Vol. 23

FLORA OF KENYA

EDITORS-IN-CHIEF
QING-FENG WANG   GEOFFREY MWACHALA

· Rubiaceae ·

YA-DONG ZHOU
CAI-FEI ZHANG
GUANG-WAN HU
GEOFFREY MWACHALA

图书在版编目(CIP)数据

肯尼亚植物志.23,茜草科:英文/周亚东等主编.—武汉:湖北科学技术出版社,2023.9
ISBN 978-7-5706-2586-4

Ⅰ.①肯… Ⅱ.①周… Ⅲ.①茜草科－植物志－肯尼亚－英文 Ⅳ.①Q948.542.4

中国国家版本馆CIP数据核字(2023)第110507号

Flora of Kenya. Rubiaceae
First Edition 2023

Published by Hubei Science & Technology Press
No. 268 Xiongchu Street, Hongshan District, Wuhan City, Hubei Province, 430070, China
ISBN:978-7-5706-2586-4

Printed in the People's Republic of China

肯尼亚植物志.23,茜草科
KENNIYA ZHIWUZHI 23 QIANCAOKE

责任编辑:曾紫风		
责任校对:王　璐		封面设计:胡　博

出版发行:湖北科学技术出版社	
地　　址:武汉市雄楚大街268号(湖北出版文化城B座13—14层)	
电　　话:027-87679468	邮　编:430070
印　　刷:湖北金港彩印有限公司	邮　编:430040

787×1092　　1/16	29.25 印张　　500千字
2023年9月第1版	2023年9月第1次印刷
定　　价:398.00元	

审图号:GS(2021)2658号

(本书如有印装问题,可找本社市场部更换)

Editorial Committee

Scientific Advisors
De-Yuan HONG [Institute of Botany, Chinese Academy of Sciences (CAS)]
Peter Hamilton RAVEN (Missouri Botanical Garden)

Editors-in-Chief
Qing-Feng WANG (Wuhan Botanical Garden/Sino-Africa Joint Research Center, CAS)
Geoffrey MWACHALA (National Museums of Kenya)

Members of Editorial Committee
Geoffrey MWACHALA (National Museums of Kenya)
Itambo MALOMBE (National Museums of Kenya)
Paul Mutuku MUSILI (National Museums of Kenya)
Emily WABUYELE (National Museums of Kenya/ Kenyatta University)
Peris Wangari KAMAU (National Museums of Kenya)
Paul Muigai KIRIKA (National Museums of Kenya)
Robert Wahiti GITURU (Jomo Kenyatta University of Agriculture and Technology)
Qing-Feng WANG (Wuhan Botanical Garden/Sino-Africa Joint Research Center, CAS)
De-Zhu LI (Kunming Institute of Botany, CAS)
Hang SUN (Kunming Institute of Botany, CAS)
Zhi-Duan CHEN (Institute of Botany, CAS)
Qin-Er YANG (South China Botanical Garden, CAS)
Nian-He XIA (South China Botanical Garden, CAS)
Shou-Zhou ZHANG (Shenzhen Fairy Lake Botanical Garden)
Guang-Wan HU (Wuhan Botanical Garden/Sino-Africa Joint Research Center, CAS)

Secretaries
Neng WEI (Wuhan Botanical Garden/Sino-Africa Joint Research Center, CAS)
Veronicah Mutele NGUMBAU (National Museums of Kenya)

Figure and Plate Editor
Jing TIAN (Wuhan Botanical Garden/Sino-Africa Joint Research Center, CAS)

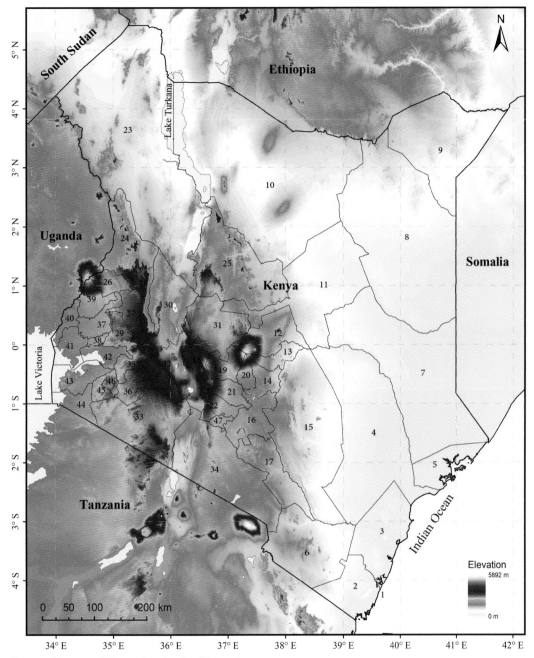

Topographic map of Kenya showing the 47 counties

1. Mombasa
2. Kwale
3. Kilifi
4. Tana River
5. Lamu
6. Teita Taveta
7. Garissa
8. Wajir
9. Mandera
10. Marsabit
11. Isiolo
12. Meru
13. Tharaka-Nithi
14. Embu
15. Kitui
16. Machakos
17. Makueni
18. Nyandarua
19. Nyeri
20. Kirinyaga
21. Murang'a
22. Kiambu
23. Turkana
24. West Pokot
25. Samburu
26. Trans Nzoia
27. Uasin Gishu
28. Elgeyo-Marakwet
29. Nandi
30. Baringo
31. Laikipia
32. Nakuru
33. Narok
34. Kajiado
35. Kericho
36. Bomet
37. Kakamega
38. Vihiga
39. Bungoma
40. Busia
41. Siaya
42. Kisumu
43. Homa Bay
44. Migori
45. Kisii
46. Nyamira
47. Nairobi

Introduction

The Republic of Kenya, with a land area of 582,646 km^2 and a 536 km coastline, is located in eastern Africa. The equator and the Great Rift Valley go through its territory on east-west direction and north-south direction respectively and intersect here. It is bordered by Somalia to the east, by Tanzania to the south, by Uganda to the west, by Ethiopia and South Sudan to the north, and facing the Indian Ocean to the southeast. The whole territory of Kenya belongs to the tropical monsoon area, but its climate varies from warm and humid tropical climate in coastal region to temperate tropical savanna climate inland due to the impact of the high altitude, and then to arid climate in the north and northeast parts of the county. The country has two rainy seasons with the "long rains" season occurring from March/April to May/June and the "short rains" from October to November/December. The annual rainfall has a declining trend from southeast to northwest.

The coastal plain occupies the southeastern part of Kenya, while most of the rest is a plateau with average elevation of 1,500 m. The Great Rift Valley splits the plateau into eastern and western parts with depth of 450–1,000 m and width of 50–100 km. There are many lakes and volcanoes distributed along the Valley. The northern part of Kenya is arid with desert and semi-desert areas occupying about 56% of the territory. Mount Kenya which is the second highest mountain of Africa, is located in the center of the country. Its summit has an altitude of 5,199 m above sea level and is covered with snow all year around. Besides Mount Kenya, there are Aberdares and Mount Elgon both with altitude more than 4,000 m and habouring well developed montane forests. In coastal area, there are Shimba Hills Forest, Arabuko-Sokoke Forest and some forest fragments, which are famous for the richness of endemic plants. The savannah is one of the most important ecosystems in Kenya and covers the remaining area and is characterized by *Adansonia digitata* and *Vachellia* species as well as succulents. The special biodiversity in Kenya is a result of its diverse climate and geography. Three of the 34 hotspots of biodiversity in the world are represented in Kenya including the eastern Afromontane, the Horn of Africa and the Coastal Forests of eastern Africa.

The *Flora of Kenya*, launched in 2015, will consist of 31 volumes. It aims to document the approximately 7,000 vascular plant species found in Kenya. This publication builds upon the well-established *Flora of Tropical East Africa* (FTEA 1952–2012), which covers Kenya, Uganda, and Tanzania. The *Flora of Kenya* incorporates recent field collections, taxonomic updates, revisions, and includes color photos to facilitate plant identification.

In this book, the classification of lycopods and ferns follows the Pteridophyte Phylogeny

Group (PPG I) system [PPG. 2016. A community-derived classification for extant lycophytes and ferns. J. Syst. Evol., 54: 563–603]. The classification of gymnosperms adheres to the system proposed by Yang et al. (2022) [Yang Y., et al. 2022. Recent advances on phylogenomics of gymnosperms and an updated classification. Plant Divers., 44: 340–350]. The orders and families of flowering plants are classified according to the Angiosperm Phylogeny Group IV (APG IV) system [Angiosperm Phylogeny Group. 2016. An update of the Angiosperm Phylogeny Group classification for the orders and families of flowering plants: APG IV. Bot. J. Linn. Soc., 181: 1–20]. Future volumes will incorporate updates to these classification systems.

The classification of infrafamiliar categories will adopt the latest advancements from systematic studies. Species delimitations will be based on the most recent revisions and the examination of materials conducted by the authors. It is important to note that new taxa should be published elsewhere before appearing in the *Flora of Kenya*. If necessary, new synonyms and typifications will be included. Each family, genus, and species will be accompanied by brief descriptions. A glossary aimed at interpreting specialized terms used in morphological descriptions will be provided, focusing on specific plant families such as Asteraceae, Orchidaceae, Poaceae, and Rubiaceae.

The distribution of each taxon in Kenya can be summarized as follows:
- Northern: County 10.
- Northwestern: County 23.
- Western: Counties 24, 26–30, 32–33, 35–46.
- Central: Counties 12–22, 25, 31, 47, and western corner of county 11.
- Eastern: Counties 7–9 and the majority of counties 4 and 11.
- Southern: Counties 6 and 34.
- Coastal: Counties 1–3, 5, and the eastern corner of county 4.

Representative specimens will be cited for each species, and a distribution map will be provided based on all consulted specimens and/or observations. Illustrations and photographs will be utilized to highlight the key morphological characteristics and interspecific differences of the plants.

The Editorial Committee of *Flora of Kenya*
9th June, 2023

Preface

Rubiaceae Juss., which is named after the genus *Rubia* L., has an estimated total number of 13,000 species in 627 genera, with 3 or 4 subfamilies and over 40 tribes, thus summing up to approximately 4% of all flowering plants making it the fourth largest angiosperm family in the world. This family has a huge diversity ranging from weedy herbs to massive rainforest trees with a large array of flower types and pollinators, and different fruit and seed types that require a myriad of ways of dispersal. Plants in this family have been documented in almost all types of habitats around the world, and adapted to a range of environmental conditions with the largest diversity found in the warm sub-tropic climates.

Sub-Saharan Africa is known for its abundant of species of Rubiaceae, with the East African region particularly the mountains and coastal forests of Kenya and Tanzania being recognized for their rich diversity of Rubiaceae species, including rare and endemic ones. Rubiaceae making it the 6th largest plant family in Kenya after Fabaceae, Poaceae, Asteraceae, Acanthaceae and Cyperaceae.

In total, there are 265 species, 23 subspecies, and 46 varieties of Rubiaceae in Kenya, belonging to 83 genera, 36 tribes and 3 subfamilies. Cultivated species are not included. The three genera with the highest number of species are *Psychotria* L. (21 species), *Pavetta* L. (19), and *Oldenlandia* L. (16). The collection points cover almost the whole country, including some reserves and national parks. The average species richness of Rubiaceae in each county is 41 species, and the five counties with the highest richness are Kwale (136 species), Kilifi (121), Teita Taveta (102), Meru (79) and Tana River (71). The four counties with the lowest species richness less than 10 species are Nyamira (9 species), Siaya (8), Vihiga (7) and Busia (5). There is a total of 36 endemic species (including infraspecies) of Rubiaceae in Kenya, and this is 13.14 % of the total number of Rubiaceae species that grow in Kenya.

The 23rd volume of the *Flora of Kenya* absorbed and drew on most of the views and information of the volumes of Rubiaceae of FTEA, and also adopted the latest research results of this family. County distribution information is provided for each species, which will make the book useful to conservation initiatives by the people and government of Kenya.

Acknowledgements

The publication of this Rubiaceae volume, as the inaugural volume of the *Flora of Kenya*, is the result of fruitful scientific cooperation between Chinese and Kenyan researchers.

This work is strongly supported by many scientific research institutions, including Wuhan Botanical Garden of CAS, Institute of Botany of CAS, Kunming Institute of Botany of CAS, South China Botanical Garden of CAS, Fairylake Botanical Garden of Shenzhen, Chenshan Botanical Garden of Shanghai, National Museums of Kenya, Jomo Kenyatta University of Agriculture and Technology, Kenya Wildlife Service, Kenya Forest Service, Royal Botanic Gardens Kew, and Royal Botanic Garden Edinburgh.

The color photos in this volume were captured by many photographers from China and Kenya. They are Qing-Feng WANG (QFW), Guang-Wan HU (GWH), Ya-Dong ZHOU (YDZ), Zhi-Xiang ZHONG (ZXZ), Sheng-Wei WANG (SWW) and Neng WEI (NW) from Wuhan Botanical Garden of CAS, Veronicah Mutele NGUMBAU (VMN) from National Museums of Kenya, Bing LIU (BL) from Institute of Botany of CAS, Cheng LIU (CL) and Tao DENG (TD) from Kunming Institute of Botany of CAS, Zheng-Wei WANG (ZWW) from Chenshan Botanical Garden of Shanghai, Shu-Dong ZHANG (SDZ) from Liupanshui Normal University. Much appreciation to Nan JIA (NJ) from Wuhan Botanical Garden of CAS for preparing all the illustrations of this volume.

The publication of this volume was supported by the fund from the Sino-Africa Joint Research Center, CAS, China (Y323771W07 and SAJC201322) and the National Natural Science Foundation of China (31800176 and 32370217).

Contents

Rubiaceae 001
1. Subfam. **Rubioideae** Verdc. 009
   1. Trib. **Urophylleae** Bremek. ex Verdc. 010
      1. **Pauridiantha** Hook. f. 010
   2. Trib. **Lasiantheae** Bremer & Manen 013
      2. **Lasianthus** Jack 013
   3. Trib. **Craterispermeae** Verdc. 017
      3. **Craterispermum** Benth. 017
   4. Trib. **Psychotrieae** Cham. & Schltdl. 020
      4. **Eumachia** DC. 020
      5. **Chassalia** Comm. ex Poir. 023
      6. **Geophila** D. Don 032
      7. **Psychotria** L. 035
   5. Trib. **Knoxieae** Benth. & Hook. f. 063
      8. **Chamaepentas** Bremek. 064
      9. **Triainolepis** Hook. f. 066
      10. **Dirichletia** Klotzsch 069
      11. **Parapentas** Bremek. 072
      12. **Dolichopentas** Kårehed & B. Bremer 074
      13. **Phyllopentas** (Verdc.) Kårehed & B. Bremer 078
      14. **Pentas** Benth. 083
      15. **Rhodopentas** Kårehed & B. Bremer 092
      16. **Pentanisia** Harv. 096
      17. **Otomeria** Benth. 101
   6. Trib. **Spermacoceae** Bercht. & J. Presl 105
      18. **Kohautia** Cham. & Schltdl. 106
      19. **Pentanopsis** Rendle 112
      20. **Conostomium** Cufod. 114
      21. **Pentodon** Hochst. 117
      22. **Agathisanthemum** Klotzsch 120
      23. **Dibrachionostylus** Bremek. 123
      24. **Cordylostigma** Groeninckx & Dessein 125
      25. **Edrastima** Raf. 129
      26. **Scleromitrion** (Wight & Arn.) Meisn. 131
      27. **Oldenlandia** L. 134
      28. **Diodia** Gronov. 155

    29. **Mitracarpus** Zucc. 157
    30. **Diodella** Small 160
    31. **Spermacoce** L. 163
    32. **Richardia** L. 173
  7. Trib. **Anthospermeae** Cham. & Schlecht. 177
    33. **Anthospermum** L. 177
  8. Trib. **Paederieae** DC. 182
    34. **Paederia** L. 183
  9. Trib. **Rubieae** Baill. 185
    35. **Rubia** L. 186
    36. **Galium** L. 188

2. Subfam. **Cinchonoideae** Raf. 203
  10. Trib. **Hymenodictyeae** Razafim. & B. Bremer 204
    37. **Hymenodictyon** Wall. 204
  11. Trib. **Naucleeae** Kostel. 208
    38. **Mitragyna** Korth. 208
    39. **Breonadia** Ridsdale 211
    40. **Uncaria** Schreb. 214
    41. **Nauclea** L. 217
  12. Trib. **Guettardeae** DC. 219
    42. **Guettarda** L. 219

3. Subfam. **Ixoroideae** Raf. 223
  13. Trib. **Mussaendeae** Benth. & Hook. f. 224
    43. **Heinsia** DC. 224
    44. **Pseudomussaenda** Wernham 228
    45. **Mussaenda** L. 228
  14. Trib. **Crossopterygeae** F. White ex Bridson 236
    46. **Crossopteryx** Fenzl 236
  15. Trib. **Ixoreae** Benth. & Hook. f. 238
    47. **Ixora** L. 239
  16. Trib. **Vanguerieae** A. Rich. ex Dumort. 243
    48. **Keetia** E.P. Phillips 244
    49. **Afrocanthium** (Bridson) Lantz & B. Bremer 250
    50. **Psydrax** Gaertn. 256
    51. **Bullockia** (Bridson) Razafim., Lantz & B. Bremer 265
    52. **Pyrostria** Comm. ex Juss. 271
    53. **Canthium** Lam. 275
    54. **Rytigynia** Blume 280

- 55. **Fadogia** Schweinf. 291
  - 56. **Multidentia** Gilli 293
  - 57. **Vangueriopsis** Robyns 296
  - 58. **Vangueria** Juss. 298
- 17. Trib. **Coffeeae** DC. 314
  - 59. **Coffea** L. 315
  - 60. **Calycosiphonia** Pierre ex Robbr. 320
  - 61. **Empogona** Hook. f. 323
  - 62. **Tricalysia** A. Rich. ex DC. 328
- 18. Trib. **Octotropideae** Bedd. 333
  - 63. **Cremaspora** Benth. 333
  - 64. **Galiniera** Delile 337
  - 65. **Didymosalpinx** Keay 340
  - 66. **Lamprothamnus** Hiern 342
  - 67. **Kraussia** Harv. 345
  - 68. **Feretia** Delile 348
  - 69. **Polysphaeria** Hook. f. 351
- 19. Trib. **Sherbournieae** Mouly & B.Bremer 355
  - 70. **Mitriostigma** Hochst. 356
  - 71. **Oxyanthus** DC. 359
- 20. Trib. **Pavetteae** A. Rich. ex Dumort. 365
  - 72. **Rutidea** DC. 365
  - 73. **Leptactina** Hook. f. 370
  - 74. **Cladoceras** Bremek. 373
  - 75. **Pavetta** L. 375
  - 76. **Tennantia** Verdc. 399
  - 77. **Tarenna** Gaertn. 401
  - 78. **Coptosperma** Hook. f. 406
- 21. Trib. **Gardenieae** A. Rich. ex DC. 414
  - 79. **Gardenia** J. Ellis 414
  - 80. **Heinsenia** K. Schum. 421
  - 81. **Rothmannia** Thunb. 423
  - 82. **Aidia** Lour. 431
  - 83. **Catunaregam** Wolf 433

Glossary 437
Abbreviations 440
New Typifications 440
New Synonyms and Nomenclatural Novelties 442
Index 443

Rubiaceae

Small to large trees, shrubs or less often annual or perennial herbs, rarely lianas or climbers, unarmed or sometimes armed. Raphides (calcium oxalate crystals) present or absent. Leaves opposite, less often whorled or decussate, entire or rarely lobed to denticulate or toothed, sometimes with bacterial nodules and domatia; stipules interpetiolar and infrequently fused to adjacent petioles, sometimes sheath-like, entire or divided into lobes or fimbriae. Inflorescences terminal, axillary or pseudo-axillary, cymose, paniculate, fasciculate, or rarely spiciform or capitate, few- to many-flowered or occasionally reduced to a solitary flower; bracteate, ebracteate, or bracts sometimes reduced, rarely enlarged. Flowers bisexual or rarely unisexual, homostylous or quite often heterostylous. Calyx gamosepalous; tube mostly fused to the inferior ovary; limb usually developed, truncate, toothed or lobed. Corolla small to large, gamopetalous, white or coloured; tube funnel-form, salver-form, campanulate, sometimes with a very long and slender lower part; lobes imbricate or valvate, sometimes contorted, spreading to somewhat reflexed. Stamens usually inserted variously in corolla throat; anthers basi- or dorsifixed, introrse. Ovary 1–many-locular, placentation axile or parietal; ovules 1–many in each locule; style simple, usually long and narrow; stigmas 1–2(–10)-lobed, lobes capitate, linear, spatulate, clavate. Fruits small to quite large, simple, capsular, berrylike or drupaceous, indehiscent or infrequently dehiscent, sometimes united into a syncarp. Seeds 1–many, very small to large, variously ellipsoid, lenticular, flattened, oblanceoloid, angled, or plano-convex, smooth or rarely winged.

About 13,000 species found all around the world especially in the tropical and subtropical regions. Three subfamilies, 21 tribes, and 83 genera occur in Kenya.

Key to genera

1a. Erect or climbing herbs, often adhesive due to prickles and harsh hairs; leaves with leaflike stipules in whorls of 4–6 or rarely many; ovary 1–2-1ocular, with single ovule in each locule 2
1b. Herbs, shrubs, trees or lianas, not adhesive; leaves usually paired, or rarely 3–6-whorled, with stipules developed between each pair; ovary 1–many-locular, with single to many ovules in each locule 3
2a. Leaf-blades large, ovate to lanceolate, with petiole very well developed; flowers 5-merous .... 35. *Rubia*
2b. Leaf-blades small, linear to lanceolate, or rarely ovate, sessile to shortly petiolate; flowers 4-merous .... 36. *Galium*
3a. Climbing herbs, scandent shrubs or lianas 4
3b. Erect, decumbent or procumbent herbs, shrubs or trees 11
4a. Stems 4-angled, with recurved spines 5

4b. Stems occasionally 4-angled; spines straight or absent 6
5a. Flowers in completely spherical heads; fruit a fusiform capsule; seeds winged...... 40. *Uncaria*
5b. Flowers in dense corymbs or subcapitate; fruit a globose berry; seeds unwinged ....74. *Cladoceras*
6a. Several calyx-lobes on each inflorescence developed into a stalked white to colored, membranous, stipitate calycophylls .... 45. *Mussaenda*
6b. No calyx-lobes developed into enlarged calycophylls 7
7a. Plants evil-smelling; fruits flattened .... 34. *Paederia*
7b. Plants not evil-smelling; fruits not flattened 8
8a. Raphides present; flowers heterostylous; stigma bifid .... 5. *Chassalia* (*C. cristata*)
8b. Raphides absent; flowers not heterostylous; stigmatic knob fusiform or cylindrical 9
9a. Stipules with broad base and several fimbriae or a single lobe; stigmatic knob fusiform ....72. *Rutidea*
9b. Stipules lanceolate, triangular or ovate, or with truncate to triangular base and a keeled lobe; stigmatic knob cylindrical.. 10
10a. Leaves chartaceous to subcoriaceous, rarely coriaceous; calyx-limb dentate or slightly lobed ....48. *Keetia*
10b. Leaves usually subcoriaceous to coriaceous; calyx-limb a dentate to repand rim, usually much smaller .... 50. *Psydrax*
11a. Herbs, woody herbs or rarely subshrubs; raphides present; corolla usually valvate 12
11b. Subshrubs, shrubs or rarely herbaceous shoots from a woody rootstock; raphides present or absent; corolla valvate, contorted or sometimes imbricate 37
12a. Ovules solitary in each locule 13
12b. Ovules 2–many in each locule 21
13a. Flowers usually unisexual .... 33. *Anthospermum*
13b. Flowers bisexual 14
14a. Flowers usually 5-merous 15
14b. Flowers usually 4-merous, or rarely 3–6-merous 17
15a. Creeping herbs, rooting at nodes; fruit a drupe, orange or red .... 6. *Geophila*
15b. Erect or trailing herbs, not rooting at nodes; fruit a capsule or drupe 16
16a. Raphides present; leaves paired; flowers several in terminal or pseudo-axillary, capitate or spike-like inflorescences .... 16. *Pentanisia*
16b. Raphides absent; leaves 3–4-whorled; flowers solitary or few in axillary cymes ....55. *Fadogia*
17a. Ovary 3–6-locular; stigmas 3–4; capsules 3–4-coccous .... 32. *Richardia*
17b. Ovary 2(–3)-locular; stigmas 1–2; capsules with 2 valves or 2 cocci, or circumscissile....... 18
18a. Fruits circumscissile .... 29. *Mitracarpus*
18b. Fruits indehiscent or open by longitudinal slits or 2-coccous 19
19a. Fruits capsular with 2 valves or with 2 cocci, usually dehiscent .... 31. *Spermacoce*
19b. Fruits with 2 cocci, somewhat indehiscent 20

20a. Seeds distinctly and characteristically lobed .... 28. *Diodia*
20b. Seeds not lobed .... 30. *Diodella*
21a. Flowers mostly 4-merous; calyx-lobes usually equal 22
21b. Flowers mostly 5-merous; calyx-lobes sometimes unequal 30
22a. Inflorescences axillary or flowers solitary, axillary 23
22b. Inflorescences terminal or terminal and axillary 25
23a. Corolla-tube over 20 mm long .... 20. *Conostomium*
23b. Corolla-tube less than 10 mm long 24
24a. Corolla-tube glabrous; stamens and styles both exserted .... 26. *Scleromitrion*
24b. Corolla-tube with a ring of hairs in the throat; stamens and styles not both exserted ....
.... 27. *Oldenlandia*
25a. Corolla-tube narrowly cylindrical; anthers and stigmas included 26
25b. Corolla-tube cylindrical or funnel-shaped; anthers and/or stigmas usually exserted 27
26a. Stigma 2-lobed, lobes filiform .... 18. *Kohautia*
26b. Stigma unlobed, ovoid or cylindrical .... 24. *Cordylostigma*
27a. Capsule opening both septicidally and loculicidally 28
27b. Capsule opening loculicidally, sometimes tardily dehiscent 29
28a. Corolla-tube bearded inside; style not shortly bifid at the apex .... 22. *Agathisanthemum*
28b. Corolla-tube glabrous inside; style shortly bifid at the apex .... 23. *Dibrachionostylus*
29a. Corolla-tube often with a ring of hairs in the throat .... 27. *Oldenlandia*
29b. Corolla-tube glabrous .... 25. *Edrastima*
30a. Calyx-lobes usually equal; ovary 2-locular .... 21. *Pentodon*
30b. Calyx-lobes usually unequal; ovary 2–10-locular 31
31a. Inflorescences in cymose heads, developed into a long simple spike in fruit ..... 17. *Otomeria*
31b. Flowers in more complicated inflorescences 32
32a. Creeping or decumbent herbs .... 11. *Parapentas*
32b. Erect herbs, woody herbs or subshrubs 33
33a. Calyx-lobes subequal, linear, subulate, narrowly triangular or slightly spathulate at the apices 34
33b. Calyx-lobes unequal, linear, subulate, spathulate or 1–3-foliaceous 35
34a. Corolla-tube slender, up to 13 cm long .... 12. *Dolichopentas*
34b. Corolla-tube cylindric, less than 2.8 cm long .... 13. *Phyllopentas*
35a. Inflorescences terminal, cymose, (1–)3–several-flowered; corolla white ..... 8. *Chamaepentas*
35b. Inflorescences terminal or axillary, dense or lax, several- to many-flowered; corolla white or brightly coloured 36
36a. Flowers distinctly vermilion-scarlet .... 15. *Rhodopentas*
36b. Flowers white, mauve, blue or pink, rarely red and then of a deeper crimson shade ....
.... 14. *Pentas*
37a. Leaf-blades with distinct bacterial nodules, scattered or sometimes restricted to the mid-rib 38

37b. Leaf-blades without distinct bacterial nodules 39
38a. Rhaphides present; flowers usually heterostylous; corolla-lobes usually 5, valvate in bud; stigma bifid .... 7. *Psychotria*
38b. Raphides absent; flowers not heterostylous; corolla-lobes 4, contorted in bud; stigma very shortly 2-lobed.. 74. *Pavetta*
39a. Raphides absent; inflorescences congested into spherical heads; flowers small 40
39b. Raphides present or absent; inflorescences not as above; flowers small or large 42
40a. Ovaries and fruitlets connate; fruit a syncarp .... 41. *Nauclea*
40b. Ovaries and fruitlets free 41
41a. Leaves opposite, broadly elliptic to obovate; stipules large, membranous, broadly elliptic .... 38. *Mitragyna*
41b. Leaves 3–4-whorled, lanceolate or oblong-lanceolate; stipules smaller, bifid, triangular......... .... 39. *Breonadia*
42a. Inflorescences elongated, spicate to racemose .... 37. *Hymenodictyon*
42b. Inflorescences not as above 43
43a. Leaves 3–6(-many)-whorled, heath-like; flowers always unisexual........... 33. *Anthospermum*
43b. Leaves usually opposite, larger; flowers usually bisexual 44
44a. Raphides present 45
44b. Raphides absent 53
45a. Ovule solitary in each locule 46
45b. Ovules 2–many in each locule 50
46a. Ovary 2-1ocular .... 3. *Craterispermum*
46b. Ovary 2–12-locular 47
47a. Stipules entire, often triangular; inflorescences axillary or supra-axillary ....2. *Lasianthus*
47b. Stipules often divided or rarely entire; inflorescences always terminal 48
48a. Stipules entire or lobed, not becoming corky; fruits not dehiscent .... 7. *Psychotria*
48b. Stipules entire, often becoming corky; fruits more or less dehiscent 49
49a. Corolla-tube often slightly curved and winged; lobes often winged .... 5. *Chassalia*
49b. Corolla-tube and lobes not as above .... 4. *Eumachia*
50a. Stipules entire or fringed; fruits fleshy .... 1. *Pauridiantha*
50b. Stipules fimbriate; fruits dry 51
51a. Calyx-lobes equal, linear, linear-lanceolate or narrowly triangular; ovary 2-locular............... .... 19. *Pentanopsis*
51b. Calyx-lobes unequally 5–7-toothed or developed into an oblique dilated reticulated shallowly concave entire or lobed lamina; ovary 2–10-locular 52
52a. Calyx-lobes unequally 5–7-toothed .... 9. *Triainolepis*
52b. Calyx-lobes developed into oblique dilated reticulated shallowly concave lamina .... 10. *Dirichletia*
53a. Flowers without secondary pollen presentation; seeds with little or no albumen.................... .... 42. *Guettarda*

53b. Flowers often with secondary pollen presentation; seeds with albumen 54
54a. Some calyx-lobes always enlarged, foliaceous or developed into membranous calycophylls 55
54b. Calyx-lobes equal or slight unequal 57
55a. Some calyx-lobes enlarged, subfoliaceous .... 43. *Heinsia*
55b. Some calyx-lobes developed into a stalked white to colored, membranous, stipitate calycophylls 56
56a. Flower buds bearing 5 apical filiform appendages; fruits dehiscing at the apex .... 44. *Pseudomussaenda*
56b. Flower buds without appendages; fruits indehiscent .... 45. *Mussaenda*
57a. Fruit a capsule, loculicidally splitting into 2 valves; seeds winged .... 46. *Crossopteryx*
57b. Fruit an indehiscent drupe; seeds unwinged 58
58a. Plants always armed with spines 59
58b. Plants unarmed 64
59a. Calyx-lobes conspicuous, leafy, 8–16 mm long, much exceeding the corolla-tube .... 58. *Vangueria* (*V. pallidiflora*)
59b. Calyx-lobes not conspicuous, or if so not exceeding the corolla-tube 60
60a. Plants scandent, some axillary branchlets reduced to recurved spines .... 74. *Cladoceras*
60b. Plants not scandent; spines straight or slightly recurved 61
61a. Flowers solitary, supra-axillary; corolla-tube funnel-shaped, 4.5–7.5 cm long .... 65. *Didymosalpinx*
61b. Flowers solitary or few to several in fascicles or cymes; corolla-tube cylindrical, much samller 62
62a. Stipules glabrous inside; inflorescences terminal or on short spurs .... 83. *Catunaregam*
62b. Stipules pubescent or hairy inside; inflorescences axillary 63
63a. Stipules pubescent inside; corolla-lobes acute or shortly apiculate .... 53. *Canthium*
63b. Stipules hairy inside; corolla-lobes acute to long apiculate or with a filiform appendage .... 54. *Rytigynia*
64a. Ovary 1-locular, with numerous ovules on 2–9 parietal placentas; fruits usually with a thick fibrous, woody or leathery wall 65
64b. Ovary 2(–6)-locular, with single to many ovules in each locule; fruits not as above 66
65a. Flowers 5–12-merous; solitary or in few-flowered clusters; ovary with numerous ovules on 2–9 parietal placentas .... 79. *Gardenia*
65b. Flowers 5-merous, often solitary and pendent; ovary with numerous ovules on 2 parietal placentas .... 81. *Rothmannia*
66a. Petiole articulate; flowers 4(–5)-merous; ovary 2-locular; ovule single in each locule .... 47. *Ixora*
66b. Petiole not articulate; flowers 4–8(–12)-merous; ovary 1–5(–6)-locular; ovule single to numerous in each locule 67
67a. Flowers usually small, 4–5(–6)-merous; ovary 2–5(–6)-locular; ovule single in each locule,

	usually pendulous from near the apex 68
67b.	Flowers small to very large, 4–8(–12)-merous; ovary 1–2(–3)-locular; ovule single to numerous in each locule, erect or attached by middle to the septum, rarely pendulous 76
68a.	Inflorescences umbellate or 1-flowered, entirely enclosed in bud by paired connate bracts..... .... 52. *Pyrostria*
68b.	Inflorescences various, never enclosed by paired bracts 69
69a.	Corolla-lobes linear-lanceolate, 2–4 cm long, much exceeding tube .... 57. *Vangueriopsis*
69b.	Corolla-lobes not as above 70
70a.	Subshrubby herbs or single-stemmed shrubs from a woody rootstock, up to 2 m tall ....56. *Multidentia*
70b.	Shrubs or small trees, up to 20 m tall 71
71a.	Stipules glabrous within; style usually at least twice as long as corolla-tube; stigmatic knob cylindric, about twice as long as wide .... 50. *Psydrax*
71b.	Stipules hairy or glaborous within; style usually much less than twice as long as corolla-tube; stigmatic knob mostly as broad as long 72
72a.	Ovary 2-locular; stigmatic knob 2-lobed; fruits 2-seeded 73
72b.	Ovary 2–5(–6)-locular; stigmatic knob 2–5-lobed; fruits 1–5-seeded 75
73a.	Calyx with a well-developed tubular part; pyrenes very thickly woody, strongly irregularly ridged with lines of dehiscence apparent ....56. *Multidentia*
73b.	Calyx without conspicuous tubular part; pyrenes not as above 74
74a.	Flowers unisexual, (4–)5–6-merous, functional males 1–20 in fascicules, functional females usually solitary... 51. *Bullockia*
74b.	Flowers bisexual, 4–5-merous, few to many in pedunculate cymes .... 49. *Afrocanthium*
75a.	Leaves usually with domatia; corolla-tube glabrous or hairy, but not bearded at the throat; fruits less than 30 mm in diameter .... 54. *Rytigynia*
75b.	Leaves without domatia; corolla-tube bearded at the throat; fruits up to 50 mm in diameter . ... 58. *Vangueria*
76a.	Inflorescences mostly axillary; ovary 2-locular with 1-several ovules in each locule; style-arms divergent; seeds few, never held together in a matrix 77
76b.	Inflorescences terminal, lateral or occasionally axillary; ovary 1–2(–4)-locular with 1–numerous ovules in each locule; style-arms seldom divergent; seeds many, sometimes held together in a matrix 86
77a.	Flowers 4–8(–12)-merous; ovary 2-locular; seed coat more or less entire, or with a distinct longitudinal ventral invagination 78
77b.	Flowers (4–)5(–6)-merous, rarely 6–7(–8)-merous; ovary 1–2-locular; seed coat has a very distinct fingerprint-like or sometimes reticulate pattern invagination 81
78a.	Ventral surface of seed with a distinct longitudinal invagination (typical "coffee-beans") .... 59. *Coffea*
78b.	Ventral surface of seed without invagination 79
79a.	Flowers (4–)5(–6)-merous; anthers with a conspicuous ribbon-like appendage .... 61. *Empogona*

79b. Flowers 4–8(–12)-merous; anthers with or without an inconspicuous apical appendage 80
80a. Corolla-lobes longer than tube; ovules (seeds) 1–2 per locule; fruits 9–13 mm in diameter .... 60. *Calycosiphonia*
80b. Corolla-lobes shorter than tube; ovules (seeds) 1–12 per locule; fruits usually less than 7 mm in diameter, rarely up to 10 mm in diameter .... 62. *Tricalysia*
81a. Ovule single in each locule 82
81b. Ovules few in each locule 84
82a. Flowers in terminal and axillary panicles; corolla-lobes 6–9; style narrowly club-shaped, densely pubescent .... 66. *Lamprothamnus*
82b. Flowers usually sessile in bracteate axillary clusters; corolla-lobes 4–5; style slender or filiform, pubescent or just hairy at the extreme base 83
83a. Fruits ovoid, crowned with persistent and sometimes beaked calyx .... 63. *Cremaspora*
83b. Fruits globose, not crowned with persistent calyx .... 69. *Polysphaeria*
84a. Flowers sessile and appearing terminal on short leafless spurs or sometimes pedicellate in pairs from the axils of young leaves .... 68. *Feretia*
84b. Flowers few to many in axillary pedunculate panicles or cymes 85
85a. Flowers pedicellate in pedunculate axillary cymes; stigmatic club deeply bifid, each lobe with 5 membranous ciliate wings .... 64. *Galiniera*
85b. Flowers many in axillary, lax, pedunculate panicles; stigmatic club long, fusiform, with 10 membranous ciliate wings .... 67. *Kraussia*
86a. Ovule single in each locule 87
86b. Ovules usually more than one, sometimes many in each locule 88
87a. Scandent shrubs, with hairy stems and opposite lateral branches; leaf-blades without bacterial nodules; stigma entire or rarely 2–3-lobed .... 72. *Rutidea*
87b. Shrubs or small trees, never scadent; leaf-blades sometimes with bacterial nodules; stigma very shortly 2-lobed .... 75. *Pavetta*
88a. Flowers usually large, with corolla-tube over 1 cm and up to 13 cm long 89
88b. Flowers small, with corolla-tube usually less than 6 mm long 92
89a. Calyx-lobes leaf-like, up to 2 cm long; fruits longitudinally ribbed .... 73. *Leptactina*
89b. Calyx-lobes not leaf-like, small; fruits not ribbed 90
90a. Leaf-blades often red when young or with the nerves red beneath; fruits subglobose, 1.1–1.3 cm in diameter .... 80. *Heinsenia*
90b. Young leaf-blades not as above; fruits often ellipsoid or fusiform, over 1.7 cm long 91
91a. Flowers few to several in subcapitulate cymes; bracts leaf-like, up to 2 cm long................... .... 70. *Mitriostigma*
91b. Flowers subsessile to pedicellate in axillary, compact to lax racemes or panicles; bracts not leaf-like, less than 1 cm long .... 71. *Oxyanthus*
92a. Flowers 5–6-merous, in sessile few-flowered clusters at the ends of short shoots; ovules 3(–4) in each locule .... 76. *Tennantia*
92b. Flowers (4–)5-merous, in terminal or pseudoaxillary, few- to many-flowered corymbose,

sessile or pedunculate; ovules 1 to many in each locule 93

93a. Fruits not crowned by the persistent calyx-limb; seed 1(–2), reticulate .... 78. *Coptosperma*

93b. Fruits mostly crowned by the persistent calyx-limb; seeds (1–)several to numerous, not reticulate 94

94a. Inflorescences of terminal corymbose; corolla-tube cylindrical to funnel-shaped, lobes oblong .... 77. *Tarenna*

94b. Inflorescences pseudoaxillary, lateral at nodes with subtending scalelike leaves; corolla-tube funnel-shaped, lobes ovate, margins revolute .... 82. *Aidia*

1. Subfam. **Rubioideae** Verdc.

Shrubs, herbs, or rarely trees, with raphides. Flowers usually heterostylous. Stipules entire, bifid, or often fimbriate. Corolla aestivation always valvate. Ovary 1–12-locular, usually with 1 to several ovules in each locule. Fruits dry or fleshy, dehiscent or indehiscent. Seeds with albumen.

Nine tribes and 36 genera occur in Kenya.

Key to tribes

1a. Ovule solitary in each locule 2
1b. Ovules 1–many in each locule 7
2a. Ovary 2-locular 3
2b. Ovary 2–12-locular 6
3a. Shrubs or trees; stipules entire, connate to form a tube .... Trib. 3. Craterispermeae
3b. Herbs, climbers or rarely subshrubs; stipules always leaf-like and whorled with leaves 4
4a. Usually foetid smelling climbers; leaves opposite .... Trib. 8. Paederieae
4b. Plants not evil-smelling; leaves and leaf-like stipules in whorls of 4–8 or more 5
5a. Flowers unisexual or hermaphrodite; ovules attached to the base of the ovary-locules ....
 .... Trib. 7. Anthospermeae
5b. Flowers hermaphrodite; ovules affixed to the septum, amphitropous .... Trib. 9. Rubieae
6a. Stipules entire, often triangular; inflorescences axillary or supra-axillary ... Trib. 2. Lasiantheae
6b. Stipules often divided or rarely entire; inflorescences always terminal ...... Trib. 4. Psychotrieae
7a. Stipules entire or fringed; fruit fleshy .... Trib. 1. Urophylleae
7b. Stipules fimbriate; fruits dry 8
8a. Calyx-lobes usually unequal; ovary 2–10-locular .... Trib. 5. Knoxieae
8b. Calyx-lobes usually equal; ovary 2(or rarely 3–4)-locular .... Trib. 6. Spermacoceae

1. Trib. **Urophylleae** Bremek. ex Verdc.

Subshrubs, shrubs or small trees, with raphides. Stipules entire or fringed. Flowers homostylous or heterostylous. Ovary 2–many-locular, with numerous ovules in each locule. Fruit a berry, 2–many-locular, with many seeds.

One genus occurs in Kenya.

1. **Pauridiantha** Hook. f.

Shrubs or small trees. Leaves opposite or whorled, shortly petiolate; blades long-acuminate, often with domatia; stipules triangular to ovate, entire. Inflorescences axillary or terminal, trichotomously corymbose or subumbellate, sessile or pedunculate, 1–many-flowered. Flowers heterostylous. Calyx-tube short, dentate or lobed. Corolla salver-shaped, white to cream; tube short, funnel-shaped or cylindrical, upper half hairy inside; lobes triangular, ovate or oblong-lanceolate, glabrous inside. Stamens included or slightly exserted in short-styled flowers. Ovary 2–3-locular; ovules numerous in each locule; stigmas 2, globose. Fruit a globose berry. Seeds numerous; endosperm oily.

About 50 species confined in tropical Africa and Madagascar; only one species in Kenya.

1. **Pauridiantha paucinervis** (Hiern) Bremek., Bot. Jahrb. Syst. 71: 212. 1940; F.T.E.A. Rubiac. 1: 153. 1976; K.T.S.L.: 526. 1994. ≡ *Urophyllum paucinerve* Hiern, Fl. Trop. Afr. 3: 74. 1877. —Type: Equatorial Guinea, Bioko (Fernando Po), 1219 m, Nov. 1860, *G. Mann 577* [holotype: K (K000172965)] (Figure 1; Plate 1)

≡ *Urophyllum holstii* K. Schum., Pflanzenw. Ost-Afrikas, C: 379. 1895. ≡ *Pauridiantha holstii* (K. Schum.) Bremek., Bot. Jahrb. Syst. 71: 212. 1940. ≡ *Pauridiantha paucinervis* (Hiern) Bremek. subsp. *holstii* (K. Schum.) Verdc., Kew Bull. 30: 283. 1975; F.T.E.A. Rubiac. 1: 153. 1976. —Types: Tanzania, E. Usambara Mts., Lutindi to Urwald, Jule 1893, *C. Holst 3277* [syntype: B (destroyed); lectotype: K (K000172969), **designated here**; isolectotypes: M (M010 6415)]; Tanzania, Kilimanjaro, Shira Plateau, *G. Volkens 1940* [syntype: B (destroyed); isosyntype: K]

Shrub or small tree, up to 12 m tall. Leaves opposite, shortly petiolate; blades oblong-elliptic or -lanceolate, 3.5–15.5 cm × 1–5.5 cm, apex acuminate, base cuncate;

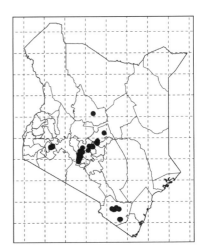

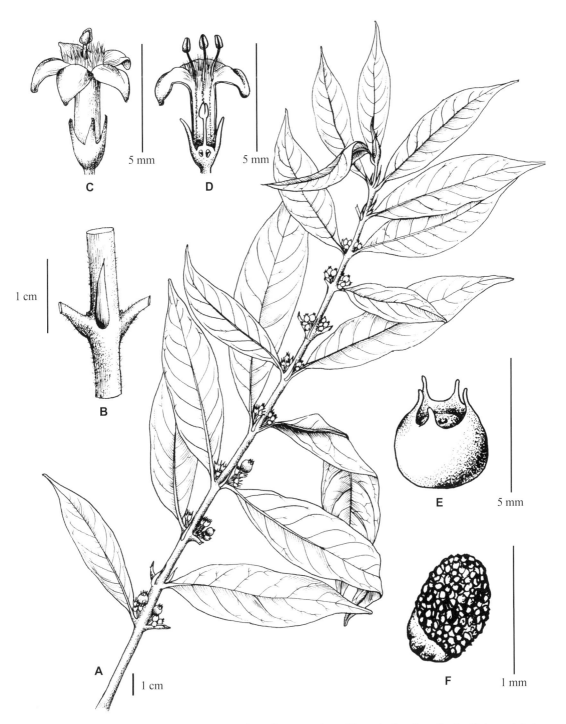

Figure 1 *Pauridiantha paucinervis*. A. fruiting branch; B. portion of branch showing stipule; C. long-styled flower; D. longitudinal section of short-styled flower, showing the stamens, style and ovary; E. fruit; F. seed. Drawn by NJ.

Plate 1 A–E. *Pauridiantha paucinervis*. Photo by GWH (A, C) and YDZ (B, D, E).

stipules lanceolate, up to 1.5 cm, adpressed pubescent. Inflorescences axillary, shortly pedunculate, 1–3-flowered, bracts lanceolate, 1.5–2.5 mm long. Flowers heterostylous. Calyx with tube ca. 1 mm long; lobes lanceolate, up to 2.8 mm long. Corolla salver-shaped, white or cream, tube 2–5 mm long; lobes oblong-lanceolate, 1.5–3 mm long. Stamens as long as the corolla-lobes in short-styled flowers, just included in long-styled flowers. Ovary 2–3-locular, with numerous ovules in each locule. Style 3–5 mm long in long-styled flowers, 1.5–3 mm long in short-styled flowers; stigmas 2, globose. Fruit a subglobose berry, 3–5 mm long, 2.5–4 mm across. Seeds orange-brown, ellipsoid.

Distribution: Western, central, and southern Kenya. [Tropical Africa and Madagascar].

Habitat: Forests; 1200–2500 m.

Kericho: southwest of Mau Forest, *Kerfoot 4883* (EA, K). Kiambu: Kikuyu Escarpment Forest, *Faden et al. 74/1327* (K). Kirinyaga: south Mount Kenya, *Gillett & Mathew 19101* (EA, K). Meru: Nyambeni Hills, *SAJIT Z0146* (HIB); east Mount Kenya, *Luke 383* (EA). Murang'a: Kieni Forest, *Beentje & Mungai 2892* (EA). Nyandarua: south Kinangop, *Gilbert & Thulin 1039* (EA). Nyeri: Ragati Forest, *Dyson 458* (EA). Samburu: Mathew's Range, *Luke 14155* (EA). Teita Taveta: Mbololo Forest, *Faden et al.* 364 (EA, K); Mount Kasigau, *Luke et al. 5349* (EA, K).

2. Trib. **Lasiantheae** Bremer & Manen

Subshrubs, shrubs, or small trees. Stipules entire, often triangular. Ovary 2–12-locular, with a single ovule in each locule, erect from the base. Fruits drupaceous, not compound or aggregated, often blue to black, with 2–12 pyrenes.

Only one genus occurs in Kenya.

2. **Lasianthus** Jack

Shrubs or rarely small trees, sometimes foetid, unarmed. Leaves opposite, distichous, usually papery to leathery, base acute to rounded or cordate, apex acuminate, acute or cuspidate; stipules interpetiolar, persistent or deciduous, triangular or lanceolate. Inflorescences 1–several-flowered fascicles, axillary, sessile or pedunculate; bracts persistent, small. Flowers bisexual or rarely unisexual, sessile or pedicellate. Calyx-tube subglobose, ovoid, oblong, campanulate or urceolate; limb (3–)4–6(–8)-dentate or lobed, persistent. Corolla white or pink, funnel-form or salver-form; tube densely hairy at the throat; lobes 4–6(–7), spreading, erect or reflexed, usually hairy inside. Stamens 4–6, inserted in corolla throat; filaments short; anthers linear or oblong, included or exserted. Ovary 3–12-locular; ovules 1 in each locule, erect from the base. Style short or elongate; stigma 3–10-lobed, linear, lanceolate or obtuse. Fruits blue, pink, purple, white or black; pyrenes 3–12, segment-shaped or pyriform, 1-seeded. Seeds narrowly oblong, black, with abundant endosperm.

About 180 species in tropical Asia, Australia, tropical America and Africa, ca. 23 species in continental Africa and three species in Madagascar and adjacent islands; only one species in Kenya.

1. **Lasianthus kilimandscharicus** K. Schum., Pflanzenw. Ost-Afrikas, C: 396. 1895; F.T.E.A. Rubiac. 1: 142. 1976; K.T.S.L.: 520. 1994. —Type: Tanzania, Kilimanjaro, above Schira, 2500 m, Mar. 1894, *G. Volkens 1949* [lectotype: K (K000172649), designated by P. L. Jannerup in Nordic J. Bot. 23(6): 675. 2006; isolectotypes: BM (BM000602108), G (G00014579)] (Figure 2; Plate 2)

= *Lasianthus kilimandscharicus* K. Schum. subsp. *glabrescens* Jannerup, Nordic J. Bot. 23: 677, figs. 3a, 14. 2006, **syn. nov.** —Type: Malawi, Southern Region, Mt. Mulanje, Litchenya Plateau, 1800 m, 4 Nov. 1986, *J.D. & E.G. Chapman 8188* [holotype: FHO; isotypes: MAL, MO (MO-1056965)]

Shrub or small tree, up to 7.5 m tall, with smooth grey bark. Leaves opposite, blades oblong, narrowly elliptic to lanceolate, 9–20 cm × 2–7 cm, above glabrous and below sparsely hairy on veins or rarely hairy, narrowly acuminate at the apex, cuneate, rounded or oblique at the base; petioles up to 1.8 cm long; stipules narrowly to broadly triangular, 1.5–6 mm × 1–4.5 mm. Flowers in compact, subsessile, axillary clusters, heterostylous; bracts ovate to lanceolate, 1–4 mm × 0.5–2 mm, hairy. Calyx pinkish white or tinged purple; tube turbinate, 0.5–2 mm long; lobes 3–4, triangular or rounded, 0.5–3 mm × 0.5–2 mm. Corolla white; tube cylindrical, 2.5–5 mm × 1–2 mm; lobes 4–5, ovate-oblong, 2–4 mm × 1.5–2 mm, sparsely hairy to glabrous outside, densely hairy inside; anthers included or slightly exserted. Ovary 4–6-locular; style 2.5–5.5 mm long; stigma-lobes 4–5, oblong or subcapitate. Fruits greenish blue, globose; pyrenes 4–6, brown, 2.5–4 mm × 1.5–2.5 mm, with a tip or hooked adaxally.

Distribution: Western, central, and southern Kenya. [East Africa from Kenya to Zimbabwe].

Habitat: Forests; 1200–2500 m.

Note: There is no obvious difference between the two types, *G. Volkens 1949* and *J.D. & E.G. Chapman 8188*. It is very difficult to distinguish the subspecies of this widespread species, just using the characteristics of indumentum on leaves and young twigs. Our field investigations also found some individuals with intermediate characters between the two subspecies.

Bomet: southwestern Mau Forest Reserve, *Mass Geesteranus 5746* (K). Kericho: Sambret catchment, *Kerfoot 3872* (EA, K). Kiambu: Gatamayu Forest, *Beentje 2722* (EA). Kirinyaga: south Mount Kenya, *Perdue & Kibuwa 8295* (EA, K). Meru: Nyambeni Hills, *SAJIT Z0144* (HIB). Murang'a: Kieni Forest, *Kokwaro & Mathenge 2990* (EA). Nakuru: Endabarra, *Bally B4845* (EA). Nyandarua: south Kinangop, *Gibert & Thulin 1038* (EA). Samburu: Mathew's Range, *Luke 14229*

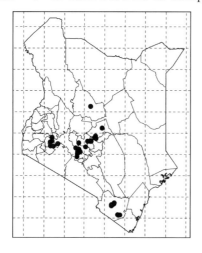

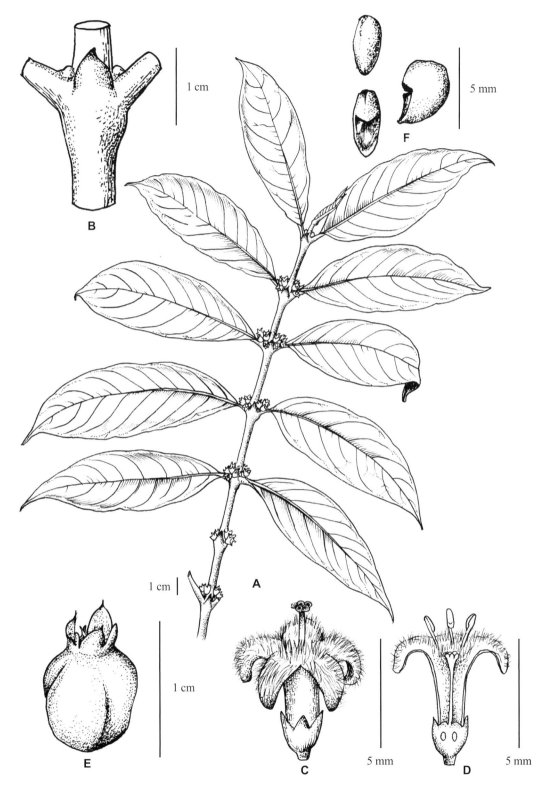

Figure 2 *Lasianthus kilimandscharicus*. A. flowering branch; B. portion of branch showing the stipule; C. long-styled flower; D. longitudinal section of short-styled flower, showing the stamens, style and ovary; E. fruit; F. pyrenes. Drawn by NJ.

Plate 2 A–E. *Lasianthus kilimandscharicus*. Photo by YDZ.

(EA, K). Nyeri: Ragati Forest Station, *Dyson 460* (EA). Teita Taveta: Mbololo Hill Forest, *Muasya et al. 480* (EA); Mount Kasigau, *Kimberly 605* (EA). Tharaka-Nithi: Chogoria Forest, *SAJIT 002953* (EA).

3. Trib. **Craterispermeae** Verdc.

Shrubs or trees, mostly with yellow-green foliage. Stipules entire, connate to form a tube. Flowers heterostylous. Ovary 2-1ocular, with a single pendulous ovule in each locule. Fruit a berry with one seed.

One genus occurs in Kenya.

3. **Craterispermum** Benth.

Shrubs or trees. Leaves opposite, yellow-green, glabrous, petiolate; blades mostly oblong or elliptic, coriaceous, net-veined; stipules entire, connate to form a tube. Inflorescences usually supra-axillary; peduncles short and stout or less often long and slender, strongly compressed; bracteoles present. Flowers hermaphrodite, heterostylous. Calyx-tube turbinate; limb cupular, truncate, sinuate or 5-toothed. Corolla funnel-shaped or shortly salver-shaped; tube short, hairy in throat; lobes 5, ovate, spreading. Stamens 5, inserted on the throat of the corolla, more or less exserted; anthers linear-oblong, dorsifixed. Ovary 2-locular, with a single pendulous ovule in each locule; style filiform; stigma bifid. Fruit a subglobose berry, 1-seeded.

About 16 species in tropical Africa, Madagascar and the Seychelles; only one species in Kenya.

1. **Craterispermum schweinfurthii** Hiern, Fl. Trop. Afr. 3: 161. 1877; F.T.E.A. Rubiac. 1: 162. 1976; K.T.S.L.: 510. 1994. —Type: South Sudan, Khor Bodo (Boddo), by streams, 10 Feb. 1870, *G. Schweinfurth 2935* [holotype: K (K000379487); isotype: M (M0106329)] (Figure 3; Plate 3)

Shrub or tree, up to 15 m tall, glabrous. Leaves, usually yellow-green when dry, blades elliptic, oblong or obovate, 5–18 cm × 2–8 cm, coriaceous, with fine net-veining, obtuse to shortly acuminate at the apex, cuneate at the base; stipules entire, 2.5–5 mm long, connate to form a tube. Inflorescences supra-axillary, several-flowered; peduncles stout, compressed, up to 1 cm long; bracts and bracteoles triangular, keeled, ca. 1.5 mm long. Calyx-tube 0.7–1.5 mm long, limb slightly toothed. Corolla white, sweet-scented; tube 3.5–6 mm long, hairy inside at the throat; lobes oblong-lanceolate to ovate, 3–6 mm long. Stamens 5, more or less exserted; anthers linear-oblong, dorsifixed. Ovary 2-locular, with a single pendulous ovule in each locule; style exserted or included, 7–7.5 mm long in long-styled flowers, 2.5–4 mm long in short-styled flowers; stigma bifid, lobes 1–2.5 mm long.

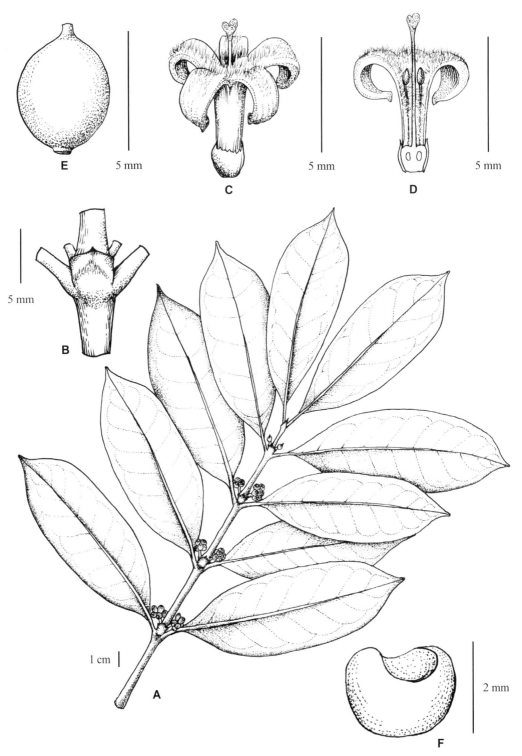

Figure 3 *Craterispermum schweinfurthii*. A. fruiting branch; B. portion of branch showing the stipules; C. long-styled flower; D. longitudinal section of long-styled flower, showing the stamens, style and ovary; E. fruit. F. seed. Drawn by NJ.

Plate 3 A–F. *Craterispermum schweinfurthii*. Photo by GWH.

Fruit a berry, subglobose or ellipsoid, dark purple when ripe, 5–6 mm long. Seeds bowl-shaped.

Distribution: Western and coastal Kenya. [Tropical Africa].

Habitat: Lowlands or upland evergreen forests or woodlands; 20–1650 m.

Kakamega: Kakamega Forest, *SAJIT 006779* (HIB). Kwale: Majoreni Area, *Luke & Luke 3797* (EA, K).

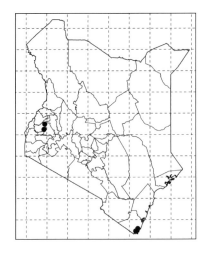

4. Trib. **Psychotrieae** Cham. & Schltdl.

Shrubs, trees or rarely herbs. Stipules divided or rarely entire. Flowers heterostylous. Ovary 2–8-locualr, with single erect ovule in each locule. Fruit a drupe or berry, fleshy, with 1-seeded pyrenes. Seeds often with horny endosperm.

Four genera occur in Kenya.

1a. Creeping herbs .... 6. *Geophila*
1b. Subshrubs, shrubs or rarely erect herbs 2
2a. Stipules entire or lobed, not becoming corky; fruit not dehiscent .... 7. *Psychotria*
2b. Stipules entire, often becoming corky; fruit more or less dehiscent 3
3a. Corolla-tube often slightly curved and winged; lobes often winged .... 5. *Chassalia*
3b. Corolla-tube and lobes not as above .... 4. *Eumachia*

4. **Eumachia** DC.

Shrubs. Stems usually 2-ribbed, with pale corky bark. Leaves opposite or 3–4-whorled; nodules absent; domatia present, small; stipules entire or 2(–many)-fid. Flowers heterostylous, 5-merous, several in terminal head-like or paniculate inflorescences; bracts and bracteoles very small or absent. Calyx-tube very short, with truncate or toothed limb. Corolla yellow or white; tube cylindrical, hairy at the throat; lobes triangular to elliptic-lanceolate. Stamens included or slightly exserted. Ovary 2-locular, with single erect ovule per locule; style with 2 stigma lobes. Fruit a drupe with 2 pyrenes. Seeds pale; endosperm not ruminate.

A genus of 83 species widely distributed in tropics of Africa, America, Asia, Australia, New Guinea, and Pacifc Islands; one species with two varieties in Kenya.

Figure 4 *Eumachia abrupta* var. *abrupta*. A. fruiting branch; B. longitudinal section of short-styled flower, showing the stamens, style and ovary; C. long-styled flower; D. seed. Drawn by NJ.

Plate 4 A–C. *Eumachia abrupta* var. *abrupta*. Photo by VMN.

1. **Eumachia abrupta** (Hiern) J.H. Kirkbr., J. Bot. Res. Inst. Texas 9: 76. 2015. ≡ *Psychotria abrupta* Hiern, Fl. Trop. Afr. 3: 205. 1877. ≡ *Chazaliella abrupta* (Hiern) E.M.A. Petit & Verdc., Kew Bull. 30: 268. 1975; F.T.E.A. Rubiac. 1: 117. 1976; K.T.S.L.: 508. 1994. —Types: Mozambique, Shigogo, Dec. 1860, *J. Kirk s.n.* [lectotype: K (K000412356), designated by C.M. Taylor et al. in Candollea 72: 295. 2017]; Mozambique, Shiramba Dembe, 14 Apr. 1860, *J. Kirk s.n.* [syntype: K (K000412357)]

Small shrub, up to 4.5 m tall. Stems with whitish-grey corky bark, internodes usually with 2 longitudinal keels. Leaves opposite or sometimes 3-whorled, shortly petiolate; blades elliptic to ovate-lanceolate, up to 20 cm × 10 cm, acute to acuminate at the apex, cuneate at the base; nodules absent; domatia reduced to small white tufts; stipules ovate or triangular, ca. 2 mm long, bifid or with several teeth. Flowers in terminal head-like clusters, 5-merous. Calyx-tube oblong-conic, ca. 1 mm long; limb very shallow. Corolla bright yellow; tube ca. 2.8 mm long; lobes ca. 1.5 mm long, spreading. Stamens included or slightly exserted. Ovary 2-locular, with single erect ovule per locule; style ca. 1.8 mm long in short-styled flowers, ca. 3.5 mm long in long-styled flowers. Fruit an ellipsoid drupe, 6–10 mm long; pyrenes 2, pale, 6–7 mm long. Seeds pale brown, compressed, ca. 5 mm long.

1a. Leaf-blades very thin, up to 20 cm × 10 cm .... a. var. *abrupta*
1b. Leaf-blades rather thick, less than 6 cm × 3 cm .... b. var. *parvifolia*

a. var. **abrupta** (Figure 4; Plate 4)

Leaf-blades very thin, large, up to 20 cm × 10 cm, usually narrowly acuminate.

Distribution: Coastal Kenya. [Ethiopia to south tropical Africa].

Habitat: Coastal evergreen forests or thickets; up to 500 m.

Kilifi: Kaya Kambe, *Robertson & Luke*

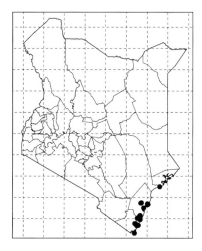

4801 (EA, K); Arabuko-Sokoke Forest, *Laugridge 91* (EA); Mangea Hill, *Luke & Robertson 1818* (EA). Kwale: Mwachi Forest Reserve, *Robertson & Luke 6175* (EA); Muhaka Forest, *Gillett 21052* (EA, K); Shimba Hills, *Magogo & Glover 375* (EA, K). Tana River: Nairobi Ranch, *Festo & Luke 2391* (EA, K).

b. var. **parvifolia** (Verdc.) C.M. Taylor, Candollea 72(2): 295. 2017. ≡ *Chazaliella abrupta* (Hiern) E.M.A. Petit & Verdc. var. *parvifolia* Verdc., Kew Bull. 30(2): 268. 1975; F.T.E.A. Rubiac. 1: 119. 1976; K.T.S.L.: 508. 1994. —Type: Kenya, Kilifi, Arabuko-Sokoke Forest, Mar. 1930, *R.M. Graham K2339* [holotype: K (K000318 634); isotypes: EA (EA000001777, EA000001778, EA000001779)]

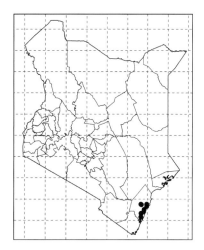

Leaf-blades rather thick, small, less than 6 cm × 3 cm, mostly blunt or subacute.

Distribution: Coastal Kenya. [Mozambique and Tanzania].

Habitat: Coastal evergreen forests; 50–450 m.

Kilifi: Mida Forest, *Padwa 69/18* (EA); Kaya Kivara, *Robertson & Luke* 4713 (EA).

5. **Chassalia** Comm. ex Poir.

Subshrubby herbs, small trees, or rarely lianas. Leaves opposite or rarely 3-whorled, usually thin, glabrous, sometimes with domatia. Stipules interpetiolar, sometimes united into a sheath, entire or with 2 short fimbriae. Inflorescences terminal, paniculate, often with subcapitate elements; bracts small. Flowers sessile or pedicellate, bisexual, 4–5-merous, usually distylous.

Calyx-tube mostly ovoid or oblong, somewhat ribbed. Buds often winged. Corolla white, pink or purple; tube cylindrical, often slightly curved and winged; lobes often winged. Ovary 2-locular; ovules solitary in each locule; stigmas 2, linear. Fruits pink, purple to black, succulent; pyrenes 2, plano-convex. Seeds compressed orbicular, endosperm fleshy.

About 110 species, mostly in tropical Africa, Madagascar and Mascarenes, few in tropical Asia; six species in Kenya.

1a. Lianas ....1. *C. cristata*
1b. Sub-shrubby herbs, shrubs, or small trees...2
2a. Corolla-tube more than 1.2 cm long; wings of buds, corolla-tube and lobes very evident.......3
2b. Corolla-tube 0.4–1.2 cm long; wings of buds, corolla-tube and lobes absent or not very evident 4
3a. Leaf-blades elliptic, apex acute; petioles often less than 2.5 cm long; stipules entire.................
.... 2. *C. umbraticola*
3b. Leaf-blades elliptic-oblanceolate, apex distinctly acuminate; petioles up to 4 cm long; stipules 2-lobed at the apex...3. *C. discolor*
4a. Corolla-tube 0.5–1.2 cm long; fruits didymous .... 6. *C. kenyensis*
4b. Corolla-tube 0.4–0.6 cm long; fruits not didymous...5
5a. Stipules often joined to form a sheath, 3–4 mm long; peduncles 2–6 cm long ....
.... 5. *C. subochreata*
5b. Stipules not joined, 1–2 mm long; peduncles 6–17 mm long ....4. *C. parvifolia*

1. **Chassalia cristata** (Hiern) Bremek., Bull. Jard. Bot. État Bruxelles 22: 104. 1952; F.T.E.A. Rubiac. 1: 122. 1976; K.T.S.L.: 507. 1994. ≡ *Psychotria cristata* Hiern, Fl. Trop. Afr. 3: 205. 1877. —Types: South Sudan, Khor Atiziri (Atazilly) near Rikkete's village, 28 Feb. 1870, *G. Schweinfurth 3159* [lectotype: K (K0004 30014), **designated here**]; D.R. Congo, Khor Dyagbe near Wando's village, 3 Mar. 1870, *G. Schweinfurth II-7* [syntype: K (K000 430013)]; D.R. Congo, Munsa (Munza), 31 Mar. 1870, *G. Schweinfurth 3463* [syntype: BM (BM000903534)] (Figure 5; Plate 5A–D)

Liana up to 6 m long. Leaf-blades elliptic-oblong, 4.5–15 cm × 1.5–6.5 cm, acuminate at the apex, cuneate at the base; petioles 0.5–2 cm long; stipules united into a sheath, up to 3.5 mm long, becoming corky. Inflorescences terminal, paniculate, up to 7.5 cm long; peduncles up to 2.2 cm long; pedicels obsolete; bracts and bracteoles small. Calyx-tube ovoid, very short; limb-tube up to 0.8 mm long; teeth broadly triangular, up to 0.5 mm long. Corolla pink,

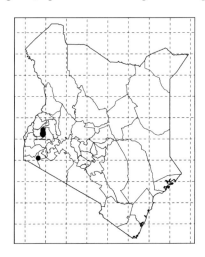

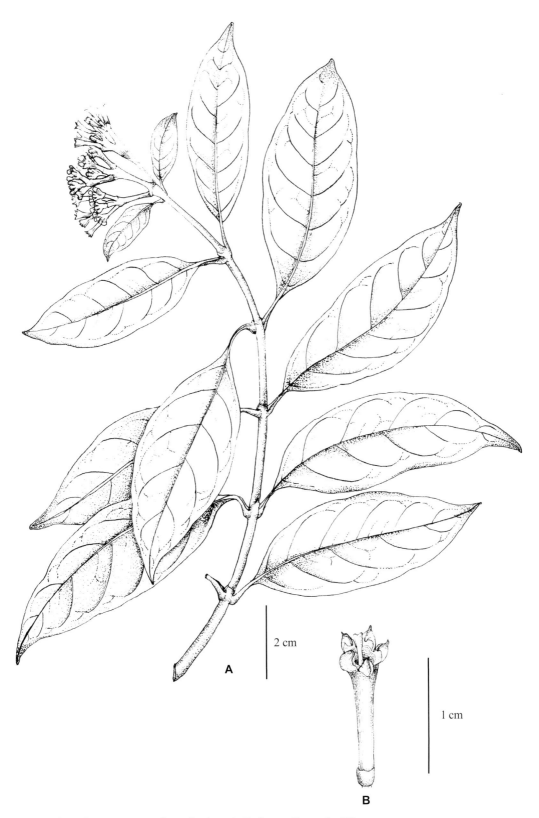

Figure 5 *Chassalia cristata*. A. a flowering branch; B. flower. Drawn by NJ.

white or purplish white outside, often white inside with a yellow ring at the throat; tube up to 1 cm long, slightly curved; lobes triangular-oblong, ca. 2 mm long. Styles dimorphic, 4.5–8.5 mm long. Fruits black or purple, subglobose, 5–6 mm long; pyrenes semi-globose, 4–5 mm in diameter. Seeds 3.5–5 mm long.

Distribution: Western Kenya. [Tropical Africa].

Habitat: Rainforests; 1350–1650 m.

Kagamega: Kakamega Forest, *SAJIT 005055* (HIB). Kisii: southern Nyanza, *Plaizier 1322* (EA).

2. **Chassalia umbraticola** Vatke., Oesterr. Bot. Z. 25: 230. 1875; F.T.E.A. Rubiac. 1: 123. 1976; W.F.E.A.: 150. 1987; K.T.S.L.: 508. 1994. —Type: Tanzania, Zanzibar Island, Sept. 1873, *J.M. Hildebrandt 1158* [holotype: B (destroyed); lectotype: K (K 000319116), **designated here**; isolectotype: BM (BM000903530)] (Plate 5E–G)

Shrub or rarely sub-shrubby herb, up to 4.5 m tall. Leaf-blades elliptic, 3–16 cm × 1.5–6.5 cm, acute or slightly acuminate at the apex, cuneate at the base; petioles 0.5–2.5 cm long, stipules broad, 1–1.5 mm long, undivided, sometimes becoming corky. Inflorescences terminal; peduncles up to 4 cm long; pedicels absent or very short; bracts very small. Calyx-tube oblong-ovoid, up to 2 mm long, slightly ribbed; limb-tube 0.2–0.5 mm long; lobes triangular, 0.1–0.5 mm long. Corolla cream to white; tube 1.5–2 cm long; lobes lanceolate, 5–7 mm long, winged. Styles dimorphic, 1–2 cm long. Fruits black, subglobose, slightly compressed, 4–7 mm in diameter.

Distribution: Coastal Kenya. [East Africa, from Kenya to Mozambique].

Habitat: Coastal forests; up to 400 m.

Kilifi: Arabuko-Sokoke Forest, *SAJIT 004655* (HIB); Kaya Jibana, *Robertson & Luke 4511* (EA); Jilore Forest, *Simpson 51* (EA). Kwale: Shimba Hills, *SAJIT 005471* (HIB); Gongoni Forest, *SAJIT 005563* (HIB); Muhaka Forest, *Brenan et al. 14530* (EA, K). Lamu: Boni Forest Reserve, *Luke & Robertson 1518* (EA, K). Mombasa: Mowesa, *Graham K1705* (EA, K). Tana River: Nairobi Ranch, *Festo & Luke 2484* (EA, K); Mto Moni, *Robertson & Luke 5419* (EA).

3. **Chassalia discolor** K. Schum., Bot. Jahrb. Syst. 34: 339. 1904; F.T.E.A. Rubiac. 1: 126. 1976; K.T.S.L.: 508. 1994. —Type: Tanzania, Lushoto Usambara Mountains, 16 Dec. 1895, *J. Buchwald 102* [lectotype: K (K000318654), **designated here**; isolectotype: BM (BM000903531)]

Shrub up to 6 m tall. Leaf-blades elliptic- to ovate-oblanceolate, 4–15.5 cm × 1.5–7 cm, narrowly acuminate at the apex, cuneate at the base; petiole up to 4 cm long; stipules triangular, 3–4 mm long, 2-lobed at the apex usually becoming corky. Inflorescences terminal, paniculate; peduncles up to 2 cm long; pedicels obsolete; bracts small. Calyx-tube oblong, 1–

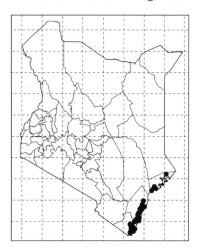

Plate 5 A–D. *Chassalia cristata*; E–G. *C. umbraticola*. Photo by GWH (A, C–F), TD (B) and VMN (G).

1.8 mm long, ribbed; limb-tube 0.7–1.2 mm long; lobes linear to triangular, 0.5–4 mm long. Corolla white suffused pink; tube slightly curved, 1–2 cm long; lobes elliptic-lanceolate, 2–4 mm long, winged. Styles dimorphic, 7–15 mm long. Fruits white, reddish or purple-black, subglobose, 5–6 mm in diameter; pyrenes subglobose.

subsp. **taitensis** Verdc., Kew Bull. 30: 275. 1975; F.T.E.A. Rubiac. 1: 127. 1976. —Type: Kenya, Taita Hills, Ngangao Forest, 1800–1925 m, 7 Jule 1969, *R.B. Faden & A. Evans 69/877* [holotype: K (K000353077); isotype: EA (EA000001774)] (Figure 6; Plate 6A, B)

Calyx-lobes linear to linear-triangular, 2.5–4 mm long.
Distribution: Southern Kenya. [Endemic].
Habitat: Moist evergreen forests; 1400–1900 m.
Note: The other two subspecies of *Chassalia discolor* K. Schum., subsp. *discolor* and subsp. *grandifolia* Verdc. only occur in Tanzania and both with short calyx-lobes less than 2.5 mm long.

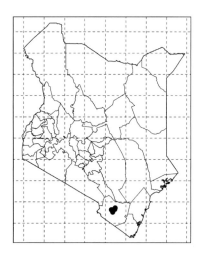

Teita Taveta: Ngangao Forest, *SAJIT 006342* (HIB); Chawia Forest, *Mwachala et al. 1057* (EA).

4. **Chassalia parvifolia** K. Schum., Bot. Jahrb. Syst. 28: 103. 1901; F.T.E.A. Rubiac. 1: 130. 1976; K.T.S.L.: 508. 1994. —Types: Tanzania, SE. Uluguru Mts., Kikurungu, *F.L. Stuhlmann 9253* [holotype: B (destroyed)]; Tanzania, Iringa, Mufindi, Kigogo Forest Reserve, 1810 m, 30 Dec. 1988, *R. Gereau, J. Lovett & C.M. Taylor 2758* [neotype: MO (MO-279863), **designated here**] (Plate 6 C–G)

Shrub to small tree, up to 7.5 m tall, with 2-ribbed or ridged stems. Leaf-blades elliptic to obovate-oblanceolate, 1–13 cm × 0.5–5 cm, abruptly acuminate at the apex, cuneate at the base; petioles short; stipules minute. Inflorescences terminal, paniculate; peduncles up to 1.7 cm long. Calyx-tube squarish, ca. 0.8 mm long; limb-tube very short, with ovate-triangular teeth. Corolla white, often tinged or tipped with pink; tube 4–6 mm long; lobes ovate, ca. 2 mm long. Fruits ovoid or ellipsoid, 4.5–5 mm long; pyrenes pale, 4–5 mm long.
Distribution: Southern Kenya. [D.R. Congo to east tropical Africa and west Mozambique].

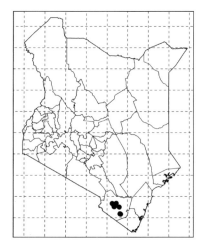

Habitat: Evergreen forests; 1300–1950 m.
Teita Taveta: Ngangao Forest, *SAJIT*

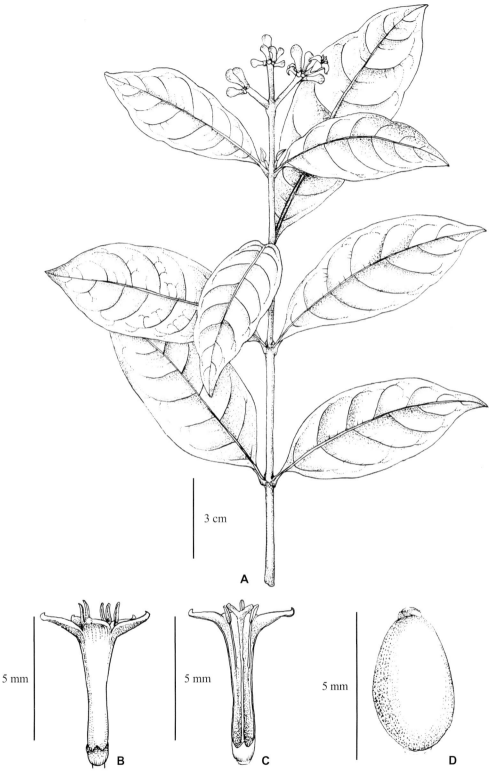

Figure 6 *Chassalia discolor* subsp. *taitensis*. A. flowering branch; B. flower; C. longitudinal section of flower, showing the stamens and style; D. fruit. Drawn by NJ.

Plate 6 A, B. *Chassalia discolor* subsp. *taitensis*; C–G. *C. parvifolia*. Photo by GWH (A, B, D–G) and BL (C).

006344 (HIB); Chawia Forest, *SAJIT 006382* (HIB); Mbololo Forest, *De Block et al. 471* (EA)

5. **Chassalia subochreata** (De Wild.) Robyns, Fl. Spermatophyt. Parc Nat. Albert 2: 367. 1947; F.T.E.A. Rubiac. 1: 131. 1976; K.T.S.L.: 508. 1994. ≡ *Psychotria subochreata* De Wild., Pl. Bequaert. 2: 428. 1924. —Type: D.R. Congo, between Masisi and Walikale, 31 Dec. 1914, *J. Bequaert 6433* [holotype: BR (BR0000008106674); isotype: BR (BR0000008828606)]

Shrub or small tree, up to 9 m tall. Leaf-blades oblanceolate to narrowly elliptic, 3.5–18 cm × 1.5–5.5 cm, narrowly acuminate at the apex, cuneate-attenuate at the base; petioles up to 3.5 cm long; stipules joined to form a sheath, 3–4 mm long. Inflorescences terminal, paniculate; peduncles 2–6 cm long. Calyx-tube ovoid, 0.8–1 mm long; limb-tube very short, with minute triangular teeth. Corolla white; tube 5–6 mm long; lobes oblong-elliptic, ca. 4.5 mm long. Styles dimorphic, 2.5–6 mm long. Fruits black, ellipsoid, 4.5–6.5 mm long; pyrenes pale, ca. 5 mm long.

Distribution: Western Kenya. [D.R. Congo to east tropical Africa].

Habitat: Evergreen forests; 1650–2550 m.

Bomet: southwestern Mau Forest Reserve, *Maas Geesteranus 5772* (K, L); Kericho: Sambret Catchment, *Kerfoot EAH11418* (EA).

6. **Chassalia kenyensis** Verdc., Kew Bulletin 30: 279. 1975; F.T.E.A. Rubiac. 1: 132. 1976; K.T.S.L.: 508. 1994. —Type: Kenya, Limuru, Limuru Girls High School, 2300 m, 22 Feb. 1970, *R.B. Faden & A. Evans 70/72* [holotype: K (K000318992); isotype: EA (EA 000002925)]

Shrub up to 2 m tall, occasionally somewhat scandent. Leaf-blades elliptic, 5–15 cm × 1.5–5 cm, acuminate at the apex, cuneate at the base; petioles up to 1.5 cm long; stipules 1.5–2 mm long, entire. Inflorescences terminal, paniculate; peduncles 1.5–3 cm long. Calyx-tube oblong, 0.5–1 mm long; limb-tube very short; lobes broadly triangular, ca. 0.5 mm long. Corolla white, tipped with purple; tube slightly curved, 0.5–1.2 cm long; lobes oblong-lanceolate, 3–4 mm long. Styles dimorphic, 3–7 mm long. Fruits reddish to purple, subglobose, 4–5 mm in diameter; pyrenes semiglobose, ca. 4 mm long.

Distribution: Central Kenya. [Tanzania].

Habitat: Evergreen forests; 1650–2300 m.

Kiambu: near Limuru Girls School, *Faden & Evans 70/72* (EA, K). Kirinyaga:

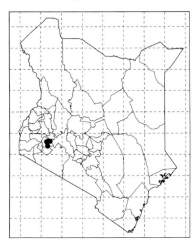

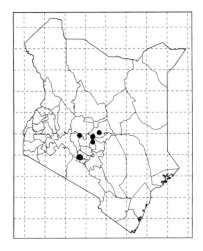

south Mount Kenya, *Gardner 1167* (EA). Laikipia: Kangari Tea Estate, *Luke 444* (EA). Meru: northeast Mount Kenya, *Rammell 1083* (EA, K).

6. **Geophila** D. Don

Perennial herbs; stems creeping, rooting at nodes. Leaves opposite, long petiolate; blade usually broadly ovate to cordate; stipules interpetiolar, entire or 2-lobed. Inflorescences terminal or pseudoaxillary, cymose to capitate, 1–several-flowered, pedunculate, bracteate. Flowers bisexual, sometimes heterostylous, sessile or subsessile. Calyx-tube obovoid, limb very short, 4–7-lobed. Corolla white, cylindrical or funnel-form; lobes 4–7. Stamens 4–7, inserted in corolla tube corolla-tube, included or slightly exserted; filaments filiform; anthers dorsifixed. Ovary 2-locular with single erect ovule in each locule; style slender; stigma 2-lobed. Fruit a drupe, orange or red, globose to ellipsoid; pyrenes 2. Seed 1 in each pyrene; testa membranous; endosperm corneous.

About 29 species widespread in tropical Africa, Madagascar, Asia, central, north, and south America; two species in Kenya.

1a. Inflorescences 1(–3)-flowered; bracts 1–2, lanceolate .... 1. *G. herbacea*
1b. Inflorescences several-flowered; bracts several, leafy, obovate, rounded-elliptic or rhomboid .. .. 2. *G. obvallata*

1. **Geophila herbacea** (Jacq.) K. Schum., Nat. Pflanzenfam. 4(4): 119. 1891. ≡ *Psychotria herbacea* Jacq., Enum. Syst. Pl. 16. 1760. —Type: Rheede, Hort. Malab. 10: t. 21. 1690 [lectotype, designated by Howard, Fl. Lesser Antilles 6(3): 416. 1989] (Figure 7; Plate 7A–D)

Creeping herb, up to 30 cm long, rooting at nodes, densely appressed hairy. Leaf-blades ovate to suborbicular with cordate base, 1–5 cm across; petioles up to 12 cm long, pubescent; stipules small, broadly ovate. Inflorescences 1(–3)-flowered; bracts 1–2, lanceolate, 2.5–4 mm long. Calyx-tube obconic, 1–2 mm long; limb-tube 0.8–1.2 mm long, with 4 lanceolate lobes 1.5–3 mm long. Corolla white; tube cylindrical, 0.5–1.3 mm long; lobes 4, ovate-lanceolate, 3–9 mm long. Stamens 4, inserted in corolla-tube. Drupes red, subglobose, 5–12 mm in diameter; pyrenes 2, 3.5–4 mm long. Seed 1 in each pyrene, ca. 3 mm long.

Distribution: Western, central, and coastal Kenya. [Tropical and subtropical Old World to Pacific].

Habitat: Evergreen forests; up to 1600 m.

Note: *Geophila herbacea* (Jacq.) K. Schum. had been always treated as a synonym of

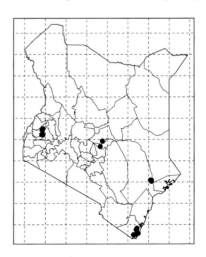

Figure 7 *Geophila herbacea*. Drawn by NJ.

Geophila repens (L.) I.M. Johnst (e.g., F.T.E.A. Rubiac. 1: 110) with the lectotype collected from the new world and a much wider pan-tropical distribution range. A recent molecular study has shown that the old world and new world samples formed different clades. For more details see I.M. Lim & K.M. Wong in K.M. Wong et al., Fl. Singapore 13: 92–95. 2019.

Kakamega: Kakamega Forest, *SAJIT 005051* (HIB); Malava Forest, *Faden & Evens 69/2043* (EA). Kwale: Shimba Hills, *SAJIT 006148* (HIB); Gongoni Forest Reserve, *Robertson & Luke 5941* (EA, K). Meru: Lower Imenti Forest, *Faden & Faden 74/889* (EA, K). Tana River: Primate Reserve, *Luke & Robertson 1157* (EA); Baomo North Forest, *Luke et al. TPR194* (EA). Tharaka-Nithi: Chuka Forest, *SAJIT VK0052* (HIB).

2. Geophila obvallata (Schumach.) Didr., Vidensk. Meddel. Naturhist. Foren. Kjøbenhavn 1854: 186. 1854; F.T.E.A. Rubiac. 1: 112. 1976. ≡ *Psychotria obvallata* Schumach., Beskr. Guin. Pl. 111. 1827. —Type: Ghana, *P. Thonning s.n.* [holotype: C]

Creeping herb, up to 60 cm long, rooting at nodes. Leaves triangular-ovate to ovate-reniform, 0.8–4(–9) cm × 0.6–5.5 cm, apex acute to rounded, base cordate; petiole up to 10 cm long; stipules elliptic, very small. Inflorescences several-flowered, enclosed in a whorl of bracts; bracts several, leafy, obovate, rounded-elliptic or rhomboid, up to 1.5 cm long. Calyx-lobes subulate, linear-lanceolate, narrowly triangular or distinctly spathulate, up to 6.5 mm long. Corolla white; tube funnel-shaped, 3–6.5 mm long; lobes ovate-oblong, 1.2–3 mm long. Styles 0.3–1 mm long. Berries black, purple or blue, 4–8 mm long; pyrenes

Plate 7 A–D. *Geophila herbacea*; E, F. *G. obvallata* subsp. *ioides*. Photo by GWH (A, C–F) and NW (B).

half-ovoid, 4–4.7 mm long. Seeds brown, ca. 3 mm long.

subsp. **ioides** (K. Schum.) Verdc., Kew Bull. 30: 267. 1975; F.T.E.A. Rubiac. 1: 113. 1976. ≡ *Geophila ioides* K. Schum., Pflanzenw. Ost-Afrikas C: 392. 1895. —Type: Mozambique, Quelimane, Jan. 1889, *F.L. Stuhlmann 711* [holotype: B (destroyed); lectotype: HBG (HBG521614), **designated here**] (Plate 7E, F)

Leaves rounded-reniform to ovate. Calyx-lobes linear or linear subulate or very slightly spathulate at the apex, 1.5–3.5 mm long. Corolla-tube 3–4 mm long.

Distribution: Coastal Kenya. [Burundi, Mozambique, and Tanzania].

Habitat: Coastal evergreen forests; up to 500 m.

Note: There are four subspecies recorded in *Geophila obvallata* (Schumach.) Didr.: subsp. *ioides* (K. Schum.) Verdc. distributed in Kenya; subsp. *obvallata* occurs in west and central Africa, from Guinea to D.R. Congo; subsp. *involucrata* (Schweinf. ex Hiern) Verdc. occurs in central Africa from Sudan to D.R. Congo; subsp. *pilosa* Figueiredo only occurs in Angola.

Kilifi: Mangea Hill, *Luke & Robertson 986* (EA). Kwale: Shimba Hills, *Magogo & Glover 1018* (EA, K); Gongoni Forest Reserve, *Robertson & Luke 6338* (EA).

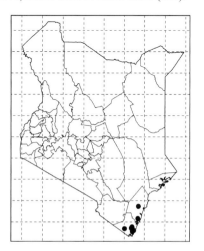

7. **Psychotria** L.

Small trees, shrubs, subshrubs, lianas, or rarely herbs. Leaves opposite or rarely whorled, either with domatia or with bacterial nodules scattered throughout the leaf-blades; stipules either entire or fimbriate, usually deciduous. Flowers mostly small, 4–5(–6)-merous, bisexual, heterostylous, sessile or pedicellate, in terminal or axillary, capitate or paniculate inflorescences. Calyx-limb minutely lobed. Corolla funnel-form to tubular, inside glabrous or variously pubescent. Ovary 2-locular; ovule single in each locule. Fruits drupaceous, fleshy, ellipsoid, ovoid, or subglobose; pyrenes 1–2, each with single seed. Seed testa usually reddish brown; endosperm fleshy or corneous, sometimes ruminate.

A very large genus with more than 1800 species, widely distributed in the tropics and subtropics of America, Asia, Africa, Madagascar and Pacific islands. In total, there are 21 known species occur in Kenya which are listed in the key and described below. *Psychotria* sp. B [*Faden et al. 72/211* (EA)] and sp. C [*Glover et al. 1411* (EA)] which were recorded by B. Verdcourt in FTEA as suspicious species need more investigation. Another specimen [*SAJIT 004565* (HIB)] collected from Taita Hills does not match any species in this genus.

1a. Inflorescences a dense head, surrounded by an involucre of separate or fused bracts 2
1b. Inflorescences lax, not surrounded by an involucre 4
2a. Calyx-lobes less than 4 mm long .... 19. *P. tanganyicensis*
2b. Calyx-lobes more than 5 mm long 3
3a. Leaf-blades beneath glabrous to sparely pubescent, but not ferruginous .... 20. *P. ceratoloba*
3b. Leaf-blades beneath densely ferruginous pubescent .... 21. *P. mildbraedii*
4a. Nodules scattered on leaves 5
4b. Nodules absent or inconspicuous on mid-rib 9
5a. Corolla-tube glabrous inside .... 18. *P. alsophila*
5b. Corolla-tube hairy inside 6
6a. Peduncles distinctly winged .... 10. *P. lauracea*
6b. Peduncles not winged 7
7a. Young stems and leaves usually pubescent; the component of inflorescences often quite dense
    .... 17. *P. kirkii*
7b. Young stems and leaves always glabrous; the component of inflorescences more often lax 8
8a. Young branches glabrous; leaves glabrous, always more than 10 cm long .... 16. *P. faucicola*
8b. Young branches glabrous to pubescent; leaves glabrous to slightly pubescent, less than 9 cm
    long .... 15. *P. punctata*
9a. Stipules obviously bilobed 10
9b. Stipules entire or very shortly bifid or toothed at the apex 13
10a. Shrubs; leaves always very small, less than 8 cm long 11
10b. Shrubs or trees; leaves always large, up to 23 cm long 12
11a. Flowers in panicles; calyx-limb 1.5–2 mm long .... 11. *P. amboniana*
11b. Flowers in fascicles; calyx-limb 0.5–0.8 mm long .... 14. *P. holtzii*
12a. Tree up to 24 m tall; fruit scarcely grooved .... 9. *P. mahonii*
12b. Shrub or small tree, up to 7.5 m tall; fruit distinctly grooved .... 7. *P. fractinervata*
13a. Domatia present 14
13b. Domatia absent 17
14a. Corolla 10–12 mm long; fruit blue .... 2. *P. crassipetala*
14b. Corolla less than 9 mm long; fruit red 15
15a. Stipules elliptic to ovate, 1–3.5 cm long .... 5. *P. orophila*
15b. Stipules triangular to ovate, mostly less than 1 cm long 16
16a. Corolla yellow .... 3. *P. bagshawei*
16b. Corolla white .... 1. *P. capensis*
17a. Leaves with ferruginous hairs on venation beneath .... 4. *P. pseudoplatyphylla*
17b. Leaves glabrous or nearly so, but never with ferruginous hairs 18
18a. Stipules 25–30 mm long .... 6. *P. taitensis*
18b. Stipules less than 11 mm long 19
19a. Fruit rounded, 11–16 mm in diameter .... 8. *P. petitii*
19b. Fruit ellipsoid, 6–8 mm long 20

20a. Calyx-limb 1–1.5 mm long .... 12. *P. leucopoda*
20b. Calyx-limb 0.5–0.8 mm long .... 13. *P. schliebenii*

1. **Psychotria capensis** (Eckl.) Vatke, Oesterr. Bot. Z. 25: 230. 1875; F.T.E.A. Rubiac. 1: 39. 1976. ≡ *Logania capensis* Eckl., S. Afr. Quart. Joun. 1: 371. 1830. —Type: South Africa, in the woods of the Zuurberg, near Bontjes River, 1000 ft (not localized)

= *Grumilea riparia* K. Schum. & K. Krause, Bot. Jahrb. Syst. 39: 560. 1970 ≡ *Psychotria riparia* (K. Schum. & K. Krause) E.M.A. Petit, Bull. Jard. Bot. État Bruxelles 34: 43. 1964; F.T.E.A. Rubiac. 1: 38. 1976; K.T.S.L.: 537. 1994. ≡ *Psychotria capensis* (Eckl.) Vatke subsp. *riparia* (K. Schum. & K. Krause) Verdc., Fl. Zambes. 5(1): 13. 1989; F.T.E.A. Rubiac. 3: 924. 1991. —Type: Tanzania, Morogoro, Liwale R., *W. Busse 557* [holotype: B (destroyed); lectotype: EA, designated by E.M.A. Petit in Bull. Jard. Bot. État Bruxelles 34: 56. 1964; isolectotype: BR (BR0000009887824)]

Shrub or small tree, up to 5 m tall. Leaf-blades ovate to obovate, 3–20 cm × 1–10 cm, oblong, rounded to acuminate at the apex, narrowly cuneate to rounded at the base; nodules absent; domatia often present; petioles up to 4 cm long; stipules triangular to ovate, 3–8 mm long. Flowers many in much-branched panicles, with peduncle up to 15 cm long. Calyx-tube obconical, 1–2 mm long; limb cupuliform, 1–1.5 mm long, with triangular lobes. Corolla-tube 2–5 mm long; lobes oblong-triangular, 1.5–3 mm long. Fruit red, subglobose or ovoid, 5–7 mm long.

1a. Young branches and leaves glabrous or nearly so .... a. var. *capensis*
1b. Young branches and leaves densely ferruginous pubescent .... b. var. *puberula*

a. var. **capensis** (Plate 8A, B)

Young branches and leaves glabrous or nearly so.

Distribution: Central, eastern, southern, and coastal Kenya. [Eastern and southern Africa].

Habitat: Riverine or dry forest, hillsides, bushland or thickets of roadsides; up to 1800 m.

Embu: Mashamba, *Robertson 2040* (EA). Isiolo: Meru National Park, *Gillett 20166* (EA). Kajiado: Emali Hill, *Beentje 2522* (EA). Kiambu: Thika, *Faden 66121* (EA). Kilifi: Malindi, *Mwadime 039* (EA). Kitui: Endau, *Mbonge 28* (EA). Kwale: Shimba Hills, *Magogo & Glover 228* (K); Mrima Hill, *Robertson MDE75* (EA). Lamu: Witu Forest, *Robertson & Luke 5481* (EA). Machakos: Kindaruma Dam, *Gillett & Faden 18237* (EA). Makueni: Chyulu Hills, *Mumiukha 411* (EA). Teita Taveta: Sagala Forest, *SAJIT 005381* (HIB). Tana River: Mbia Forest, *Luke &*

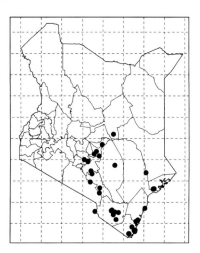

Robertson 1232 (EA). Tharaka-Nithi: Kijegge Hill, *Beentje & Powys 4072* (EA).

b. var. **puberula** (E.M.A. Petit) Verdc., Fl. Zambes. 5(1): 14. 1989; F.T.E.A. Rubiac. 3: 924. 1991. ≡ *Psychotria riparia* (K. Schum. & K. Krause) E.M.A. Petit var. *puberula* E.M.A. Petit, Bull. Jard. Bot. État Bruxelles 34: 46. 1964; K.T.S.L.: 537. 1994. —Type: Kenya, Kitui Boma, 1158 m, 18 Jan. 1942, *P.R.O. Bally 1529* [holotype: K (K000311717); isotypes: BR (BR0000009888234), G (G00350047)] (Plate 8C, D)

Young branches and leaves densely ferruginous pubescent.

Distribution: Central, southern, and coastal Kenya. [East Africa, from Ethiopia to Mozambique].

Habitat: Coastal forests, dry forests or hillsides; up to 1200 m.

Kilifi: Malindi, *Polhill & Paulo 802* (EA). Kwale: Shimba Hills, *SAJIT V0309* (HIB); Dzombo Hill, *Robertson et al. MDE210* (EA); Buda Mafisini Forest, *Drummond & Hemsley 3951* (EA). Machakos: Moboloni, *Bally B8530* (EA). Makueni: Boma, *Bally B1529* (EA). Teita Taveta: Taita Hills, *Mwachala et al. in EW 779* (K).

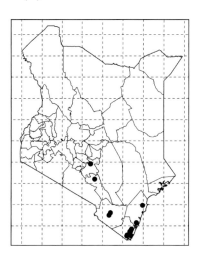

2. **Psychotria crassipetala** E.M.A. Petit, Bull. Jard. Bot. État Bruxelles 34: 191. 1964; F.T.E.A. Rubiac. 1: 40. 1976; K.T.S.L.: 534. 1994. —Type: Kenya, Taita Hills, NNE of Ngerenyi, Ngangao, 1850 m, 15 Sept. 1953, *R.B. Drummond & J.H. Hemsley 4336* [holotype: K (K000311716); isotypes: BR (BR0000008852243), EA (EA000001820)]

Shrub or small tree, up to 10 m tall. Leaf-blades elliptic to broadly elliptic or obovate, 6–15 cm × 3–9 cm, rounded to shortly acuminate at the apex, cuneate at the base, glabrous; nodules absent; domatia present; petioles up to 5 cm long, glabrous; stipules ovate-triangular, ca. 5 mm long. Flowers heterostylous, 5-merous, many in much-branched panicles. Calyx-tube subglobose, ca. 1.5 mm long, glabrous; lobes ca. 1 mm long. Corolla white; tube funnel-shaped, 4–5 mm long, glabrous inside; lobes markedly thick, ca. 6 mm long. Fruits greenish blue, subglobose, ca. 8 mm in diameter.

Distribution: Southern and coastal Kenya. [Tanzania].

Habitat: Montane forests or hillsides; 50–1900 m.

Lamu: Utwani Forest, *Dale K3831*

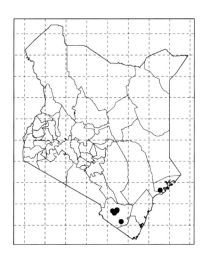

(EA). Teita Taveta: Mbololo Hill, *Wakanene & Mwangangi 440* (EA); Mount Kasigau, *Muasya & Medley 681* (EA); Ngangao Forest, *Faden & Evans 69/875* (EA).

3. **Psychotria bagshawei** E.M.A. Petit, Bull. Jard. Bot. État Bruxelles 34: 55. 1964; F.T.E.A. Rubiac. 1: 43. 1976; K.T.S.L.: 534. 1994. —Type: Uganda, Toro, Isunga, 8 Jule 1906, *A.G. Bagshawe 1092* [lectotype: BM (BM000903373), designated by E.M.A. Petit in Bull. Jard. Bot. État Bruxelles 34: 56. 1964] (Plate 8E, F)

Shrub or small tree, up to 3 m tall. Leaf-blades elliptic, 4–20 cm × 2–8 cm, acute or shortly acuminate at the apex, narrowly cuneate at the base, glabrous; petioles up to 5 cm long; stipules ovate to triangular, 5–10 mm long, shortly ferruginous-puberulous. Flowers heterostylous, 5-merous, many in much-branched panicles, the individual clusters always congested. Calyx-tube obconical, ca. 1 mm long, glabrous; limb cupuliform, ca. 1 mm long. Corolla-tube 2.5–3.5 mm long; lobes oblong, 2.5–3 mm long. Fruits red, subglobose, 4.5–7 mm in diameter.

Distribution: Western and central Kenya. [D.R. Congo, Tanzania, and Uganda].

Habitat: Evergreen forests; 1550–1700 m.

Kakamega: Kakamega Forest, *SAJIT 006786* (HIB). Kiambu: Limuru, *Fukuoka K143* (EA). Nandi: Yala River Bridge, *Gillett 16721* (EA, K).

4. **Psychotria pseudoplatyphylla** E.M.A. Petit, Bull. Jard. Bot. État Bruxelles 34: 90. 1964; F.T.E.A. Rubiac. 1: 48. 1976; K.T.S.L.: 537. 1994. —Type: Tanzania, SE side of Kilimanjaro, 1800 m, 16 Dec. 1933, *H.J. Schlieben 4362* [holotype: BR (BR0000008855145); isotypes: BM (BM000903359), G (G00342436), HBG (HBG521153), LISC (LISC002676), S (S-G-5123), M, P] (Plate 9A, B)

Shrub or small tree, up to 4 m tall. Leaf-blades elliptic to obovate, 6–22 cm × 3–11.5 cm, obtuse to shortly acuminate at the apex, cuneate at the base, glabrous above, ferruginous pubescent on the nerves beneath; nodules absent; domatia sometimes present; petioles up to 6 cm long; stipules elliptic, 1.7–3.5 cm long. Flowers heterostylous, in much-branched panicles. Calyx-tube conic, ca. 0.5 mm long; limb cupuliform, ca. 0.6 mm long. Corolla-tube 4.5–5 mm long; lobes ovate, 2–2.5 mm long. Fruits ellipsoid or subglobose, 4.5–10 mm in diameter.

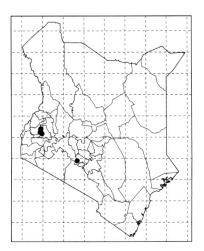

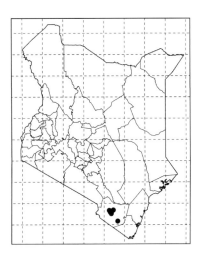

Plate 8 A, B. *Psychotria capensis* var. *capensis*; C, D. *P. capensis* var. *puberula*; E, F. *P. bagshawei*. Photo by GWH (A, B, E, F) and VMN (C, D).

Distribution: Southern Kenya. [Tanzania].
Habitat: Forests; 1150–2000 m.

Teita Taveta: Ngangao Forest, *SAJIT 004516* (HIB); Mount Kasigau, *Hemp 5367* (K); Mbololo Forest, *Block et al. 492* (K).

5. **Psychotria orophila** E.M.A. Petit, Bull. Jard. Bot. État Bruxelles 34: 92. 1964; F.T.E.A. Rubiac. 1: 49. 1976; K.T.S.L.: 536. 1994. —Type: Kenya, Masailand, near Nairobi, *C.F. Elliott 27* [lectotype: K (K000412275), designated by E. Petit in Bull. Jard. Bot. État Bruxelles 34: 93. 1964] (Plate 9C–F)

Shrub or small tree, up to 10 m tall. Leaf-blades elliptic or oblong-elliptic, 7–18 cm × 3.5–10 cm, obtuse or slightly acuminate at the apex, cuneate at the base, glabrous above, glabrous beneath or pubesent on nerves; petioles up to 3 cm long; nodules absent; domatia hairy; stipules ovate or ovate-elliptic, 1–2 cm long. Flowers heterostylous, in much-branched panicles; peduncles 2–7(–15) cm long. Calyx-tube obconic, ca. 1 mm long; limb cupuliform, 1–1.5 mm long. Corolla-tube 2.5–5 mm long; lobes oblong-elliptic, 1.8–4 mm long. Fruits ellipsoid, 6–7 mm long.

Distribution: Northern, western, central, and southern Kenya. [Ethiopia, Tanzania, and Uganda].

Habitat: Montane forests; 1550–2600 m.

Note: The only collection from northern Narok, *Glover et al. 1411* (EA), was identified as *Psychotria* sp. B in F.T.E.A. and K.T.S.L., while the only obviously distinction between *Glover et al. 1411* (EA) and *P. orophila* E.M.A. Petit was that the calyx-limb of the former (ca. 2.5 mm long) was much longer than the later (1–1.5 mm long). We thought this variation in *Glover et al. 1411* (EA) was not a stable feature, because of some shorter calyx-limbs also found on the same specimens. Here, we treated *P.* sp. B as *P. orophila*.

Baringo: Lembus Forest, *Buch 61/69* (EA). Bungoma: Mount Elgon, *Hedberg 277* (K). Kajiado: Ngong Hills, *Khayota 62* (EA). Kakamega: Kakamega Forest, *SAJIT 004687* (HIB). Kericho: Tinderet Forest Reserve, *Geesteranus 5425* (K); Mau Forest, *Kerfoot 1605* (EA). Kiambu: Kieni Forest, *Beentje 2844* (EA); Tigoni Forest, *Birmie 683* (EA). Kirinyaga: Castle Forest Station, *Spjut & Ensor 2991* (EA). Makueni: Chyulu Hills, *Bally 1094* (EA). Marsabit: Mount Kulal, *Herlocker H-438* (EA); Mount Nyiru, *Luke 13998* (EA). Meru: Nyambeni Forest, *Polhill & Verdcourt 269* (EA); Marimba Forest, *Verdcourt & Polhill 2984* (EA). Nakuru: Eburru Forest, *Luke et al. 8836* (K). Nyandarua: Aberdare National Park, *Croat 28393* (K). Samburu: Ndoto Mountains, *Bytebier & Kirika 36* (EA); Mount Nyiru, *Kerfoot 1953* (EA). Tharaka-Nithi: Chogoria Forest, *SAJIT 001349* (HIB). Trans Nzoia: Saiwa Swamp, *Beentje 3032* (EA).

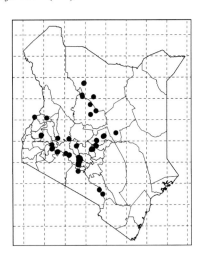

6. **Psychotria taitensis** Verdc., Kew Bull. 30: 248. 1975; F.T.E.A. Rubiac. 1: 50. 1976; K.T.S.L.: 538. 1994. —Type: Kenya,

Plate 9 A, B. *Psychotria pseudoplatyphylla*; C–F. *P. orophila*. Photo by GWH (A, E), ZWW (B) and YDZ (C, D, F).

Mt. Kasigau, pipeline road from Rukanga, 1400–1600 m, 6 Feb. 1971, *R.B. Faden et al. 71/153* [holotype: EA (EA000001870); isotypes: EA (EA000001867, EA000001868, EA000001869), K (K000311715), MO (MO-2049832)]

Small understorey tree. Leaf-blades elliptic or oblong-elliptic, 7.5–14 cm × 3.5–7 cm, very shortly bluntly acuminate or subacute at the apex, cuneate at the base; petioles up to 3.5 cm long; stipules large, broadly elliptic to round, 2.5–3 cm long. Flowers in branched inflorescences, each component capitate; peduncles up to 7 cm long. Calyx-tube obconic to semigloboses, 0.7–1.2 mm long; limb-lobes ca. 1 mm long. Corolla-tube ca. 2.5 mm long; lobes ovate-lanceolate, ca. 2 mm long. Fruits red, rounded, 7–8 mm in diameter.

Distribution: Southern Kenya. [Endemic].
Habitat: Forests; 1450–1500 m.

Teita Taveta: Mount Kasigau, *Luke et al. 4151* (EA).

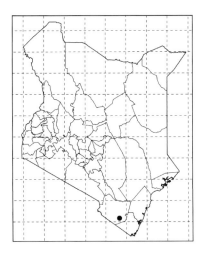

7. **Psychotria fractinervata** E.M.A. Petit, Bull. Jard. Bot. État Bruxelles 34: 127. 1964; F.T.E.A. Rubiac. 1: 50. 1976; K.T.S.L.: 535. 1994. —Type: Tanzania, Kilimanjaro, Marangu, Oct. 1893, *G. Volkens 1120* [lectotype: K (K000412274), designated by E. Petit in Bull. Jard. Bot. État Bruxelles 34: 128. 1964; isolectotypes: BM (BM000903351), BR (BR0000008853448), HBG (HBG521183)] (Figure 8; Plate 10A–C)

Shrub or small tree, up to 7.5 m tall. Leaf-blades narrowly elliptic, oblong-elliptic or ovate, 6–20 cm × 2–8.5 cm, acute to acuminate at the apex, cuneate at the base; domatia present; nodules absent; petioles up to 2.5 cm long; stipules obovate, 1–2 cm long, bifid at the apex. Flowers heterostylous, in much-branched panicles; peduncles up to 10 cm long. Calyx-tube hemispherical, 1–2 mm long; limb cupuliform, 2–3 mm long; lobes irregularly triangular, 0.5–1 mm long. Corolla-tube 4.5–7 mm long; lobes oblong, 3–6 mm long. Fruits ellipsoid to globose, 5–7 mm long.

Distribution: Northern, western, and central Kenya. [Tanzania].
Habitat: Upland forests; 1450–2600 m.

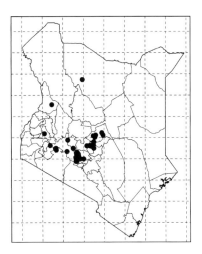

Embu: Irangi, *Ossent 170* (EA). Kericho: Sambret Catchment, *Kerfoot 4088* (EA). Kiambu: Kieni Forest, *Beentje 2917* (EA); Limuru, *Luke et al. 8370* (EA). Kirinyaga:

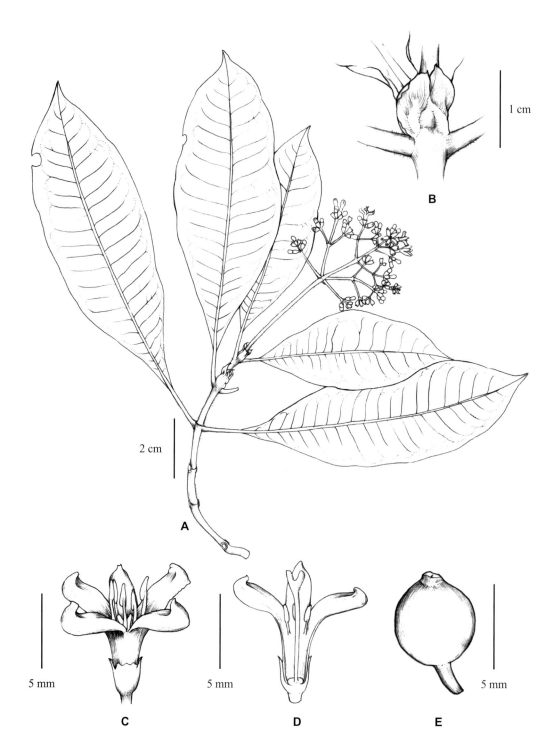

Figure 8 *Psychotria fractinervata*. A. flowering branch; B. portion of branch showing the stipule; C. short-styled flower; D. longitudinal section of long-styled flower, showing the stamens and style; E. fruit. Drawn by NJ.

Kamweti Forest, *Beentje 3867* (K); Castle Forest Reserve, *Tarrow JKCAT383* (EA). Marsabit: Mount Kulal, *Luke et al. 10822* (EA). Meru: Nyambene Hills, *SAJIT Z0170* (HIB); Marimba, *Verdcourt & Polhill 2985* (EA). Nakuru: Eburru Forest, *Luke et al. 8924* (EA). Nandi: Kapsabet, *SAJIT 006596* (HIB). Nyandarua: Aberdares Mountains, *Gardner K2392* (EA). Nyeri: Ragati Forest, *Beentje 3849* (K). Tharaka-Nithi: Mount Kenya, *SAJIT 002942* (HIB). West Pokot: Cherangani Hills, *SAJIT 006903* (HIB).

8. **Psychotria petitii** Verdc., Kew Bull. 30: 249. 1975; F.T.E.A. Rubiac. 1: 54. 1976; K.T.S.L.: 537. 1994. —Type: Kenya, Taita Hills, 8 km NNE of Ngerenyi, Ngangao, 1850 m, 15 Sept. 1953, *R.B. Drummond & J.H. Hemsley 4343* [holotype: K (K000412272); isotypes: BR (BR0000008855084), EA (EA000001817), K (K000412273)]

Shrub or small tree, up to 9 m tall. Leaf-blades elliptic to oblong-elliptic, 4.5–13.5 cm × 2.2–5 cm, shortly acuminate at the apex, cuneate at the base; nodules and domatia absent; petioles up to 3 cm long; stipules ovate elliptic, 7–10 mm long. Flowers heterostylous, 5-merous, in branched panicles, each component dense; peduncles up to 6 cm long. Calyx-tube subglobose, ca. 1 mm long; limb ca. 1 mm long, lobes oblong to triangular, up to 1 mm long. Corolla-tube ca. 4 mm long; lobes ovate-triangular, ovate-elliptic, ca. 2 mm long. Fruits globose, 1–1.6 cm in diameter.

Distribution: Southern Kenya. [Endemic].
Habitat: Forests; 1450–2200 m.

Teita Taveta: Ngangao Forest, *SAJIT 006348* (HIB); Mbololo Hill, *Faden et al. 72/265* (EA); Vuria Hill, *Beentje et al. 1127* (K).

9. **Psychotria mahonii** C.H. Wright, Bull. Misc. Inform. Kew 1906(4): 106. 1906; F.T.E.A. Rubiac. 1: 58. 1976; K.T.S.L.: 536. 1994. —Type: Malawi, Likangala stream, cultivated at Kew, *J. Mahon 597-1898* [holotype: K (K000412302)]

Large tree, up to 15(–24) m tall. Leaf-blades narrowly elliptic, oblong-elliptic or obovate, 3–23 cm × 1.5–10 cm, acuminate at the apex, cuneate to rounded at the base, glabrous to slightly pubescent or distinctly ferruginous hairy beneath; nodules absent; domatia hairy; petioles up to 3.5 cm long; stipules obovate, 0.4–1.7 cm long; bifid at the apex. Flowers heterostylous, (4–)5-merous, in much-branched panicles; peduncles 1.5–8 cm long. Calyx-tube obconic, ca. 1 mm long; limb cupuliform, 1–1.5 mm long. Corolla-tube 4–6 mm long; lobes elliptic-oblong, 2.5–3.5 mm long. Fruits red, subglobose, 5–6 mm in diameter, slightly grooved.

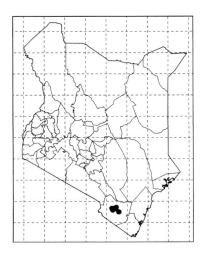

1a. Leaf-blades small, less than 16 cm × 9 cm, glabrous to slightly pubescent, but never ferruginous... a. var. *puberula*
1b. Leaf-blades large, up to 22 cm × 10 cm, glabrous to distinctly ferruginous hairy beneath.........
.... b. var. *pubescens*

a. var. **puberula** (E.M.A. Petit) Verdc., Kew Bull. 30: 253. 1975; F.T.E.A. Rubiac. 1: 60. 1976; K.T.S.L.: 536. 1994. ≡ *Psychotria megistosticta* (S. Moore) E.M.A. Petit var. *puberula* E.M.A. Petit, Bull. Jard. Bot. État Bruxelles 34: 116. 1964. —Type: Tanzania, Njombe, Ruhudji R., Lupembe, Sept. 1931, *H.J. Schlieben 1168A* [holotype: BR (BR 0000008853974); isotypes: BM (BM000 903353), G, K (K000412271), M (M010 6247)] (Plate 10D, E)

Leaf-blades small, less than 16 cm × 9 cm, with 8–13 pairs of lateral nerves, glabrous to slightly pubescent, but never ferruginous.

Distribution: Western and central Kenya. [East Africa].

Habitat: Montane forests; 1800–2450 m.

Bomet: Sotik, *Bally B7835* (EA). Kiambu: Kieni Forest, *Beentje & Mungai 2901* (EA). Kirinyaga: Kamweti Forest, *Faden et al. 71/870* (EA). Murang'a: Kimakia Forest, *Tayler 15145* (EA). Nandi: North Nandi Forest, *SAJIT 006632* (HIB). Nyandarua: Aberdares Mountains, *Wimberk K1513* (EA). Nyeri: Kagochi, *Kerfoot 1497* (EA). Tharaka-Nithi: Chogoria, *Kirika et al. KMK10* (EA). Trans Nzoia: near Kitale, *Welster 8830* (EA).

b. var. **pubescens** (Robyns) Verdc., Kew Bull. 30: 254. 1975; F.T.E.A. Rubiac. 1: 61. 1976; K.T.S.L.: 536. 1994. ≡ *Grumilea bequaertii* De Wild. var. *pubescens* Robyns, Bull. Jard. Bot. État Bruxelles 17: 96. 1943. —Type: D.R. Congo, Kivu, between Lubenga and Sake, Feb. 1932, *J. Lebrun 5019* [lectotype: BR(BR0000008854230), designated by O. Lachenaud in Opera Bot. Belg. 17: 147. 2019; isolectotype: BR (BR0000008852861)] (Plate 10F, G)

Leaf-blades large, up to 22 cm × 10 cm, with 7–22 pairs of lateral nerves, glabrous to distinctly ferruginous hairy beneath.

Distribution: Western and central Kenya. [D.R. Congo, Tanzania, and Uganda].

Habitat: Montane or riverine forests;

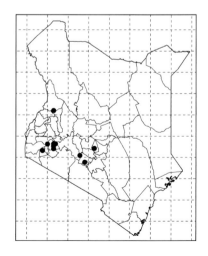

Plate 10 A–C. *Psychotria fractinervata*; D, E. *P. mahonii* var. *puberula*; F, G. *P. mahonii* var. *pubescens*. Photo by GWH.

1600–2350 m.

Bomet: Southwest Mau Forest, *Geesteranus 5749* (K). Kericho: Sambret Catchment, *Kerfoot 4337* (EA). Kiambu: Kijabe, *Buch 60/128* (EA). Kirinyaga: Kimaru Tea Estate, *Gilbert & Gilbert 6005* (EA, K). Kisii: *Napier 3004* (EA, K). Nairobi: Karura, *Hutchins s.n.* (EA). West Pokot: Cherangani, *Tweedie 2801* (EA).

10. **Psychotria lauracea** (K. Schum.) E.M.A. Petit, Bull. Jard. Bot. État Bruxelles 34: 129. 1964; F.T.E.A. Rubiac. 1: 61. 1976; K.T.S.L.: 535. 1994. ≡ *Grumilea lauracea* K. Schum., Pflanzenw. Ost-Afrikas, C: 392. 1895. —Type: Tanzania, Kilimanjaro, Marangu, 15 Dec. 1893, *G. Volkens 1393* [lectotype: BR (BR0000008853820), designated by E. Petit in Bull. Jard. Bot. État Bruxelles 34: 131. 1964; isolectotypes: E (E00193745), PRE (PRE0594731-0), HBG (HBG521175), K (K000412268)] (Figure 9; Plate 11A, B)

Shrub or small tree, up to 6 m tall. Leaves oblong-elliptic or ovate lanceolate, 5–22 cm × 3–10.5 cm, obtusely acuminate or rounded at the apex, cuneate to rounded at the base, with or without nodules; stipules ovate or oblong, 1–2 cm, long, bifid at the apex. Flowers sweet-scented, heterostylous, in much-branched pyramidal panicles; peduncles 4–12 cm long, glabrous, winged. Calyx-tube obconic, ca. 0.8 mm long; limb cupuliform, subtruncate. Corolla-tube funnel-shaped, 2–3 mm long; lobes ovate, 2–2.5 mm long. Fruits red, subglobose, 4–6 mm in diameter.

Distribution: Western, central, and coastal Kenya. [D.R. Congo, Rwanda, Tanzania, Uganda, and Zambia].

Habitat: Forests; up to 1900 m.

Embu: *Bally 356* (K). Kilifi: Pangani Rocks, *Luke 1837* (EA). Kwale: Shimba Hills, *SAJIT 005508* (HIB); Gongoni Forest Reserve, *Robertson & Luke 5975* (EA); Mrima Hill, *Faden 70/236* (EA). Lamu: Witu Forest, *Robertson & Luke 5486* (EA). Meru: Nyambeni Hills, *Polhill & Verdcourt 276* (EA). Migori: Bukeria, *Napier 2938* (EA). Narok: Endama, *Glover et al. 1890* (EA); Lolgorien, *Kuchar et al. 5531* (EA). Samburu: Kijegge Hill, *Kamau 190* (EA). Teita Taveta: Taita Hills, *Mwachala et al. 779* (EA); Mount Kasigau, *Medley 647* (EA); Mbololo Hill, *Burney & Burney T315* (EA).

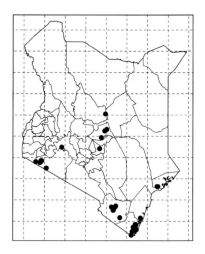

11. **Psychotria amboniana** K. Schum., Pflanzenw. Ost-Afrikas, C: 390. 1895; F.T.E.A. Rubiac. 1: 77. 1976; W.F.E.A.: 155. 1987; K.T.S.L.: 534. 1994. —Types: Tanzania, Tanga, Amboni, Jule 1893, *C. Holst 2716* [syntype: B (destroyed); lectotype: K (K000284036), **designated here**; isolectotype: HBG (HBG521214)] & *C. Holst. 2885a* [syntype: B (destroyed)] (Figure 10; Plate 11C–F)

= *Psychotria amboniana* K. Schum. var. *velutina* (E.M.A. Petit) Verdc., Kew Bull. 30: 260. 1975; F.T.E.A. Rubiac. 1: 79. 1976

≡ *Psychotria albidocalyx* K. Schum. var. *velutina* Petit., Bull. Jard. Bot. État

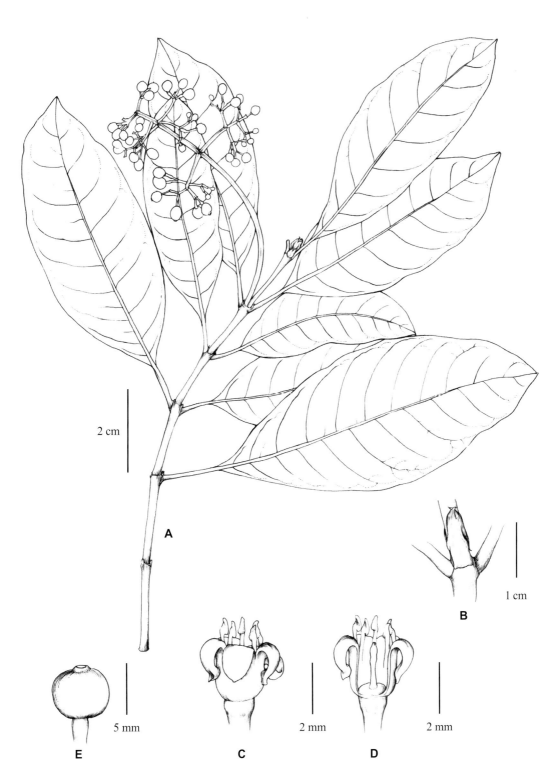

Figure 9 *Psychotria lauracea*. A. fruiting branch; B. portion of branch showing the stipule; C. flower; D. longitudinal section of short-styled flower, showing the stamens and style; E. fruit. Drawn by NJ.

Plate 11 A, B. *Psychotria lauracea*; C–F. *P. amboniana*. Photo by GWH (A–C, F), BL (D) and ZXZ (E).

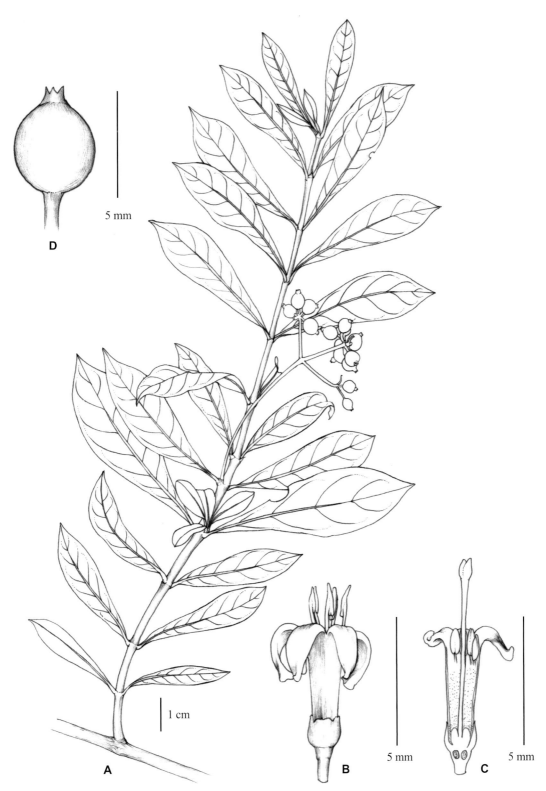

Figure 10 *Psychotria amboniana*. A. a fruiting branch; B. short-styled flower; C. longitudinal section of long-styled flower, showing the stamens, style and ovary; D. fruit. Drawn by NJ.

Bruxelles 36: 83. 1966. —Type: Kenya, Kwale, Mazeras, *R.M. Graham* in *F.D. 2337* [holotype: K; isotypes: EA (EA000001872), G (G00350034)]

= *Psychotria albidocalyx* K. Schum., Abh. Preuss. Akad. Wiss. 1894: 21. 1894. —Type: Kenya, Kwale, Bome, 15 Mar. 1902, *T. Kässner 307* [holotype: BM (BM000903599); isotype: K (K000284057)]

Shrub 3 m tall. Leaves oblong-elliptic to narrowly obovate, 1.5–8 cm × 0.2–3.3 cm, acute to shortly acuminate at the apex, very narrowly cuneate at the base, petioles up to 3(–15) mm long; domatia present; nodules inconspicuous on mid-rib; stipules 1–2 mm long, bifid at the apex. Flowers heterostylous, many in rather congested glabrous panicles. Calyx-tube campanulate, 0.5–1 mm long; limb cupuliform, 1.2–2 mm long. Corolla-tube 4–6 mm long; lobes narrowly oblong, 3–4 mm long, appendiculate at the apex. Fruits subglobose to ovoid, 4–6 mm in diameter.

Distribution: Coastal Kenya. [Tanzania].

Habitat: Lowland forests; up to 450 m.

Kilifi: Mangea Hill, *Luke & Robertson 600B* (EA); Arabuko-Sokoke Forest, *Langridge 101* (EA); Kaya Kivara, *Robertson & Luke 4768* (EA). Kwale: Shimba Hills, *SAJIT 005469* (HIB); Chuna Forest, *Luke & Robertson 573* (EA); Gongoni Forest, *Luke et al. 8352* (EA). Lamu: Kiunga Region, *Muchiri 505* (EA); Hindi to Kiunga, *Luke & Robertson 1451* (EA); Witu Forest, *Gillet 20421* (EA). Mombasa: *Linder 2660* (K). Tana River: south of Bfunde, *Robertson & Luke 5339* (EA); Nairobi Ranch, *Festo & Luke 2489* (EA).

12. **Psychotria leucopoda** E.M.A. Petit, Bull. Jard. Bot. État Bruxelles 36: 84. 1966; F.T.E.A. Rubiac. 1: 79. 1976; K.T.S.L.: 536. 1994. —Type: Tanzania, E. Usambara Mts., Sigi R., 12.8 km below Amani, 457 m, 29 Dec. 1956, *B. Verdcourt 1748* [holotype: BR (BR0000008853929); isotypes: EA (EA 000001871), K (K000412262), PRE (PRE 0594732-0)] (Plate 12A)

Shrub or small tree, up to 3 m tall. Leaf-blades elliptic to narrowly oblong-elliptic, 3.5–18 cm × 1–7 cm, acuminate at the apex, cuneate at the base; glabrous or ciliate along mid-rib; nodules inconspicuous; petioles up to 2 cm long; stipules ovate-triangular, 3–6 mm long. Flowers heterostylous, in branched panicles; peduncles up to 6 cm long. Calyx-tube ca. 1 mm long; limb deeply cupuliform, 1–1.7 mm long. Corolla-tube 3.5–5 mm long; lobes oblong-lanceolate, 2.5–3 mm long.

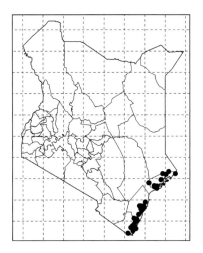

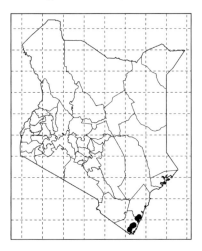

Fruits ellipsoid, 6–8 mm long.

Distribution: Coastal Kenya. [Tanzania].

Habitat: Lowland forests; up to 400 m.

Kilifi: Kaya Rabai, *Luke et al. 6295* (EA). Kwale: Shimba Hills, *SAJIT 006010* (HIB); Mkongani North Forest, *Luke & Robertson 513* (EA).

13. **Psychotria schliebenii** E.M.A. Petit, Bull. Jard. Bot. État Bruxelles 36: 86. 1966; F.T.E.A. Rubiac. 1: 80. 1976; K.T.S.L.: 537. 1994. —Type: Tanzania, Ulanga, Mahenge, 900–1000 m, 8 Feb. 1932, *H.J. Schlieben 1733* [holotype: BR (BR0000008854001); isotypes: B (B100160776), BM (BM000903597), G (G 00350042), HBG (HBG521132), LISC (LISC 002677), K (K000284113), M (M0106257), S (S-G-5129), P]

Shrub, up to 3 m tall. Leaf-blades narrowly elliptic, oblong-elliptic or obovate, 4.5–18 cm × 1.5–9 cm, acuminate at the apex, cuneate at the base, glabrous; nodules often on midrib, inconspicuous; petiole up to 8 cm long; stipules ovate-triangular, 3–5 mm long. Flowers heterostylous, in much-branched panicles; peduncles very short or up to 7 cm long. Calyx-tube ovoid, ca. 1 mm long; limb cupuliform, very short, lobes triangular or oblong, up to 1 mm long. Corolla-tube 3–4 mm long; lobes triangular-ovate, 2–3 mm long. Fruits red, ellipsoid, 6.5–8 mm long.

1a. Peduncles usually very short .... a. var. *sessilipaniculata*
1b. Peduncles up to 7 cm long .... b. var. *parvipaniculata*

a. var. **sessilipaniculata** E.M.A. Petit, Bull. Jard. Bot. État Bruxelles 36: 84. 1966; F.T.E.A. Rubiac. 1: 81. 1976; K.T.S.L.: 537. 1994. —Type: Tanzania, Mvomero, Turiani, Nov. 1953, *S.R. Semsei 1499* [holotype: K (K000412261); isotypes: BR (BR0000008854087), EA (EA000001788)]

Leaf-blades 7–18 cm × 2.5–9 cm. Peduncles less than 4 mm long.

Distribution: Coastal Kenya. [Tanzania].

Habitat: Lowland forests; up to 450 m.

Kwale: Shimba Hills, *Magogo & Glover 14* (EA); Gongoni Forest, *Robertson & Luke 5939* (EA). Tana River: Tana River National Primate Reserve, *Luke & Robertson 1165* (EA).

b. var. **parvipaniculata** E.M.A. Petit, Bull.

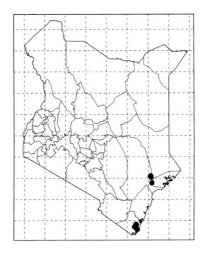

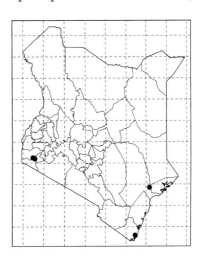

Jard. Bot. État Bruxelles 36: 87. 1966; F.T.E.A. Rubiac. 1: 81. 1976; K.T.S.L.: 537. 1994. —Type: Kenya, Kwale, Gazi, Gogoni Forest, *I.R. Dale in F.D. 3570* [holotype: BR (BR000 0008854018), isotype: K (K000311712)]

Leaf-blades 3–13 cm × 1–3.5 cm. Peduncles developed, up to 7 cm long.

Distribution: Western and coastal Kenya. [Endemic].

Habitat: forests; up to 1350 m.

Homa Bay: Kabuoch Forest, *Jarret 41* (EA). Kwale: Gongoni Forest, *Luke et al. 8348* (EA). Migori: Kuja River, *Glasgow 46/5* (EA).

14. **Psychotria holtzii** (K. Schum.) E.M.A. Petit, Bull. Jard. Bot. État Bruxelles 36: 90. 1966; F.T.E.A. Rubiac. 1: 83. 1976; K.T.S.L.: 535. 1994. ≡ *Anthospermum holtzii* K. Schum., Bot. Jahrb. Syst. 34: 340. 1904. —Type: Tanzania, Dar es Salaam, Mogo Forest, *A. Engler 2137, 2161 & 2186* [syntypes: B (destroyed)]; ibid., *C. Holtz 301/02* [neotype: EA, designated by E. Petit in Bull. Jard. Bot. État Bruxelles 36: 90. 1966] (Plate 12B, C)

= *Psychotria holtzii* (K. Schum.) E.M.A. Petit var. *pubescens* Verdc., Kew Bull. 30: 261. 1975; F.T.E.A. Rubiac. 1: 84. 1976. —Type: Tanzania, ca. 25.6 km west of Dar es Salaam on Morogoro Road, Oct. 1957, *J.R. Welch 397* [holotype: K; isotype: EA (EA000001787)]

Shrub, up to 3 m tall. Leaf-blades narrowly elliptic to oblong, 1–4 cm × 0.5–1.5 cm, acute or sub-obtuse at the apex, cuneate at the base; petioles up to 2.5 mm long; stipules prominently bilobed, 1–2 mm long. Flowers heterostylous, in sessile few-flowered subumbellate inflorescences. Calyx-tube obconic, ca. 0.8 mm long, pubescent; limb very short, 0.5–0.8 mm long. Corolla-tube 3–3.5 mm long; lobes oblong-elliptic, 1.7–3 mm long, appendiculate at the apex. Fruits crimson-red, subglobose, ca. 5 mm in diameter, crowned by the persistent calyx.

Distribution: Coastal Kenya. [Tanzania].
Habitat: Lowland forests; up to 500 m.

Kilifi: Mangea Hill, *Luke & Robertson 322* (EA). Kwale: Shimba Hills, *SAJIT 005507* (HIB); Gongoni Forest, *Luke & Robertson 2370* (EA); Mrima Hill, *Verdcourt 1884* (EA).

15. **Psychotria punctata** Vatke, Oesterr. Bot. Z. 25(7): 230. 1875; F.T.E.A. Rubiac. 1: 89. 1976; K.T.S.L.: 537. 1994. —Type: Tanzania, Zanzibar Island, Sept. 1873, *J.M. Hildebrandt 1136* [holotype: W; isotypes: BM (BM000903589), K (K000284234)]

Shrub, up to 3 m tall. Leaf-blades elliptic to ovate-elliptic, 3–13 cm × 1–6.3 cm, obtuse, rounded or emarginate, rarely acute at the apex, cuneate at the base, glabrous or puberulous; petioles up to 2 cm long; stipules ovate-triangular, up to 2.4 mm long, slightly bifid at the apex. Flowers sweet-scented, heterostylous, in lax panicles. Calyx-tube turbinate, 0.5–1 mm long; limb cupuliform, 0.6–1 mm long. Corolla-tube 3.5–5.5 mm long; lobes oblong-elliptic, 2.5–3.5 mm long. Fruits red, subglobose, 5–6 mm in diameter.

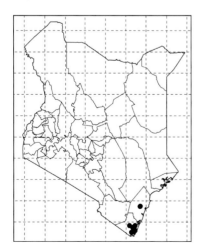

Plate 12 A. *Psychotria leucopoda*; B, C. *P. holtzii*. Photo by GWH.

1a. Plant glabrous; leaf-blades up to 13 cm × 6.3 cm..a. var. *punctata*
1b. Young stems sparsely to densely pubescent but soon glabrous; leaf-blades less than 9 cm × 2.5 cm 2
2a. Leaf-blades very small, less than 3.5 cm × 2.5 cm .... b. var. *minor*
2b. Leaf-blades narrowly oblanceolate, up to 9 cm × 2.2 cm.. c. var. *tenuis*

a. var. **punctata** (Plate 13A, B)

Plant glabrous; leaf-blades thick, up to 13 cm × 6.3 cm.

Distribution: Coastal Kenya. [Tanzania and Comoros Islands].

Habitat: Bushlands or forests; up to 250 m.

Kilifi: Gede Forest, *Gerhardt & Steiner 262* (EA); Malindi, *Robertson 7645* (EA). Kwale: Kaya Waa, *SAJIT 005576* (HIB); Funzi Island, *SAJIT 005600* (HIB); Gazi Bay, *SAJIT 006238* (HIB). Lamu: Kiunga, *Gillespie 80* (K); Witu Forest, *Faden et al. 74/1135* (EA).

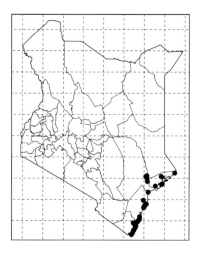

Mombasa: English Point, *Napier 3252* (EA). Tana River: Tana River Primate Reserve, *Luke et al. TPR29* (EA); Kurawa, *Polhill & Paulo 632* (EA).

b. var. **minor** E.M.A. Petit, Bull. Jard. Bot. État Bruxelles 36: 112. 1966; F.T.E.A. Rubiac. 1: 90. 1976; K.T.S.L.: 537. 1994. —Type: Kenya, Samburu to Mackinnon Road, 3 Aug. 1953, *R.B. Drummond & J.H. Hemsley 4070* [holotype: K; isotypes: BR (BR0000008853493), EA (EA000001799), S (S-G-5124), SRGH (SRGH0106688-0)]

Young stems sparsely to densely pubescent but soon glabrous; leaf-blaes very small, less than 3.5 cm × 2.5 cm.

Distribution: Coastal Kenya. [Tanzania].
Habitat: Bushlands; up to 350 m.

Kilifi: Dakabuka Hill, *Dale K1074* (EA). Kwale: between Samburu and Mackinnon Road, *Drummond & Hemsley 4070* (EA).

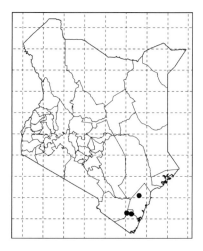

c. var. **tenuis** E.M.A. Petit, Bull. Jard. Bot. État Bruxelles 36: 112. 1966; F.T.E.A. Rubiac. 1: 90. 1976; K.T.S.L.: 537. 1994. —Type: Kenya, Kilifi, Arabuko-Sokoke Forest, Aug. 1936, *I.R. Dale* in *F.D. 3535* [holotype: BR (BR0000008853509); isotypes: EA (EA 000001798), K]

Young stems sparsely to densely

pubescent but soon glabrous; leaf-blades narrowly oblanceolate, up to 9 cm × 2.2 cm.

Distribution: Coastal Kenya. [Endemic].

Habitat: Bushlands and forests; up to 100 m.

Kilifi: Arabuko-Sokoke Forest, *Luke & Robertson 2605* (EA); Kaya Kauma, *Robertson & Luke 5701* (EA).

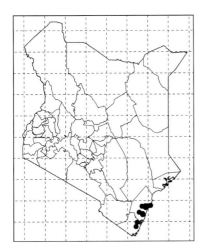

16. **Psychotria faucicola** K. Schum., Bot. Jahrb. Syst. 34: 336. 1904; F.T.E.A. Rubiac. 1: 90. 1976; K.T.S.L.: 535. 1994. —Type: Tanzania, E. Usambara Mts., Amani, *A. Engler 606* [holotype: B (destroyed)]; ibid., 15 Sept. 1903, *O. Warnecke* in *Herb. Amani 474* [neotype: K (K000284651), designated by E. Petit in Bull. Jard. Bot. État Bruxelles 36: 124. 1966; isoneotypes: BM (BM000903582), E (E00193726), EA (EA000001797), P] (Plate 13C, D)

Shrub or subshrub, up to 3.6 m tall. Leaf-blades broadly to narrowly elliptic, 10–25 cm × 4–12 cm, glabrous on both surfaces, acute to shortly acuminate at the apex, cuneate at the base; nodules usually numerous; petioles up to 9 cm long, glabrous; stipules 4–13 mm long. Flowers heterostylous, in much-branched panicles. Calyx-tube turbinate, ca. 1 mm long; limb cupuliform, 0.7–1 mm long, with obsolete lobes. Corolla-tube 2–2.8 mm long; lobes oblong-elliptic, 1.5–2.3 mm long. Fruits red, subglobose, 5–6.5 mm in diameter.

Distribution: Coastal Kenya. [Tanzania].

Habitat: Forests; up to 450 m.

Kilifi: Gedi Forest, *Faden et al. 71/680*

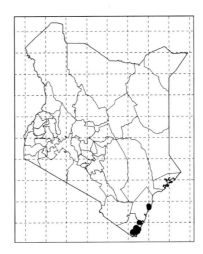

(EA). Kwale: Shimba Hills, *SAJIT V0353* (HIB); Gongoni Forest, *Robertson & Luke 5966* (EA); Mrima Hill, *Faden et al. 70/242* (EA).

17. **Psychotria kirkii** Hiern, Fl. Trop. Afr. 3: 206. 1877; F.T.E.A. Rubiac. 1: 92. 1976; K.T.S.L.: 535. 1994. —Type: Mozambique, near Morumballa, 1858, *J. Kirk 9* [holotype: K (K000412294)] (Plate 13E, F)

= *Psychotria kirkii* Hiern var. *nairobiensis* (Bremek.) Verdc., Kew Bull. 30: 262. 1975; F.T.E.A. Rubiac. 1: 94. 1976; K.T.S.L.: 535. 1994. ≡ *Psychotria nairobiensis* Bremek., J. Bot. 71: 278. 1933. —Type: Kenya, Nairobi, 22 Sept. 1916, *Dawson 490* [holotype: K].

= *Psychotria kirkii* Hiern var. *volkensii* (K. Schum.) Verdc., Kew Bull. 30: 262. 1975; F.T.E.A. Rubiac. 1: 94. 1976; K.T.S.L.: 535.

1994. ≡ *Psychotria volkensii* K.Schum., Pflanzenw. Ost-Afrikas, C: 390. 1895. —Type: Tanzania, Kilimanjaro, Murangu, Jule 1893, *G. Volkens 604* [syntype: B (destroyed); lectotype: K (K000284613), designated by E. Petit in Bull. Jard. Bot. État Bruxelles 36: 114. 1966; isolectotype: BM (BM000903585)]

= *Psychotria kirkii* Hiern var. *tarambassica* (Bremek.) Verdc., Kew Bull. 30: 263. 1975; F.T.E.A. Rubiac. 1: 95. 1976; K.T.S.L.: 535. 1994. ≡ *Psychotria tarambassica* Bremek., J. Bot. 71: 280. 1933. —Type: Kenya, Baringo, Kamasia, Tarambas Forest, Nov. 1930, *I.R. Dale* in *F.D. 2436* [holotype: K; isotype: EA (EA000001792, EA000001793)]

= *Psychotria kirkii* Hiern var. *hirtella* (Oliv.) Verdc., Kew Bull. 28: 321. 1973; F. T.E.A. Rubiac. 1: 96. 1976; K.T.S.L.: 535. 1994. ≡ *Psychotria hirtella* Oliv., Trans. Linn. Soc. London, Bot. 2: 336. 1887. —Type: Tanzania, Kilimanjaro, Oct. 1884, *H.H. Johnston s.n.* [holotype: K (K000412258); isotype: BM (BM000903583)]

Shrub or subshrub, up to 6 m tall. Young stems and leaves usually pubescent. Leaf-blades ovate, oblong-elliptic or lanceolate, 2–18 cm × 0.5–9 cm, acute to acuminate at the apex, cuneate at the base; domatia absent; nodules numerous, scattered on the lamina; petioles up to 2.5 cm long; stipules ovate-triangular, 2.5–11 mm long, bifid at the apex. Flowers heterostylous, in panicles or umbels, the component often quite dense. Calyx-tube 0.5–1 mm long; limb cupuliform, 0.7–1.5 mm long; lobes obsolete. Corolla-tube cylindric, 2.5–6 mm long; lobes oblong to elliptic, 2–5 mm long, often reflex. Fruits red, subglobose, 5–7 mm in diameter.

Distribution: Widespread in Kenya. [Tanzania and Uganda].

Habitat: Forests, thickets, bushlands, hillsides with outcropping rocks or wooded grasslands; up to 2600 m.

Baringo: Kabarnet, *SAJIT 006539* (HIB). Kiambu: Thika, *Malombe & Kirika 36* (EA). Kilifi: Malindi, *Kimeu KEFRI641* (EA). Kitui: Mutomo Hill, *Gillet 18558* (EA); Nuu Forest, *Kuchar & Msafiri 14917* (EA). Laikipia: Rumuruti, *Carter & Stannard 335* (K). Machakos: Mua Hills, *Gillett 16211* (EA). Makueni: Kithembe Hill, *Mwangangi 2190* (EA); Nzaui Hill, *Kokwaro 1843* (EA). Marsabit: Mount Kulal, *Hepper & Jaeger 6863* (EA); Mount Marsabit, *Herlocker & Gaqar 541* (EA); Furroli, *Gillet 13891* (K). Meru: Lake Nkuga, *SAJIT 003814* (HIB). Nairobi: Nairobi Arboretum, *Williams 480* (EA). Nakuru: Nguruman Hills, *Ndegwa 936* (EA). Samburu: Lolokwi, *Curry & Glen 83* (EA); Maralal, *Gilbert et al. 5127* (EA); Mathew's Range, *Ichikawa 673* (EA). Teita Taveta: Taita Hills, *Mwachala et al.* in *EW 534* (EA); Mount Kasigau, *Luke et al. 4224* (K). Tharaka-Nithi: Kijegge Hill, *Beentje & Powys 4083* (EA). Turkana: Kuwalath Lowdar, *Paulo 1044* (EA). Uasin Gishu: Eldoret, *Williams 92* (EA). Wajir: Gurar, *Osiri 1* (EA, K). West Pokot: Mount Sekerr, *Agnew et al. 10391* (EA).

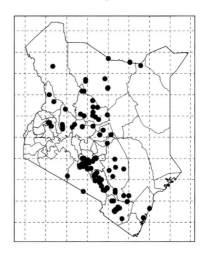

Plate 13 A, B. *Psychotria punctata* var. *punctata*; C, D. *P. faucicola*; E, F. *P. kirkii*. Photo by GWH.

18. **Psychotria alsophila** K. Schum., Pflanzenw. Ost-Afrikas, C: 390. 1895; F.T.E.A. Rubiac. 1: 99. 1976; K.T.S.L.: 534. 1994. —Type: Tanzania, W. Usambara Mts., Kwa Mshuza, Aug. 1893, *C. Holst 9065* [holotype: B (destroyed); lectotype: HBG (HBG521216), designated by E. Petit in Bull. Jard. Bot. État Bruxelles 36: 128. 1966; isolectotypes: BR (BR0000008851772), K (K 000412253), P (P00553323), US (US00516692)]

Shrub or rarely small tree, up to 6 m tall. Leaves narrowly elliptic or elliptic-lanceolate, 2–12 cm × 0.6–4.5 cm, subacute or narrowly rounded at the apex, cuneate at the petioles; petioles up to 0.5 cm long; stipules 4–8 mm long, with 2 acuminate lobes. Flowers heterostylous, several in lax panicle, with each branch congested and subcapitate; peduncles up to 4 cm long; pedicels 0–3 mm long. Calyx-tube ovoid, ca. 1 mm long; limb cupuliform, ca. 0.8 mm long. Corolla white, glabrous outside and inside; tube 4–7 mm long; lobes ovate-elliptic, 2.7–3 mm long. Fruits red, rounded, 5–7.5 mm in diameter.

Distribution: Southern Kenya. [Tanzania].
Habitat: Forests; 1700–1900 m.
Teita Taveta: Ngangao Forest, *Faden et al. 72/213* (EA).

19. **Psychotria tanganyicensis** Verdc., Kew Bull. 30: 258. 1975; F.T.E.A. Rubiac. 1: 75. 1976; K.T.S.L.: 538. 1994, non *Psychotria macrophylla* Ruiz & Pav. (1799). ≡ *Uragoga macrophylla*, K.Krause, Bot. Jahrb. Syst. 39: 569. 1907. —Type: Tanzania, E. Usambaras Mts., Gonja Mt., 5 Oct. 1905, *A. Engler* in *Herb. Amani 3366* [holotype: B (destroyed); lectotype: EA (EA000001875), **designated here**] (Plate 14A–D)

Small shrub, up to 3 m tall. Leaf-blades elliptic to obovate, 16–32 cm × 7–14 cm, acuminate at the apex, cuneate at the base, glabrous or with ferruginous hairy beneath; petioles 2–3.5 cm long; stipules elliptic-oblong to obovate, 1.5–2.5 cm long, bifid at the apex. Inflorescences mostly many-flowered, terminal, head-like, sessile or pedunculate; main bracts paired at the base forming an involucre; pedicels 1–2 mm long. Calyx-tube ovoid-obconic, 1.2–2 mm long; limb-lobes ovate or obovate, ca. 2 mm long. Corolla cream or white; tube cylindric below, funnel-shaped at the apex, 6–7 mm long; throat hairy; lobes ovate, 2.5–3 mm long. Fruits purple or blue, ellipsoid, 7–8.5 mm long.

Distribution: Coastal Kenya. [Tanzania].
Habitat: Forests; 300–450 m.
Kwale: Shimba Hills, *SAJIT 006160* (HIB).

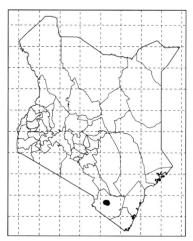

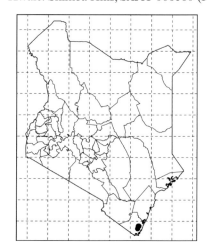

20. **Psychotria ceratoloba** (K. Schum.) O. Lachenaud, Opera Bot. Belg. 17: 187. 2019. ≡ *Uragoga ceratoloba* K. Schum., Bot. Jahrb. Syst. 28: 105. 1899. —Type: D.R. Congo, du Kasaï, entre Ipamu et Pangu, *Vanderyst 9513* [neotype: BR, designated by O. Lachenaud in Opera Bot. Belg. 17: 187. 2019] (Plate 14E, F)

= *Psychotria peduncularis* (Salisb.) Steyerm. var. *ciliato-stipulata* Verdc., Kew Bull. 30: 689. 1976; F.T.E.A. Rubiac. 1: 74. 1976. —Type: D.R. Congo, Lesse, 9 May 1914, *J. Bequaert 4188* [holotype: BR (BR0000008855022); isotype: BR (BR0000008855138)]

Shrub or shrubby herb, up to 1.2 m tall. Leaf-blades ovate to narrowly ovate or lanceolate, 9–26.5 cm × 4–13.5 cm, acute to abruptly acuminate at the apex, cuneate to rounded at the base, beneath glabrous to sparsely pubescent; petioles 1.5–6.5 cm long; stipules oblong, ovate to rounded, 1–2.2 cm long, bifid at the apex. Flowers usually numerous in involucrate capitate inflorescences; peduncles 0.5–1.5 cm long. Calyx-tube ellipsoid, 2–3 mm long; limb-lobes obsolate. Corolla white; tube cylindric at the base, funnel-shaped at upper part, 4.5–6.5 mm long. Fruit blue, ellipsoid, 7–10 mm long.

Distribution: Western Kenya. [D.R. Congo, Ethiopia, Sudan, and Uganda].

Habitat: Forests; 1550–1850 m.

Kakamega: Kakamega Forest, *SAJIT 006708* (HIB). Kericho: *Napier 6099* (K).

21. **Psychotria mildbraedii** (K. Krause) O. Lachenaud, Opera Bot. Belg. 17: 192. 2019. ≡ *Uragoga mildbraedii* K. Krause, Wiss. Ergebn. Deut. Zentr.-Afr. Exped., Bot. 2: 337. 1911. —Types: D.R. Congo, Forestier Central, Aruwimi entre Bomili et Pangapres Bafwayabu, 1908, *J. Mildbraed 3258* [holotype: B (destroyed); isotype: BR (BR0000008858030, fragment)]; Yangambi, au pied des falaises de l'Isalowe, *J. Louis 11306* [epitype: BR (BR0000021335600), designated by O. Lachenaud in Opera Bot. Belg. 17: 192. 2019; isoepitypes: B, BR (BR 0000021335617), K, MO, P, WAG, YA]

= *Cephaelis suaveolens* Hiern, Fl. Trop. Afr. 3: 224. 1877. ≡ *Cephaelis peduncularis* Salisb. var. *suaveolens* (Hiern) Hepper, Kew Bull. 16: 156. 1962. ≡ *Psychotria peduncularis* (Salisb.) Steyerm. var. *suaveolens* (Hiern) Verdc., Kew Bull. 30: 257. 1975; F.T.E.A. Rubiac. 1: 73. 1976. —Type: Sudan, Jur, Orel, 7 May 1869, *G. Schweinfurth 1736* [lectotype: K (K000412282), designated by F.N. Hepper in Kew Bull. 16: 156. 1962; isolectotypes: P (P00551335), BM, K (K000412282), M (M0187151)]

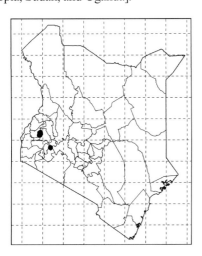
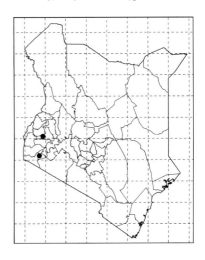

Plate 14 A–D. *Psychotria tanganyicensis*; E, F. *P. ceratoloba*. Photo by ZXZ (A–D) and GWH (E, F).

Shrub, up to 4.5 m tall. Leaves obovate-elliptic or lanceolate, 2.5–16 cm × 3–9 cm, acute to abruptly acuminate at the apex, cuneate to rounded at the base, sparely or densely ferruginous pubescent beneath at least on the main venation; petioles 1–3.5 cm long; stipules obovate-oblong to lanceolate, 1–1.5 cm long, bifid at the apex. Flowers usually numerous in involucrate capitate inflorescences; peduncles 1–4 cm long. Calyx-lobes very small, glabrous. Fruits ellipsoid, 5–8 mm long.

Distribution: Western Kenya. [Tropical Africa].

Habitat: Riverine forests; 1600–1650 m.

Kisii: Kuja River, *Glasgow 46/2* (EA). Nandi: Kaimosi, *Bamps 6454* (EA).

5. Trib. **Knoxieae** Benth. & Hook. f.

Herbs, subshrubs, shrubs or rarely small trees. Stipules fimbriate and colleter-tipped. Flowers heterostylous with unequal calyx-lobes. Ovary 2–10-locular, with solitary to many ovules in each locule. Fruits dry, dehiscent or indehiscent, or sometimes drupaceous. Ten genera occur in Kenya.

1a. Ovule solitary in each locule .... 16. *Pentanisia*
1b. Ovules 2–many in each locule 2
2a. Subshrubs, shrubs or rarely herbaceous shoots from a woody rootstock 3
2b. Herbs, woody herbs or rarely subshrubs 4
3a. Calyx-limb unequally 5–7-toothed .... 9. *Triainolepis*
3b. Calyx-limb developed into an oblique dilated reticulated shallowly concave entire or lobed lamina .... 10. *Dirichletia*
4a. Inflorescences in cymose heads, and developed into a long simple spike in fruit .... 17. *Otomeria*
4b. Flowers in more complicated inflorescences 5
5a. Creeping or decumbent herbs, often rooting at the nodes .... 11. *Parapentas*
5b. Erect herbs, woody herbs or subshrubs 6
6a. Calyx-lobes subequal, linear, subulate, narrowly triangular or slightly spathulate at the apices 7
6b. Calyx-lobes unequal, linear, subulate, spathulate or 1–3-foliaceous 8
7a. Corolla-tube slender, up to 13 cm long .... 12. *Dolichopentas*
7b. Corolla-tube cylindric, less than 2.8 cm long .... 13. *Phyllopentas*
8a. Inflorescences terminal, cymose, (1–)3–several-flowered; corolla white .... 8. *Chamaepentas*
8b. Inflorescences terminal or axillary, dense or lax, several- to many-flowered; corolla white or brightly coloured 9
9a. Flowers distinctly vermilion-scarlet .... 15. *Rhodopentas*
9b. Flowers white, mauve, blue or pink, only rarely red and then of a deeper crimson shade .... 14. *Pentas*

8. **Chamaepentas** Bremek.

Perennial herbs or subshrubs. Leaves opposite, petiolate or sessile; stipules often divided into 2–several filiform to subulate colleter-tipped segments. Inflorescences terminal, cymose, (1–)3–several-flowered. Flowers often white, hermaphrodite, dimorphic or not. Calyx-tube campanulate or urceolate; lobes 5, unequal, linear to subulate or spathulate. Corolla-tube very long, narrowly tubular, the throat often densely hairy inside; lobes 5, lanceolate- to ovate-oblong, elliptic or obovate. Stamens included, the upper part of the corolla-tube always widened at the point of inclusion. Ovary 2-locular, with numerous ovules in each locule; style exserted, tomentose above; stigma bilobed. Capsule ovoid to oblong, often beaked, dehiscent at the apex, sometimes separating into 2 cocci. Seeds many, minute, brown.

Six species restricted in East Africa; one species in Kenya.

1. **Chamaepentas hindsioides** (K. Schum.) Kårehed & B. Bremer, Taxon 56: 1075. 2007. ≡ *Pentas hindsioides* K. Schum., Bot. Jahrb. Syst. 34: 330. 1904; F.T.E.A. Rubiac. 1: 193. 1976; U.K.W.F.F.: 225. 2013. —Types: Tanzania, W. Usambara Mts., Sakare-Manka, *A. Engler 1056* [holotype: B (destroyed)]; Tanzania, Pare, Usangi, 1892 m, May 1928, *A.E. Haarer 1362* [neotype: EA; isoneotype: K (K000319689), designated by B. Verdcourt in Bull. Jard. Bot. État Bruxelles 23: 278. 1953] (Figure 11; Plate 15)

Woody herb or shrub, up to 1.8 m tall. Leaves ovate-lanceolate to elliptic, 3–11 cm × 1–4.5 cm, acute or somewhat acuminate at the apex, cuneate at the base, hairy; petioles 0.5–2.3 cm long; stipules with 2–5 linear setae up to 2–5 mm long, the colleters conspicuous, orange or white, ca. 0.5 mm long. Inflorescences terminal, cymose, several-flowered. Flowers white, sweet smelling. Calyx-tube 3–4 mm long; lobes linear-spathulate to oblanceolate or subulate, 0.8–2 cm long. Corolla-tube 2.5–5.5 cm long; lobes elliptic or ovate-oblong to oblong-lanceolate, 0.6–1.2 cm long. Stamens included. Style exserted for 0–3 mm, minutely tomentose. Capsules 6.5–10 mm long, glabrous to hairy. Seeds very numerous, ca. 1 mm long.

Distribution: Central and southern Kenya. [Tanzania].

Habitat: Bushlands or forest margins; 1400–1950 m.

Makueni: Nzaui Forest, *Agnew et al. 8356* (EA). Teita Taveta: Mbololo Forest, *Beentje et al. 1009* (EA); Ngangao Forest, *Mwachala et al. 1049* (EA); Mount Kasigau, *Bally B13573* (EA).

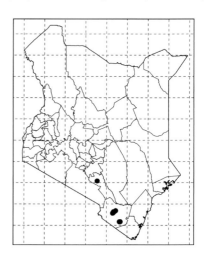

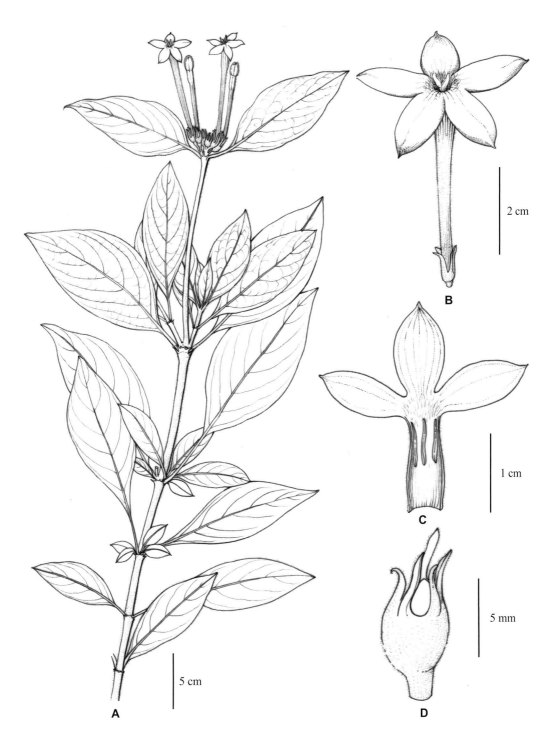

Figure 11 *Chamaepentas hindsioides*. A. flowering branch; B. flower; C. longitudinal section of the upper part of corolla, showing the stamens; D. fruit. Drawn by NJ.

Plate 15 A–C. *Chamaepentas hindsioides*. Photo by GWH (A, B) and BL (C).

9. **Triainolepis** Hook. f.

Shrubs or small trees. Leaves lanceolate to ovate or oblong-elliptic; stipules ovate or ovate-triangular, 3–5-lobed. Inflorescences terminal, corymbose; bracts very small. Flowers hermaphrodite, heterostylous, (4–)5-merous. Calyx-tube campanulate, the limb cupular, unequally 5–7-toothed. Corolla white; tube hairy at the throat; lobes lanceolate, hairy outside, glabrous inside. Stamens exserted or included. Ovary 4–10-locular; ovules 2(–3) in each locule; style exserted or included; stigma-lobes 4–10, filiform. Drupe red, globose or depressed globose, containing a woody or bony 4–10-celled putamen. Seeds ellipsoid, compressed, with membranous testa and fleshy albumen.

A small genus with 13 species occurring in Madagascar, Comoros Islands, Aldabra Group and the coast of east Africa; only one species in Kenya.

1. **Triainolepis africana** subsp. **hildebrandtii** Hook. f., Gen. Pl. 2: 126. 1873; F.T.E.A. Rubiac. 1: 149. 1976; K.T.S.L.: 548. 1994. —Type: Mozambique, Rovuma Bay, Mar. 1861, *J. Kirk s.n.* [holotype: K (K000286841)]

Shrub or rarely a small tree, up to 6 m tall. Leaves elliptic, narrowly ovate or elliptic-lanceolate, 3–12 cm × 1–6 cm, narrowly acuminate at the apex, cuneate at the base, pubescent or glabrous except for veins beneath; petioles 0.5–1.7 cm long; stipules ovate or triangular, 1–4 mm long, often with 3 unequal subulate fimbriae, colleter-tipped. Inflorescences terminal, corymbose; peduncle up to 2.2 cm long; pedicels very short. Calyx-tube 1–1.5 mm long; limb-tube up to 2.5 mm long; lobes up to 3 mm long. Corolla white; tube campanulate, up to 1 cm long; lobes 3.5–5.5 mm long. Style 1–1.5 cm long in long-styled flowers, 4–5.5 mm long in short-styled flowers; stigma 6–8-lobed. Drupe dark red, subglobose, 4–6 mm in diameter when dry; putamen depressed-subglobose, 4–5 mm in diameter, grooved. Seeds narrowly oblong-ellipsoid, 2–2.4 mm long.

subsp. **hildebrandtii** (Vatke) Verdc., Kew Bull. 30: 282. 1975; F.T.E.A. Rubiac. 1: 150. 1976; K.T.S.L.: 548. 1994. ≡ *Triainolepis hildebrandtii* Vatke, Oesterr. Bot. Z., 25: 230–231. 1875. —Type: Tanzania, Zanzibar,

Plate 16 A–E. *Triainolepis africana*. Photo by BL (A–C) and GWH (D, E).

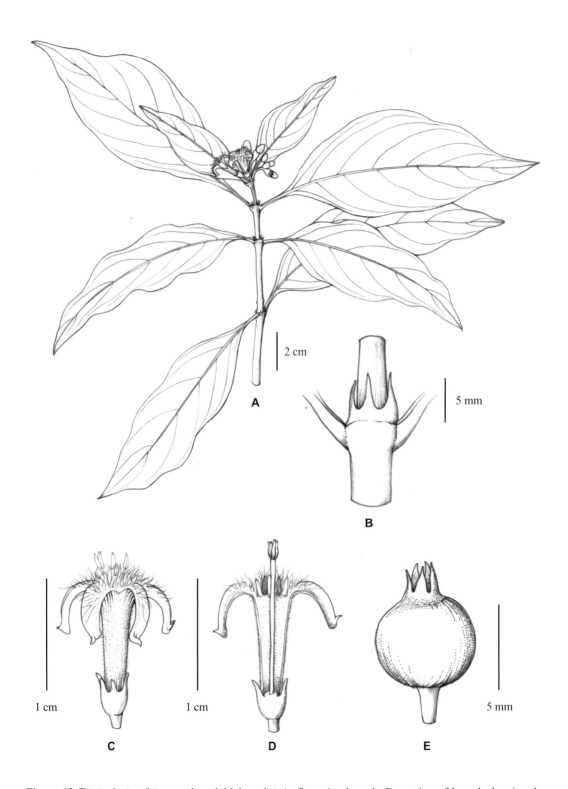

Figure 12 *Triainolepis africana* subsp. *hildebrandtii*. A. flowering branch; B. portion of branch showing the stipule; C. short-styled flower; D. longitudinal section of long-styled flower, showing the stamens and style; E. fruit. Drawn by NJ.

Sept. 1873, *J.M. Hildebrandt 1126* [holotype: B (destroyed); lectotype: K (K000430041), **designated here**] (Figure 12; Plate 16)

Leaves glabrous except for veins beneath.

Distribution: Coastal Kenya. [Aldabra Group, Comoros Islands, Madagascar, Malawi, Mozambique, Tanzania, and Zambia].

Habitat: Coastal bushlands; up to 100 m.

Note: *Triainolepis africana* subsp. *africana* occurs in Tanzania and Mozambique, with leaf-blades distinctly pubescent above, and sparsely to densely pubescent beneath.

Kilifi: Gede Forest, *Gerhardt & Steiner 26* (EA); Watamu, *Kimeu et al. KEFRI671* (EA, K). Kwale: Diani, *SAJIT 005450* (HIB); Diani Forest, *Gillett & Kibuwa 19850* (EA, K). Lamu: Manda Island, *Brathy et al. 119* (EA). Mombasa: *Graham 2124* (EA, K). Tana River: Nairobi Ranch, *Festo & Luke 2355* (EA).

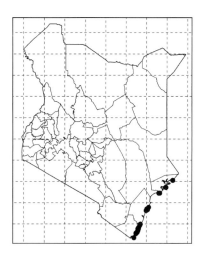

10. **Dirichletia** Klotzsch

Shrubs. Leaves opposite, shortly petiolate; stipules with 3–5 linear or filiform colleter-tipped setae. Inflorescences terminal, corymbose, few-many-flowered. Flowers hermaphrodite, dimorphic. Calyx-tube small, campanulate; limb developed into an oblique dilated reticulated shallowly concave entire or lobed lamina. Corolla-tube slender, usually hirsute inside; lobes 4–5, oblong or ovate. Stamens 5, included or exserted. Ovary 2-locular; ovules many in each locule; style filiform; stigma bipartite. Fruits oblong, costate, attenuate at the base, dicoccous.

A small genus with 5 species occurring in East Africa from Ethiopia to Zimbabwe, also found in Somalia and Yemen (Socotra Island); one species in Kenya.

1. **Dirichletia glaucescens** Hiern, Fl. Trop. Afr. 3: 51. 1877; C.P.K.: 378. 2016. ≡ *Carphalea glaucescens* (Hiern) Verdc., Kew Bull. 28: 424. 1974; W.F.E.A.: 150. 1987; K.T.S.L.: 506. 1994. —Type: Somalia, Tola River, Apr. 1873, *J. Kirk s.n.* [holotype: K (K000414224)] (Figure 13; Plate 17)

= *Dirichletia asperula* K. Schum., Pflanzenw. Ost-Afrikas, C: 378. 1895. —Type: Kenya, Teita Taveta, Ndi Mt. *J.M. Hildebrandt 2595* [holotype: B (destroyed); lectotype: K (K000414222), **designated here**; isolectotype: BM]

= *Dirichletia ellenbeckii* K. Schum., Bot. Jahrb. Syst. 33: 336. 1903. —Types: Ethiopia, Gobelle Valley, *H. Ellenbeck 1053A* [syntype: B (destroyed)]; Ethiopia, Arussi-Galla, Buchar, *H. Ellenbeck 2013* [syntype: B (destroyed)]; Kenya, Mandera, Karro Gudda, *H. Ellenbeck 2169a* [syntype: B (destroyed)]

Much-branched shrub up to 3 m high. Leaf-blades, elliptic to lanceolate or narrowly obovate, 1–8.5 cm × 0.5–2.5 cm, acute to

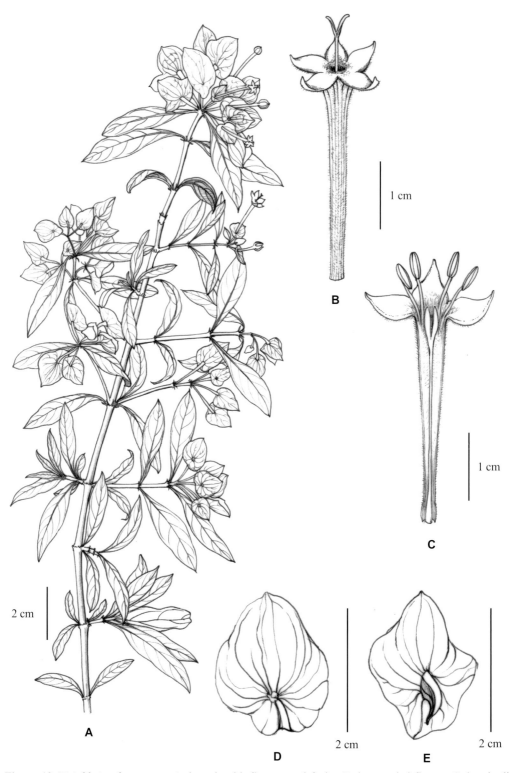

Figure 13 *Dirichletia glaucescens*. A. branch with flowers and fruits; B. long-styled flower; C. longitudinal section of short-styled flower, showing the stamens and style; D. vertical view of fruit, showing the enlarged calyx-limb; E. dorsal view of fruit. Drawn by NJ.

Plate 17 A–C. *Dirichletia glaucescens*. Photo by QFW.

obtuse at the apex, cuneate at the base; petioles up to 5 mm long; stipules with 3–5 setae 1–5 mm long. Inflorescences densely cymose, terminal, with peduncles 0.1–1.5 cm long. Pedicels 1.5–9 mm long. Calyx-tube obconic to cylindrical, 1.5–3 mm long, hairy; limb white, greenish or pinkish, elliptic, 0.5–2 mm long, enlarged in fruit, 1.3–2.8 cm × 0.8–2.3 cm. Corolla white or pale pink, with tube 1.2–3.5 cm long; lobes ovate, 3.5–6 mm long. Style filiform; stigma 2(–4)-lobed. Fruits obconic, often curved, 4–10 mm long, strongly ribbed. Seeds obconic-fusiform, 2.3–2.7(–4) mm long.

Distribution: Northern central, eastern, southorn, and coastal kenya. [Ethiopia, Somalia, and Tanzania].

Habitat: Deciduous woodlands, bushlands, and thickets; 200–1500 m.

Garissa: Modo Gash-Garissa, *Stannard & Gilbert 1030* (EA). Isiolo: Garba Tula, *Adamson s.n.* (EA). Kajiado: Kitui-Kibwezi Road, *Bally 13149* (EA, K). Kilifi: Malindi, *Wood 1396* (EA). Kitui: ca. 5 km NNW of

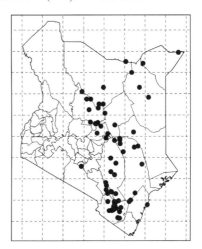

Voo, *Kuchar 8941* (EA). Makueni: Mtito Andei, *Verdcourt 1170* (EA, K). Mandera: Dandu, *Gillett 13024* (EA, K). Marsabit: Sato 704 (EA). Meru: Meru National Park, *Ament & Magogo 383* (EA, K). Samburu: near Ndoto Mountains, *Gilbert & Gachathi 5268* (EA, K). Teita Taveta: Mount Kasigau, *Muasya & Medley 707* (EA); near Manyani, *Luke 81* (EA). Tana River: ca. 7 km west of Hatama Corner, *Gillett & Gachathi 20512* (EA, K); ca. 22 km from Garissa, *Stannard & Gilbert 1082* (EA, K). Tharaka-Nithi: Tana-Rojeweru River, *Gillett 18896* (EA, K). Wajir: Dela, *Dale K725* (EA, K).

11. **Parapentas** Bremek.

Procumbent perennial herbs. Leaves opposite, petiolate; stipule-sheath divided into several subulate colleter-tipped lobes. Inflorescences terminal or pseudo-axillary, 1–3-flowered. Flowers white or lilac, hermaphrodite, isostylous or heterostylous. Calyx-tube campanulate; lobes 5, subequal, linear or linear-oblong. Corolla-tube narrowly tubular, glabrous or sparsely to densely pilose within throat; lobes 5, elliptic, minutely papillate inside. Stamens and style included or well exserted. Ovary 2-locular; ovules many in each locule. Capsules obconic or globose, loculicidally dehiscent. Seeds many, brown, angular.

A small genus with three species restricted to tropical Africa; only one species in Kenya.

1. **Parapentas battiscombei** Verdc., Bull. Jard. Bot. État 23: 54. 1953; F.T.E.A. Rubiac. 1: 225. 1976; U.K.W.F.F.: 222. 2013. —Type: Kenya, E. Mt. Kenya Forest, *E. Battiscombe 695* [holotype: EA (EA000001581); isotype: K (K000311692)] (Figure 14; Plate 18)

Perennial herb, with procumbent stems up to 30 cm long, rooting at the nodes. Leaves ovate or elliptic-ovate, 0.6–4.5 cm × 0.3–2.5 cm, bluntly acute at the apex, cuneate to rounded at the base; petioles up to 2 cm long, pubescent; stipule-sheath 1–1.5 mm long, divided into 3–5 filiform fimbriae 1–4 mm long, with scarcely evident colleters. Inflorescences terminal or pseudo-axillary, 1–3-flowered. Calyx-tube subglobose, very short, hairy; lobes unequal or subequal, oblong to narrowly spathulate, 2–6 mm long. Corolla white, pale blue or lilac; tube 1.4–2.3 cm long, densely hairy within the throat; lobes oblong-elliptic, 5–8 mm long. Style exserted 2–5 mm in long-styled flowers, 1.4 cm long in short-styled flowers, glabrous; stigma-lobes filiform 1–2.5 mm long. Capsules brown, oblong-ovoid, 3–4 mm long.

Distribution: Central and southern Taveta. [Endemic].

Habitat: Evergreen forests; 1150–2200 m.

Meru: Nyambeni Hills, *Lavranos 17256* (EA). Teita Taveta: Mount Kasigau, *Luke 9431*

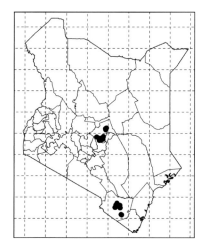

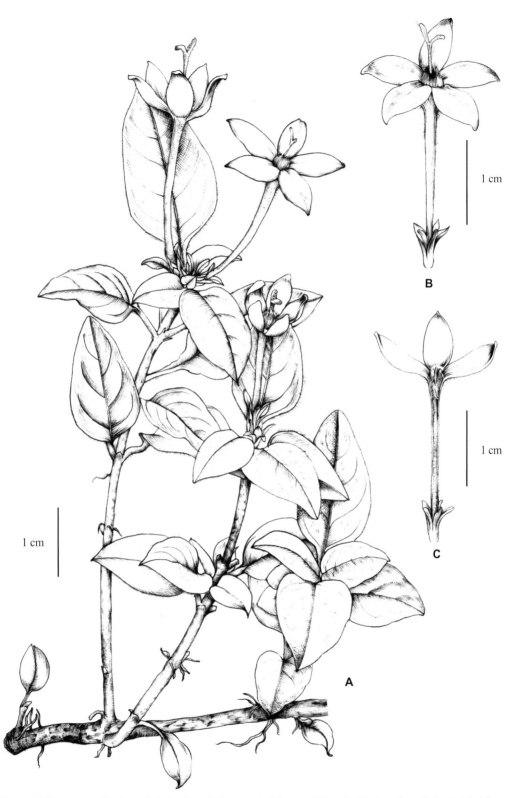

Figure 14 *Parapentas battiscombei*. A. plant; B. long-styled flower; C. longitudinal section of short-styled flower, showing the stamens and style. Drawn by NJ.

Plate 18 A, B. *Parapentas battiscombei*. Photo by YDZ.

(EA); Chawia Forest, *Beentje et al. 828* (EA, K); Ngangao Forest, *Barry 8768* (EA, K). Tharaka-Nithi: Chuka Forest, *SAJIT VK0052* (EA); Chogoria Track of Mount Kenya, *SAJIT 003984* (HIB).

12. **Dolichopentas** Kårehed & B. Bremer

Herbs, sometimes with a woody rootstock. Leaves opposite or often 3–5-whorled; stipules with 1–7 linear or deltoid setae. Inflorescences terminal, cymose, several-branched, few- to many-flowered. Flowers not dimorphic, white. Calyx-tube ovoid or oblong-ovoid, glabrous to hairy; lobes usually 5, subequal, linear to deltoid or narrowly triangular. Corolla-tube very long, dilated at the apex, pubescent outside, throat densely hairy; lobes linear, elliptic-lanceolate, elliptic or oblong-obovate. Stamens entirely included. Style always exserted, tomentose with white papillae. Fruits oblong to obovoid-obtriangular, ribbed. Seeds minute, brown.

A small genus with four species restricted in tropical Africa; two species in Kenya.

1a. Shrubby herb with 2–3 main stems from a woody rootstock; corolla-tube usually short, 2–4.2 cm long, lobes linear; fruit 3–6.5 mm in diameter..1. *D. longiflora*

1b. Herb, usually with single stem; corolla-tube 3–13 cm long, lobes deltoid to filiform; fruits 1–1.5 cm in diameter... 2. *D. decora*

1. **Dolichopentas longiflora** (Oliv.) Kårehed & B. Bremer, Taxon 56: 1076. 2007. ≡ *Pentas longiflora* Oliv., Trans. Linn. Soc. London, Bot. 2: 335. 1887; F.T.E.A. Rubiac. 1: 195. 1976; W.F.E.A.: 155. 1987; U.K.W.F.F: 225. 2013. —Type: Tanzania, Kilimanjaro, Oct. 1884, *H.H. Johnston s.n.* [holotype: K (K 000394968); isotype: BM (BM000902857)] (Plate 19)

= *Pentas longiflora* Oliv. f. *glabrescens* Verdc., Bull. Jard. Bot. État Bruxelles 23: 286. 1953. —Type: Kenya, S. Kavirondo, Kisii, Sept. 1933, *E.R. Napier* in *C.M. 10183* [holotype: EA (EA000001767)]

Shrubby herb to 2 m tall, with 2–3 main stems from a woody rootstock. Leaves opposite or 3-whorled; blades lanceolate or rarely ovate-lanceolate, 5–15 cm × 0.7–3 cm, acute at the apex, narrowed at the base, glabrous to velvety or pubescent; petiole up to 1 cm long; stipules with 3–7 linear setae 0.1–1.3 cm long. Inflorescences terminal, cymose, several-branched, 20–100-flowered; bracteoles filamentous, up to 3 mm long. Calyx-tube glabrescent to velvety, 1.5–2 mm long; lobes subequal, linear, 3–7.5 mm long. Corolla-tube 2–4.2 cm long, dilated at the apex, pubescent outside; lobes oblong or elliptic, 3–6 mm long; throat hairy. Stamens entirely included. Style exserted 1–6 mm, tomentose with white scaly papillae. Fruits oblong, depressed, 3–6.5 mm long, ribbed, shortly pubescent.

Distribution: Western, central, and southern Kenya. [Burundi, D.R. Congo, Ethiopia, Malawi, Rwanda, Tanzania, and Uganda].

Habitat: Grasslands, bushlands, thickets, and forest edges; 1050–2450 m.

Baringo: Perkerra, *Verdcourt 1203* (EA). Bomet: Abossi (Trans Mara), *Glover et al. 2303* (EA). Bungoma: Mount Elgon, *Lugard 144* (EA). Elgeyo-Marakwet: Katimok Forest, *Dale 2406* (EA). Homa Bay: near Magenche, *Plaizier 1336* (EA). Kajiado: Chyulu Hills, *Gilbert 6196* (EA). Kericho: east of Ngoina Tea Estate, *Perdue & Kibuwa 9341* (EA). Kiambu: Kikuyu Escarpment Forest, *Perdue & Kibuwa 8254* (EA). Kisii: *Napier s.n.* (EA). Kisumu: Muhoroni, *Johansen s.n.* (LD). Makueni: Nzaui Hill, *Agnew et al. 8404* (EA). Migori: ca. 2 km north of Nyamarambe, *Vuyk 303* (EA, WAG). Murang'a: Kiru, *Waiganjo JWW44* (EA). Nakuru: Rongai, *Someren s.n.* (EA). Nandi: North Nandi Forest, *SAJIT 006606* (EA, HIB). Narok: Masai Mara Game Reserve, *Kuchar 11619* (EA). Nyamira: Ngoina, *Ossent 640* (EA). Nyandarua: Mount Lolderodo, *Gardner 2012* (EA). Teita Taveta: Mount Kasigau, *Joana s.n.* (EA). Trans-Nzoia: Saiwa Swamp, *Birnie 401* (EA). Uasin Gishu: *Hervey 1380* (EA). West Pokot: Sekerr, *Agnew*

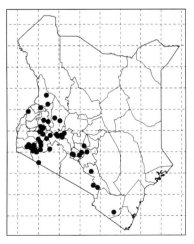

Plate 19 A–G. *Dolichopentas longiflora*. Photo by GWH.

et al. 10342 (EA).

2. **Dolichopentas decora** (S. Moore) Kårehed & B. Bremer, Taxon 56: 1075. 2007. ≡ *Pentas decora* S. Moore, J. Bot. 48: 219. 1910. —Type: D.R. Congo, Luendarides, 28 Jan. 1908, *T. Kässner 2419* [holotype: BM (BM000902856); isotype: K (K000394984)]

Herb 0.3–1.5 m tall; stems usually single, mostly somewhat woody, glabrous to pubescent. Leaves 3–5-whorled or rarely opposite; blades elliptic or elliptic-lanceolate, 3.9–13 cm × 0.9–5.6 cm, subacute or rounded at the apex, rounded to cuneate at the base, glabrous to densely velvety; petioles obsolete; stipules with 1–5 ciliate less than 1 cm long setae. Inflorescences terminal, 1–4-branched, each cluster with ca. 30 or more flowers. Calyx glabrous to hairy; tube 2–5 mm long; lobes deltoid to filiform, 3–11 mm long. Flowers white, sweet-scented. Corolla-tube 3–13 cm long, scarcely dilated at the apex; throat densely hairy with long hairs; lobes linear-oblong to elliptic-lanceolate, 0.5–1.8 cm long, acute. Stamens entirely included. Style exserted up to 1.8 cm, densely tomentose with white papillae; stigma-lobes elliptic, 1–3 mm long, thickened. Fruits obovoid-obtriangular, 1–1.5 cm long, prominently ribbed, pubescent. Seeds brown, ovoid, 1–1.5 mm long.

1a. Leaves mostly glabrous; corolla-tube up to 12 cm long .... a. var. *decora*
1b. Leaves glabrescent to velutinous; corolla-tube often 5–6 cm long .... b. var. *triangularis*

a. var. **decora**

Leaves mostly glabrous. Corolla-tube longer, up to 12 cm long.

Distribution: Western Kenya. [Angola, Central Africa, D.R. Congo, Ethiopia, Malawi, Nigeria, South Sudan, Tanzania, Uganda, and Zambia].

Habitat: Grasslands, bushlands, thickets, and forest edges; 1050–2450 m.

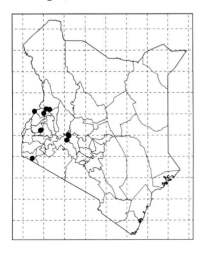

Bungoma: Mount Elgon, *Jack 25* (EA). Kakamega: Kakamega Forest, *Kokwaro 3762* (EA). Migori: Suna, *Rayner s.n.* (EA). Nakuru: Solai Escarpment, *Gardner 1470* (EA). Trans-Nzoia: Kitale Grassland Research Station, *Williams 281* (EA). West Pokot: Kapenguria, *Napier 1991* (EA).

b. var. **triangularis** (De Wild.) Kårehed & B.Bremer, Taxon 58: 317 2009. ≡ *Pentas triangularis* De Wild., Repert. Spec. Nov. Regni Veg. 13: 139. 1914. ≡ *Pentas decora* S. Moore var. *triangularis* (De Wild.) Verdc., Bull. Jard. Bot. État Bruxelles 23: 291. 1953. —Type: D.R. Congo, Katanga, Lualaba Kraal, Dec. 1912, *H.A. Homblé 934* [holotype: BR (BR0000008850775); isotype: BR (BR0000008850768, BR0000008850782), MO (MO-391717)]

Leaves glabrescent to velutinous. Corolla-tube often 5–6 cm long.

Distribution: Western Kenya. [Central Africa, D.R. Congo, Ethiopia, Malawi, Nigeria, Tanzania, and Uganda].

Habitat: Grasslands or woodlands; 1300–2100 m.

Trans-Nzoia: near Kitale, *Tweedie 1189* (K); Elgon, *Lugard 9* (K).

13. **Phyllopentas** (Verdc.) Kårehed & B. Bremer

Herbs, subshrubs or shrubs, always covered with ferruginous hairs at least when dry. Leaves opposite, blades elliptic, ovate or oblong-ovate, acuminate at the apex, rounded to cuneate at the base; stipules with 3–10 filiform or subulate segments. Inflorescences axillary or terminal, 3–several-branched, the cymes few- to many-flowered. Flowers dimorphic, white to mauve. Calyx-lobes usually 5, subequal, filiform or subulate, or one enlarged into an ovate petaloid stipitate lamina. Corolla pubescent outside, throat densely hairy; lobes elliptic-lanceolate. Stamens obviously exserted in short-styled flowers. Style included or exserted. Fruit a capsule, glabrous or pubescent. Seeds minute, many.

A small genus with 12 species restricted in tropical Africa and Madagascar; two species in Kenya.

1a. Corolla-tube short, up to 1.4 cm long .... 1. *P. schimperi*
1b. Corolla-tube long, 1.5–2.8 cm long .... 2. *P. elata*

1. **Phyllopentas schimperi** (Hochst.) Y.D. Zhou & Q.F. Wang, Phytotaxa 435(3): 252. 2020. ≡ *Mussaenda schimperi* Hochst., Unio Itin. 38. 1840. ≡ *Pentas schimperi* (Hochst.) Wieringa, Blumea 53: 567. 2008. ≡ *Phyllopentas schimperiana* (Vatke) Kårehed & B. Bremer, Taxon 56(4): 1076. 2007, *nom. illeg.* ≡ *Pentas schimperiana* (A. Richard) Vatke, Linnaea 40: 192. 1876, *nom. illeg.*; F.T.E.A. Rubiac. 1: 187. 1976; U.K.W.F.F.: 225. 2013. —Type: Ethiopia, ad latus boreali montis Scholoda, 26 Oct. 1837, *G.H.W. Schimper I: 38* [lectotype: WAG (WAG0003029), designated by J.J. Wieringa in Blumea 53: 567. 2008; isolectotypes: BR (BR0000008849946, BR0000008358363 & BR0000008850041), GH (GH01154976), HAL (HAL0113665), L (L0057932), HBG (HBG521292), HOH (HOH009711), MPU (MPU021496), REG (REG000627), S (S05-9970 & S05-9973), TUB (TUB004516 & TUB004517), K, MO, P, W] (Figure 15; Plate 20)

= *Pentas thomsonii* Scott-Elliot, J. Linn. Soc., Bot. 32: 435. 1896. —Types: Kenya, Laikipia, Sept. 1884, *J. Thomson s.n.* [lectotype: K (K000394970), designated by B. Verdcourt in F.T.E.A. Rubiac. 1: 188. 1976; isolectotype: BM]; Nandi Hills, *G.F. Scott-Elliot 6954* [syntype: K?]

Shrub or woody herb up to 3(–5) m tall, covered with ferruginous hairs at least when dry. Leaves ovate, ovate-lanceolate or ovate-oblong, 6–21 cm × 1.8–8 cm, acute to acuminate at the apex, cuneate to round at the base; petioles up to 1.5 cm long; stipules large, with 6–10 linear-lanceolate brown hairy setae. Inflorescences terminal, several-branched. Calyx-tube hairy, 1.5–3 mm long; lobes subequal, linear, up to 1.2 cm long, hairy

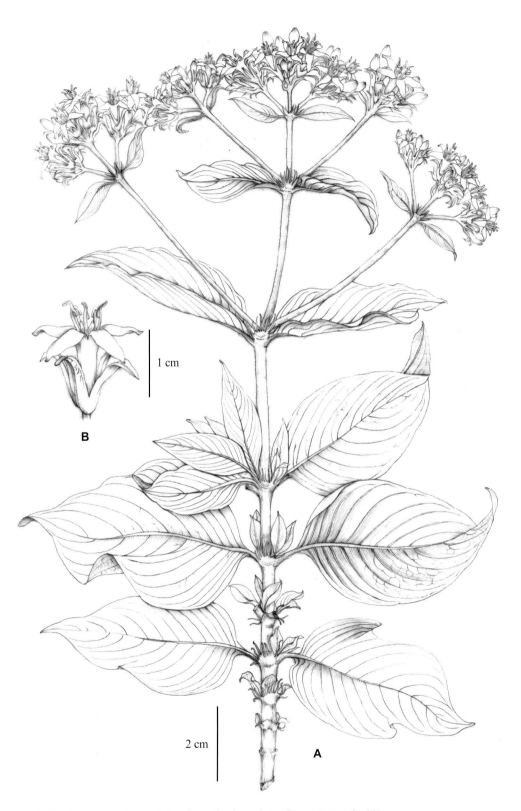

Figure 15 *Phyllopentas schimperi*. A. a flowering branch; B. flower. Drawn by NJ.

Plate 20 A–H. *Phyllopentas schimperi*. Photo by NW.

outside. Corolla white, often tinged pink, glabrous or pubescent; tube up to 1.4 cm long, dilated at the apex. Style exserted for 0.5–5 mm long in long-styled flowers. Capsules 4–6 mm long.

Distribution: Western, central, and southern Kenya. [Burundi, Cameroon, D.R. Congo, Equatorial Guinea, Ethiopia, Malawi, Rwanda, Sao Tome, Tanzania, Uganda, and Zambia].

Habitat: Grasslands, bushlands, bamboo or evergreen forests; 1400–3000 m.

Baringo: Tugen Hills, *Tweedie 3922* (K). Bungoma: Mount Elgon, *Lugard 2974* (EA). Kajiado: Loita Hills, *Fayad 131* (EA). Kericho: Tinderet Forest, *Perdue & Kibuwa 9186* (EA, K). Kisii: ca. 4 km after Mogumo Market, *Plaizier 1278* (WAG). Marsabit: Mount Kulal, *Kamau et al. GBK12* (EA). Nakuru: Mau Forest, *Jex-Blake 3081* (EA, K). Samburu: Mount Nyiru, Newbould 3467 (K). Teita Taveta: Mbololo Forest, *Beentje 953* (EA). Trans-Nzoia: west slope of Mount Elgon, *SAJIT PR0120* (EA, HIB); Kitale Museum Nature Trail, *Beentje 2969* (EA). West Pokot: Cherangani Hills, *Dale 3266* (EA, K).

2. **Phyllopentas elata** (K. Schum.) Kårehed & B. Bremer, Taxon 56: 1076. 2007. ≡ *Pentas elata* K. Schum., Pflanzenw. Ost-Afrikas, C: 377. 1895; F.T.E.A. Rubiac. 1: 188. 1976. —Type: Tanzania, Moshi, Himo, Mar. 1894, *G. Volkens 1822* [holotype: B (destroyed); lectotype: BM (BM000902860), **designated here**] (Plate 21)

Shrub 3–5 m tall, covered with ferruginous hairs at least when dry. Leaves elliptic, 6–18 cm × 2–6.5 cm, shortly acuminate at the apex, cuneate at the base; petioles up to 1.5 cm long; stipules with 4–9 filiform 3–5 mm long segments. Inflorescences axillary and terminal; flowers sweet-scented; peduncles up to 3 cm long. Calyx-tube 1.6–2 mm long, pubescent; lobes subequal, subulate or slightly spathulate at the apices, 5–10 mm long. Corolla white, 1.5–2.8 cm long; lobes elliptic, 5–6 mm long. Capsules 5 mm long.

Distribution: Central Kenya. [Tanzania].

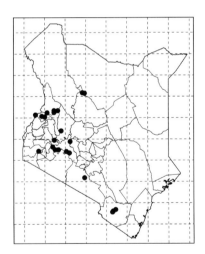
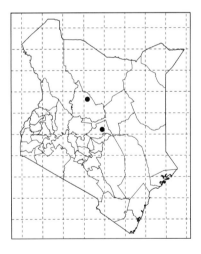

Habitat: Evergreen forests; 2200–2300 m.

Meru: Nyambene Hills, *SAJIT Z0174* (HIB). Samburu: Ndoto Mountains, *Cameron 269* (EA).

Plate 21 *Phyllopentas elata*. Photo by YDZ.

14. **Pentas** Benth.

Perennial herbs or subshrubs, with a fibrous or woody rootstock. Leaves opposite or rarely 3–4-whorled, hairy on both surfaces, short- to long-petiolate; stipules divided into 2–many filiform colleter-tipped segments. Inflorescences terminal or axillary, dense or lax, several- to many-flowered. Flowers usually dimorphic. Calyx-tube ovoid or globose; lobes usually 5, unequal, 1–3-foliaceous, enlarged. Corolla white or brightly coloured; tube shortly cylindrical to narrowly tubular, hairy in the throat; lobes ovate or oblong. Stamens included or exserted. Ovary 2-locular, ovules numerous in each locule. Capsules subglobose to ovoid or obovoid, opening at the apex into 4 valves or sometimes separating into 2 cocci. Seeds minute, brownish.

A small genus with 16 species widely distributed in North Africa, tropical Africa, north part of South Africa (Natal) and Madagascar, also widely cultivated in other parts of the world; six species in Kenya.

1a. Corolla-tube 3–9 cm long .... 6. *P. suswaensis*
1b. Corolla-tube less than 4 cm long 2
2a. Corolla-tube 1–4 cm long .... 5. *P. lanceolata*
2b. Corolla-tube usually no more than 10 mm long, rarely up to 11 mm long 3
3a. Prophytic herb with numerous short stems from a thick woody rootstock; corolla-tube 4–5 mm long .... 4. *P. arvensis*
3b. Herbs with 1–3 stems from a taproot; corolla-tube 4–11 mm long 4
4a. Corolla white or blue; tube 4–5 mm long .... 3. *P. pubiflora*
4b. Corolla white, reddish or purple; tube usually more than 7 mm long 5
5a. Petioles 1–7 cm long; corolla white .... 1. *P. micrantha*
5b. Petioles 0–1 cm long; corolla reddish or purple, more rarely white .... 2. *P. zanzibarica*

1. **Pentas micrantha** Baker, J. Linn. Soc., Bot. 21: 408. 1885; F.T.E.A. Rubiac. 1: 202. 1976. —Type: Madagascar, Tanala Forest, Dec. 1883, *R. Baron 3292* [holotype: K (K000046869)]

Herb, up to 90 cm tall. Leaf-blades elliptic to ovate or ovate-oblong, 5.5–14.5 cm × 1.5–6.5 cm, acute at the apex, cuneate at the base, glabrescent or sparsely pubescent on both sides; petioles 1–4(–7) cm long; stipules with 4–7 setae up to 6 mm long. Inflorescences of a lax terminal corymbs, several- to many-flowered. Flowers white or lavender-blue. Calyx-tube 1–1.5 mm long; lobes unequal, the largest one lanceolate, 3–10 mm long, the rest ones minute. Corolla-tube 6.5–10 mm long; lobes ovate, 1.8–3 mm long. Style slightly exserted. Capsules obtriangular to oblong-ovoid, 2.5–5 mm long.

subsp. **wyliei** (N.E. Br.) Verdc., Bull. Jard. Bot. État Bruxelles 23: 308. 1953; F.T.E.A. Rubiac. 1: 203. 1976. ≡ *Pentas wyliei* N.E. Br., Bull. Misc. Inform. Kew 1901: 123. 1901. —Type: South Africa, Zululand, Ungoya, 3 Apr. 1899, *J. Wylie* in *herb. J.M.*

Wood 7590 [holotype: K (K000414198); isotypes: NH (NH0008006-0), PRE (PRE 0151950-0), US (US00137513)]

Lower petioles mostly 1–4(–7) cm long. Corolla-tube 7–9.5(–11) mm long.

Distribution: Coastal Kenya. [East Africa from Kenya to northern South Afirca].

Habitat: Lowland forests; up to 800 m.

Note: *Pentas micrantha* subsp. *micrantha* only occurs in Madagascar.

Kilifi: Kaya Jibana, *Luke & Luke 4316* (EA). Kwale: Gongoni Forest, *Luke & Luke 9038* (EA).

2. **Pentas zanzibarica** (Klotzsch) Vatke, Oesterr. Bot. Z. 25: 232. 1875; F.T.E.A. Rubiac. 1: 203. 1976; W.F.E.A.: 155. 1987; U.K.W.FF: 226. 2013. ≡ *Pentanisia zanzibarica* Klotzsch, Naturw. Reise Mossambique 1: 286. 1861. —Types: Tanzania, Zanzibar, *Peters s.n.* [holotype: B (destroyed)]; Tanzania, Zanzibar, Sept. 1873, *J.M. Hildebrandt 1128* [neotype: W, designated by B. Verdcourt in Bull. Jard. Bot. État Bruxelles 23: 321. 1953; isoneotypes: BM, K (K000414215)]

Herb or shrubby herb, up to 1.8 m tall, with a woody rootstock. Leaves opposite, blades lanceolate to ovate, 4–14.5 cm × 1.4–5.8 cm, base cuneate, apex acute; petioles 0–1 cm long; stipules with several 4–9(–14) mm long setae. Inflorescences head-like, terminal or axillary; peduncles up to 15 cm long. Calyx-tube hairy, 1–1.3 mm long; lobes unequal, 1–9 mm long. Corolla blue, lilac, mauve or rarely whitish or red; tube 4–10 mm long; lobes oblong-elliptic, 1.5–6.5 mm long. Capsules pubescent, 2–5.5 cm long, beaked.

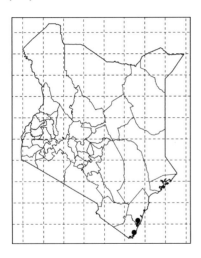

1a. Leaves small, 4–11.5 cm × 1.4–3.7 cm..a. var. *zanzibarica*
1b. Leaves large, 11–14.5 cm × 5.2–5.8 cm..b. var. *tenuifolia*

a. var. **zanzibarica** (Plate 22A, B)

Leaves small, 4–11.5 cm × 1.4–3.7 cm.

Distribution: Western, central, southern, and coastal Kenya. [Tropical East Africa].

Habitat: Grasslands or forest margins; up to 2600 m.

Homa Bay: Solko Kasuanga, *Kirika 251* (EA). Kajiado: Chyulu Hills, *Luke & Luke 7370* (EA); Mount Suswa, *Glover 4138* (EA); Ol Ndonyo Orok Hill, *Nyakundi 211* (EA).

Kiambu: Kijabe, *Verdcourt 3236* (EA); Makongo Forest Reserve, *Mutangah & Musila 4326* (EA). Kwale: Shimba Hills, *Luke et al. 8319* (EA); Mrima Forest, *Verdcourt 1867* (EA); Dzombo Hill, *Robertson et al. 345* (EA). Makueni: Kikoko Hill, *Mwangangi 1668* (EA). Meru: Meru National Park, *Mwangangi & Fosberg 652* (EA). Nairobi: Nairobi City Park, *Mwangangi & Kamau 3973* (EA). Nakuru: Menengai, *Mwangangi 203* (EA); Ol Longonot Estate, *Kerfoot 358* (EA); Mau

Escarpment, *Sikes 327* (EA). Narok: Olarro Camp, *Luke & Luke 7313* (EA); Ndunyangero Hill, *Glover et al. 1607* (EA); Noolpopong, *Kuchar & Msafiri 6866* (EA). Nyeri: Iganjo Kiburi Githi Location, *Kibue 213* (EA). Samburu: Leroghi, *Leakey 8566* (EA). Teita Taveta: Mount Kasigau, *Muasya 364* (EA); Bura, *Mwachala et al. EW1254* (EA). Trans-Nzoia: Endebess, *Irwin et al. 9641* (EA).

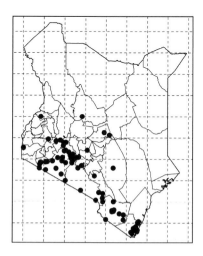

b. var. **tenuifolia** Verdc., Kew Bull. 30: 288. 1975; F.T.E.A. Rubiac. 1: 205. 1976. —Type: Kenya, Trans-Nzoia, Elgon, Endebess, Sept. 1943, *M.V.B. Webster* in *Herb. Amani 9644* [holotype: EA (EA000001500)]

Leaves large, 11–14.5 cm × 5.2–5.8 cm.

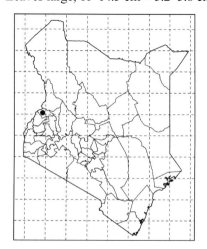

Distribution: Western Kenya. [Endemic].
Habitat: Forests; ca. 1800 m.

Trans-Nzoia: Endebess, *Webster* in *Herb. Amani 9644* (EA).

3. **Pentas pubiflora** S. Moore, J. Linn. Soc., Bot. 38: 254. 1908; F.T.E.A. Rubiac. 1: 206. 1976; U.K.W.F.F.: 225. 2013. —Type: Uganda, Mt. Ruwenzori, 1903, *A.F.R. Wollaston s.n.* [holotype: BM (BM000902848); isotypes: BM (BM000902847), MO (MO-391822)] (Plate 22C)

Herb, with stems sometimes woody at the base, up to 1.5(–3) m tall. Leaf-blades ovate to lanceolate, (3–)9.5–14.5 cm × (1–)3–5 cm, acute at the apex, cuneate at the base, pubescent on both sides; petioles up to 3 cm long; stipules with 5–9 filiform setae up to 1.2 cm long. Inflorescences of terminal corymbs; peduncles 2–4.5 cm long. Flowers white, rarely tinged pale blue or pinkish. Calyx-tube ca. 1.5 mm long; lobes unequal, one foliaceous, ovate, 2–6.5 mm long, the rest ones minute. Corolla-tube funnel-shaped, 4–5 mm long; lobes oblong-lanceolate, 2–3 mm long. Stamens included or slightly exserted. Capsules 2–4.5 mm long, ribbed.

Distribution: Western and southern Kenya.

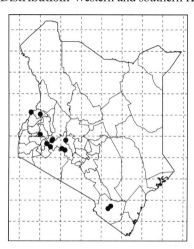

Plate 22 A, B. *Pentas zanzibarica* var. *zanzibarica*; C. *P. pubiflora*. Photo by LB (A, B) and YDZ (C).

[D.R. Congo and Uganda].

Habitat: Grasslands or forest margins; 1400–2500 m.

Bomet: southwest Mau Forest, *Kerfoot 4339* (EA). Kericho: Londiani, *Graham 1033* (EA). Nakuru: Eburu Forest, *Luke et al. 8917* (EA). Nandi: *Battiscombe K1206* (EA). Teita Taveta: Mbololo Forest, *Wakanene et al. 385* (EA). Trans-Nzoia: Kitale Museum, *Ekkens DB133* (EA).

4. **Pentas arvensis** Hiern, Fl. Trop. Afr. 3: 47. 1877; F.T.E.A. Rubiac. 1: 208. 1976; U.K.W.F.F.: 225. 2013. —Type: Sudan, Mittuland, Derago, Jan. 1870, *G. Schweinfurth 2775* [holotype: K (K000394974); isotypes: BM (BM000902875), GOET (GOET010421), M (M0106433), P (P00539248 & P00539249)]

Erect herb up to 50 cm tall, with a woody rootstock. Leaves paired or rarely 3-whorled; blades elliptic to lanceolate, 2.5–6 cm × 0.5–2 cm, bluntly acute or obtuse at the apex, cuneate at the base, hairy on both sides; petioles

0–2 mm long; stipules with 3–5 linear or deltoid 3–4(–9) mm long lobes. Inflorescences laxly globose, few- to many-flowered. Flowers white or pinkish. Calyx-tube 1–1.3 mm long; lobes unequal, the largest one oblong, up to 3 mm long, the rest ones minute. Corolla-tube 4–6 mm long; lobes oblong or elliptic, 1.8–3 mm long; stamens included or exserted; style slightly exserted. Capsules subglobose, 2.5–4 mm long.

Distribution: Western Kenya. [Tropical Africa].

Habitat: Grasslands or forest margins; 1500–1600 m.

Kakamega: Kakamega Forest, *Carroll II8* (EA). Kisumu: Nyanza Basin, *Battiscombe 653* (EA).

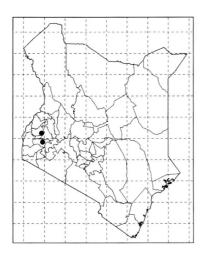

5. **Pentas lanceolata** (Forssk.) Deflers, Voyage au Yemen: 142. 1889; F.T.E.A. Rubiac. 1:208. 1976; W.F.E.A.: 155. 1987; U.K.W.F.F.:225. 2013; C.P.K.: 384. 2016. ≡ *Ophiorrhiza lanceolata* Forssk., Fl. Aegypt.-Arab. 42. 1775. —Type: Yemen, Hadie Mts., Mar. 1763, *P. Forsskal s.n.* [holotype: C; isotypes: BM (BM 000945087), S (S05-9977 & S05-9978)]

= *Pentas ainsworthii* Scott-Elliot, J. Linn. Soc., Bot. 32: 433. 1896. —Type: Kenya, Kitui, Ukamba, 1893, *G.F. Scott-Elliot 6437* [holotype: K (K000414212); isotype: BM (BM000902876)]

Herb or subshrub up to 1.3 m tall, hairy. Leaves ovate, lanceolate, ovate-lanceolate or elliptic, 3–13 cm × 1–6 cm, acute at the apex, cuneate at the base; petioles up to 5 cm long; stipules with 3–9(–14) setae. Inflorescences with terminal and axillary components combined into a single cluster. Calyx-tube 1–3 mm long; lobes very unequal, the largest one lanceolate, 0.5–1.3 cm long, the rest ones minute. Flowers mauve to white. Corolla-tube 1–4 cm long, dilated at the apex; lobes oblong-ovate to elliptic, 0.3–1 cm long; throat hairy within; style exserted or included. Fruits obtriangular, 4–6 mm long; beak 1–2 mm long.

1a. Corolla predominantly white, more rarely tinged lilac or pink, tube 2–4 cm long..a. var. *lanceolata*
1b. Corolla usually pink, lilac, mauve or magenta, rarely white, tube 1–2.2 cm long....................2
2a. Inflorescences congested in flowering and fruiting..b. var. *leucaster*
2b. Inflorescences with branches lax and spicate in fruiting .... c. var. *nemorosa*

a. var. **lanceolata** (Figure 16; Plate 23A–C)

Corolla predominantly white, more rarely tinged lilac or pink; tube 2–4 cm long.

Distribution: Northern, northwestern, western, central, and southern Kenya. [Tropical East Africa, Sudan, South Sudan, Ethiopia and also in Arabia].

Habitat: Grasslands, bushlands, or forest margins; 1300–2400 m.

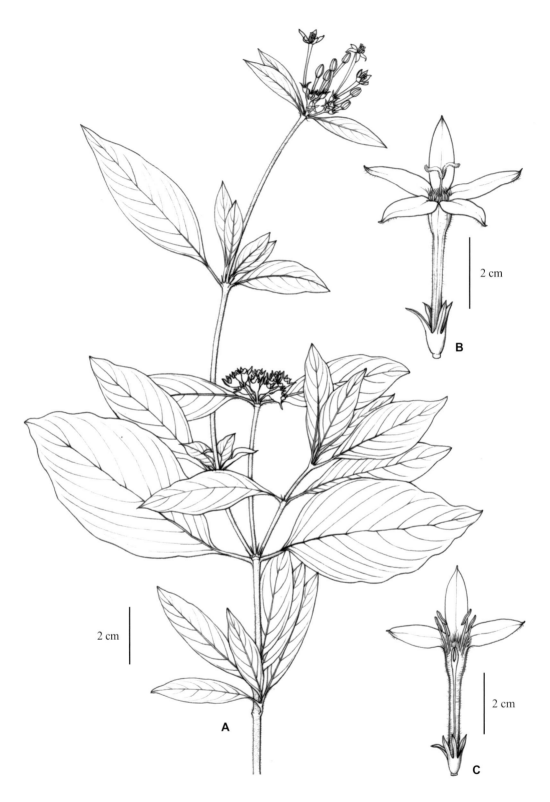

Figure 16 *Pentas lanceolata* var. *lanceolata*. A. flowering branch; B. long-styled flower; C. longitudinal section of short-styled flower, showing the stamens and style. Drawn by NJ.

Kiambu: Ngubi Forest Reserve, *Perdue & Kibuwa 8246* (EA). Kajiado: Ngong Hills, *Beentje 1772* (EA); Loitokitok, *Ibrahim 665* (EA). Kitui: Mumoni Hill, *Malombe & Muasya 868* (EA). Makueni: near Ithemboni, *Mwangangi 828* (EA). Marsabit: Marsabit Forest, *Faden 68/517* (EA); Mount Kulal, *Luke 1082* (EA). Nairobi: Karura Forest, *Kuchar 4445* (EA). Narok: Iligeri, *Kuchar 10891* (EA). Nyandarua: north Kinangop, *Kuchar 12305* (EA). Nyeri: west Mount Kenya, *SAJIT 002052* (HIB). Samburu: Mount Nyiru, *Kerfoot 2070* (EA). Teita Taveta: Ngangao Forest, *Beentje 2133* (EA). Turkana: Moruassigar, *Newbould 7238* (EA).

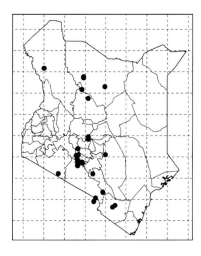

b. var. **leucaster** (K. Krause) Verdc., Bull. Jard. Bot. État Bruxelles 23: 347. 1953; F.T.E.A. Rubiac. 1: 210. 1976. ≡ *Pentas leucaster* K. Krause, Wiss. Ergebn. Deut. Zentr.-Afr. Exped., Bot. 2: 312. 1911. —Type: Rwanda, Lake Mohasi, 1907, *J. Mildbraed 444* [holotype: B (destroyed); lectotype: BR (BR0000008850713, fragment only), designated by B. Verdcourt in Bull. Jard. Bot. État Bruxelles 23: 348. 1953] (Plate 23D, E)

Corolla usually pink, lilac, mauve or magenta, rarely white; tube 1–2.2 cm long. Inflorescences congested in flower and fruit.

Distribution: Widespread in Kenya. [East Africa].

Habitat: Grasslands or bushlands; 1000–2400 m.

Bungoma: west Mount Elgon, *Jack 5* (EA). Elgeyo-Marakwet: Kapsowar, *Hepper & Field 4974* (EA). Kakamega: Kakamega Forest, *Verdcourt 1640* (EA). Kiambu: Muguga Forest, *Kokwaro & Kabuye 337* (EA). Kisii: east of Ikoba, *Vuyk 263* (EA). Kisumu: Songhor, *Havey 10562* (EA). Machakos: Ol Donyo Sabuk, *Lind & Napper 5507* (EA).

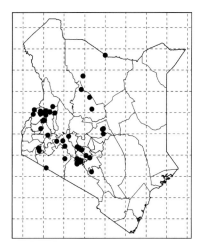

Makueni: Kilungu, *Mwangangi 2119* (EA). Marsabit: Furroli, *Gillett 13903B* (EA). Meru: Nyambeni Hills, *Luke et al. 1283B* (EA). Murang'a: Aberdares Mountains, *Gardner 14798* (EA). Nairobi: Nairobi City Park, *Kahurananga & Kiilu 2989* (EA). Nakuru: Njoro, *Monro 53* (EA). Narok: Noolpopong, *Kuchar & Musafiri 6808* (EA). Nyandarua: Kinangop, *Brown 1158* (EA). Samburu: Mathew's Range, *Luke 14113* (EA). Trans-Nzoia: Saiwa Swamp National Park, *Kirika 98* (EA). Uasin Gishu: Kipkarren, *Brodhurst-*

Hill 142 (EA). West Pokot: Cherangani Hills, *Bridson 103* (EA).

c. var. **nemorosa** (Chiov.) Verdc., Bull. Jard. Bot. État Bruxelles 23: 349. 1953; F.T.E.A. Rubiac. 1: 211. 1976; W.F.E.A.: 155. 1987. ≡ *Pentas parvifolia* Hiern var. *nemorosa* Chiov., Lav. Reale Ist. Bot. Modena 6: 52. 1935. —Type: Kenya, Aberdare Mts., Tusu, 25 Nov. 1910, *G. Balbo 678* [lectotype: FT (FT003374), designated by B. Verdcourt in Bull. Jard. Bot. État Bruxelles 23: 349. 1953; isolectotype: FT (FT003374)] (Plate 23F, G)

Corolla usually pink, lilac, mauve or magenta, rarely white; tube 1–2.2 cm long. Inflorescences with branches lax and spicate in fruiting.

Distribution: Central and western Kenya. [Ethiopia and Tanzania].

Habitat: Forest margins; 1200–2400 m.

Embu: Inamindi Valley, *Beecher 359* (EA). Kajiado: Emali Hills, *Beentje 2503* (EA). Kiambu: Limuru, *Verdcourt 357* (EA). Kirinyaga: near Thiba Fishing Camp, *Starreuski 178/51/1* (EA). Kisii: near Kisii, *Vuyk & Breteler 168* (EA). Kitui: Muumoni Hills, *Kirika et al. NMK578* (EA). Laikipia: Nanyuki, *Wildins 7467* (EA). Makueni: Kilungu, *Mwangangi 2124* (EA). Meru: Ngaia Forest, *Luke et al. 7281* (EA). Murang'a: Aberdare Mountains, *Battiscombe 14815* (EA). Nairobi: Karura Forest, *Mwangangi 241* (EA). Nyeri: Aberdare National Park, *De Block & Stieperaere 502* (EA). Samburu: Mathew's Range, *Kerfoot 1143* (EA). Tharaka-Nithi: Chuka Forest, *Beentje 2218* (EA). West Pokot: Sekerr, *Agnew et al. 10351* (EA).

6. **Pentas suswaensis** Verdc., Kirkia 5: 274. 1966; F.T.E.A. Rubiac. 1: 212. 1976; U.K.W.F.F.: 225. 2013. —Type: Kenya, Mt. Suswa, Aug. 1952, *B. Verdcourt 710* [holotype: EA (EA000002975)]

Subshrub or woody herb up to 1.2 m tall; stems densely pubescent. Leaves opposite or 3-whorled; blades elliptic, 3–6 cm × 1.5–3.5 cm, subacute or obtuse at the apex, cuneate at the base, densely hairy; petioles 2–6 mm long; stipules with 5 setae, 2–5 mm long. Inflorescences few-flowered. Calyx-tube 2 mm long, densely pubescent, strongly ribbed; lobes unequal, oblong, elliptic or subspathulate, 3–7 mm long. Corolla white or greenish cream; tube (3–)5.5–9 cm long, pubescent outside; lobes 4–5, oblong, 6–

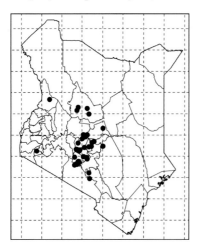
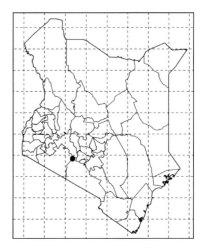

Plate 23 A–C. *Pentas lanceolata* var. *lanceolata*; D, E. *P. lanceolata* var. *leucaster*; F, G. *P. lanceolata* var. *nemorosa*. Photo by GWH (A, C) and YDZ (B, F, G) and NW (D, E).

10 mm long. Style exserted 1 cm in long-styled flowers, tomentellous. Capsule oblong, 6–8 mm long, strongly ribbed, densely pubescent.

Distribution: Southern Kenya. [Endemic].

Habitat: Bushlands; 1800–2000 m.

Kajiado: Mount Suswa, *Glover & Oledonet 4453* (EA).

15. **Rhodopentas** Kårehed & B. Bremer

Shrubs or subshrubs with woody erect or scrambling stems. Leaf-blades ovate-lanceolate, ovate-oblong or elliptic-oblong, acute or acuminate at the apex, cuneate at the base, glabrescent to sparsely hairy above, pubescent to velvety beneath, petiolate; stipules with 3–9 linear setae. Inflorescences terminal or axillary, dense to lax. Calyx-tube obovoid, glabrous to velvety; lobes 5–6, unequal, foliaceous. Flowers dimorphic. Corolla distinctly vermilion-scarlet; tube dilated at the apex, glabrous to hairy outside; lobes linear-oblong, oblong-lanceolate or elliptic. Stamens included or well exserted. Capsule oblong to obovoid, ribbed, eventually dividing into 2 cocci and showing little apical dehiscence.

A small genus with two species in East Africa from Ethiopia to Zambia and Malawi, both in Kenya.

1a. Leaves large, 3.5–15 cm × 1.7–6.5 cm, with petioles up to 2 cm long .... 1. *R. bussei*
1b. Leaves small, 1.2–10.5 cm × 0.3–0.8(–2) cm, with very short petioles less than 6 mm long .... 2. *R. parvifolia*

1. **Rhodopentas bussei** (K. Krause) Kårehed & B. Bremer, Taxon 56: 1076. 2007. ≡ *Pentas bussei* K. Krause, Bot. Jahrb. Syst. 43: 134. 1909; F.T.E.A. Rubiac. 1: 200. 1976. —Type: Tanzania, Lindi, Rondo Plateau, 600 m, May 1903, *W. Busse 2628* [holotype: B (destroyed); lectotype: K (K000394964), **designated here**; isolectotypes: BR (BR0000008850669), EA (EA000001880)] (Figure 17; Plate 24A–C)

Shrub or shrubby herb up to 4 m tall. Leaf-blades ovate-lanceolate or ovate-oblong, 3.5–15 cm × 1.7–6.5 cm, acute or acuminate at the apex, cuneate at the base, sparsely shortly hairy above, pubescent to velvety beneath; petiole up to 2 cm long; stipules with 3–9 linear setae 0.4–1.5 cm long. Inflorescences terminal and axillary, dense or lax, up to 8 cm wide, many-flowered. Calyx-tube obovoid, ca. 1.5 mm long, glabrous or velvety; lobes very unequal, foliaceous, 1–3 much enlarged, narrowly lanceolate, 0.5–1.8 cm long, the rest small, linear, deltoid or lanceolate, 1.5–7 mm long. Corolla vermilion-scarlet; tube 0.7–2 cm long, dilated at the apex, glabrous to hairy outside; lobes narrowly oblong, elliptic or oblong-lanceolate, 2.5–12 mm long; stamens entirely included. Capsules oblong to obovoid, 3–6 mm long, 10-ribbed.

Distribution: Central and coastal Kenya. [Burundi, D.R. Congo, Malawi, Somalia, Tanzania, and Zambia].

Habitat: Grasslands, bushlands, woodlands, and dry evergreen forests; up to 1800 m.

Kilifi: Brachystegia Forest, *Spjut & Ensor 2615* (EA, K); Gede Forest, *Gerhardt & Mariette 143* (EA). Kitui: Mutha, *Joana 7412*

Figure 17 *Rhodopentas bussei*. A. a flowering branch; B. short-styled flower. Drawn by NJ.

(K). Kwale: Shimba Hills, *SAJIT V0217* (HIB); Buda Forest, *Kokwaro 3932* (EA). Lamu: Boni Forest, *Luke & Robertson 1483* (EA); ca. 2 km north of Hindi, *Gillett 20336* (EA, K). Makueni: Kiboko Research Station, *Muriithi 114* (EA). Mombasa: Bamburi, *Haller 37* (EA).

6 mm long. Inflorescences corymbose, terminal. Calyx-tube obovoid, ca. 1.5 mm long; lobes 5–6, unequal, 1.5–12 mm long. Corolla scarlet, rarely pink and white; tube 0.7–2.2 cm long; lobes oblong to linear-oblong, 2.5–10 mm long. Capsules ribbed, oblong to obovoid, 3–6 mm long.

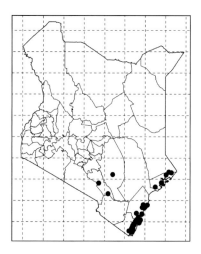

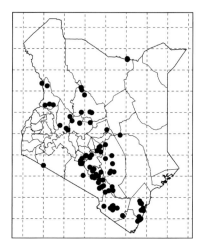

2. **Rhodopentas parvifolia** (Hiern) Kårehed & B. Bremer, Taxon 56: 1076. 2007; C.P.K.: 385. 2016. ≡ *Pentas parvifolia* Hiern, J. Linn. Soc., Bot. 16: 262. 1877; F.T.E.A. Rubiac. 1: 201. 1976; W.F.E.A.: 155. 1987; U.K.W.F.F.: 225. 2013. —Type: Kenya, Mombasa, Mar. 1876, *J.M. Hildebrandt 1994* [holotype: BM (BM000902853); isotypes: K, W] (Plate 24D–F)

= *Pentas parvifolia* Hiern f. *spicata* Verdc., Bull. Jard. Bot. État 23: 306. 1953; F.T.E.A. Rubiac. 1: 202. 1976. —Type: Kenya, Chyulu foothills, 15 May 1938, *P.R.O. Bally 735* [holotype: EA (EA000001760)]

Erect subshrub up to 2.5 m high. Leaf-blades opposite, elliptic-lanceolate, ovate or oblong-elliptic, 1.2–10.5 cm × 0.3–2 cm, acute at the apex, cuneate at the base; petioles 0–6 mm long; stipules with several setae up to

Distribution: Widespread in Kenya. [Ethiopia, Tanzania, and Uganda].

Habitat: Grasslands, thickets, or dry evergreen forests; up to 2400 m.

Baringo: Bogoria, *Vincens 266* (EA). Embu: Kiang'ombe Hill, *Kirika et al. NMK568* (EA, K). Kajiado: Enkorika, *Kuchar et al. 7952* (EA). Kiambu: Kuraiha Farm near Juja Town, *Malombe & Kirika 58* (EA). Kilifi: Mangea Hill, *Luke & Robertson 1778* (EA). Kwale: Kaya Bombo, *Luke 3481* (EA, K). Laikipia: Vaso Narok River, *Carter & Stannard 329* (EA, K). Machakos: Katumani, *Thomas 712* (EA, K). Makueni: Kithembe Village, *Mwangangi 870* (EA, K). Kitui: Makongo Forest Reserve, *Mwangangi 4647* (EA). Marsabit: Kakala, *Gillett 12797* (EA). Meru: Meru National Park, *Hamilton 279* (EA). Mombasa: Bamburi Beach, *Napier 3338* (EA). Murang'a: Thika Falls, *Napier 5631*

Plate 24 A–C. *Rhodopentas bussei*; D–F. *R. parvifolia*. Photo by BL (A, B, D–F) and GWH (C).

(EA). Nairobi: Nairobi National Park, *Faden et al. 74/551* (EA, K). Narok: Masai Mara Game Reserve, *Kuchar 4395* (EA). Samburu: Gurika Hill, *Carter & Stannard 518* (EA). Teita Taveta: Bura River, *Mwachala 205* (EA). Tana River: Athi River Falls, *Gedye 3582* (EA). West Pokot: Napau Hills, *Padure 115* (EA, K).

16. **Pentanisia** Harv.

Annual, perennial herbs or subshrubs. Leaves opposite or rarely 3-whorled; stipules with bases connate with the petiole, fimbriate. Inflorescences terminal or pseudo-axillary, capitate or spike-like, few- to many-flowered. Flowers (3–)4–5-merous, usually blue, dimorphic. Calyx-tube ovoid or rectangular; lobes with 1–3 enlarged and often leaf-like, the rest small. Corolla-tube narrowly cylindrical; lobes (3–)4–5, ovate to oblong; throat densely hairy. Ovary 2–5-locular; ovule single in each locule. Style filiform; stigma divided into 2–5 filiform lobes equal in number of locule. Fruits dry or slightly fleshy, globular. Seeds small, compressed.

A genus with about 15 species confined to tropical Africa and Madagascar; five species in Kenya.

1a. Slender annual herb; flowers 4-merous .... 1. *P. parviflora*
1b. Perennial herbs; flowers usually 5-merous 2
2a. Ovary 2-locular or rarely abnormally 3-locular; stigma bifid 3
2b. Ovary 3–5-locular; stigma 3–5-lobed 4
3a. Pyrophytic herb from a distinctly woody rootstock; leaves glabrous or rarely shortly hairy .....
.... 2. *P. schweinfurthii*
3b. Subprostrate herb with rootstock not distinctly woody; leaves pubescent .... 3. *P. foetida*
4a. Corolla-tube 0.8–2.3 cm long .... 4. *P. ouranogyne*
4b. Corolla-tube 2.7–4 cm long .... 5. *P. longituba*

1. **Pentanisia parviflora** Stapf ex Verdc., Kew Bull. 6: 383. 1952. ≡ *Paraknoxia parviflora* (Stapf ex Verdc.) Bremek., Kew Bull. 8: 439. 1953; F.T.E.A. Rubiac. 1: 166. 1976; U.K.W.F.F.: 226. 2013. —Type: Uganda, Teso, Serere, 1100 m, May 1932, *P. Chandler 672* [holotype: EA (no seen); isotypes: BM (BM000903206), BR (BR0000008850195), K (K000379457)]

Annual hairy herb, up to 38 cm tall. Leaves opposite, sessile to shortly petiolate; blades narrowly elliptic, lanceolate to linear or linear-oblanceolate, 1.3–4.2 cm × 0.2–1.7 cm, acute at the apex, cuneate at the base, stipules with 3–7 deltoild segments 1–2 mm long from a short base. Inflorescences terminal, head-like, sessile. Calyx-tube short; lobes 3–4, small or sometimes with a foliaceous. Corolla white or tinged bluish mauve; tube narrowly funnel-shaped, up to 4 mm long; lobes 3–4, oblong, 1.3–1.5 mm long. Ovary 2-locular; ovules solitary in each locule; style filiform; stigma bifid, lobes 0.6–0.8 mm long. Fruits ovoid, indehiscent, 1.3–1.8 mm long, hairy. Seeds 0.8–0.9 mm long.

Distribution: Western Kenya. [Tropical Africa].

Habitat: Grasslands, bushlands, or old cultivations; 1300–2200 m.

Nandi: Chemase, *Tallantire 159* (EA). Trans-Nzoia: Kitale Grassland Research Station, *Verdcourt 2454* (EA, K). Uasin Gishu: Kipkarren, *Brodhurst-Hill 446* (EA).

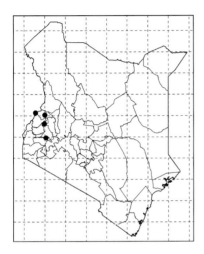

2. **Pentanisia schweinfurthii** Hiern, Fl. Trop. Afr. 3: 131. 1877; F.T.E.A. Rubiac. 1: 168. 1976; U.K.W.F.F.: 226. 2013. —Type: Sudan, Dar Fertit, south from the Gudjo, 29 Jan. 1871, *G. Schweinfurth 8* [holotype: K; isotypes: BM (BM000903215), W]

Perennial pyrophytic herb, up to 24 cm tall, with a distinctly woody rootstock. Leaves opposite, elliptic or elliptic-obovate, 0.3–6 cm × 0.2–2 cm, acute to obtuse at the apex, cuneate at the base, glabrous or rarely shortly hairy; petiole up to 2 mm long; stipules with 2–4 deltoid lobes or 1 trifid lobe, 1–4.5(–7) mm long. Inflorescences capitate or branched and spike-like. Calyx-tube squarish; lobes unequal, the longest 1–3.5 mm long, the rest minute. Corolla bright blue, white, pale lilac or purple; tube 0.6–1.3 cm long, glabrous or pubescent with short white hairs; throat densely hairy; lobes ovate-oblong to ovate-lanceolate, 2–6 mm long. Style exserted; stigma lobes linear. Fruits ovoid, 1.5–2.5 mm long, 2-locular. Seeds broadly elliptic, ca. 2 mm long, concavo-convex.

Distribution: Western Kenya. [Tropical Africa].

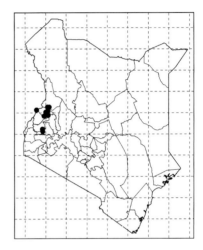

Habitat: Grasslands or woodlands; 1500–2500 m.

Bungoma: Kiminini, *Mainwaring 2666* (EA). Kakamega: Kakamega Forest, *Gilbert 6878* (EA, K). Trans-Nzoia: Saiwa Swamp National Park, *Hooper & Townsend 1444* (EA, K). Uasin Gishu: Hoey's Bridge, *Mainwaring 1551* (EA). Vihiga: Kaimosi, *Archer 242* (EA). West Pokot: Cherangani Hills, *Webster 8835* (EA).

3. **Pentanisia foetida** Verdc., Kew Bulletin 6: 381. 1952; F.T.E.A. Rubiac. 1: 169. 1976; U.K.W.F.F.: 226. 2013. —Type: Kenya, Limuru, Tigoni Dam, 1770 m, 4 Mar. 1951, *R.W. Rayner 421* [holotype: EA (EA000001780); isotypes: BM (BM 000903207), BR (BR0000008828620), FI, LISC (LISC002639), K (K000311702 & K

000311703), P, U] (Plate 25A)

Perennial stink herb, with several subprostrate hairy stems, up to 60 cm long. Leaves ovate to ovate-lanceolate, 3–8 cm × 1.5–3 cm, acute at the apex, cuneate at the base, pubescent on both surfaces; petioles up to 1.2 cm long; stipules with 5–7 subequal segments 2–5.5(–9) mm long, from a base 2–4 mm long. Inflorescences terminal or pseudo-axillary, capitate or later extending and becoming spikes up to 5 cm long. Flowers lilac, white, pinkish red or purple, dimorphic. Calyx-tube ovoid, 1–1.5 mm long; lobes 5, 1 foliaceous, 4–6 mm long. Corolla-tube 5–10 mm long, pubescent outside, hairy inside the throat; lobes 5–6, oblong-lanceolate, 1.5–3 mm long. Style exserted; stigma bifid, lobes filiform. Fruits ovoid, brownish black, ca. 2.5 mm long. Seeds brown, elliptic, ca. 1.2 mm long.

Distribution: Central Kenya. [Tanzania].

Habitat: Grasslands or forest edges; 1700–2400 m.

Embu: Thiba, *Copley 391* (K). Kiambu: Tigoni Dam, *Rayner 421* (EA). Kirinyaga: south Mount Kenya, *Kabuye 55* (EA). Meru: east slope of Mount Kenya, *SAJIT 003803* (HIB); Marimba Forest, *Verdcourt & Polhill 2992* (EA, K). Nyandarua: Aberdare Mountains National Park, *Block & Stieperaere 507* (EA).

4. **Pentanisia ouranogyne** S. Moore, J. Bot. 18: 4. 1880; F.T.E.A. Rubiac. 1: 172. 1976; W.F.E.A.: 155. 1987; U.K.W.F.F: 226. 2013; C.P.K.: 383. 2016. —Type: Kenya, Kitui, Apr. 1877, *J.M. Hildebrandt 2754* [holotype: BM (BM000903202); isotypes: JE (JE00000328), K, M (M0106368 & M0106369), W] (Figure 18; Plate 25B–D)

Perennial herb up to 60 cm tall, with a slender rootstock; stems usually with spreading white hairs. Leaves opposite, hairy, lanceolate to linear-lanceolate, 1.5–10 cm × 0.3–2.7 cm, base cuneate, apex acute; petioles up to 1 cm long, adnate to the stipular sheath; stipules membranous, with several bristly hairy setae 2–9.5 mm long. Inflorescences head-like, terminal, with peduncles 0–16 cm long. Calyx hairy, with leaf-like lobe 3–10 mm long. Corolla bright blue, rarely pink, with tube 0.7–2.3 cm long; lobes 2–5.5 mm long. Ovary 2–5-locular; style slender; stigma 2–5-lobed, 1–2 mm long. Fruits reddish-brown, globose, woody, 2.5–5 mm in diameter. Seeds pale-brown, elliptic, ca. 3 mm long.

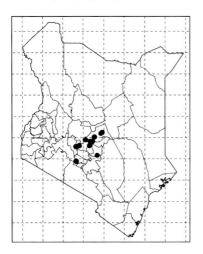

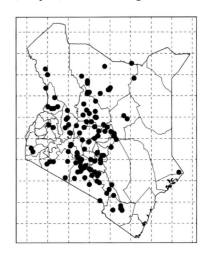

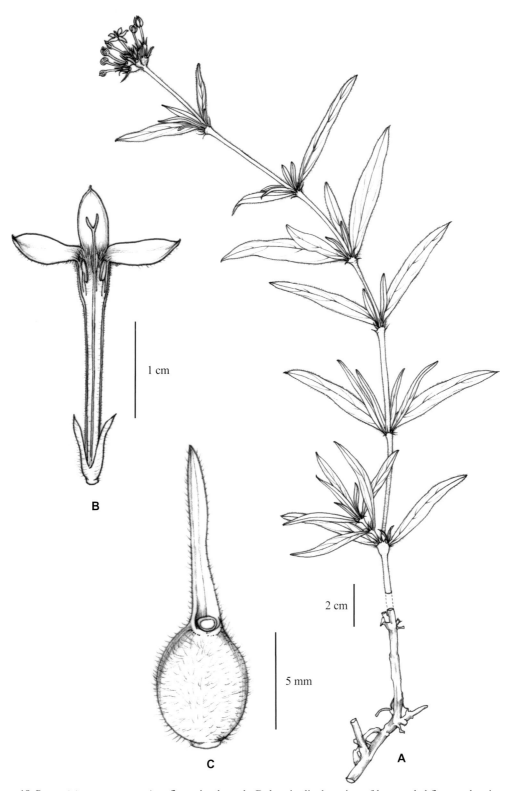

Figure 18 *Pentanisia ouranogyne*. A. a flowering branch; B. longitudinal section of long-styled flower, showing stamens and style; C. fruit. Drawn by NJ.

Plate 25 A. *Pentanisia foetida*; B–D. *P. ouranogyne*. Photo by YDZ.

Distribution: Widespread in Kenya. [Ethiopia, Somalia, Tanzania, and Uganda].

Habitat: Grasslands or woodlands,, also in old cultivations; 300–2450 m.

Baringo: near Chepkesin, *Bonnefille & Riollet 39* (EA). Embu: Mwea National Reserve, *Mwangangi et al. 4964* (EA). Garissa: Mkondoni, *Edwards E142* (EA). Homa Bay: Lambwe Valley, *Makin 67* (EA). Isiolo: near Veterinary Primary School, *Kimani 277* (EA). Kajiado: *Kokwaro 3695* (EA). Kiambu: Thika, *Faden 67351* (EA). Kirinyaga: Cotton Research Farm, *Robertson 1686* (EA). Kitui: Yatta, *Seki JKCAT1564* (EA). Laikipia: Ol Ari Nyiro Ranch, *Muasya 1547* (EA). Machakos: Lukenya, *Agnew & Musumba 5336* (EA). Makueni: Sultan Hamud, *Georgiadis 6* (EA). Marsabit: Marsabit National Reserve, *SAJIT Z0316* (HIB). Meru: Meru National Park, *Lambrecht 11* (EA). Nairobi: Mbagathi, *Napier 87* (EA). Nakuru: near Lake Nakuru, *Kutilek 118* (EA). Narok: near Masai Mara Game Reserve, *Kimeu et al. KEFRI429* (EA). Nyeri: near Karura Farm, *Williams 312* (EA). Samburu: Lopet Plateau, *Carter & Stannard 464* (EA, K). Teita Taveta: Irima Hill, *Kokwaro 2631* (EA). Tharaka-Nithi/Meru: Meru National Park, *Ament & Magogo 184* (EA). Turkana: *Ohta 54* (EA). Uasin Gishu: Eldoret, *Tweedie 31* (EA). Wajir: Moyale, *Gillett 12950* (EA). West Pokot: Sigor, *Roberts 3* (EA).

5. **Pentanisia longituba** (Franch.) Oliv., Unkn. Horn of Afr., App.: 319. 1888; F.T.E.A. Rubiac. 1: 174. 1976. ≡ *Knoxia longituba* Franch., Sert. Somal.: 32. 1882. —Type: Somalia, *G. Révoil 53* [holotype: P(P03947630)]

Perennial herb, up to 60 cm tall. Leaves lanceolate or oblong-lanceolate to linear-lanceolate, 2.5–7.5 cm × 0.4–2.3 cm, acute at the apex, cuneate at the base, sessile or very shortly petiolate; stipules with several setae, 0.8–4 mm long. Inflorescences terminal, densely capitate. Flowers white or pale blue. Calyx-tube 2–4 mm long; the foliaceous lobe lanceolate, 0.6–2 cm long. Corolla-tube 2.5–4.3 cm long; lobes oblong-elliptic, 3–7.5 mm long. Ovary 3–5-locular; style exserted; stigma 3–5-fid, lobes filiform, 1–2 mm long. Fruits very pale brown, globose, 4.5–5.5 mm in diameter.

Distribution: Northern Kenya. [Ethiopia and Somalia].

Habitat: Open deserts or bushlands; 900–1050 m.

Marsabit: Furroli, *Gillett 13806* (BR, K).

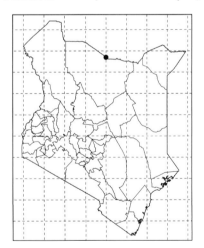

17. Otomeria Benth.

Erect ascending or procumbent, annual or perennial herbs. Leaves opposite; stipules with base divided into several narrow segments. Flowers 5-merous, in cymose heads, developed into a long simple spike in fruiting. Calyx-tube campanulate, ovoid or elongate-oblong; lobes unequal, 1–3 foliaceous. Corolla-tube elongated, slender, densely hairy at the throat; lobes 5, spreading, ovate to orbicular, narrowing to the base. Stamens 5, inserted at the throat of the corolla, exserted or included; anthers linear-oblong. Ovary 2-locular, with ovules numerous in each locule; style exserted or included. Capsules oblong, compressed, ribbed. Seeds small, reticulate.

A small genus of 8 species widely distributed in tropical Africa; two species in Kenya.

1a. Petioles 0–0.6 cm long; flowers usually scarlet, isostylous, with styles always exserted............ .... 1. *O. elatior*
1b. Petioles up to 2 cm long; flowers pale pink or white with a blue or maroon central spot, heterostylous, with style exserted or included...2. *O. oculata*

1. **Otomeria elatior** (A. Rich. ex DC.) Verdc., Bull. Jard. Bot. État 23: 18. 1953; F.T.E.A. Rubiac. 1: 214. 1976; W.F.E.A.: 154. 1987; U.K.W.F.F: 226. 2013. ≡ *Sipanea elatior* A. Rich., Mem. Fam. Rubiac.: 196. 1830. —Type: Angola, *da Silva s.n.* [holotype: P (P00539269)]

Erect or rarely straggling herb, up to 3 m tall. Leaf-blades ovate-elliptic to lanceolate or rarely linear, 1.5–9.5 cm × 0.7–3.2 cm, acute at the apex, rounded or cuneate at the base, glabrous, pubescent to densely hairy on both side; petioles 0–6 mm long; stipules with 1–3 linear setae, 1–5 mm long. Inflorescences spike-like, 1–6.5 cm long, elongated up to 37 cm long in fruiting. Flowers usually scarlet, isostylous. Calyx-tube ca. 2 mm long; lobes unequal, 1–3-foliaceous, lanceolate, up to 2.4 cm long, others 1–4 mm long. Corolla-tube slender, up to 2.7 cm long, densely hairy at the throat; lobes ovate, orbicular or elliptic-spathulate, 0.5–1.8 cm long. Style slightly exserted; stigmas elliptic. Fruits oblong, compressed, up to 1.2 cm long, strongly ribbed.

Distribution: Western Kenya. [Tropical Africa].

Habitat: Swampy places in wooded grasslands; 900–2550 m.

Kakamega: Mumias, *Battiscombe 668* (EA, K). Kisii: Lolgorien, *Napier 2916* (EA). Uasin Gishu: Burnt Forest, *Webster 8834* (EA).

2. **Otomeria oculata** S. Moore, J. Bot. 18: 4. 1880; F.T.E.A. Rubiac. 1: 215. 1976; W.F.E.A.: 154. 1987; U.K.W.F.F.: 226. 2013. —Type: Kenya, Kitui, Apr. 1877, *J.M. Hildebrandt 2756* [holotype: BM (BM000839221); isotypes: K (K000414234), M (M0106431), W] (Figure 19; Plate 26)

Erect or subshrubby herb, up to 60 cm tall. Leaf-blades lanceolate, narrowly rhomboid or rarely ovate, 2.5–6.5(–8.5) cm × 0.7–2.3 cm,

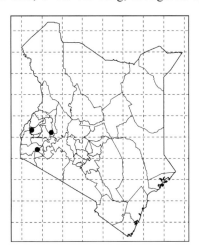

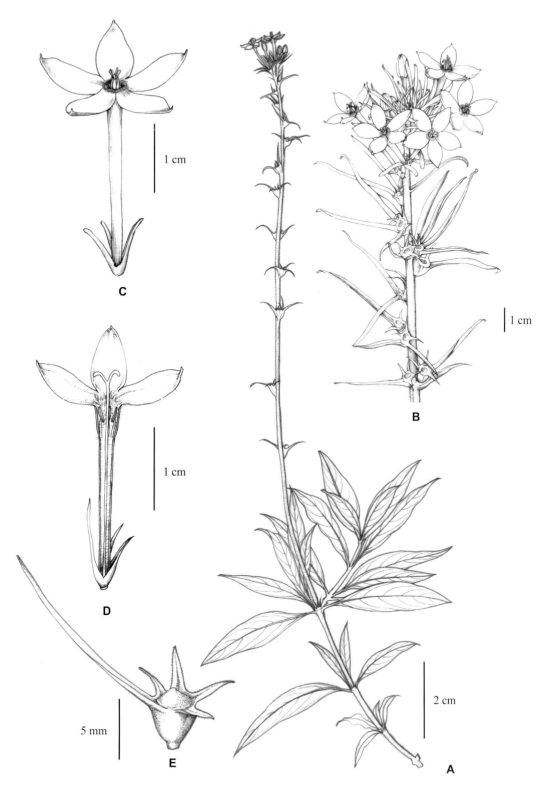

Figure 19 *Otomeria oculata*. A. branch with flowers and fruits; B. inflorescence; C. short-styled flower; D. longitudinal section of long-styled flower; E. fruit. Drawn by NJ.

Plate 26 A–D. *Otomeria oculata*. Photo by NW (A, B) and GWH (C, D).

acute at the apex, narrowly cuneate at the base, hairy both sides; petioles up to 2 cm long; stipules with 1–3 flat 1.5–3 mm long segments. Inflorescences capitate at first but soon becoming a spike 5–32 cm long. Flowers pale pink or white with a blue or maroon central spot, heterostylous. Calyx-tube 1–2 mm long; lobes unequal, 1–3 linear, up to 2.5 cm long, others 1–5 mm long. Corolla-tube slender, 1.8–3.2 cm long; lobes narrowly to broadly elliptic, 0.5–1 cm long; style slightly exserted or included. Fruits obtriangular-oblong, 3.5–6 mm long, ribbed.

Distribution: Northern, northwestern, western, and central Kenya. [Ethiopia and Uganda].

Habitat: Grasslands or bushlands; 500–1700 m.

Baringo: Lesurut Island, *Luke 659* (EA). Isiolo: Veterinary Primary School, *Kimani 259* (EA). Kitui: Endau Forest, *SAJIT MU0114* (HIB); Katumba Hill, *Kuchar 9058* (EA). Laikipia: Lomasa on Ewaso Ng'iro River, *Faden 24/85* (EA). Machakos: Ol doinyo Sabuk, *Verdcourt & Bally 841* (EA, K); Nguungi Hill, *Kimani 188* (EA). Marsabit: Mount Kulal, *Synnott 1766* (EA); Sololo, *Gillett 13673* (EA, K). Meru: Meru National Park, *Ament & Magogo 276* (K); Shaba, *Sutton 20* (EA). Murang'a: Mitubir, *Archer 33* (EA). Samburu: Wamba, *Newbould 2915* (K); Mount Ndoto, *Hepper & Jaeger 7244* (EA, K). Tana River: Kora National Park, *Mungai et al. 128/83* (EA, K). Tharaka-Nithi: north of Ishiara on road to Meru, *Gilbert et al. 5708* (EA, K).

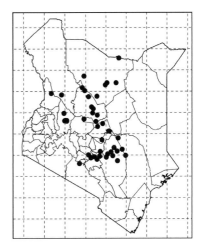

6. Trib. **Spermacoceae** Bercht. & J. Presl

Herbs or subshrubs. Stipules fimbriate. Ovary 2(or rarely 3–4)-locular, with single to numerous ovules in each locule. Fruits dry, capsular or dividing into dehiscent or indehiscent cocci.

Fifteen genera occur in Kenya.

1a. Ovule solitary in each locule..2
1b. Ovules 2–many in each locule 6
2a. Ovary 3–6-locular; stigmas 3–4; capsules 3–4-coccous .... 32. *Richardia*
2b. Ovary 2(–3)-locular; stigmas 1–2; capsules with 2 valves or 2 cocci, or circumscissile...........3
3a. Fruits circumscissile.. 29. *Mitracarpus*
3b. Fruits indehiscent or open by longitudinal slits or 2-coccous...4
4a. Fruits capsular with 2 valves or with 2 cocci, usually dehiscent .... 31. *Spermacoce*

4b. Fruits with 2 cocci, somewhat indehiscent 5
5a. Seeds distinctly and characteristically lobed .... 28. *Diodia*
5b. Seeds not lobed .... 30. *Diodella*
6a. Flowers mostly 4-merous 7
6b. Flowers mostly 5-merous 8
7a. Annual or short-lived perennial herbs .... 21. *Pentodon*
7b. Shrubs or subshrubs .... 19. *Pentanopsis*
8a. Inflorescences axillary or flowers solitary, axillary 9
8b. Inflorescences terminal or terminal and axillary 11
9a. Corolla-tube over 20 mm long .... 20. *Conostomium*
9b. Corolla-tube less than 10 mm long 10
10a. Corolla-tube glabrous; stamens and styles both exserted .... 26. *Scleromitrion*
10b. Corolla-tube with a ring of hairs in throat; stamens and styles not both exserted .... 27. *Oldenlandia*
11a. Corolla-tube narrowly cylindrical; anthers and stigmas included 12
11b. Corolla-tube cylindrical or funnel-shaped; anthers and/or stigmas usually exserted 13
12a. Stigma 2-lobed, lobes filiform .... 18. *Kohautia*
12b. Stigma unlobed, ovoid or cylindrical .... 24. *Cordylostigma*
13a. Capsules opening both septicidally and loculicidally 14
13b. Capsules opening loculicidally, sometimes tardily dehiscent 15
14a. Corolla-tube bearded inside; style not shortly bifid at the apex .... 22. *Agathisanthemum*
14b. Corolla-tube glabrous inside; style shortly bifid at the apex .... 23. *Dibrachionostylus*
15a. Corolla-tube often with a ring of hairs in throat .... 27. *Oldenlandia*
15b. Corolla-tube glabrous .... 25. *Edrastima*

18. **Kohautia** Cham. & Schltdl.

Annual or perennial herbs or rarely subshrubs. Leaves sessile, opposite, blades linear to narrowly elliptic-lanceolate; stipules with 2-several fimbriae or rarely reduced to a simple lobe. Flowers not heterostylous, in terminal panicle-, corymb- or head-like inflorescences. Calyx-lobes 4(–5), small, equal. Corolla-tube narrowly cylindrical; lobes 4(–5), narrowly linear to broadly elliptic. Stamens always included or only the anther-tips exserted. Ovary 2-locular; ovules numerous; style always included; stigma-lobes 2, filiform. Capsules crowned by the permanent calyx lobes, globose or ellipsoid, scarcely beaked, loculicidally splitting at the apex. Seeds numerous, angular.

About 30 to 40 species widely distributed in Africa, Madagascar, the Arabian peninsula, Iran, India, Pakistan and tropical Asia; three species in Kenya.

1a. Corolla scarlet-pink, pink or crimson-purple; capsules oblong-ellipsoid..............1. *K. coccinea*
1b. Corolla white, yellowish, blue or lilac; capsules spherical 2
2a. Flowers always paired along the branchlets of inflorescence; corolla-tube less than 5 mm long ....2. *K. aspera*
2b. Flowers always single along the branchlets of inflorescence; corolla-tube more than 10 mm long..3. *K. caespitosa*

1. **Kohautia coccinea** Royle, Ill. Bot. Himal. Mts.: 241, t. 53, f. 1. 1839; F.T.E.A. Rubiac. 1: 235. 1976; U.K.W.F.F.: 223. 2013. —Type: India, Himalaya, Budraj, *H.B. Royle s.n.* [holotype: LIV] (Figure 20)

Annual herb, up to 45(–70) cm tall. Leaf-blades linear or linear-lanceolate, 1–8 cm × 0.1–0.8 cm, acute at the apex, narrowed at the base; stipule with 2–6 filiform fimbriae 1.5–5 mm long. Inflorescences many-flowered, spike- or raceme-like. Flowers paired at each node. Calyx-tube ellipsoid, 1.2–1.5 mm long; lobes linear, 1.8–5.2(–7) mm long. Corolla usually pink, scarlet-pink or crimson-purple, narrowly cylindrical, 2–6 mm long; lobes (3–)4(–5), oblong-elliptic, 1.5–4.5(–7.5) mm long. Style 1.2–1.5 mm long; stigma 1–1.2 mm long. Capsules oblong-ellipsoid, 3–5.5 mm long, scabrid to sparsely papillose. Seeds darkbrown, irregularly angular, ca. 0.3 mm long.

Distribution: Northern, western, central, and southern Kenya. [Tropical Africa, also in North India].

Habitat: Grasslands, woodlands, roadsides, or in cultivated fields; 800–2400 m.

Baringo: Sabatia, *Graham 3025* (K). Homa Bay: Mbita Point, *Dissemond 68* (EA). Kajiado: Chyulu Hills, *Luke & Luke 9489* (EA). Kericho: east of Londiani, *Gillett 19315* (EA). Kisii: *Napier s.n.* (EA). Marsabit: Marsabit National Reserve, *Bally & Smith 14795* (EA, K). Nairobi: Nairobi National Park, *Mbuvi 450* (EA). Nyeri: Nanyuki, *Moreau & Moreau 92* (EA,K). Kiambu: Langata, *Archer 255* (EA). Nakuru: Ol Longonot Estate, *Kerfoot 3575* (EA). Narok: llgeri, *Kuchar 10843* (EA). Samburu: near Poror, *Bally 8563* (K). Trans-Nzoia: Kitale, *Birnie 411* (EA). Uasin Gishu: Kipkarren, *Brodhurst-Hill & Napier 134* (K). West Pokot: south of Cherangani, *Symes 209* (EA).

2. **Kohautia aspera** (B. Heyne ex Roth) Bremek., Verh. Kon. Ned. Akad. Wetensch., Afd. Natuurk., Sect. 2, 48(2): 113. 1952; F.T.E.A. Rubiac. 1: 241. 1976; U.K.W.F.F.: 223. 2013. ≡ *Hedyotis aspera* B. Heyne ex Roth, Nov. Pl. Sp.: 94. 1821. — Type: India, *B. Heyne s.n.* [neotype: K-W (K001110087), designated by I.M. Turner in Taxon 70(2): 393. 2021]

Annual herb up to 40 cm tall. Leaf-blades linear to narrowly elliptic, 1.5–5 cm × 0.5–

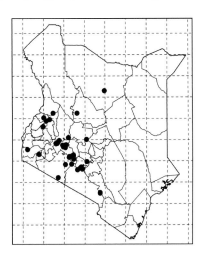

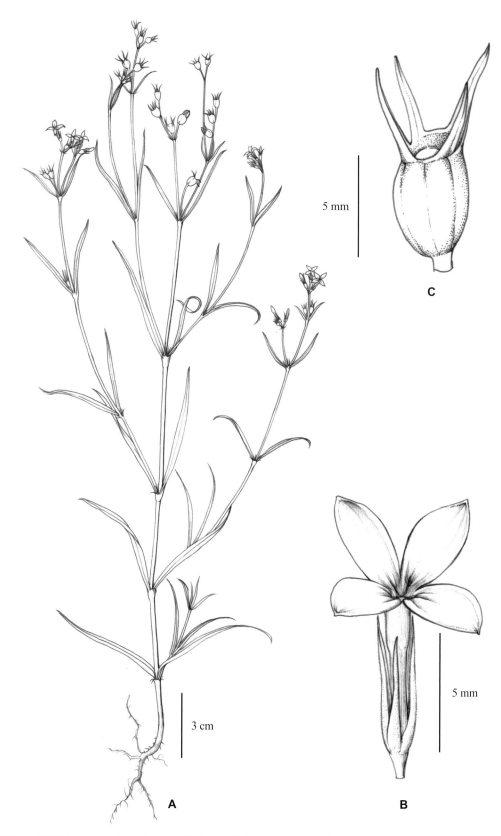

Figure 20 *Kohautia coccinea*. A. plant; B. flower; C. fruit. Drawn by NJ.

4 mm, acute at the apex, narrowed to the base; stipules with 2 fimbriae 0.5–2 mm long. Inflorescences lax, the flowers mostly in pairs well spaced on the rhachis. Calyx-tube ellipsoid, 1–1.2 mm long, papillate; lobes linear-lanceolate, 0.7–2 mm long. Corolla white, brownish, greenish, pale blue or purple; tube narrowly cylindrical, 2.5–4.7 mm long; lobes oblong, 1–1.2 mm long. Style ca. 0.6 mm long; stigma-lobes 1.2–1.4 mm long. Capsule oblong-subglobose, 2.5–4 mm long. Seeds pale brown, angular, 0.4–0.6 mm long.

Distribution: Northern, western, centeral, and southern Kenya. [Africa, Arabia, Australia, and India].

Habitat: Grassland, open bushlands, and thickets; 750–1950 m.

Baringo: Chemolingot Borehole Area, *Timberlake 469* (EA). Isiolo: Buffalo Springs, *Napper 584* (EA). Kajiado: Chyulu Plains, *Luke & Luke 9480* (EA). Kirinyaga: Mwea Plains, *Robertson 1588* (EA, K). Laikipia: Rumuruti, *Hepper & Jaeger 6621* (K). Makueni: Kilima Kiu, *Jex-Blake s.n.* (EA). Marsabit: ca. 15 km south on road from Turbi, *Kuchar 9127* (EA). Nairobi: Nairobi National Park, *Verdcourt & Polhill 3153* (EA). Trans-Nzoia: northeast Elgon, *Tweedie 4092* (EA).

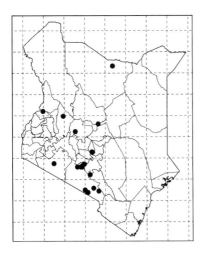

3. **Kohautia caespitosa** Schnizl., Flora 25, Beibl. 1(10): 145. 1842; F.T.E.A. Rubiac. 1: 238. 1976; W.F.E.A.: 152. 1987; U.K.W.F.F: 223. 2013. —Type: Sudan, Cordofani, Arasch-Cool, 7 Oct. 1839, *K.G.T. Kotschy 138* [holotype: W; isotypes: E (E00193644), G (G00014495 & G00014496), HBG (HBG521379), K (K000414254), L (L.2923226), LD (LD1410160), M (M0106455 & M0106456), MPU (MPU021358 & MPU022789), P (P00539319 & P00539321), S (S14-15793), TUB (TUB004502 & TUB004503)]

Annual, perennial or sometimes subshrubby herb, up to 90 cm tall. Leaf-blades linear to linear-lanceolate or narrowly elliptic, 0.5–8 cm × 0.3–5(–12) mm, acute to narrowly apiculate at the apex, narrowed to the base; stipules with filiform fimbriae 0.5–3 mm long. Inflorescences often trichotomous or dichasial. Flowers always solitary at each node. Calyx-tube ovoid, ca. 1 mm long, verruculose, scabridulous or glabrous; lobes lanceolate to triangular, 0.5–2 mm long. Corolla white, grey, buff, yellowish, pink or lilac, the tube often purplish and the lobes ochraceous, always pale inside, glabrous or papillate outside; tube narrowly cylindrical, 0.8–1.4 cm long; lobes linear-oblong to narrowly elliptic, 2–6 mm long. Style 1.5–4 mm long. Capsules subglobose or ovoid, 1.5–4 mm long, verruculose, hispidulous or glabrous.

1a. Ovary and capsules verruculose; corolla papillose outside .... a. var. *caespitosa*
1b. Ovary and capsules not verruculose, always grabrous; corolla glabrous outside 2
2a. Corolla robust, lobes pale-yellow inside, 1–1.8 mm wide .... b. var. *kitaliensis*
2b. Corolla small, lobes white inside, 0.5–0.7 mm wide .... c. var. *amaniensis*

a. var. **caespitosa**

Corolla glabrous outside. Ovary and capsules not verruculose, always grabrous.

Distribution: Eastern Kenya. [Arabia and northeast Africa].

Habitat: Bushlands; 200 m.

Wajir: ca. 80 km southwest of Wajir, *Bally & Smith B14494* (EA).

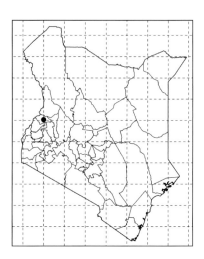

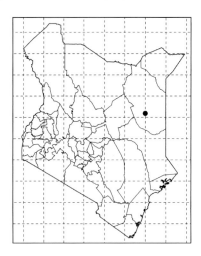

b. var. **kitaliensis** Verdc., Kew Bull. 30: 290. 1975; F.T.E.A. Rubiac. 1: 239. 1976. —Type: Kenya, Trans Nzoia, Kitale, 1890 m, Apr. 1963, *E.M. Tweedie 2595* [holotype: K (K000311695)]

Corolla glabrous outside, robust, lobes pale-yellow inside, 1–1.8 mm wide. Ovary and capsule not verruculose, always grabrous.

Distribution: Western Kenya. [Endemic].

Habitat: Grasslands or bushlands; 1800–1900 m.

Trans-Nzoia: Kitale, *Bogdan 3676* (EA).

c. var. **amaniensis** (K. Krause) Bremek., Verh. Kon. Akad. Wetensch., Afd. Natuurk., Sect. 2, 48(2): 108. 1952; F.T.E.A. Rubiac. 1: 239. 1976. ≡ *Oldenlandia amaniensis* K. Krause, Bot. Jahrb. Syst. 43: 129. 1909. —Type: Tanzania, Tanga, Totohoon, near Moa, 14 Aug. 1906, *K. Braun* in *Herb. Amani 1383* [lectotype: EA (EA000001577), **designated here**; isolectotype: EA (EA000001576)] (Plate 27)

Corolla glabrous outside. Ovary and capsules not verruculose, always grabrous.

Distribution: Widespread in Kenya. [Northeast and tropical east Africa].

Habitat: Grasslands, bushlands, or old cultivations and other disturbed places; up to 1800 m.

Baringo: Lake Bogoria National Reserve, *Mwachala et al. 627* (EA); Campi ya Samaki, *Ossent 661* (EA). Garissa: ca. 19 km from Garissa on Hagadera Road, *Brenan et al. 14756* (EA, K). Isiolo: *Newbould 2874* (K). Kilifi: south of Ras Kitua, *Frazier 2209* (EA,

Plate 27 A–D. *Kohautia caespitosa* var. *amaniensis*. Photo by ZWW (A, C), BL (B) and GWH (D).

K); Watamu, *van der Hagen & Simpson 96* (EA); Arabuko-Sokoke Forest, *Jeffery 262* (EA). Kitui: Nuu Hill, *Mwachala et al. 563* (EA). Kwale: Matuga, *Robertson 3485* (EA, K); Titanium Power Line, *Nyange 0384* (K). Laikipia: Doldol, *Kirika et al. 03/37/07* (EA). Lamu: Kiunga Island, *Gillespie 255* (K); Shella Sand Dunes, *Greenway & Rawlins 8915* (EA, K). Machakos: Mua Hills, *Gillett 16205* (EA, K). Makueni: Malivani Hill, *Malombe 8-1679* (EA). Mandera: ca. 12 km south of El Wak on Wajir Road, *Gilbert & Thulin 1658* (EA, K). Marsabit: ca. 12 km southwest of Marsabit near Karsadera, *Carter & Stannard 636* (EA, K); East Rudolf National Park, *Kuchar 9148* (EA). Meru: Meru National Park, *Ament 571* (EA, K). Mombasa: Likoni, *Johansen 102* (EA). Nakuru: Naivasha, *Seldon in E.A.H.1184* (EA, K). Narok: south of Lorsuate Hills, *Kuchar 13824* (EA). Samburu: Mathew's Range, *Newbould 3162* (EA, K). Teita Taveta: west of Bura River, *Mwachala et al. 323* (EA); Buchuma, *Kiambati 1* (EA). Tana River: Kurawa, *Polhill & Paulo 555* (EA, K); near Leka, *Gillett 16451* (EA, K). Turkana: Lokori, *Mathew 6199* (EA, K); Karasuk, *Lye 9152* (K). Uasin Gishu: Kipkarren, *Napier 108* (K).

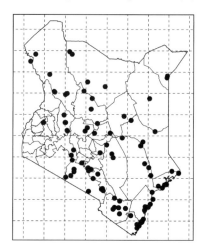

19. **Pentanopsis** Rendle

Shrubs or subshrubs. Leaves opposite or crowded on short shoots, linear-subulate to lanceolate; stipules usually sheathing, united with petiole, with fimbriate margin, persistent. Flowers 4-merous, fragrant, heterostylous, solitary or few together terminating short shoots. Calyx-tube obovoid; lobes narrowly triangular, linear-lanceolate or linear. Corolla white; tube slender; lobes oblong-elliptic to lanceolate. Stamens inserted in corolla-tube; anthers oblong, included or rarely slightly exserted in short-styled flowers. Ovary 2-locular; ovules numerous in each locule; style shortly exserted in long-styled flowers, included in short-styled flowers; stigmas bifid, filiform. Capsules obovoid or ellipsoid, with a short beak, loculicidally dehiscent. Seeds flat, elliptic.

A small genus with two species confined to north-eastern tropical Africa; only one species in Kenya.

1. **Pentanopsis fragrans** Rendle, J. Bot. 36: 29. 1898; F.T.E.A. Rubiac. 1: 246. 1976; K.T.S.L.: 531. 1994. —Type: Somalia, Wagga Mt., 1897, *E. Lort Phillips s.n.* [holotype: BM (BM000902893)] (Figure 21)

Small shrub, up to 2.5 m tall. Leaves opposite or crowded on short shoots; blades linear to

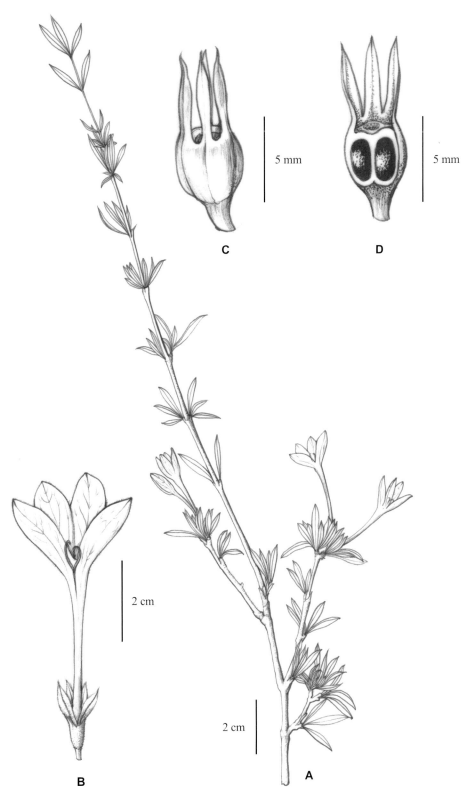

Figure 21 *Pentanopsis fragrans*. A. flowering branch; B. flower; C. fruit; D. longitudinal section of fruit. Drawn by NJ.

lanceolate, 6–40 mm × 0.8–5 mm, acute and mucronulate at the apex; margins usually revolute; stipule-sheath 2–7 mm long, with few fimbriae up to 1 mm long. Calyx-tube obovoid, 1.5–2.5 mm long; lobes 4, narrowly triangular to linear, 2–11 mm long. Corolla white or tinged purplish outside; tube 1.8–3.7 cm long; lobes 4, oblong-elliptic to lanceolate, 5–15 mm long. Style 1.6–2.2 cm long in short-styled flowers and 2.1–3 cm long in long-styled flowers. Anthers 1–2 mm long in long-styled flowers, 2.1–3 mm long in short-styled flowers. Capsule 4.5–7 mm long. Seeds 1.5–2 mm long.

Distribution: Eastern Kenya. [Ehiopia and Somalia].

Habitat: Deciduous or evergreen bushlands, usually on rocky grounds; 1450–1500 m.

Mandera: Dandu, *Gillett 12789* (EA, K).

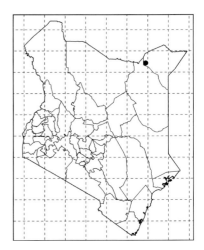

20. **Conostomium** Cufod.

Erect perennial herbs or subshrubs. Leaves opposite, sessile, linear to lanceolate; stipules sheathing the stem, truncate, often fimbriate. Flowers 4-merous, in terminal corymbose, spike- or raceme-like inflorescences, or solitary to several fascicled in leaf-axils. Calyx-lobes linear-subulate. Corolla salver-shaped; tube slender; lobes valvate in bud. Stamens always included in the tube or only the tips of the anthers exserted. Ovary 2-locular, ovules numerous in each locule; style usually exserted or rarely included; stigma-lobes oblong. Capsule ovoid or subglobose, with a distinct beak, loculicidally dehiscent. Seeds angular.

A small genus with 5 species confined to east and south Africa; two species in Kenya.

1a. Leaves linear or narrowly lanceolate, narrowed to the base; flowers pedicellate, with small corolla-tube less than 6 cm long ....1. *C. longitubum*
1b. Leaves lanceolate, subcordate at the base; flowers sessile, with corolla-tube 9–15 cm long ......
....2. *C. quadrangulare*

1. **Conostomium longitubum** (Beck) Cufod., Nuovo Giorn. Bot. Ital., n.s., 55: 85. 1948. ≡ *Oldenlandia longituba* Beck, Harrar Leipzig, App.: 461. 1888. —Type: Ethiopia, Harrar, *D.K. von Hardegger s.n.* [holotype: W (1886-0011267)] (Figure 22)

= *Conostomium kenyense* Bremek., Verh. Kon. Akad. Wetensch., Afd. Natuurk., Sect. 2, 48(2): 132. 1952; F.T.E.A. Rubiac. 1: 245. 1976; W.F.E.A.: 150. 1987; U.K.W.E.F: 223. 2013. —Type: Kenya, Samburu, Lorogi, 29 Sept. 1935, *D.G.B. Leakey 35* in *F.D. 3466* [holotype: K; isotype: EA (EA000001575)]

= *Conostomium camptopodum* Bremek.,

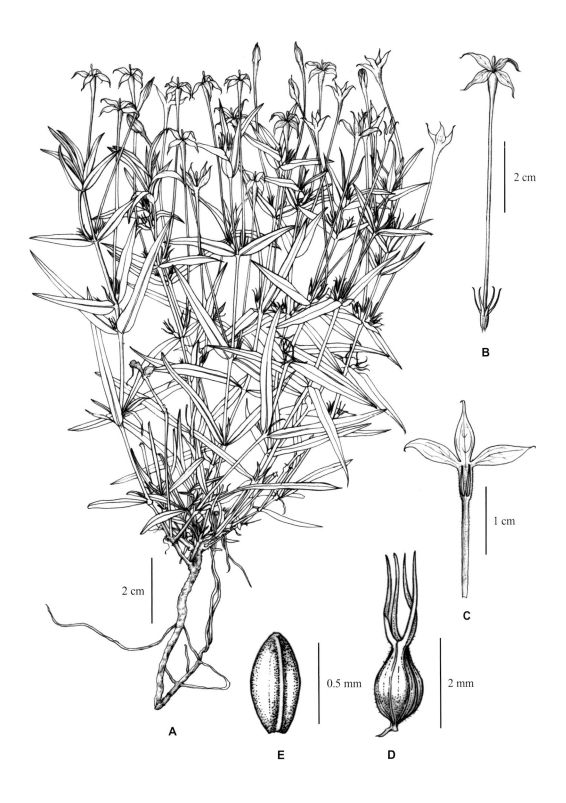

Figure 22 *Conostomium longitubum*. A. plant; B. flower; C. longitudinal section of the upper part of corolla, showing the stamens; D. fruit; E. seed. Drawn by NJ.

Verh. Kon. Ned. Akad. Wetensch., Afd. Natuurk., Sect. 2, 48(2): 133. 1952. —Type: Kenya, Turkana, Nataparin, Jule 1932, *A.M. Champion 136* [holotype: K (K000311693)]

= *Conostomium microcarpum* Bremek., Verh. Kon. Ned. Akad. Wetensch., Afd. Natuurk., Sect. 2, 48(2): 133 1952. —Type: Kenya, Turkana, edge of Turkana Desert, 29 Jule 1938, *I.B. Pole-Evans & J. Erens 1558* [holotype: K (K000319148); isotypes: E (E00193637), PRE (PRE0593517-0)]

Perennial herb or subshrub, up to 2.5 m tall; young stems 4-angular, glabrous to scabrid or shortly pubescent. Leaves opposite, sessile, linear to lanceolate, 6–70 mm × 2–7 mm, acute at the apex, narrowed to the base, margins usually revolute; stipule-sheath up to 2.5 mm long, mostly with a few short fimbriae. Flowers pedicellate, several in spike-like, raceme-like or corymbose inflorescences. Calyx-tube ovoid, up to 2 mm long; lobes linear to lanceolate, 1–6 mm long. Corolla white or cream, often tinged with blue, purple or green outside; tube 2–6 cm long; lobes oblong-lanceolate, 5–15 mm long, acute. Capsule subglobose, 1.5–3 mm long, with a beak 1.5–2 mm long. Seeds narrowly ellipsoid, 0.6–0.8 mm long.

Distribution: Northern, northwestern, central, and eastern Kenya. [Ethiopia and Somalia].

Habitat: Deciduous or evergreen bushlands, usually on rocky grounds; 100–1900 m.

Baringo: Chemolingot Hill, *Timberlake 294* (EA). Embu: ca. 2 km northwest of Kamburu Bridge on Tana River, *Robertson 1721/B* (EA). Garissa: Hagadera, *Hale B2554* (K). Isiolo: *Adamson 685* (EA, K). Kitui: *Gillett & Faden 18252* (EA, K). Machakos: Maboloni, *Bally 8374* (EA, K). Marsabit: Kalacha, *Luke et al. 10871* (K). Samburu: Gurika Hill near Lesirikan, *Carter & Stannard 514* (EA, K). Turkana: Lokori to Kailongol, *Mathew 6794* (EA, K). Wajir: Dadaab Wajir Road, *Gillett 21258* (EA, K). West Pokot: ca. 1 km below Turkwell Gorge, *Lye 9010* (K).

2. **Conostomium quadrangulare** (Rendle) Cufod., Nuovo Giorn. Bot. Ital. n.s., 55: 85. 1948; F.T.E.A. Rubiac. 1: 243. 1976; W.F.E.A.: 151. 1987; U.K.W.F.F: 223. 2013. ≡ *Pentas quadrangularis* Rendle, J. Bot. 34: 127. 1896. —Type: Ethiopia, Lake Stephanie, 27 June 1895, *A. Donaldson-Smith s.n.* [holotype: BM (BM000902942)]

Perennial herb up to 0.6 m tall, with 4-ribbed stems from a woody root. Leaves opposite,

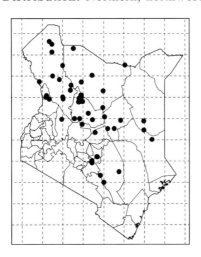

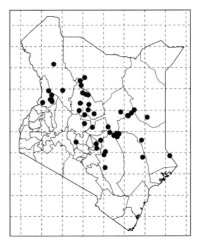

sessile, linear to lanceolate, 1.8–6.8 cm × 0.2–1.9 cm, narrowly tapering acute at the apex, narrowed to subcordate at the base; stipule-sheath truncate, 0.5–1.5 mm long, fringed with hairs. Flowers axillary, sessile, usually many in leafy spike-like inflorescences. Calyx-tube ellipsoid, 3–4 mm long; lobes 0.5–1.3 cm long. Corolla white or pinkish; tube slender, up to 15 cm long; lobes ovate or triangular-ovate, 1–2 cm long. Style exserted; stigma-lobes oblong, up to 5.5 mm long. Capsules ovoid, 5–7 mm long, 8-ribbed, crowned by the persistent calyx-lobes, beaked. Seeds trigonous-ellipsoid, 0.6–0.8 mm long.

Distribution: Northern, northwestern, central, and eastern Kenya. [Ethiopia, South Sudan, and Uganda].

Habitat: Deciduous bushlands or grasslands; 50–1900 m.

Embu: Siakago, *Bertaina 11* (EA, K). Garissa: ca. 5 km northeast of Garissa on road to Wajir, *Brenan et al. 14718* (EA, K). Isiolo: north of Mount Kenya, *Linsen & Giesen 94* (EA). Kitui: ca. 5 km southwest of Ikanga, *Kuchar 8959* (EA). Laikipia: Doidoi, *Kirika et al. 03/35/07* (EA). Marsabit: Mount Kulal, *Lamprey et al. ALPK23* (EA, K). Meru: Meru National Park, *Gillett & Hamilton 214* (EA). Nyeri: northeast Aberdares, *Dowson 549* (K). Samburu: Mathew's Range, *Ichikawa 659* (EA). Tana River: Kora National Reserve, *van Sameren 914* (EA). Turkana: Loupe, *Mathew 6429* (EA, K). Wajir: Dadaab-Wajir road, *Brenan et al. 14844* (EA, K). West Pokot: near Sigor, *Kenya N. Mus. 2nd 1974 Exped. 166* (EA).

21. **Pentodon** Hochst.

Annual or short-lived perennial glabrous herbs. Leaves opposite, sessile; stipules sheathing, 1–5-fimbriate. Inflorescences lax, terminal or axillary. Flowers small, hermaphrodite, dimorphic or not. Calyx-tube obconic or campanulate; lobes 5, equal, very narrowly triangular. Corolla-tube narrowly funnel-shaped, throat hairy; lobes 5, ovate-triangular. Stamens exserted or included. Ovary 2-locular; ovules numerous in each locule; style filiform, glabrous; stigma 2-lobed, lobes filiforms. Capsules campanulate or oblong, with slightly raised beak, loculicidally dehiscent. Seeds small, numerous, brown, angular.

A small genus with only two species widespread in Africa and extending to Arabia and Madagascar; one species in Kenya.

1. **Pentodon pentandrus** Vatke, Oesterr. Bot. Z. 25(7): 231. 1875; F.T.E.A. Rubiac. 1: 263. 1976; U.K.W.F.F.: 225. 2013. —Type: Ghana, *P. Thonning s.n.* [holotype: C; isotype: S (S-G-3017)] (Figure 23; Plate 28)

= *Pentodon pentandrus* Vatke var. *minor* Bremek., Verh. Kon. Akad. Wetensch., Afd. Natuurk., Sect. 2, 48(2): 179. 1952; F.T.E.A. Rubiac. 1: 265. 1976. **syn. nov.** —Type: South Africa, Natal, Durban, *F. Krauss 332* [holotype: TUB; isotypes: BM (BM000902964), K (K0 00414341), MO (MO-2049451 & MO-391712)]

Annual or short-lived perennial herb; stems decumbent or ascending, up to 90 cm long. Leaves opposite, sessile; blades linear-lanceolate, to elliptic, 1.3–8 cm × 0.3–2.7 cm, subacute to sharply acute at the apex, rounded

Figure 23 *Pentodon pentandrus*. A. plant; B. long-styled flower; C. longitudinal section of short-styled flower, showing stamens and style. Drawn by NJ.

to cuneate at the base; stipule-sheath up to 3 mm long with fimbriae 0.5–3 mm long. Inflorescences lax, axillary. Calyx-tube 0.5–1.5 mm long; limb-tube 0.2–1 mm long; lobes 0.5–1.5 mm long. Corolla white, pink or bluish; tube narrowly to widely funnel-shaped, 1.5–4.5 mm long; lobes ovate-triangular, 1–3 mm long. Style 1–3.5 mm long; stigma-lobes 0.8–2 mm long. Capsule 2–4 mm long, crowned by the persistent calyx-lobes. Seeds black, angular, ca. 0.3 mm long.

Distribution: Northwestern, western, central, eastern, southern, and coastal Kenya. [Arabian Peninsula and Africa].

Habitat: Swamp, lake, and river margins; up to 2000 m.

Note: Heterostyly in this species is not a stable trait in different populations, which could make it difficult to identify certain specimens. At the same time, the size of the capsule and leaves are also not distinguishing characters of different varieties. We suggest

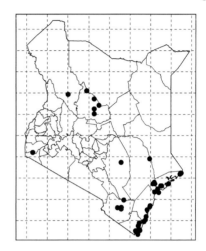

Plate 28 A, B. *Pentodon pentandrus*. Photo by GWH.

merging the two varieties into one, treating *Pentodon pentandrus* Vatke var. *minor* Bremek. as a synonym of *P. pentandrus*. In addition, we found that there are unisexual female flowers in the field (Plate 28B) and some flowers have three stigmas.

Garissa: Tana River, Bura, *Njoroge et al. 79* (K). Homa Bay: shores of Lake Victoria, *Hanid & Kiniaruh 739* (EA). Kilifi: Magangani Area, *Kirika et al. NMK746* (EA); Arabuko-Sokoke Forest, *Simpson 332* (EA). Kitui: Mount Endau, *Ossent 616* (EA). Kwale: Ngongoni Forest, *SAJIT 006226* (HIB); Shimba Hills, *Luke et al. 6028* (EA). Lamu: Badar Water Pan, *Gilbert & Kuchar 5891* (EA, K); Sinyori, *Sangai 970* (EA). Mombasa: Mombasa Island, *Drummond & Hemsley 1048* (EA). Samburu: Mathew's Range, *Kerfoot 2557* (K). Teita Taveta: Sagala Forest, *SAJIT 005353* (HIB). Tana River: up to 1 km south of Garsen, *Gillett & Kibuwa 19908* (EA, K); Shirikisho, *Robertson & Luke 5430* (EA). Turkana: Kerio River at Lokori, *Mwangangi 1393* (EA).

22. **Agathisanthemum** Klotzsch

Perennial herbs or subshrubs. Leaves opposite, sessile, narrowly lanceolate to ovate or oblong; stipules 3–13-fimbriated. Inflorescences cymose, corymbose or head-like, many-flowered. Flowers usually heterostylous, 4-merous. Calyx-lobes keeled. Corolla shortly tubular, densely hairy at the throat. Ovary 2(–3)-locular; ovules numerous on peltate placentas; style filiform, shortly hairy; stigma-lobes subglobose or ovoid. Capsules subglobose, produced into a conical beak. Seeds numerous, angular.

A small genus with 5–6 species confined to tropical Africa and the Comoros Islands; two species in Kenya.

1a. Flowers in laxly corymbose inflorescences; calyx-lobes lanceolate, 1.4–3.2 mm long.............. .... 1. *A. bojeri*
1b. Flowers in dense subglobose inflorescences; calyx-lobes linear-lanceolate, 3.3–9 mm long...... .... 2. *A. globosum*

1. **Agathisanthemum bojeri** Klotzsch, Naturw. Reise Mossambique 1: 294. 1861; F.T.E.A. Rubiac. 1: 254. 1976; W.F.E.A.: 150. 1987. —Types: Tanzania, Zanzibar, 1833, *W. Bojer s.n.* [syntype: B (destroyed); lectotype: K (K000414400), **designated here**]; Zanzibar, *W. Peters s.n.* [syntype: B (destroyed)] (Plate 29)

Perennial herb or subshrub up to 1.5 m tall. Leaves subsessile, narrowly lanceolate to elliptic or ovate-oblong, 1.2–5.8 cm × 0.1–2 cm, subacute to acute or apiculate at the apex, cuneate at the base; stipule-sheath pubescent, 0.5–4.5 mm long with (2–)5 fimbriae 1–5.8 mm long. Inflorescences laxly corymbose, many-flowered. Flowers heterostylous. Calyx-tube 0.9–1.7 mm long, glabrous or pubescent, limb tube 0.2–0.8 mm long; lobes lanceolate, 1.4–3.2 mm long. Corolla white; tube 1–2.8 mm long; lobes ovate-oblong, 1.5–2.7 mm. Style, 0.7–4 mm long; stigma lobes very short. Capsules 1–2.8 mm long. Seeds black, ca.

Plate 29 A–D. *Agathisanthemum bojeri*. Photo by GWH (A, B, D) and BL (C).

0.3 mm long, ovoid-trigonous, reticulate.

Distribution: Southern and coastal Kenya. [Eastern to southern Africa].

Habitat: Grasslands or woodlands; up to 1350 m.

Garissa: Mkondoni Water Hole, *Gilbert & Kuchar 5922* (EA, K). Kilifi: Arabuko-Sokoke Forest, *Gillett & Kibuwa 20029* (EA); Dida Swamp, *Greiling 66* (EA); ca. 8 miles south of Jilore Forest Station, *Perdue & Kibuwa 10054* (EA, K). Kwale: Shimba Hills, *SAJIT 006050* (HIB); Cha Simba Forest, *Drummond & Hemsley 1088* (EA, K); Mwaluganje Elephant Sanctuary, *Mwadime et al. 135* (EA). Lamu: Mambosasa Forest Station, *Sangai 978* (EA); ca. 8 km along road from Mkunumbi to Witu, *Kuchar 18917* (EA). Mombasa: Likoni, *Jex-Blake s.n.* (EA). Teita Taveta: Laghonyi, Bura, *Mwachala et al. in EW 3144* (EA). Tana River: Kanwe-Mayi Pool, *Hooper & Townsend 1180* (EA, K).

1839, *G.H.W. Schimper 512* [lectotype: P (P03953647), designated here; isolectotypes: BM, G (G00004268, G00004269, G00004270), K (K000414401, K000414403), M (M0106463, M0106464), MO (MO-391280), P (P03953650 left plant, P03953652, P03953648), TUB (TUB004494), WAG (WAG0002943)]; Ethiopia, Kouaietha, *Quartin Dillon s.n.* [syntype: P (P03953650 right plant)]

Erect herb 1.2 m tall, with a woody rootstock. Leaves subsessile, elliptic, oblong-elliptic or oblong-lanceolate, 1–8 cm × 0.6–2.5 cm, acute at the apex, cuneate at the base; stipule-sheath triangular, 2–9 mm long, with (0–)3–5 linear fimbriae 4–9 mm long. Flowers not heterostylous, in dense subglobose inflorescences 1–1.4 cm in diameter. Calyx-tube 1.2–2.1 mm long, glabrous or pubescent; limb-tube 0.8–1.8 mm long; lobes linear-lanceolate, 3.3–9 mm long. Corolla white, cream, pale yellow, blue, lilac, pink, purple or

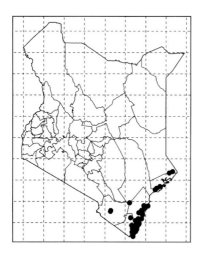

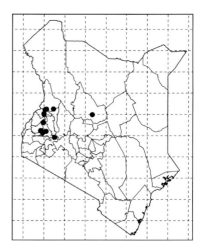

2. Agathisanthemum globosum (Hochst. ex A. Rich.) Klotzsch, Naturw. Reise Mossambique 294. 1861; F.T.E.A. Rubiac. 1: 258. 1976; U. KW.F.F.: 224. 2013. ≡ *Hedyotis globosa* Hochst. ex A. Rich., Tent. Fl. Abyss. 1: 360. 1847. —Types: Ethiopia, Teeli (Dochli), 20 Nov.

mauve; tube 1.8–4 mm long; lobes triangular, 2.2–4.1 mm long. Style 3.7–6.5 mm long, exserted; stigma-lobes 0.4–0.7 mm long. Capsules compressed-ellipsoid, ca. 2.4 mm long, crowned by the persistent calyx-lobes, mostly pubescent. Seeds dark purplish brown, ovoid-trigonous, ca.

0.6 mm long.

Distribution: Western Kenya. [Eastern and central Africa].

Habitat: Grasslands or woodlands; 900–2400 m.

Kakamega: Kakamega Forest, near Forest Station, *Gilbert & Mesfin 6652* (EA, K). Kericho: Tinderet Forest, *Irwin 239* (EA, K). Nandi: Kaimosi, *Rogers 728* (K). Samburu: Mathew's Range, *Bronner 963* (EA). Trans-Nzoia: Kitale Museum Nature Trail, *Mungai 134/84* (EA). Uasin Gishu: Kipkarren River, *Brodhurst-Hill 487* (EA). West Pokot: Kapenguria, *Napier 1968* (EA).

23. **Dibrachionostylus** Bremek.

Perennial glabrous herbs with 4-ribbed or winged stems. Leaves opposite, sessile, linear to lanceolate; stipules shortly sheathing the stem, with several fimbriae. Inflorescences terminal, corymbiform, dense, many-flowered. Flowers small, heterostylous, 4-merous. Calyx-tube ovoid; lobes narrowly triangular. Corolla-tube cylindrical; lobes oblong-ovate. Stamens slightly longer than corolla-lobes in short-styled flowers, or as far as the throat in long-styled flowers. Ovary 2-locular; ovules numerous in each locule. Style excluding; stigma-lobes 2, ellipsoid. Capsules subglobose, shortly beaked. Seeds sharply trigonous.

A monotypic genus confined to central Kenya.

1. **Dibrachionostylus kaessneri** (S. Moore) Bremek., Verh. Kon. Akad. Wetensch., Afd. Natuurk., Sect. 2, 48(2): 164. 1952; F.T.E.A. Rubiac. 1: 259. 1976; U.K.W.F.F.: 224. 2013. ≡ *Oldenlandia kaessneri* S. Moore, J. Bot. 43: 249. 1905. —Type: Kenya, Nairobi, 6 Sept. 1902, *T. Kässner 957* [holotype: BM (BM000902925); isotypes: K (K000352615), MO (MO-716947)] (Figure 24)

Perennial glabrous herb, up to 40 cm tall. Leaves sessile, linear to lanceolate, 1.5–5.5 cm × 0.2–0.8 cm, acute at the apex, cuneate at the base; stipule-sheath very short, with 5–7 up to 5 mm long fimbriae. Inflorescences terminal, corymbiform, with peduncles up to 6 cm long. Flowers very small, heterostylous. Calyx-tube ovoid; lobes 4, up to 1 mm long. Corolla white or pale mauve; tube cylindrical, 1–2 mm long; lobes 4, up to 2 mm long. Ovary 2-locular; ovules numerous in each locule. Style 3.7–4 mm long in long-styled flowers, ca. 1.5 mm long in short-styled flowers. Capsule subglobose, 1.2–1.8 mm long, shortly beaked. Seeds pale yellow-brown, sharply trigonous.

Distribution: Central Kenya. [Endemic].

Habitat: Grasslands or rocky places; 1100–1850 m.

Embu: Emberre, *Graham 21052* (K). Kiambu: Thika, *Luke et al. 16114* (K). Kiambu:

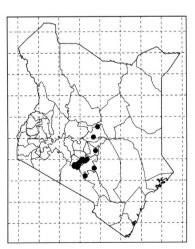

Figure 24 *Dibrachionostylus kaessneri*. A. plant; B. portion of branch showing the stipule; C. long-styled flower; D. longitudinal section of short-styled flower, showing the stamens, style and ovary; Drawn by NJ.

Campus of Jomo Kenyatta College of Agriculture and Technology, *Seki JKCAT-74* (EA); northeast of Ruiru, *Perdue & Kibuwa 8185* (EA, K). Machakos: Yatta Plateau, *Gillett et al. 23957* (EA, K). Makueni: Kilima Kiu, *Ritchie s.n.* (EA). Meru: eastern Meru, *Ward 2564* (K). Nairobi: Nairobi National Park, *Verdcourt & Polhill 3145* (EA, K).

24. **Cordylostigma** Groeninckx & Dessein

Annual or perennial herbs. Leaves opposite, sessile, linear, linear-lanceolate or elliptic-lanceolate; stipules sheathing the stem, mostly with 2–8(–11) fimbriae. Flowers 4(–5)-merous, isostylous, few to many in terminal cymose inflorescences. Calyx-tube subglobose; lobes narrowly triangular to lanceolate or oblong-lanceolate. Corolla-tube narrowly cylindrical, hairy or papillate at the throat inside; lobes elliptic, ovate-elliptic or oblong. Style ending in a single ovoid or cylindrical stigma. Capsules almost hemispherical, subglobose or oblong. Seeds subglobose, ellipsoid, angular or irregularly ovoid.

A small genus with nine species, mainly in tropical and southern Africa, also in Madagascar; four species in Kenya.

1a. Decumbent herb with a woody rhizomatous rootstock .... 1. *C. prolixipes*
1b. Erect herb without a woody rhizomatous rootstock 2
2a. Corolla-tube glabrous at the throat .... 2. *C. virgatum*
2b. Corolla-tube densely bearded at the throat 3
3a. Corolla scarlet or vermilion-red; lobes broadly elliptic, 4.5–9 mm long .... 3. *C. obtusilobum*
3b. Corolla lilac, pink, red, blue or white; lobes elliptic, usually less than 4.5 mm long ....
.... 4. *C. longifolium*

1. **Cordylostigma prolixipes** (S. Moore) Groeninckx & Dessein, Taxon 59(5): 1466. 2010. ≡ *Oldenlandia prolixipes* S. Moore, J. Bot. 43: 351. 1905. ≡ *Kohautia prolixipes* (S. Moore) Bremek., Verh. Kon. Akad. Wetensch., Afd. Natuurk., Sect. 2, 48(2): 67. 1952; F.T.E.A. Rubiac. 1: 231. 1976. ≡ *Oldenlandia pedunculata* K. Schum. & K. Krause, Bot. Jahrb. Syst. 39: 519. 1907, *nom. superfl.* —Type: Kenya, Kwale, Duruma, 23 Mar. 1902, *T. Kässner 442* [holotype: BM (BM000902940); isotype: K (K000311696)]

Perennial decumbent herb, up to 35 cm tall, with a woody rhizomatous rootstock. Leaves opposite, sessile, linear to lanceolate, 1–3 cm × 0.2–1 cm, acute at the apex, rounded to cuneate at the base; stipule-sheath 1–3 mm, with 1–5

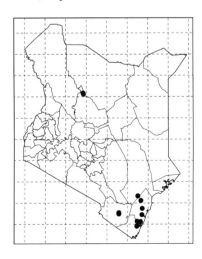

fimbriae up to 3 mm long. Inflorescences usually with elongated peduncles up to 12 cm long, few-flowered. Calyx-tube subglobose; lobes oblong-lanceolate, up to 2.5 mm long, keeled. Corolla purple-red, with tube 5–6 mm long; lobes oblong, 3–5.5 mm long. Style 1 mm long; stigma 1.2 mm long. Capsule hemispherical, 2–3 mm long. Seeds trigonous.

Distribution: Central, southern, and coastal Kenya. [Endemic].

Habitat: Grasslands or rocky places; 100–2400 m.

Kilifi: near Ganze, *Robertson 4373* (EA). Kwale: Maluganji Forest, *Robertson & Luke 6055* (EA, K). Samburu: Mount Nyero, *Blake 16* (K). Teita Taveta: Sagala Hills, *Faden & Evans 71/43* (EA, K). Tana River: Galana Ranch, *Field 56* (EA).

2. **Cordylostigma virgatum** (Willd.) Groeninckx & Dessein, Taxon 59(5): 1466. 2010. ≡ *Hedyotis virgata* Willd., Sp. Pl. 1: 567. 1797. ≡ *Kohautia virgata* (Willd.) Bremek., Verh. Kon. Akad. Wetensch., Afd. Natuurk., Sect. 2, 48(2): 77. 1952. —Type: Guinea, *Anonymous s.n.* [holotype: B (B-W 02596)]

Annual or perennial herb, up to 60 cm tall, usually with numerous erect branches. Leaves opposite, sessile, linear to lanceolate or lanceolate-elliptic, 1–3 cm × 0.5–7 mm, apex acute and shortly aristate; stipule-sheath 1–2.5 mm long, with 2–6 reflexed filiform fimbriae up to 6 mm long. Flowers several in lax terminal cymes. Calyx-tube obovate-subglobose; lobes narrowly triangular. Corolla variously coloured; tube 2.5–7 mm long, glabrous or very slightly papillose at the throat; lobes ovate-triangular to ovate-elliptic, 1–2.5 mm long. Style and stigma together as long as the narrow tube. Capsules hemispherical to subglobose, 1–2.5 mm long.

Distribution: Central and coastal Kenya. [Tropical and southern Africa].

Habitat: Open grasslands, bushlands, or woodlands; up to 1700 m.

Note: Not seen the holotype deposited at B, Bremek. [in Verh. Kon. Akad. Wetensch., Afd. Natuurk., Sect. 2, 48(2): 77. 1952] supposed the specimen *P. Thonning s.n.* collected from Guinea [S (S-G-3020)] which was later misinterpreted as the designation of isoneotype [Groeninckx et al., Taxon 59(5): 1467. 2010].

Kiambu: Ruiru, *Napier 380* (K). Kilifi: Arabuko-Sokoke Forest, *Luke 3026* (EA, K). Kwale: Shimba Hills, *Magogo & Glover 352* (EA, K). Kajiado: outside Nairobi National Park, *Faden et al. 74/554* (EA, K). Tana River: Nairobi Ranch, *Festo & Luke 2382* (EA).

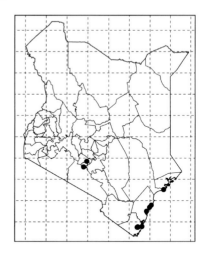

3. **Cordylostigma obtusilobum** (Hiern) Groeninckx & Dessein, Taxon 59(5): 1466. 2010 ≡ *Oldenlandia obtusiloba* Hiern, Fl. Trop. Afr. 3: 56. 1877 ≡ *Kohautia obtusiloba* (Hiern) Bremek., Verh. Kon. Akad. Wetensch., Afd. Natuurk., Sect. 2, 48(2): 66. 1952; F.T.E.A. Rubiac. 1: 230. 1976. —Types: Tanzania,

Bagamoyo, Kingani, Oct. 1870, *J. Kirk s.n.* [lectotype: K (K000319438), designated by Bremek., Verh. Kon. Akad. Wetensch., Afd. Natuurk., Sect. 2, 48(2): 67. 1952]; Mozambique, *J. Forbes 358* [syntype: K (K000414359)]

Annual or perennial erect herb, with 4-ribbed stems, up to 70 cm tall. Leaves opposite, sessile, linear to lanceolate, 1.5–7 cm × 1–6 mm, acute at the apex, narrowed to the base; stipule-sheath 1–2 mm long, with 2–4 fimbriae up to 5 mm long. Flowers in lax pedunculate cymes. Calyx-tube subglobose; lobes narrowly lanceolate, 3–4 mm long. Corolla scarlet; tube 8–12 mm long, densely hairy at the throat; lobes broadly elliptic, 4.5–9 mm long. Capsules almost hemispherical, 2–4 mm long.

Distribution: Coastal Kenya. [Tanzania].
Habitat: Coastal grasslands; up to 300 m.
Kilifi: Kilifi-Kaloleni Road, *Faden & Faden 74/1261* (EA, K). Kwale: Shimba Hills, *Magogo & Glover 995* (EA, K); near Kiwegu, *Luke & Robertson 2323* (EA). Lamu: Mokowe to Bodhei, *Festo & Luke 2629* (EA, K). Mombasa: Changamwe, *Napier 6291* (EA, K). Tana River: Hewani to Wema, *Robertson & Luke 5310* (EA).

4. **Cordylostigma longifolium** (Klotzsch) Groeninckx & Dessein, Taxon 59(5): 1466. 2010. ≡ *Kohautia longifolia* Klotzsch, Naturw. Reise Mossambique 1: 297. 1861; Bremek., Verh. Kon. Akad. Wetensch., Afd. Natuurk., Sect. 2, 48(2): 68. 1952; F.T.E.A. Rubiac. 1: 232. 1976. ≡ *Oldenlandia longifolia* (Klotzsch) K. Schum., Pflanzenw. Ost-Afrikas, C: 376. 1895, *nom. illeg.*, non (Schum.) DC (1830). —Types: Mozambique, Sena, *W. Peters s.n.* [holotype: B (destroyed)]; Mozambique, Gonubi Hill, 6 Apr. 1898, *R. Schlechter 12181* [neotype: K, designated by Bremek. in Verh. Kon. Ned. Akad. Wetensch., Afd. Natuurk., Sect. 2, 48(2): 68. 1952; isoneotypes: BM, BR (BR0000008849014), E (E00193 642), G (G00014483, G00014484), HBG (HBG521394), L (L.2923139), W] (Figure 25)

Annual or perennial herb, up to 70(–100) cm tall. Leaves linear to lanceolate, 5–90 mm × 1–12 mm, acute to apiculate at the apex, narrowed to the base; stipule-sheath 0.5–6 mm long , with 2–several fimbriae up to 8 mm long. Flowers in dense to very lax cymes. Calyx-tube subglobose to ovoid-elliptic; lobes narrowly triangular, 1–7 mm long. Corolla reddish, blue or white; tube 4–6.5 mm long,

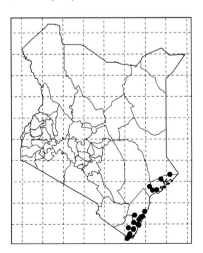

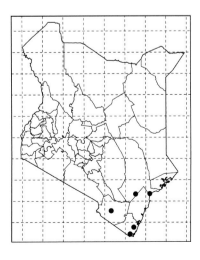

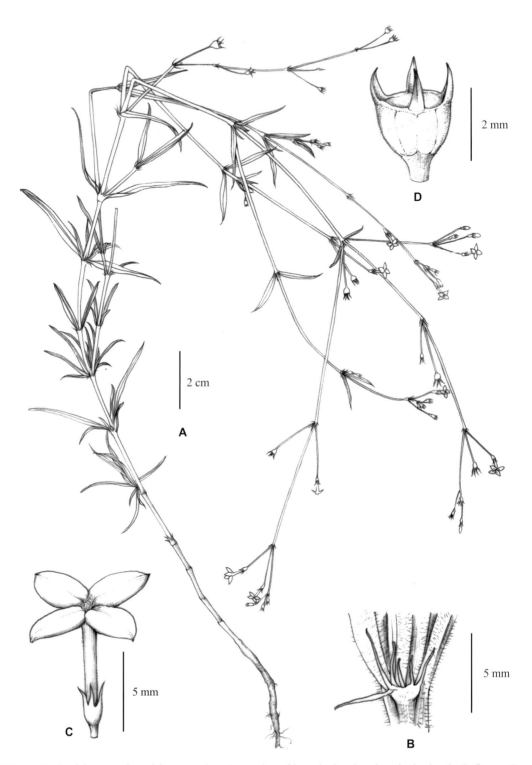

Figure 25 *Cordylostigma longifolium*. A. plant; B. portion of branch showing the stipule-sheath; C. flower; D. fruit. Drawn by NJ.

shortly hairy in throat; lobes elliptic, 2–6.5 mm long. Style 1–1.8 mm long. Capsules hemispherical to globose, mostly 1.5–3 mm in diamemter, glabrous or shortly hairy. Seeds brown, 0.4–0.5 mm long.

Distribution: Southern and coastal Kenya. [East Africa].

Habitat: Grasslands, bushlands, or old cultivations; up to 1200 m.

Kwale: Shimba Hills, *Magogo & Glover 826* (K). Teita Taveta: Kurawa, *Polhill & Paulo 603* (EA, K); west side of Bura River, *Mwachala 257* (EA). Tana River: Kurawa, ca. 30 miles south of Garsen, *Polhill & Paulo 603* (EA).

25. Edrastima Raf.

Annual or short-lived perennial herbs. Leaves opposite; stipules sheathed, produced into a 2-fid lobe. Flowers in terminal and axillary clusters, homostylous. Calyx-tube subglobose; lobes 4, equal. Corolla-tube glabrous; stigma subglobose. Capsules subglobose with a slightly raised beak. Seeds trigonous.

A small genus with five species discontinuously distributed in north America, south America, India, Southeast Asia, Tropical and subtropical Africa, and Madagascar; only one species in Kenya. It was traditionally placed in *Oldenlandia* L., while the latter was not proved to be a natural monophyletic group. Based on morphological and recent molecular evidence, the Bremekamp's *Oldenlandia* subg. *Anotidopsis* (Benth. & Hook.f.) Bremek. is recognized as the rank of genus using the earliest available generic name, *Edrastima* Raf.

1. **Edrastima goreensis** (DC.) Neupane & N.Wikstr., Taxon 64(2): 316. 2015. ≡ *Hedyotis goreensis* DC., Prodr. 4: 421. 1830. ≡ *Oldenlandia goreensis* (DC.) Summerh., Bull. Misc. Inform. Kew 1928: 392. 1928; F.T.E.A. Rubiac. 1: 279. 1976; U.K.W.F.F.: 224. 2013. —Type: Senegal, Cape Verde Peninsula, near Goree, Kounoum, 19 Mar. 1829, *G.S. Perrottet 484* [lectotype: P, designated by C.E.B. Bremekamp in Verh. Kon. Ned. Akad. Wetensch., Afd. Natuurk., Sect. 2, 48(2): 197. 1952; isolectotypes: S (S14-9343), W]

Annual or perennial herb with prostrate, decumbent or ascending stems, up to 90 cm long. Leaf-blades elliptic to ovate, 0.5–2.5 cm × 0.3–1.5 cm, obtuse or subacute at the apex, rounded or cuneate at the base, glabrescent to sparsely hairy; stipule-sheath 0.7–1(–2) mm long, produced into a 2-fid lobe 1–2 mm long, ciliate. Inflorescences terminal, sessile, many-flowered, subglobose. Flowers homostylous, 4-merous or rarely 5–6-merous. Calyx-tube subglobose, ca. 1 mm long; lobes narrowly triangular, 1–1.5 mm long. Corolla white; tube ca. 0.3 mm long; lobes ovate-triangular, 0.8–1.2 mm long. Capsules subglobose, ca. 1.5 mm long, slightly beaked.

1a. Stems glabrous or glabrescent..a. var. *goreensis*
1b. Stems sparsely to densely hairy ....b. var. *trichocarpa*

a. var. **goreensis**

Stems glabrous or glabrescent.

Distribution: Western Kenya. [Africa and Madagascar].

Habitat: Open grasslands or bushlands; 1700–2050 m.

Homa Bay: near Nyachogochogo, *Plaizier 827* (EA). Narok: Masai Mara Game Reserve, *Kuchar 11448* (EA). Trans Nzoia: Kitale, *Bogdan AB4213* (EA).

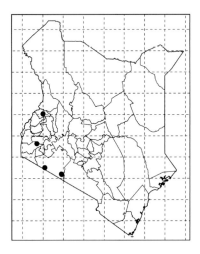

b. var. **trichocarpa** (Bremek.) Y.D. Zhou, **comb. nov.** ≡ *Oldenlandia goreensis* (DC.) Summerh. var. *trichocarpa* Bremek., Verh. Kon. Ned. Akad. Wetensch., Afd. Natuurk., Sect. 2, 48(2): 198. 1952; F.T.E.A. Rubiac. 1: 279. 1976. —Type: Togo, Lome, Maritime, 1900, *O. Warnecke 239* [holotype: K; isotypes: BR (BR0000008849441), EA (EA000001568), G, (G00014545), L (L0000424), P (P00539284)] (Plate 30)

= *Oldenlandia verticillata* Bremek. var. *trichocarpa* Bremek., Verh. Kon. Akad. Wetensch., Afd. Natuurk., Sect. 2, 48(2): 200. 1952. —Type: Tanzania, Bukoba, Nyakato, Aug. 1931, *A.E. Haarer 2060* [holotype: K (K000414324); isotype: EA (EA000001574)]

= *Oldenlandia bullockii* Bremek., Kew Bull. 13: 382. 1959. —Type: Kenya, Mumias, *Whyte s.n.* [holotype: K]

Stems sparsely to densely hairy.

Distribution: Western and coastal Kenya. [Tropical Africa].

Habitat: Forests, fringes of seasonal pools,

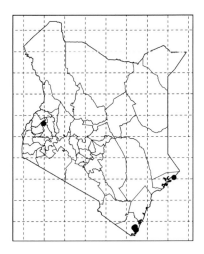

seasonal lakes in bushlands, or dam sites; up to 1850 m.

Note: south Neupane and north Wikström made the combination name of *Oldenlandia goreensis* (DC.) Neupane & N.Wikstr., based on their molecular evidence, but they did not deal with the name, *O. goreensis* var. *trichocarpa* Bremek., which is an accepted variety of *O. goreensis*. So, here we proposed the new combination.

Kwale: Shimba Hills, *Magogo & Glover 378* (EA). Lamu: Kiwayu Area, *Luke 5624* (EA). Uasin Gishu: Kipkarren, *Brodhurst-Hill 349* (EA).

Plate 30 A, B. *Edrastima goreensis* var. *trichocarpa*. Photo by GWH.

26. **Scleromitrion** (Wight & Arn.) Meisn.

Annual or perennial herbs; stems diffuse, decumbent or erect, branched at the base. Leaves opposite. Stipules membranous, with 2–7 fimbriae. Flowers homostylous, terminal or axillary panicles or solitary in each node; corolla divided equally or more than half way down; stamens and styles both exserted. Capsule subglobose, loculicidally dehiscent from the apex. Seeds many, trigonous.

A small genus with 13 species, widely distributed in tropical Africa, Madagascar, India, tropical Asia, Australia, and Pacific Islands; only one species in Kenya. The current molecular evidence shows that *Scleromitrion* clade should be recognized from *Hedyotis-Oldenlandia* complex, as this group have homostylous flowers with exserted stamens and styles. The molecular evidence also shows that the tropical species *Oldenlandia lancifolia* (Schumach.) DC. is included in the *Scleromitrion* clade. Here we move *O. lancifolia* as well as its variety into the genus *Scleromitrion* (Wight & Arn.) Meisn., and make two new combinations as below.

1. **Scleromitrion lancifolium** (Schumach.) Y.D. Zhou, **comb. nov.** ≡ *Hedyotis lancifolia* Schumach., Beskr. Guin. Pl. 72. 1827. ≡ *Oldenlandia lancifolia* (Schumach.) DC., Prodr. 4: 425. 1830; F.T.E.A. Rubiac. 1: 292. 1976. —Types: Ghana, Valley of Aquapim, *P. Thonning 210* [lectotype: C (C10003933), designated by Bremek. in Verh. Kon. Ned.

Akad. Wetensch., Afd. Natuurk., Sect. 2, 48(2): 230. 1952; isolectotypes: C (C10003937), S]; Ghana, southern part of the county, *P. Thonning 177* [syntype: C (C10003935)]; Ghana, without precious locality, *P. Thonning s.n.* [syntype: S (S-G-3012)]

Perennial straggling or prostrate herb, up to 90 cm long, stems much branched near the base, rooting at the nodes. Leaf-blades 1–8 cm × 0.2–1.8 cm, linear to linear-lanceolate, or elliptic-lanceolate, acute at the apex, cuneate at the base; stipule-sheath ca. 1 mm long, with 2–5 linear fimbriae ca. 1.5 mm long. Flowers homostylous, solitary or several in axillary panicles at the nodes; pedicels slender, up to 3 cm long. Calyx-tube 0.8–1 mm long; lobes triangular, 1–1.8 mm long. Corolla white; tube ca. 1 mm long, glabrous inside; lobes triangular, 1–2 mm long. Stamens and styles both exserted. Capsules depressed subglobose, 2–3 mm long, loculicidally dehiscent from the apex.

var. **scabridulum** (Bremek.) Y.D. Zhou, **comb. nov.** ≡ *Oldenlandia lancifolia* (Schumach.) DC. var. *scabridula* Bremek., Verh. Kon. Ned. Akad. Wetensch., Afd. Natuurk., Sect. 2, 48(2): 232. 1952; F.T.E.A., Rubiac. 1: 293. 1976. —Type: Tanzania, Njombe, upper Ruhudje R., Lupembe, *H.J.E. Schlieben 626* [holotype: K (K000414319); isotypes: BR (BR000000 5850846), G (G00014537)] (Figure 26; Plate 31)

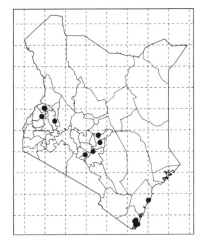

Plate 31 A–C. *Scleromitrion lancifolium* var. *scabridulum*. Photo by YDZ (A, C) and VMN (B).

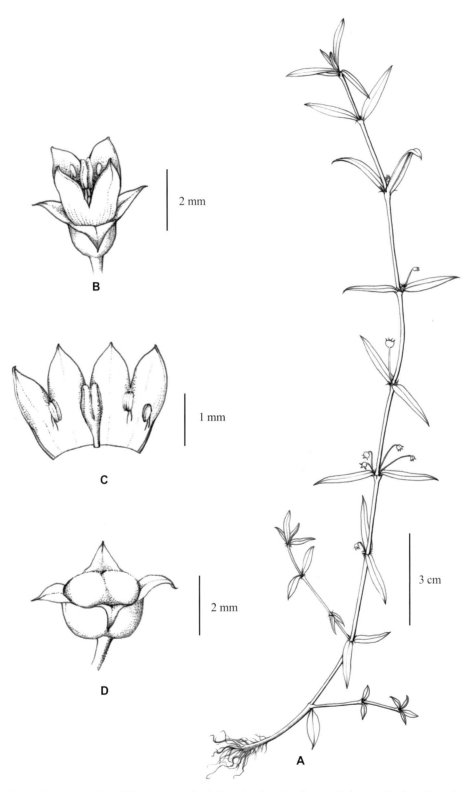

Figure 26. *Scleromitrion lancifolium* var. *scabridulum*. A. plant; B. flower; C. longitudinal section of corolla, showing stamens and style; D. fruit. Drawn by NJ.

Young stems, pedicels and leaf-blades scabridulous. Calyx-tube puberulous.

Distribution: Western, central, and coastal Kenya. [Tropics].

Habitat: Forests, riversides, or swampy grasslands; up to 2350 m.

Embu: *Lewis 5909* (K). Kiambu: north side of Thika River, *Faden 68/153* (EA). Kilifi: Gede Forest, *Gerhardt & Steiner 162* (EA). Kwale: Fihoni Swamp, *SAJIT V0547* (HIB); Shimba Hills, *Luke et al. 7459* (EA). Tharaka-Nithi: Chogoria Forest, *SAJIT 003987* (HIB). Trans Nzoia: Kitale, *Tweedie 3250* (K). Uasin Gishu: Brockley Primary School, *Gilbert & Mesfin 6445* (EA, K).

27. Oldenlandia L.

Annual or perennial herbs. Leaves opposite; stipules with 1-several fimbriae. Flowers 4(–6)-merous, bisexual, heterostylous or isostylous, in terminal or axillary, dense or lax inflorescences, sometimes fasciculate or solitary at nodes. Calyx-lobes usually 4, equal. Corolla-tube cylindrical, often hairy at throat; lobes 4. Ovary 2-locular; ovules numerous. Capsules subspherical to oblong, usually with a beak. Seeds numerous.

A large genus with about 240 species in the tropics, especially in Africa; 16 species in Kenya. Actually, the current molecular evidence shows that *Hedyotis* L., *Oldenlandia* L. and related genera (*Hedyotis-Oldenlandia* complex) are highly debated groups in the Tribe Spermacoceae Bercht. & J. Presl of Rubiaceae with no consensus to date on their generic delimitations. *Oldenlandia* L. is not a monophyletic group, and could be transferred into at least eight genera, including *Oldenlandia* s.s., *Debia* Neupane & N. Wikstrom, *Dimetia* (Wight & Arn.) Meisn., *Edrastima* Raf., *Exallage* Bremek., *Involucrella* (Benth. & Hook. f.) Neupane & N. Wikstrom, *Pentanopsis* Rendle clade, *Scleromitrion* (Wight & Arn.) Meisn., as well as some species not located in any groups, such as *Oldenlandia fastigiata* Bremek. For Kenya, *Oldenlandia lancifolia* (Schumach.) DC. has been transferred into *Scleromitrion*, *O. goreensis* has been transferred into *Edrastima*.

1a. Leaves broad, ovate, elliptic or oblong, less than 6 times as long as wide, usually abruptly narrowed into stalk 2
1b. Leaves narrow, linear to narrowly elliptic, more than 6 times as long as wide, gradually narrowed at the base and pointed at the apex 8
2a. Flowers solitary or rarely in pairs or few on axillary stalks 3
2b. Flowers many in terminal or axillary inflorescences 4
3a. Stems elongated; leaves large, up to 3.5 cm long; stipules triangular, 4–10-setose; flowers solitary or rarely in pairs or few .... 1. *O. violacea*
3b. Stems usually very short; leaves small, less than 1.3 cm long; stipules truncate, 1–3-setose; flowers solitary .... 2. *O. monanthos*
4a. Inflorescences sessile to subsessile .... 6. *O. cryptocarpa*
4b. Inflorescences with obvious peduncles, up to 3 cm long 5

5a. Flowers usually consisting of small heads in a panicle compound inflorescence .... 4. *O. johnstonii*
5b. Flowers in few- to several-flowered simple cyme 6
6a. Leaves large, 0.8–3.8 cm × 0.4–1.9 cm, with petioles 1–4 mm long .... 3. *O. friesiorum*
6b. Leaves small, 0.3–1.2 cm × 0.3–0.9 cm, with petioles up to 2 mm long .... 5. *O. rupicola*
7a. Flowers many in dense terminal sessile heads 8
7b. Flowers solitary, or few to several in fascicles at each node or in terminal or axillary cymose, umbel-like or paniculate inflorescees 9
8a. The leaves of the inflorescence nodes reduced or hidden; margins of calyx-lobes not hyaline .... 11. *O. wiedemannii*
8b. The leaves of the inflorescence nodes well developed; margins of calyx-lobes always hyaline .... 12. *O. ichthyoderma*
9a. Flowers in terminal paniculate inflorescences 10
9b. Flowers solitary, or few to several in fascicles or in cymose or umbel-like inflorescences at each node 11
10a. Annual herb, with basal leaves rosulate .... 7. *O. rosulata*
10b. Perennial herb; lowermost leaves never rosulate .... 8. *O. affinis*
11a. Plant of sand-dunes with densely hairy stems and somewhat succulent leaves .... 9. *O. richardsonioides*
11b. Not growing on sand-dunes 12
12a. Most of flowers in stalked cymes or umbel-like inflorescees at each node 13
12b. Most of flowers solitary or in fascicles at each node 14
13a. Tufted herb; corolla-tube always more than 2 mm long, lobes 2–3 mm long .... 10. *O. scopulorum*
13b. Not a tufted herb; corolla-tube less than 2 mm long, lobes 0.7–2 mm long ... 15. *O. fastigiata*
14a. Pedicels very short, less than 4 mm long .... 14. *O. acicularis*
14b. Pedicels longer, up to 2.6 cm long 15
15a. Plant prostrate; flowers small; corolla-tube shortly cylindrical, less than 1 mm long .... 16. *O. corymbosa*
15b. Plant erect; flowers large; corolla-tube narrowly tubular, up to 1.1 cm long .... 13. *O. herbacea*

1. Oldenlandia violacea K. Schum., Pflanzenw. Ost-Afrikas, C: 374. 1895. —Type: Tanzania, Kilimanjaro, Marangu, 2300 m, 15 Spet. 1893, *G. Volkens 848* [holotype: B; isotypes: BM, E (E00193621), K (K000414323)] (Plate 32A–F)

Perennial glabrous prostrate herb, up to 60 cm long, rooting at the nodes. Leaves opposite; blades lanceolate, elliptic-oblong to elliptic, 5–35 mm × 2–12 mm, acute at the apex, cuneate to rounded at the base; petioles 1.2–3 mm long; stipule-sheath 0.7–3 mm long, lobe triangular, 4–10-setose. Flowers mostly solitary at the node, sometimes paired or rarely in fascicles, 4-merous, heterostylous; pedicels up to 2.6 cm long. Calyx-tube campanulate, 1–1.5 mm long, glabrous;

Oldenlandia | 135

lobes ovate-triangular, 1–2 mm long. Corolla white; tube 3–5 mm long, densely hairy at the throat; lobes oblong-lanceolate, 1.7–4 mm long. Capsules obconic-campanulate, ca. 2 mm × 2.5 mm, glabrous.

Distribution: Western, central, and southern Kenya. [Ethiopia and Tanzania].

Habitat: Forests, streamsides, or woodlands; 1550–2750 m.

Note: *Oldenlandia violacea* K. Schum. was first described on the basis of a collection from Kilimanjaro of Tanzania. A simple comparison between *O. violacea* and another related species *O. monanthos* (Hochst. ex A. Rich.) Hiern was made by Schumann in his original description, with the statement that the former is obviously larger than the latter. Verdcourt (1976) treated *O. violacea* as a synonym of *O. monanthos* in the FTEA. In our field work from 2018 to 2019, several species of *Oldenlandia* were collected from Kenya, including some populations of *O. monanthos*, which is a prostrate herb with solitary or rarely fasciculate flowers at each node. After our critical examination of these collections, we believe that they could match perfectly with the types of *O. monanthos* and the types of *O. violacea* in morphological features respectively. So, we reinstated the independent specific status of *O. violace*. For more detail, see Zhou et al. (2022) [Zhou et al., 2022. Reinstatement of the independent specific status of *Oldenlandia violacea* (Rubiaceae) from the synonymy of *O. monanthos*. Phytotaxa, 507: 293–300].

Bomet: western Mau Forest Reserve, *Geesteranus 5766* (K). Elgeyo-Marakwet: Kibukuimet, *SAJIT 007055* (HIB). Kajiado: Ngong Hills, *Lind et al. 5748* (EA); Chyulu Hills, *Luke et al. 11567* (EA). Kericho: Sambret, *Kerfoot 2781* (EA). Nakuru: Eburru Forest, *Luke et al. 8935* (EA). Narok: Ndunyangerro, *Glover et al. 1682* (EA). Nyandarua: edge of Kinangop Plateau, *Kuchar 12337* (EA). Trans Nzoia: Mount Elgon, *Beentje 1939* (EA). Uasin Gishu: Tinderet Forest Reserve, *Geesteranus 5397* (K). West Pokot: Cherangani Hills, *SAJIT Z0075* (HIB).

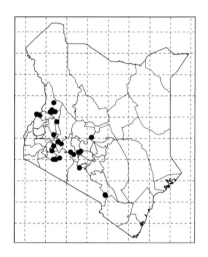

2. **Oldenlandia monanthos** (Hochst. ex A. Rich.) Hiern, Fl. Trop. Afr. 3: 60 1877; Bremek., Verh. Kon. Ned. Akad. Wetensch., Afd. Natuurk., Sect. 2, 48(2): 201. 1952, excl. syn. *O. violacea*; F.T.E.A. Rubiac. 1: 281. 1976; W.F.E.A.: 154. 1987. ≡ *Hedyotis monanthos* Hochst. ex A. Rich., Tent. Fl. Abyss. 1: 359. 1847. —Type: Ethiopia, Endeschap, 30 Jule 1838, *G.H.W. Schimper 1370* [lectotype: K (K000414332), designated by Bremek. in Verh. Kon. Ned. Akad. Wetensch., Afd. Natuurk., Sect. 2, 48(2): 202. 1952; isolectotypes: G (G00014534), LG (LG0000090028694), MO, P (P00539298); S (S14-13591), TUB (TUB004498); syntypes: Ethiopia, Choa, unknown date, *R.Quartin-Dillon & A. Petit s.n.* [P (P00539295, P00539296, P00539297)] (Plate 32G–J)

Plate 32 A–F. *Oldenlandia violacea*; G–J. *O. monanthos*. Photo by SWW (A, F), GWH (B–E, H, J) and YDZ (G, I).

Perennial prostrate herb, glabrous. Stems slender, short or elongated. Leaves oval, obtuse or subacute, 5–13 mm × 0.2–5 mm, narrowed at the base, acute at the apex; stipules shortly sheathing, truncate or 1–3-setose. Flowers 4-merous, solitary, axillary, heterostylous; pedicels up to 1.3 cm long. Calyx-tube campanulate, 1.5–3.5 mm long; lobes lanceolate, obtuse, shortly ciliate, as long as the calyx-tube. Corolla white to lilac; tube funnel-shaped, 1.5–5 mm long, hairy at the throat; lobes ovate, obtuse, 2–3 mm long. Style included or exserted, with two short spreading branches. Capsule subglobose, crowned with the erect calycine lobes.

Distribution: Western and central Kenya. [Ethiopia and Uganda].

Habitat: Upland rainforest edges, open grasslands, roadsides, or swampsides; 1700–2750 m.

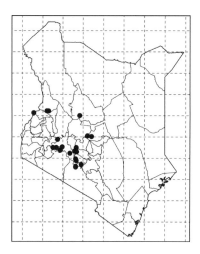

Baringo: Londiani, *Davis s.n.* (K). Bomet: Mau Forest, *Glover et al. 1441* (EA). Kajiado: Ngong Hills, *Mwangangi 2271* (EA). Kiambu: Tigoni, *Luke et al. 8369* (EA). Meru: Mount Kenya, *SAJIT Z0133* (HIB). Nairobi: *Napier 473* (K). Nakuru: Lake Naivasha, *Polhill 439B* (K). Narok: Chebil-Darakwa, *Glover et al. 1129* (EA).

Nyandarua: north Kinangop, *Kuchar 9597* (EA). Nyeri: Aberdare Mountains, *Lind 2683* (EA). Samburu: Leroghi, *Kerfoot 1148* (EA). Trans Nzoia: Mount Elgon, *Major & Cyril 494* (K). West Pokot: Cherangani Hills, *SAJIT 006820* (HIB).

3. **Oldenlandia friesiorum** Bremek., Verh. Kon. Ned. Akad. Wetensch., Afd. Natuurk., Sect. 2, 48(2): 204. 1952; F.T.E.A. Rubiac. 1: 282. 1976; W.F.E.A.: 153. 1987. —Type: Kenya, Meru, Churi River, 23 Feb. 1922, *R.E. & T.C.E. Fries 1848* [holotype: S (S-G-4342); isotypes: BR (BR0000008849496), K (K000311629)] (Plate 33A–D)

Perennial procumbent herb, up to 30 cm long, sometimes rooting at the nodes. Leaf-blades ovate or oblong-ovate, 0.5–4 cm × 0.5–2 cm, acute or very slightly acuminate at the

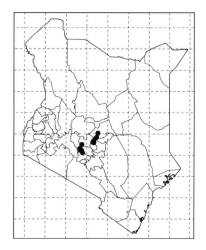

apex, rounded at the base; petioles 1–4 mm long; stipule-sheath 0.5–2 mm long, produced into a triangular lobe up to 2 mm long, bearing 5–7 filiform fimbriae 1–3 mm long. Flowers heterostylous, in few-flowered terminal or axillary cymes; peduncles up to 3 cm long; pedicels 1–3 mm long. Calyx-tube subglobose, ca. 1 mm long; lobes narrowly triangular or oblong-lanceolate, 1.8–2 mm long. Corolla white, pale lilac, blue or pink; tube ca. 2 mm long; lobes oblong, 2–3 mm long. Capsules subglobose, 2–3 mm long.

Distribution: Central Kenya. [Tanzania].

Habitat: Montane forests, bamboo zones, or *Erica* forest edges; 1850–2600 m.

Kiambu: Kame-Gakoe Road, *Kuchar 8348* (EA). Kirinyaga: Mount Kenya, Castle Camp Site to Kamweti Camp, *SAJIT 003035* (HIB). Meru: Mariba Forest, *Verdcourt & Polhill 2991* (EA). Murang'a: Kimakia Forest Reserve, *Gillett 16642* (EA). Nyandarua: Wamuhu, *Kerfoot 1388* (EA). Nyeri: Aberdare Mountains, *Agnew & Timberlake 11142* (EA, K).

4. **Oldenlandia johnstonii** (Oliv.) K. Schum. ex Engl., Abh. Königl. Akad. Wiss. Berlin 1891: 397. 1892; Bremek., Verh. Kon. Ned. Akad. Wetensch., Afd. Natuurk., Sect. 2, 48(2): 205. 1952; F.T.E.A. Rubiac. 1: 283. 1976; W.F.E.A.: 153. 1987. ≡ *Hedyotis johnstonii* Oliv., Kilima-Njaro Exped. 341. 1886. —Type: Tanzania, Kilimanjaro, *H.H. Johnston s.n.* [holotype: K]

Perennial prostrate herb, up to 1 m long, rooting at the nodes. Leaf-blades ovate or elliptic to elliptic-lanceolate 1.5–4.5 cm × 0.5–1.3 cm, acute at the apex, rounded to cuneate at the base; petioles 2–8 mm long; stipule-sheath 1–1.5 mm long, produced into a triangular lobe 2–2.5 mm long, bearing 5–9 filiform fimbriae 1–5 mm long. Flowers dimorphic, in axillary and terminal inflorescences; peduncles 0.5–3 cm long; pedicels up to 3.5 mm long. Calyx-tube obconic, ca. 1 mm long; lobes triangular-lanceolate, ca. 2 mm long. Corolla white or slightly pinkish; tube 1.2–1.5 mm long, hairy at the throat; lobes oblong, 2–2.5 mm long. Capsules ovoid or oblong-subglobose, ca. 1.5 mm long.

Distribution: Central, southern, and coastal Kenya. [Tanzania].

Habitat: Forests, thickets, dry bushlands, woodlands, grasslands, or roadsides; up to 1900 m.

Note: B. Verdcourt mentioned in FTEA that, *Oldenlandia johnstonii* (Oliv.) K. Schum. ex Engl. should be separated into two subspecies, as some specimens from coastal area of Kenya and Tanzania have ovary and capsule with short spreading hairs. But this is not a very stable trait in specimens from Kenya, and in the absence of more evidence from field work, we have retained the species as a whole.

Kajiado: Ngong Forest, *Fukuoka K176* (EA). Kiambu: Thika, *Faden 67361* (EA). Kilifi: Arabuko-Sokoke Forest, *Beentje 2308* (EA); Marafa, *Polhill & Paulo 785* (EA, K); Mangea Hill, *Luke & Robertson 1774* (EA). Kitui: Makongo Forest Reserve, *Mwangangi 4652* (EA). Kwale: Shimba Hills, *Luke & Robertson 2702* (K); Lungalunga-Msambweni Road, *Drummond & Hemsley 3863* (EA); Malungaji, *Luke & Robertson 5987* (EA). Makueni: Kithembe Hill, *Mwangangi 938* (EA).

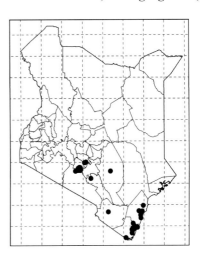

Mombasa: Mombasa Island, *Tweedie 2203* (K). Murang'a: Mitubiri, *Archer 24* (EA). Nairobi: Karura Forest, *Kahuranaga & Kibuwa 898* (EA); Nairobi National Park, *Verdcourt & Polhill 3160B* (EA). Teita Taveta: Ngangao Forest, *De Block et al. 302* (EA).

5. **Oldenlandia rupicola** (Sond.) Kuntze, Revis. Gen. Pl. 1: 293. 1891; Bremek., Verh. Kon. Ned. Akad. Wetensch., Afd. Natuurk., Sect. 2, 48(2): 208. 1952; F.T.E.A. Rubiac. 1: 284. 1976. ≡ *Hedyotis rupicola* Sond., Fl. Cap. 3: 12. 1865. —Type: South Africa, Natal, Tagoma, *W.T. Gerrard & M.J. McKen 1364* [holotype: TCD (TCD0001233); isotypes: BM, K (K000414275), S (S14-14292)] (Plate 33E–G)

Perennial procumbent herb, up to 1.4 m long, often rooting at the nodes. Leaf-blades elliptic-ovate to ovate, 3–10 mm × 2–7 mm, acute at the apex, rounded to cuneate at the base; petioles up to 2 mm long; stipule-sheath 0.5–1 mm long, with 5–7 lobes 1–3 mm long. Flowers heterostylous, solitary or more often in terminal cymes; peduncles up to 1.5 cm long. Calyx-tube obconic, 1–1.5 mm long; lobes broadly triangular, 1–4 mm long. Corolla white, pale blue, lilac or pink; tube 2–7 mm long, hairy in the throat; lobes triangular to oblong-lanceolate, 1.5–5.5 mm long. Capsule subglobose, 2–2.5 mm long.

Distribution: Southern Kenya. [Eastern and southern Africa, from Kenya to South Africa].

Habitat: Forests or hillsides, always on rock surfaces; 1650–2200 m.

Teita Taveta: Iyale Forest, *SAJIT 004544* (HIB); Vuria Forest, *SAJIT 004581* (HIB); Mount Kisagau, *Luke 4213* (EA).

6. **Oldenlandia cryptocarpa** Chiov., Res. Sci. Somalia Ital. 1: 88. 1916; Bremek., Verh. Kon. Ned. Akad. Wetensch., Afd. Natuurk., Sect. 2, 48(2): 193. 1952. —Type: Somalia, Fra Goriei ed El Magu, 1913, *G. Paoli 626* [holotype: FT (FT003301)]

= *Oldenlandia borrerioides* Verdc., Kew Bull. 30: 292. 1975; Verdc., F.T.E.A. Rubiac. 1: 287. 1976. —Type: Kenya, Tana River, 48 km south of Garsen, Kurawa, 5 Oct. 1961, *R.M. Polhill & S. Paulo 601* [holotype: K (K000311628); isotypes: BR (BR0000008849236), EA (EA000002951), FT (FT003297), PRE (PRE0594687-0)]

Annual erect herb, up to 16 cm tall. Leaf-

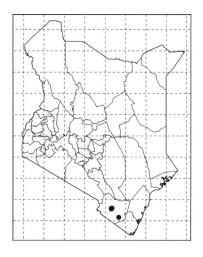
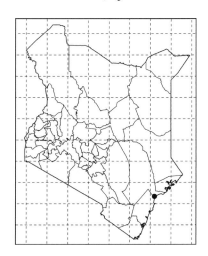

Plate 33 A–D. *Oldenlandia friesiorum*; E–G. *O. rupicola*. Photo by GWH (A), BL (B, C, E–G) and YDZ (D).

blades narrowly elliptic-lanceolate, 5–17 mm × 2–5 mm, acute at the apex, narrowed to the base; stipule-sheath ca. 2 mm long, produced into a lobe ca. 2 mm long, bearing 2 fimbriae 1–2 mm long. Flowers heterostylous, subsessile, in terminal and lateral sessile heads. Calyx-tube obovoid, ca. 1.5 mm long; lobes narrowly lanceolate, 1.5–3 mm long. Corolla white, narrowly funnel-shaped, 4–6 mm long; lobes oblong-elliptic, 2–3 mm long. Capsules obovoid, 1.6–1.8 mm long.

Distribution: Coastal Kenya. [Somalia].
Habitat: Coastal bushlands; ca. 15 m.
Tana River: Kurawa, *Polhill & Paulo 601* (EA).

7. Oldenlandia rosulata K. Schum., Bot. Jahrb. Syst. 23(3): 416. 1896; Bremek., Verh. Kon. Ned. Akad. Wetensch., Afd. Natuurk., Sect. 2, 48(2): 225. 1952; F.T.E.A. Rubiac. 1: 290. 1976. —Type: Angola, Huila, next to River Monino, Varzens, Jan. 1860, *F.M.J. Welwitsch 5320* [holotype: B (destroyed); lectotype: K (K000414292), designated by Bremek. in Verh. Kon. Ned. Akad. Wetensch., Afd. Natuurk., Sect. 2, 48(2): 225. 1952; isolectotypes: BM (BM00090 2955), G (G00014533), LISU (LISU2085 36), PRE (PRE0152010-0)]

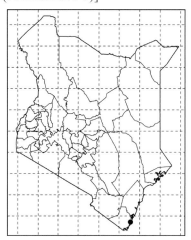

Annual erect herb, up to 15 cm tall. Basal leaves rosulate, with blades spathulate, lanceolate or elliptic, 3–7 mm × 1.3–3 mm, sometimes disappearing before maturity; stem leaves with blades linear to linear-lanceolate, 3–30 mm × 0.3–2.2 mm; stipule-sheath ca. 0.3 mm long with 2 short teeth ca. 0.5 mm long. Flowers few in terminal or axillary panicles; peduncles up to 2.8 cm long; pedicels up to 5 mm long. Calyx-tube subglobose, 0.5–0.7 mm long; lobes broadly triangular, 0.3–0.9 mm long. Corolla white to pale purple; tube 1–3.6 mm long; lobes oblong-ovate, 0.7–2 mm long. Capsules subglobose, 1–1.8 mm long, beaked.

var. **littoralis** Verdc., Kew Bull. 30(2): 293. 1975; F.T.E.A. Rubiac. 1: 291. 1976. —Type: Kenya, Kwale, Ukunda, 9 Aug. 1957, *Y.E. Symes 133* [holotype: EA (EA000001499)]

Corolla-tube ca. 1 mm long; lobes ca. 0.5 mm long. Capsules transversely oblong, compressed.

Distribution: Coastal Kenya. [Endemic].
Habitat: Grasslands; around sea level.
Kwale: Ukunda, *Symes 133* (EA).

8. Oldenlandia affinis (Roem. & Schult.) DC., Prodr. 4: 428. 1830; Bremek., Verh. Kon. Ned. Akad. Wetensch., Afd. Natuurk., Sect. 2, 48(2): 226. 1952; F.T.E.A. Rubiac. 1: 291. 1976. ≡ *Hedyotis affinis* Roem. & Schult., Syst. Veg. (ed. 15 bis) 3: 194. 1818. —Type: India, *J.G. Koenig s.n.* [holotype: B]

Perennial decumbent or straggling herb, up to 1.2 m tall, from a woody rootstock. Leaf-blades linear-lanceolate to narrowly oblong, 1–6.5 cm × 0.1–1.2 cm, acute to tapering acuminate at the apex, cuneate to rounded at the base; stipule-sheath 1 mm long, with 1–5 short fimbriae 0.2–0.5 mm long. Flowers heterostylous, in lax terminal or pseudo-axillary lax panicles; pedicels up to 1.5 cm long. Calyx-tube subglobose, 0.6–1 mm long; lobes triangular, 0.5–1 mm long. Corolla deep blue to violet; tube 3–4.5 mm long, hairy inside; lobes elliptic or oblong, 1.8–3.2 mm long. Capsules globose, 2–2.8 mm in diameter.

subsp. **fugax** (Vatke) Verdc., Kew Bull. 30: 293. 1975; F.T.E.A. Rubiac. 1: 292. 1976. ≡ *Hedyotis fugax* Vatke, Oesterr. Bot. Z. 25: 232. 1875. —Type: Tanzania, Zanzibar, *J.M. Hildebrandt 1007* [holotype: W; isotype: BM]

Plant more robust; calyx-lobes longer. Capsule more globose.

Distribution: Coastal Kenya. [Tropical Africa and Madagascar].

Habitat: Coastal forest or thickets; up to 400 m.

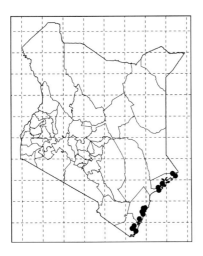

Note: *Oldenlandia affinis* subsp. *affinis* occurs in South Asia from India to Malaya.

Kilifi: Arabuko-Sokoke Forest, *Musyoki & Hansen 1019* (EA); Mida-Jilore Forest, *Gillett 20020* (EA). Kwale: Shimba Hills, *Drummond & Hemsley 1076* (EA); Titanium Base, *Nyange & Chidzinga 333* (EA). Lamu: Boni Forest, *Gillespie 364* (K); Kui Island, *Rawlins s.n.* (EA). Tana River: Nairobi Ranch, *Festo & Luke 2411* (K).

9. **Oldenlandia richardsonioides** (K. Schum.) Verdc., Kew Bull. 28: 420. 1974; F.T.E.A. Rubiac. 1: 295. 1976. ≡ *Mitratheca richardsonioides* K. Schum., Bot. Jahrb. Syst. 33: 335. 1903. —Type: Kenya, Lamu, Kiunga, *D. Riva 1650* [holotype: FI (FT003315)]

= *Oldenlandia richardsonioides* (K. Schum.) Verdc. var. *gracilis* Verdc., Kew Bull. 28: 421. 1974; F.T.E.A. Rubiac. 1: 295. 1976. —Type: Kenya/Somali Republic border, Kiunga Archipelugo, 23 Aug. 1961, *J.B. Gillespie 254* [holotype: K (K000316850)]

Annual erect herb, up to 20 cm tall; stems always much-branched from the base. Leaves sessile, slightly succulent; blades linear to lanceolate or oblanceolate, 5–40 mm × 1–7.5 mm, acute and apiculate at the apex, narrow to the base; stipule-sheath 1–2.5 mm long, with 1–5 filiform fimbriae 0.5–4.5 mm long. Flowers heterostylous, in terminal and

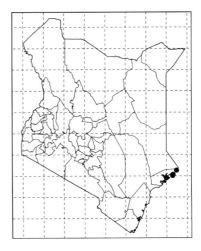

axillary dense to lax 2–9-flowered cymes; peduncles up to 15 mm long; pedicels up to 5 mm long. Calyx-tube campanulate, 0.8–1 mm long; lobes ovate-triangular, 1–3 mm long. Corolla white or pale lilac; tube 1.5–3 mm long, hairy at throat; lobes 1.5–3 mm long. Capsules campanulate, 1–2 mm long.

Distribution: Coastal Kenya. [Somalia].

Habitat: Coastal open bushlands or sandy beaches; 0–10 m.

Lamu: Kiwayu KWS/WWF Camp, *Luke 6127* (EA); Kiwayu, *Pomeroy 16762* (EA); Osine, *Greenway & Rawlins 9289* (EA).

10. **Oldenlandia scopulorum** Bullock, Bull. Misc. Inform. Kew 1932(10): 497. 1932; Bremek., Verh. Kon. Ned. Akad. Wetensch., Afd. Natuurk., Sect. 2, 48(2): 239. 1952; F.T.E.A. Rubiac. 1: 299. 1976; W.F.E.A.: 154. 1987. —Types: Kenya, Mt. Elgon, 2200 m, Oct.–Nov. 1930, *E.J. & C. Lugard 49* [holotype: K (K000316269); isotype: EA (EA000003012)]; Kenya, Mt. Elgon, 2515 m, Dec. 1930, *E.J. & C. Lugard 346* [paratypes: EA (EA000003013), K(K000316268)]

= *Oldenlandia scopulorum* Bullock var. *lanceolata* Bremek., Verh. Kon. Ned. Akad. Wetensch., Afd. Natuurk., Sect. 2, 48(2): 239. 1952. —Type: Kenya, Londiani, May 1920, *G. Lindblom s.n.* [lectotype: S (S14-14461), designated here; isolectotype: S (S14-14466)]

= *Oldenlandia filipes* Bremek., Kew Bull. 11: 169. 1956. —Type: Kenya, Londiani, 14 July 1951, *D. Davis 23* [holotype: K (K000316260); isotype: BM]

Perennial herb, up to 25 cm tall, from a woody rootstock, usually much branched near the base. Leaf-blades linear to linear-lanceolate, 5–25 mm × 1–3.5 mm, acute to acuminate at the apex, cuneate at the base; stipule-sheath 1–2 mm long, with 2–3 filiform fimbriae 1–5 mm long. Flowers heterostylous, in terminal and apparently axillary subcorymbose 2–5-flowered inflorescences; peduncles up to 3 cm long; pedicels up to 1.5 cm long. Calyx-tube campanulate, ca. 0.8 mm long; lobes triangular or lanceolate, 1–2 mm long. Corolla white, pink, lilac, mauve or pale blue; tube 1.5–3 mm long, hairy at the throat; lobes ovate-oblong, 1.5–3 mm long. Capsules subglobose, 1.5–2 mm long.

Distribution: Western, central, southern, and coastal Kenya. [Tropical East Africa].

Habitat: Grasslands, woodlands, or roadsides; 150–2550 m.

Kajiado: Ngong Forest, *van Someren B1093* (K). Kericho: Kedowa, *Napier 10184* (EA). Kiambu: Muguga, *Verdcourt 642* (EA). Kilifi: Arabuko-Sokoke Forest, *Luke & Robertson 2613* (EA). Kisumu: Onjiko, *Opiko B717* (K). Laikipia: Rumuruti, *Hepper & Jaeger 6659* (K). Machakos: Mutitu, *Kirika 615* (EA). Meru: north slope of Mount Kenya, *Rogers 398* (EA). Nairobi: Langata Barracks, *Beentje 1684* (EA). Nakuru: Lake Naivasha, *Beentje 2061* (EA); Elementeita, *Bogdan AB1207* (K); Lake Nakuru, *Kulileki 77* (EA). Narok: Eorengitok, *Glover*

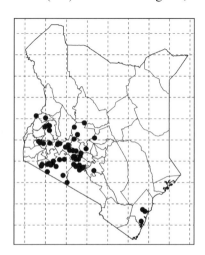

et al. 1275 (EA); ca. 10 km SSE of Narok, *Kuchar 8204* (EA); Ndunyangero Hill, *Gwynne & Samuel 1668* (K). Nyeri: near Nyeri Town, *Tallantire 676* (EA). Trans Nzoia: Kitale, *Beentje 2998* (EA). Uasin Gishu: Kipkarren, *Brodhurst-Hill 188* (EA).

11. **Oldenlandia wiedemannii** K. Schum., Bot. Jahrb. Syst. 28: 57. 1899; Bremek., Verh. Kon. Ned. Akad. Wetensch., Afd. Natuurk., Sect. 2, 48(2): 240. 1952; F.T.E.A. Rubiac. 1: 300. 1976; W.F.E.A.: 154. 1987. —Types: Tanzania, near Moshi, Wiedemann s.n. (holotype: B, destroyed); Tanzania, Moshi, May 1927, *D. Haarer 474* [neotype: K, designated by Bremek. in Verh. Kon. Ned. Akad. Wetensch., Afd. Natuurk., Sect. 2, 48(2): 241. 1952; isoneotype: EA (EA000003010)]

= *Oldenlandia kaessneri* K. Schum. & K. Krause, Bot. Jahrb. Syst. 39(3-4): 520. 1907, *nom. illeg.*, non S. Moore (1905). ≡ *Oldenlandia wiedemannii* K. Schum. var. *glabricaulis* Bremek., Verh. Kon. Ned. Akad. Wetensch., Afd. Natuurk., Sect. 2, 48(2): 241. 1952. —Type: Kenya, Sultan Hamoud, *T. Käessner 653* [holotype: B, destroyed; lectotype: K, designated by Bremek. in Verh. Kon. Ned. Akad. Wetensch., Afd. Natuurk., Sect. 2, 48(2): 241. 1952; isolectotypes: BM, MO]

Perennial erect herb or subshrub, up to 30 cm tall, always from a woody taproot. Leaf-blades linear to linear-lanceolate, 10–65 mm × 0.8–4 mm, acute at the apex, narrowly cuneate at the base; stipule-sheath 2–4 mm long, with 1–4 filiform fimbriae 2–4 mm long. Flowers heterostylous, in dense terminal heads; peduncles 0–5 mm long, but always forming a pseudo-peduncle up to 6 cm long, with the leaves at the node reduced or hidden; pedicels up to 3 mm long. Calyx-tube subglobose, ca. 1 mm long; lobes triangular to triangular-lanceolate, 1.5–3 mm long. Corolla white, pale lilac, pale pink or bluish; tube 1.5–4 mm long, hairy at the throat; lobes oblong-elliptic, 1.6–2.2 mm long. Capsules subglobose, 1–1.8 mm long.

Distribution: Western, southern, and coastal Kenya. [Tanzania].

Habitat: Forests, hillsides, woodlands, bushlands, thickets, or open grasslands scattered with trees, always in rocky places; 50–2500 m.

Kajiado: Ilpartimaro, *Kuchar & Msafiri 8000* (EA). Kilifi: Mangea Hill, *Luke 1069* (EA). Kwale: Mackinnon, *Drummond & Hemsley 4099* (EA). Laikipia: ca. 42 km northwest of Nanyuki, *Beentje 1757* (EA). Machakos: Mua Hills, *Verdcourt & Napper 2167* (EA). Makueni: Kilima Kiu, *Gillett 18368* (EA). Meru: Ngare-Ndare, *Napier 2569* (K). Nakuru: near Lake Naivasha, *Mwangangi 705* (EA). Narok: Masai Mara Game Reserve, *Kuchar & Msafiri 6142* (EA); Olarro Camp, *Luke 7315* (EA). Samburu: Lerogi Forest, *Carter & Stannard 437* (K); Sirwan, *Newbould 3418* (K). Teita Taveta: Voi-Taveta Road, *Faden74/457* (EA); Tsavo East Park, *Bally 13385* (EA).

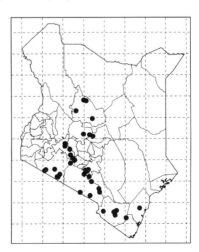

12. Oldenlandia ichthyoderma Cufod., Nuovo Giorn. Bot. Ital., n.s., 55: 83. 1948; Bremek., Verh. Kon. Ned. Akad. Wetensch., Afd. Natuurk., Sect. 2, 48(2): 242. 1952; F.T.E.A. Rubiac. 1: 302. 1976. —Types: Ethiopia, Sagan Omo, Lake Rudolf, Elolo, 5 Aug. 1939, *R. Corradi 2754* [lectotype: FT (FT003304), designated by Bremek., Verh. Kon. Ned. Akad. Wetensch., Afd. Natuurk., Sect. 2, 48(2): 243. 1952]; Ethiopia, Sagan Omo, Lake Rudolf, Elolo, 5 Aug. 1939, *R. Corradi 2759* [syntype: FT (FT003309)]

Perennial erect herb, up to 25 cm tall, usually from a woody rootstock. Leaf-blades linear-lanceolate, 15–50 cm × 1–3.5 cm, acute and apiculate at the apex, narrowed to the base; stipule-sheath 3–5 mm long, with 5–7 setae 2–5 mm long. Flowers heterostylous, in condensed sessile terminal heads, the leaves at nodes well developed. Calyx-tube turbinate, ca. 1.4 mm long; lobes ovate-lanceolate, 3–3.5 mm long, margins always hyaline. Corolla white; tube 3–3.3 mm long, hairy at the throat; lobes oblong-elliptic, 1.5–1.8 mm long. Capsules 1.7–2 mm long.

Distribution: Northern, central, and eastern Kenya. [Ethiopia and Somalia].

Habitat: Bushlands or roadsides, always on sandy soil; 50–900 m.

Isiolo: ca. 1 km north of Mado Gashi, *Gilbert & Thulin 1110* (EA, K). Marsabit: east of Lake Turkana, *Kimeu JKM-293a* (EA). Meru: Meru National Park, *Ament & Magogo 328* (K). Samburu: Mount Ololokwe, *Bally & Smith 14748* (EA). Tana River: Kora National Reserve, *van Someren 885* (EA).

13. Oldenlandia herbacea (L.) Roxb., Hort. Bengal. 11. 1814; Bremek., Verh. Kon. Akad. Wetensch., Afd. Natuurk., Sect. 2, 48(2): 244. 1952; F.T.E.A. Rubiac. 1: 305. 1976; W.F.E.A.: 153. 1987. ≡ *Hedyotis herbacea* L., Sp. Pl. 1: 102. 1753. —Type: Sri Lanka, *P. Hermann 4.19* [holotype: BM]

Annual or perennial erect, decumbent or spreading herb, up to 60 cm tall. Leaf-blades linear to linear-lanceolate or narrowly oblanceolate, 6–55 mm × 1–3.5 mm, acute at the apex, cuneate at the base; petioles not developed; stipule-sheath up to 0.5 mm long, truncate, with a few setae ca. 0.3 mm long. Flowers usually isostylous or rarely heterostylous, solitary or paired at the node; pedicels up to 3.5 cm long. Calyx-tube ovoid, 0.5–1 mm long; lobes narrowly triangular, 0.5–2.5 mm long. Corolla white, lilac or mauve; tube cylindrical, 0.2–1.1 cm long; lobes ovate, 1–3 mm long. Capsules subglobose, 2–5 mm long.

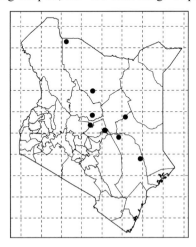

1a. Corolla-tube less than 4 mm long ....a. var. *herbacea*
1b. Corolla-tube more than 7 mm long .... b. var. *holstii*

a. var. **herbacea** (Plate 34A–C)

Corolla-tube less than 4 mm long.

Distribution: Widespread in Kenya. [Africa, Indian Subcontinent to Andaman Islands].

Habitat: Bushlands, woodlands, thickets, or grasslands, also in cultivations; up to 2200 m.

Baringo: Tugen Hills, *Vincens 112* (EA). Embu: southeast Mount Kenya, *Battiscombe K686* (EA). Kajiado: Emali, *van Someren 82* (K). Kakamega: Yala, *Agnew & Musumba 8574* (EA). Kericho: Thessalia, *Johansen 56* (EA). Kiambu: Thika, *Faden 67551* (EA). Kilifi: Arabuko-Sokoke Forest, *Faden 71/688* (EA). Kirinyaga: Thiba River Fishing Camp, *Copley B398* (K). Kisii: Namanga Ranch, *Plaizier 1089* (EA). Kitui: Mutomo Hill, *Gillett 18565* (EA). Kwale: Shimba Hills, *Magogo & Glover 319* (EA). Laikipia: Doldol, *Kirika et al. 03/37/07* (K). Machakos: Mua Hills, *Gillett 16213* (EA). Makueni: Kilima Kiu Ranch, *Gillett 2070* (EA). Mandera: Dandu, *Gillet 13140* (K). Marsabit: Moyale, *Gillett 13515* (K). Murang'a: north of Thika River, *Faden 68/213* (EA). Nairobi: Langata, *Ngeno 76* (EA). Nakuru: west of Lake Nakuru, *Verdcourt 823* (EA). Nandi: North Nandi Forest, *SAJIT 006607* (HIB). Narok: Olarro Camp, *Luke et al. 7314* (EA); Olodungoro, *Glover et al. 2121* (EA). Nyeri: Mount Kenya Forest, *Dale 3032* (EA). Samburu: Ndoto Mountains, *Gilbert et al. 5617* (EA, K). Teita Taveta: Mwandongo Forest Reserve, *Mwachala et al.* in *EW597* (EA); Ngangao Forest, *Faden et al. 230* (EA). Tana River: Katumba Hill, *Gilbert & Thulin 1733* (EA). Tharaka-Nithi: Chogoria Forest, *SAJIT 003988* (HIB). Trans Nzoia: Kitale, *Beentje 2997* (EA). West Pokot: Kongolai Escarpent, *Alobine 442* (EA).

b. var. **holstii** (K. Schum. ex Engl.) Bremek., Verh. Kon. Ned. Akad. Wetensch., Afd. Natuurk., Sect. 2, 48(2): 249. 1952; F.T.E.A. Rubiac. 1: 306. 1976. ≡ *Oldenlandia holstii* K. Schum., Pflanzenw. Ost-Afrikas, C: 376. 1895. —Type: Tanzania, Lushoto, Usambara, Kwa Mshuza, Aug. 1893, *C.H.E. Holst 8942* [holotype: B, destroyed; lectotype: K (K000414314), **designated here**; isolectotypes: BM, G (G00008893, G00014543), KFTA, M (M0106473), P (P00539292 & P00539293), S (S14-13068), US (US00137406)] (Plate 34D–F)

Corolla-tube more than 7 mm long.
Distribution: Southern Kenya. [Northeast

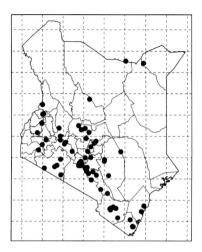
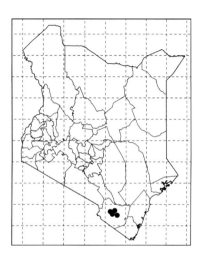

Plate 34 A–C. *Oldenlandia herbacea* var. *herbacea*; D–F. *O. herbacea* var. *holstii*. Photo by YDZ (A, C), GWH (B, D, E) and BL (F).

Africa, Ethiopia to Tanzania].

Habitat: Forests, open woodlands, or grasslands; 1300–1850 m.

Teita Taveta: near Kajire, *SAJIT 005346* (HIB); Ngangao Forest, *De Block et al. 310* (K); Mbololo Hill, *Kamau & Mwangangi 444* (EA).

14. Oldenlandia acicularis Bremek., Verh. Kon. Ned. Akad. Wetensch., Afd. Natuurk., Sect. 2, 48(2): 269. 1952; F.T.E.A. Rubiac. 1: 314. 1976. —Type: Uasin Gishu, Kipkarren, Aug. 1931, *E. Brodhurst-Hill & E.R. Napier 221* [holotype: K (K000316936); isotype: EA (EA000001600)]

Annual erect herb, up to 15 cm tall. Leaves sessile, blades linear, 10–30 cm × 0.5–0.8 mm, acuminate at the apex; stipule-sheath 1.5–3 mm long, with 2–5 setae 1–4 mm long. Flowers isostylous, 1–3 at each node, with pedicels 1–4 mm long. Calyx-tube campanulate, ca. 0.8 mm long; lobes linear-lanceolate, 1.2–1.8 mm long. Corolla white; tube ca. 1.2 mm long, hairy at the throat; lobes ovate, ca. 1.2 mm long. Capsules campanulate, laterally compressed, ca. 1.5 mm long.

Distribution: Western Kenya. [Endemic].

Habitat: Rocky pools; ca. 1500 m.

Uasin Gishu: Kipkarren, *Brodhurst-Hill 221* (EA, K).

15. Oldenlandia fastigiata Bremek., Verh. Kon. Ned. Akad. Wetensch., Afd. Natuurk., Sect. 2, 48(2): 260. 1952; F.T.E.A. Rubiac. 1: 311. 1976. —Type: Tanzania, Ulanga, Mahenge, Mbangala, *H.J.E. Schlieben 1797* [holotype: B; isotypes: BM (BM000902950), BR (BR0000008849489), G (G00014547), K (K000414310), P (P00539282), S (S14-9248)]

Annual or perennial herb, up to 60 cm long. Leaf-blade linear to linear-lanceolate, 10–60 mm × 1–5 mm, acute at the apex, narrowed to the base; petioles not developed; stipule-sheath 1–2 mm long, with 3–5 fimbriae 0.5–2.5 mm long. Flowers isostylous or heterostylous, in 3–10-flowered sessile or pedunculates axillary cymes or fascicles; peduncles up to 3 cm long; pedicels 1–4 mm long. Calyx-tube subglobose, 0.5–0.8 mm long; lobes narrowly triangular, 1–2 mm long. Corolla white or tinged blue or pink; tube 1–2 mm long, densely hairy at throat; lobes ovate-oblong, 0.7–2 mm long. Capsules depressed globose, 1–2 mm long.

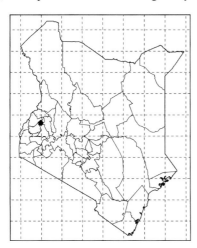

1a. Inflorescences sessile or shortly pedunculate; flowers not heterostylous.......... a. var. *fastigiata*
1b. Inflorescences distinctly pedunculate; flowers heterostylous 2
2a. Leaf-blades very thin; inflorescences lax, with the pedicels up to 5 mm long...........................
.... b. var. *pseudopenton*
2b. Leaf-blades not thin as above; inflorescences not lax, with the pedicels less than 2 mm long ...
.... c. var. *somala*

a. var. **fastigiata**

Inflorescences sessile or shortly pedunculate; flowers not heterostylous.

Distribution: Widespread in Kenya. [East Africa, from Ethiopia to Mozambique].

Habitat: Woodlands, thickets, grasslands, or sandy roadsides; up to 1800 m.

Baringo: ca. 2 km south of Muktan, *Luke BFFP896* (E.A). Kajiado: Eusso Nyiro, *Verdcourt et al. 420* (EA). Kilifi: Malindi, *Rawlins 657* (EA). Kitui: Mutomo Hill, *Gilbert 18592* (EA). Kwale: Shimba Hills, *Luke et al. 5723* (EA); Mrima Hill, *Verdcourt 1870* (EA). Laikipia: Pelagalagi River Valley, *Luke 3388* (EA). Lamu: Witu to Kipini, *Hooper & Townsend 1171* (EA). Machakos: Yatta Plateau, *Gillett 23952* (EA). Makueni: Kibwezi Forest Reserve, *Luke 14953* (EA). Mandera: ca. 6 km south of El Wak, *Gilbert & Thulin 1233* (EA). Mombasa: *Napier 3480*

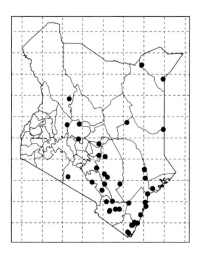

(EA). Teita Taveta: Taita Hills, *Mwachala et al. in EW623* (EA); Bura Valley, *Mungai et al. in EW1302* (EA); Aruba Dam, *Hooper & Townsend 1079* (EA). Tana River: Tana River National Primate Reserve, *Luke et al. TPR 687* (K); Tana Delta, *Luke 15401* (EA). Tharaka-Nithi: Ishiara on road to Meru, *Gilbert et al. 5713* (EA). Turkana: Ewosinaikenyi, *Mwangangi 1385* (EA).

b. var. **pseudopentodon** Verdc., Kew Bull. 30(2): 299. 1975; F.T.E.A. Rubiac. 1: 313. 1976. —Type: Kenya, Tana River, 1 km south of Garsen, 15 July 1972, *J.B. Gillett & S.P. Kibuwa 19909* [holotype: K (K000311625); isotype: EA (EA000001601)]

Leaf-blades very thin. Inflorescences lax, with the pedicels up to 5 mm long.

Distribution: Central, southern, and coastal Kenya. [Endemic].

Habitat: Riparian forests or open damp grasslands; up to 1250 m.

Embu: Mwea National Reserve, *Mwangangi 4943* (EA). Teita Taveta: Voi, *Napper 968* (EA, K). Tana River: Kora National Park, *Mutangah 117* (EA).

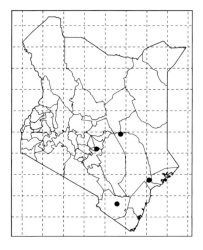

c. var. **somala** (Bremek.) Verdc., Kew Bull. 30(2): 299. 1975; F.T.E.A. Rubiac. 1: 312. 1976. ≡ *Oldenlandia somala* Chiov. ex Bremek., Verh. Kon. Akad. Wetensch., Afd. Natuurk., Sect. 2, 48(2): 251. 1952. —Type: Somalia, Bulo Aran, 12 Dec. 1935, *F. Bisi*

134 [holotype: FT (FT003302)]

Leaf-blades not thin; inflorescences not lax, with the pedicels less than 2 mm long.

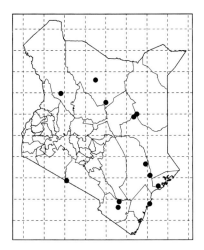

Distribution: Widespread in Kenya. [Northeast Africa, from Ethiopia, Somalia to Tanzania].

Habitat: Hillsides, bushlands, thickets, or floodplain grasslands; up to 1550 m.

Garissa: ca. 17 km east of Modo Gash, *Stannard & Gilbert 892* (EA). Kajiado: Shombole, *Greenway 232* (EA). Marsabit: Marsabit on the way to Isiolo, *Magogo 1346* (EA). Teita Taveta: Lugard Falls, *Hucks 807* (EA).

16. **Oldenlandia corymbosa** L., Sp. Pl. 1: 119. 1753; Bremek., Verh. Kon. Akad. Wetensch., Afd. Natuurk., Sect. 2, 48(2): 262. 1952; F.T.E.A. Rubiac. 1: 308. 1976; W.F.E.A.: 153. 1987. —Type: Plumier, Nov. Pl. Amer. 42, t. 36 (1703). [lectotype designated by B. Verdcourt in F.T.E.A. Rubiac. 1: 308. 1976]

Annual prostrate or erect herb, up to 30 cm long. Leaf-blades linear to narrowly elliptic, 5–35 mm × 0.5–7 mm, acute and apiculate at the apex, narrowed to the base; petioles not developed; stipule-sheath 0.5–3 mm long, with 2–5 fimbriae 0.5–1 mm long. Flowers not heterostylous, single in each node or few in pedunculate umbel-like inflorescences; peduncles up to 18 mm long; pedicels up to 15 mm long. Calyx-tube ellipsoid, 0.7–1 mm long; lobes triangular, 0.5–1.8 mm long. Corolla white or tinged blue, pink or purple; tube 0.6–1 mm long; lobes ovate to oblong, 0.5–1.2 mm long. Capsules ovoid, 1.2–2.2 mm long.

1a. Flowers single in each node ....b. var. *caespitosa*
1b. Flowers in few-flowered inflorescences 2
2a. Leaves mostly elliptic, up to 7 mm wide.. a. var. *corymbosa*
2b. Leaves mostly linear, 1–3 mm wide..3
3a. Plants mostly over 10 cm tall ....c. var. *linearis*
3b. Plants very small, mostly less than 7 cm tall... d. var. *nana*

a. var. **corymbosa** (Plate 35)

Leaves mostly elliptic, up to 7 mm wide. Flowers in few-flowered inflorescences.

Distribution: Western, central, southern, and coastal Kenya. [Tropics and subtropics of the world].

Habitat: Forest margins, open woodlands, or grasslands, often in rocky places or in old cultivations; up to 2000 m.

Kilifi: Malindi, *Bogdan 2565* (K). Kisumu: Nanga, *Turner 6714* (EA). Kwale: Msambweni, *Robertson 3321* (EA). Laikipia: Ol-Ari-Nyiro-Ranch, *Muasya 2131* (EA). Lamu: Kiunga, *Gillespie 22* (K). Narok: ca. 5 km west of new Mara Bridge, *Kuchar 4312* (EA). Teita Taveta: Mbololo Forest, *De Block et al. 500* (EA). Turkana: Kaputir Upper Turkwell, *Newbould 7380* (EA). Uasin Gishu: Kipkarren, *Brodhurst-Hill 13* (EA).

MO, NY (NY00132356)]

Flowers always single at each node.

Distribution: Widespread in Kenya. [Tropical Africa, Arabian Peninsula, Indian Subcontinent to Andaman Islands].

Habitat: Forest margins, hillsides, thickets, open bushlands, or grasslands; up to 2350 m.

Note: The specimen [*Lewis 5908* (K)]

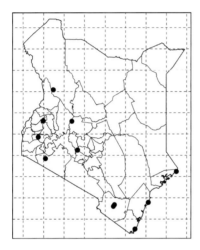

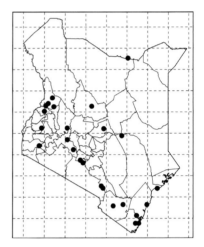

b. var. **caespitosa** (Benth.) Verdc., Kew Bull. 30(2): 298. 1975; F.T.E.A. Rubiac. 1: 310. 1976. ≡ *Oldenlandia herbacea* (L.) Roxb. var. *caespitosa* Benth. Niger Fl. 403. 1849. ≡ *Oldenlandia caespitosa* (Benth.) Hiern, Fl. Trop. Afr. 3: 61. 1877; Bremek., Verh. Kon. Akad. Wetensch., Afd. Natuurk., Sect. 2, 48(2): 262. 1952. ≡ *Hedyotis corymbosa* (L.) Lam. var. *caespitosa* (Benth.) R. Dutta & Deb, Taxon. Revis. *Hedyotis* Indian Subcont.: 151. 2004. —Type: Liberia, Cape Palmas, Jule 1841, *J.R.T. Vogel 51* [holotype: K (K000414339)]

= *Oldenlandia subtilis* S. Moore, J. Bot. 43: 249. 1905. —Type: Kenya, Kitui District, Galunka, 21 May 1902, *T. Kässner 781* [holotype: BM (BM000902952); isotypes: K,

from Ruiru of central Kenya was identified as *Oldenlandia capensis* L.f. But when we compared this specimen with the type of *O. capensis*, we thought *Lewis 5908* should be *O. corymbosa* var. *caespitosa* (Benth.) Verdc. for its pedicels always more than 3 mm long, while *O. capensis* with flowers almost sessile in its inflorescences. We did not find the specimen *Hurger s.n.* which was cited in FTEA. We doubt the occurrence of *O. capensis* in Kenya, and this needs more field investigations.

Baringo: Lake Bogoria, *Vincens 53* (EA). Kakamega: Kakamega Forest, *Verdcourt 1635* (EA). Kilifi: Arabuko-Sokoke Forest, *Luke & Robertson 2589* (EA). Kisii: ca. 19 km southwest of Kisii, *Vuyk & Bretler 237* (EA). Kwale:

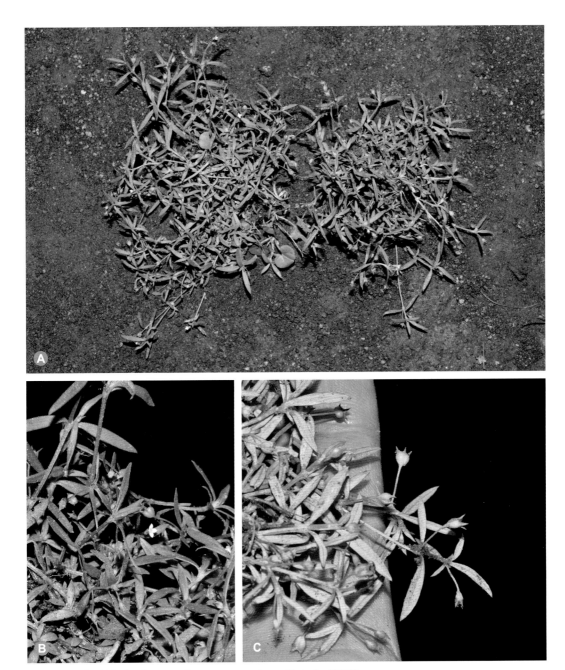

Plate 35 A–C. *Oldenlandia corymbosa* var. *corymbosa*. Photo by SWW.

Shimba Hills, *Magogo & Estes 1204* (K). Marsabit: Moyale, *Gillett 13476* (K). Meru: Nyambeni Hills, *Luke 1082* (EA). Nairobi: Nairobi National Park, *Lewis 5948* (K). Nakuru: Nakuru National Park, *Hingley 144* (EA). Samburu: Mathew's Range, *Newbould s.n.* (K). Teita Taveta: Aruba Lodge, *Kuchar & Msafiri 5682* (EA). Tana River: Tana Delta, *Luke & Hamerlynck 078* (EA). West Pokot: Cherangani Hills, *Hepper et al. 5057* (K).

c. var. **linearis** (DC.) Verdc., Kew Bull. 30(2):

296. 1975; F.T.E.A. Rubiac. 1: 309. 1976 ≡ *Oldenlandia linearis* DC., Prodr. 4: 425. 1830; Bremek., Verh. Kon. Akad. Wetensch., Afd. Natuurk., Sect. 2, 48(2): 258. 1952. —Type: Senegal, Bay of St. Louis, 1 Jan. 1830, *S. Perrottet s.n.* [holotype: G (G00014776); isotype: P]

Plants mostly over 10 cm tall. Leaves mostly linear, 1–3 mm wide. Flowers in few-flowered inflorescences.

Distribution: Western, central, southern, and coastal Kenya. [Tropical Africa, Arabian Peninsula, Indian Subcontinent to Andaman Islands].

Habitat: Hillsides, woodlands, bushlands, grasslands, or in old cultivations; up to 1850 m.

Kwale: Twiga Beach, *Napper 1357* (EA). Laikipia: ca. 42 km northwest of Nanyuki, *Beentje 1744* (EA). Murang'a: north side of Chania River, *Faden 68/214* (EA). Nairobi: Nairobi National Park, *Agnew 9417* (EA). Nakuru: northwest of Lake Nakuru, *Mwangangi 105* (EA). Teita Taveta: Tsavo East Mudanda Rock, *Faden 72/7* (EA). Trans Nzoia: Kitale, *Abugadon s.n.* (EA). West Pokot: ca. 19 km north of Kacheliba, *Leippert 5037* (EA).

d. var. **nana** (Bremek.) Verdc., Kew Bull. 30: 296. 1975; F.T.E.A. Rubiac. 1: 310. 1976. ≡ *Oldenlandia linearis* DC. var. *nana* Bremek., Verh. Kon. Akad. Wetensch., Afd. Natuurk., Sect. 2, 48(2): 259. 1952. —Type: Uganda, Mbale, Bugisu, Bulago, 27 Aug. 1932, *A.S. Thomas 320* [holotype: K (K000414312); isotype: KAW (KAW000083)]

Plants very small, mostly less than 7 cm tall. Leaves mostly linear. Flowers in few-flowered inflorescences.

Distribution: Western, central and coastal Kenya. [Tanzania].

Habitat: Thickets or roadsides; up to 2350 m.

Kajiado: Chyulu Hills, *Luke 9490A* (EA). Kwale: Diani Beach, *Luke et al. 9023* (K). Laikipia: Rumuruti, *Hepper & Jaeger 6658* (K). Meru: Nyambeni Hills, *Luke et al. 10282* (K). Nakuru: Ol Longonot, *Kerfoot 3577* (EA). Trans Nzoia: ca. 8 miles from Kitale towards Cherangani, *Symes 754* (EA). West Pokot: Sebit, *Agnew et al. 10248* (EA).

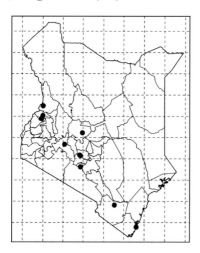
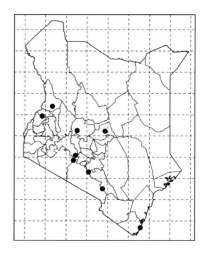

28. **Diodia** Gronov.

Annual or perennial herbs; stems erect, decumbent or prostrate, terete or mostly 4-angled. Leaves opposite or seemingly verticillate, sessile or shortly petiolate; blades linear to ovate; stipules connate and fused to petioles. Inflorescences axillary or in terminal spikes. Flowers sessile, 4-merous. Calyx-tube ellipsoid, ovoid or obconic; lobes 2–6. Corolla-tube funnel-shaped to campanulate; lobes 4–6, valvate. Stamens 4–6; anthers exserted, dorsifixed. Ovary 2(–4)-locular; ovules solitary in each locule; style filiform, exserted; stigmas 2, linear or capitate. Fruit of 2(–4) indehiscent cocci. Seeds ellipsoid-oblong.

About 50 species mostly distributed in America and Africa; one species with two varieties in Kenya.

1. **Diodia aulacosperma** K. Schum., Pflanzenw. Ost-Afrikas, C: 394. 1895; F.T.E.A. Rubiac. 1: 337. 1976; W.F.E.A.: 151. 1987. —Type: Kenya, Lamu Island, Dec. 1875, *J.M. Hildebrandt 1903* [holotype: B (destroyed); lectotype: K (K000316213), **designated here**]

Annual erect, decumbent or prostrate herb, up to 90 cm long or tall. Leaves opposite; blades elliptic-lanceolate to ovate, 0.8–5 cm × 0.3–2.5 cm, apex acute to rounded, base cuneate to truncate; petioles 1–15 mm long; stipule-sheath 2–3 mm long, with 3–8 setae 1–6 mm long. Flowers sessile, several in axillary clusters. Calyx-tube turbinate, 1–2 mm long; lobes 4–6, unequal. Corolla white, pink or mauve; tube funnel-shaped, 4–7 mm long; lobes ovate-triangular, 2–4.5 mm long. Fruit of 2 indehiscent cocci, oblong-ovoid, 2.5–4 mm long. Seeds oblong, with longitudinal grooves.

1a. Leaf-blades succulent, rounded, oblong or elliptic, not markedly narrowed to the base ....
.... a. var. *aulacosperma*
1b. Leaf-blades less succulent, ovate to elliptic, markedly narrowed to the base............................
.... b. var. *angustata*

a. var. **aulacosperma** (Figure 27; Plate 36)

Leaf-blades succulent, rounded, oblong or elliptic, not markedly narrowed to the base.
Distribution: Coastal Kenya. [Somalia and Tanzania].
Habitat: Grassland and bushland or old cultivations and other disturbed places; up to 50 m.
Kilifi: Arabuko-Sokoke Forest, *Luke & Robertson 2586B* (EA); Malindi, *Tweedie*

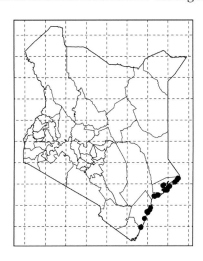

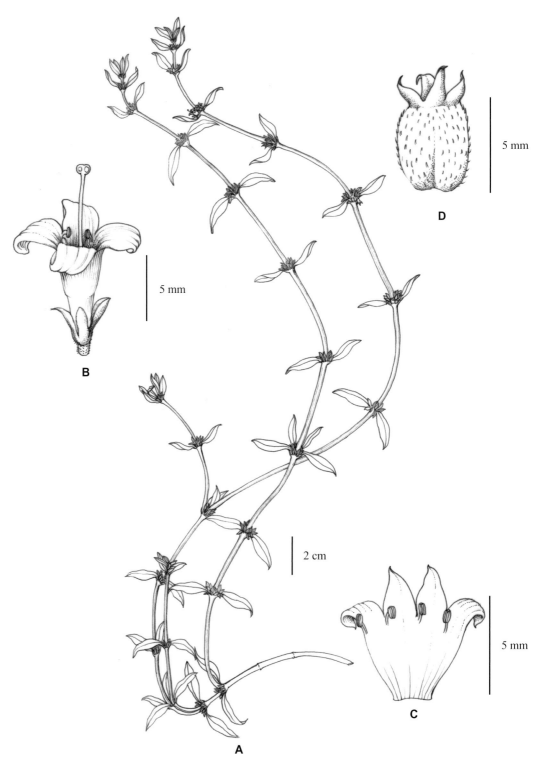

Figure 27 *Diodia aulacosperma* var. *aulacosperma*. A. plant; B. flower; C. longitudinal section of corolla, showing the stamens; D. fruit. Drawn by NJ.

Plate 36 A, B. *Diodia aulacosperma* var. *aulacosperma*. Photo by ZWW.

3177 (K). Lamu: Shela, *Bock 5* (EA); Kiwayu Area, *Luke 5647* (EA). Mombasa: *Whyte s.n.* (K). Tana River: Shekiko, *Luke & Robertson 1386* (EA).

b. var. **angustata** Verdc., Kew Bull. 30: 300. 1975; F.T.E.A. Rubiac. 1: 339. 1976. —Type: Tanzania, Tanga, Sawa, 13 Jule 1955, *H. Faulkner 1666* [holotype: K (K000316171); isotypes: BR (BR0000008829863), EA (EA 000001596), K (K000316172)]

Leaf-blades less succulent, ovate to elliptic, markedly narrowed to the base.
Distribution: Coastal Kenya. [Tanzania].
Habitat: Grasslands, bushlands, or old cultivations and other disturbed places; up to 100 m.
Kilifi: Mambrui, *Polhill & Paulo 725* (EA, K); Msabaha, *Robertson 3259* (K); Mida Creek, *Kuchar 12015* (EA). Kwale: Funzi Peninsula, *Luke & Luke 9029* (EA); Diani Beach, *Gillett 18639* (EA, K). Lamu: Sankuri Hill, *Festo et al. 2779* (EA, K); Shela, *Robertson 3779* (EA). Mombasa: English Point, *Napier 3291* (EA). Tana River: Nairobi Ranch, *Festo & Luke 2335* (EA).

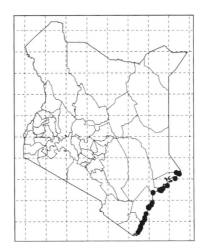

29. **Mitracarpus** Zucc.

Annual or perennial herbs, rarely subshrubs, with stem erect or prostrate, always 4-angled. Leaves opposite, subsessile or sessile, linear-lanceolate to ovate or broadly elliptic. Stipules connate with the petioles to form a fimbriated sheath, setose. Inflorescences terminal or axillary, glomerulate or capitate, several- to many-flowered. Calyx-tube obconic, obovoid or subglobose; lobes 4–5, usually unequal. Corolla salver-shaped or funnel-shaped, glabrous or pubescent in throat; lobes 4. Stamens 4, inserted in the throat, anthers included or exserted. Ovary 2(–3)-locular; ovule solitary in each locule. Style short or long, divided into 2 short linear stigmas. Fruit a thin circumscissile capsule, the upper part splitting off together with the calyx-lobes, the septum persistent. Seeds 2, oblong or globose; endosperm fleshy.

About 65 species widespread in tropical and subtropical central, north, and south America and the Antilles, one species widely naturalized in tropical Africa, Asia, Australia, and Pacific islands, and also in Kenya.

1. **Mitracarpus hirtus** (L.) DC., Prodr. 4: 572. 1830. ≡ *Spermacoce hirta* L., Sp. Pl. (ed. 2) 1: 148. 1762. —Type: Jamaica, *Herb. Linn. 125.8* [lectotype: LINN, designated by B. Verdcourt in Kew Bull. 30: 318. 1975] (Figure 28; Plate 37)

= *Mitracarpus villosus* (Sw.) DC., Prodr. 4: 572. 1830; F.T.E.A. Rubiac. 1: 375. 1976; U.K.W.F.F.: 228. 2013. ≡ *Spermacoce villosa* Sw., Prodr. 29. 1788. —Type: Jamaica, *O. Swartz s.n.* [holotype: S; isotypes: BM, SBT (SBT13430)]

Annual herb, erect or spreading, up to 40 cm tall. Leaves elliptic, 1–6 cm × 0.3–2.3 cm, subacute at the apex, cuneate at the base, margins often scabrid; stipule-sheath 1–3 mm long, divided into 1–15 fimbriae, 1–5 mm long. Inflorescences axillary, subglobose, 0.5–1.8 cm in diameter. Flowers subsessile. Calyx-tube 1–1.4 mm long; limb tube 0.1–0.4 mm long; lobes 4, 2 larger ones oblong-lanceolate, 1.3–3 mm long, and 2 smaller ones, 0.5–1.5 mm long. Corolla white; tube 1.4–1.9 mm long;

Plate 37 A, B. *Mitracarpus hirtus*. Photo by YDZ.

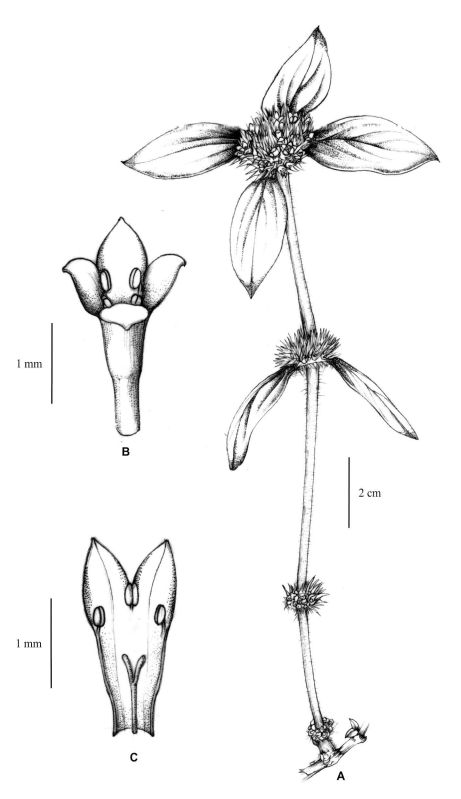

Figure 28 *Mitracarpus hirtus*. A. flowering branch; B. flower; C. longitudinal section of corolla, showing stamens and style. Drawn by NJ.

lobes ovate, 0.6–1 mm long. Style 1.1–1.6 mm long; stigma 0.3–0.5 mm long. Capsules subglobose, ca. 1 mm in diamter. Seeds pale yellow-brown, oblate-suboblong, ca. 0.8 mm long.

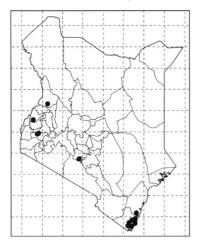

Distribution: Western, central and coastal Kenya. [Native to the Antilles and central, north, and south America; naturalized in tropical Africa, Asia, Australia, and Pacific islands].

Habitat: Disturbed areas; up to 1500 m.

Bungoma: near Uganda border, *Tweedie 2256* (K). Kilifi: Kaloleni, *Fufuoka K-262* (EA). Kisumu: ca. 4 miles north of Kisumu, *Drummond & Hemsley 4486* (EA). Kwale: Shimba Hills National Reserve, *Hiepko 2650* (EA). Nairobi: East Africa Herbarium, near Lois Leakey's Statue, *Malombe & Kimeu 1344* (EA). West Pokot: Cherangani Hills, Kacherop Forest, *SAJIT Z0008* (HIB).

30. **Diodella** Small

Annual or perennial scrambling herbs or subshrubs; stems always 4-angular and pubescent on the angles. Leaves opposite, sessile or shortly petiolate, linear, lanceolate to ovate; stipules with base united to the petiole, divided into fimbriae. Flowers mostly few in small axillary clusters. Calyx-tube ellipsoid, ovoid or obconic, pubescent to hairy; lobes 2–4, persistent. Corolla-tube funnel-shaped, glabrous or pubescent outside; lobes 4, valvate. Stamens 4, exserted; filaments inserted at the throat. Ovary 2-locular; ovules single in each locule; style exserted; stigmas 2, capitate. Fruits of 2 indehiscent cocci. Seeds 2, obovoid, oblong or compressed ellopsoid, with a broad ventral groove; endosperm corneous; embryo straight.

About 12 species widely distributed in tropical America and Africa; only one species in Kenya.

1. **Diodella sarmentosa** (Sw.) Bacigalupo & Cabral, Darwiniana 44(1): 100. 2006. ≡ *Diodia sarmentosa* Sw., Prodr. 30. 1788; F.T.E.A. Rubiac. 1: 336. 1976. —Type: Jamaica, *Swartz s.n.* [holotype: S] (Figure 29; Plate 38)

Straggling, scrambling or procumbent herb up to 3.6 m long; stems 4-angular, pubescent on the angles. Leaves elliptic, 1.8–6.3 cm × 0.7–2.8 cm, acute at the apex, narrowed to the base; petioles 1–5 mm long; stipule-sheath 1–2 mm long with lines of hairs, bearing 5–7 setae 1–7 mm long. Flowers usually few in axillary clusters at most nodes. Calyx-tube obconic, 1.5–2 mm long; lobes

4, oblong-lanceolate or narrowly triangular, 1.5–3 mm long. Corolla mauve or white; tube glabrous, funnel-shaped, 1.8 mm long; lobes triangular, 1 mm long. Ovary 2-locular; style slightly exserted, minutely papillate. Fruits oblong-ellipsoid, 3.5–5 mm long. Seeds compressed ellipsoid, 2–4 mm long, with a broad ventral groove, finely rugulose.

Distribution: Western and coastal Kenya. [Widespread in tropical Africa and America].

Habitat: Evergreen forests; up to 1600 m.

Kakamega: Kakamega Forest, *FOKP925* (EA). Kisii: ca. 17 km southeast of Kisii, *Vuyk & Breteler 128* (EA). Kwale: Shimba Hills, *SAJIT 005989* (HIB); Tiomin Kwale Mine Mukurumudzi, *Luke et al. 5899* (EA).

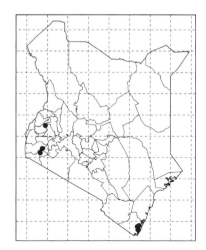

Plate 38 A, B. *Diodella sarmentosa*. Photo by VMN (A) and GWH (B).

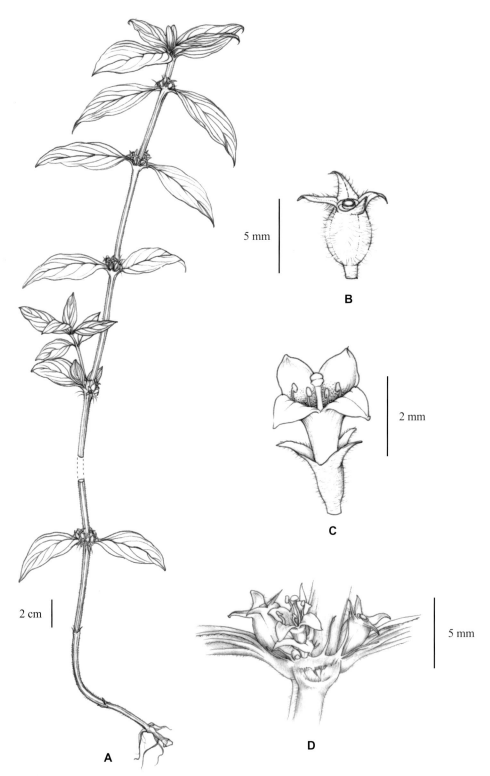

Figure 29 *Diodella sarmentosa*. A. plant; B. portion of branch showing the stipule, flower and young fruit; C. flower; D. fruit. Drawn by NJ.

31. **Spermacoce** L.

Herbs or rarely small subshrubs. Raphides present. Leaves opposite or sometimes borne on very short axillary stems and appearing whorled or fascicled, without domatia, sessile or petiolate; stipules ofen fused to petiole bases or leaves, truncate to rouded or broadly triangular, with 1 to many filiform fimbriae. Inflorescences of axillary clusters or rarely terminal capitula, several to many flowered, sessile, bracteate; bracts usually filiform-laciniate or stipuliform. Flowers bisexual, sessile or subsessile. Calyx-tube obovoid, turbinate or obconic; limb-lobes 2–4(–8), triangular, oblong or lanceolate. Corolla often white, funnel- or salver-shaped, pubescent throughout or just in throat; lobes (3–)4, valvate. Stamens 4, the filaments inserted in the tube or at the throat, included or exserted. Ovary 2-locular, ovule single in each locule; stigma capitate or 2-lobed. Fruits capsular, ellipsoid to subglobose, 2-valved. Seeds oblong, ellipsoid or ovoid.

A genus with about 280 species widespread in tropical and warm temperate regions of the world; 10 species in Kenya, including two unnamed species, *Spermacoce* sp. A [*Polhill & Paulo 930* (EA); *Greenway 10431* (EA)] and sp. B [*Bally 8546* (EA); *Robertson & Luke 6064* (EA)], which were preliminarily described in FTEA. Because we did not conduct a field investigation about these two species, they are not induded in the key and descriptions.

1a. Capsules always splitting from the base .... 10. *S. sphaerostigma*
1b. Capsules splitting from the apex 2
2a. Perennial herb with many decumbent to suberect stems from a woody usually multi-headed rootstock .... 7. *S. minutiflora*
2b. Annual or if perennial then with only a few stems from a small root stock 3
3a. Flowers in spicate or capitate heads 4
3b. Flowers not in spikes or heads, the flowering nodes well-separated 6
4a. Flower nodes often condensed into a terminal spike .... 2. *S. subvulgata*
4b. Flower nodes condensed into heads supported by numerous associated leaves 3
5a. Leaves with a pale margin and mid-rib; heads with few or no bracteoles amongst the flowers.
 .... 3. *S. chaetocephala*
5b. Leaves without a thick pale margin; heads with numerous thin bracteoles amongst the flowers
 .... 4. *S. radiata*
6a. Corolla-tube less than 3 mm long 7
6b. Corolla-tube more than 3 mm long 9
7a. Leaf-blades elliptic .... 6. *S. laevis*
7b. Leaf-blades linear-lanceolate 8
8a. Flower nodes 0.5–2 cm wide, with numerous flowers and many reddish brown setiform bracteoles .... 3. *S. chaetocephala*
8b. Flower nodes mostly small and under 1 cm wide, with far fewer flowers and less conspicuous bracteoles .... 5. *S. pusilla*

9a. Perennial herb with stems scrambling or decumbent .... 8. *S. princeae*
9b. Annual unbranched or sparsely branched herb with erect stems... 10
10a. Calyx-lobes usually 4 .... 9. *S. senensis*
10b. Calyx-lobes 2 .... 1. *S. filituba*

1. **Spermacoce filituba** (K. Schum.) Verdc., Kew Bull. 30(2): 302. 1975; F.T.E.A. Rubiac. 1: 349. 1976. ≡ *Borreria filituba* K. Schum., Bot. Jahrb. Syst. 28: 110. 1899. —Type: Kenya, Lamu, 29 Apr. 1896, *F. Thomas 213* [holotype: B (destroyed); lectotype: K (K000311691), **designated here**] (Plate 39A–C)

Annual herb, up to 60 cm tall, with 4-angled pubescent stems. Leaf-blades narrowly lanceolate, 15–55 mm × 2.5–11 mm, acute at the apex, rounded at the base, densely pubescent; stipule-sheath 2–3 mm long, with 8–10 fimbriae 6–9 mm long. Flowers congested into clusters at the nodes scattered along the main stems or lateral branches. Calyx-tube conic or ellipsoid, ca. 1.8 mm long; lobes 2, lanceolate, 2–3.5 mm long. Corolla white to violet blue; tube narrow cylindrical, (5–)10–12 mm long; lobes lanceolate, ca. 2 mm long. Capsules ellipsoid, 2.5–3 mm long.

Distribution: Central, eastern, southern, and coastal Kenya. [East Africa from Ethiopia and Somalia to Tanzania].

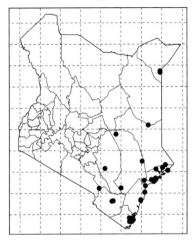

Habitat: Forest edges, bushlands, woodlands, grasslands, or roadsides; up to 1600 m.

Garissa: Boni Forest, *Kuchar 13549* (EA). Kajiado: Chyulu Hills, *Luke & Luke 10300* (EA). Kilifi: Arabuko-Sokoke Forest, *Luke & Robertson 2593* (EA). Kitui: Mutomo Hill, *SAJIT MU0055* (HIB). Kwale: Kaya Kambe, *Robertson & Luke 6279* (EA); Shimba Hills, *Magogo & Glover 715* (EA). Lamu: Kiwayu Area, *Luke & Luke 5670* (EA). Mandera: ca. 12 km south of El Wak on Wajir Road, *Gilbert et al. 1653* (EA). Teita Taveta: Ndololo, *Hucks 688* (EA). Tana River: Kora Game Reseve, *Mungai et al. 269/83* (EA). Wajir: Dadaab-Wajir Road, just north of Lagh Dera, *Gillett et al. 20636* (EA).

2. **Spermacoce subvulgata** (K. Schum.) J.G. García, Mem Junta Invest. Ultramar, 2 Ser. 6: 49. 1959; F.T.E.A. Rubiac. 1: 352. 1976; U.K.W.F.F.: 227. 2013. ≡ *Borreria subvulgata* K. Schum., Bot. Jahrb. Syst. 28: 111. 1899. —Types: Zimbabwe, near Umtali, 6 Apr. 1898, *R. Schlechter 12184* [syntype: B (destroyed); lectotype: K (K000173017), **designated here**; isolectotypes: BOL (BOL 138646), BR (BR0000008207821), HBG (HBG521827), P (P00462447), WAG (WAG 0003069)]

= *Tardavel kaessneri* S. Moore, J. Bot. 43: 250. 1905. —Type: Kenya, Machakos, Sani, 18 May 1902, *T. Kassner 749* [holotype: BM (BM000903631)]

Annual or perennial erect herb, up to

90 cm tall. Leaf-blades linear, 3–6 cm × 1–3 mm, acute at the apex, narrowed to the base; stipule-sheath 3 mm long, with 6–11 fimbriae 3–6.5 mm long. Flowering nodes often condensed into a terminal spike. Calyx-tube ca. 1.5 mm long, ellipsoid; lobes linear, 2(–4), 3–4.5 mm long. Corolla white or more rarely blue; tube narrowly funnel-shaped, 4–5 mm long; lobes lanceolate, 2.5–5 mm long. Capsules oblong-ellipsoid, 3–4 mm long.

Distribution: Southern Kenya. [Tropical Africa, from Kenya to Namibia].

Habitat: Hillsides; 1250–1300 m.

Kajiado: Chyulu Hills, *Barry 1043* (EA).

long. Calyx-tube ellipsoid, ca. 1.4 mm long; lobes 4, linear-lanceolate or slightly spathulate, 1–2 mm long. Corolla white or pale mauve; tube narrowly cylindrical with upper quarter funnel-shaped, ca. 1.3 mm long; lobes ovate-elliptic, ca. 0.8 mm long. Capsules oblong-ellipsoid, ca. 2.5 mm long.

Distribution: Western Kenya. [Tropical Africa].

Habitat: Roadsides, open bushlands, or woodlands; 1250–1700 m.

West Pokot: near Kongelai, *Tweedie 2877* (EA); near Mount Marobus, *Carter & Stannard 2* (EA).

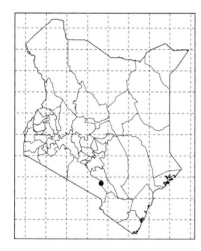

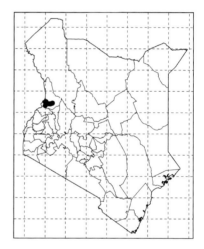

3. **Spermacoce chaetocephala** DC., Prodr. 4: 554. 1830; F.T.E.A. Rubiac. 1: 354. 1976; U.K.W.F.F.: 228. 2013. —Type: Senegal, Bakel, Sept. 1828, *Leprieur & Perrottet s.n.* [holotype: G-DC(G00667210)] (Plate 39D, E)

Annual erect herb, 60 cm tall. Leaf-blades linear to linear-lanceolate, 25–60 mm × 3–8 mm, acute at the apex, broadened at the base; stipule-sheath 1–2 mm long, with 5–7 setae up to 1.5 cm long. Flowers congested into usually quite large axillary heads; bracteoles numerous, reddish-brown, ca. 3 mm

4. **Spermacoce radiata** (DC.) Hiern, Fl. Trop. Afr. 3: 237 1877; F.T.E.A. Rubiac. 1: 356. 1976; U.K.W.F.F.: 227. 2013. ≡ *Borreria radiata* DC., Prodr. 4: 542. 1830. —Types: Senegal, Galam, 1825, *Sieber 8* [syntype: G-DC (G00667095); isosyntypes: BR (BR0000008203984), K (K000422900), M (M0106212)]; Walo, 1829, *Perrottet & Leprieur s.n.* [syntype: G-DC (G00667095)]

Annual herb, 4(–100) cm tall; stems densely covered with spreading white hairs. Leaf-blades very narrowly elliptic-lanceolate, 2–

Plate 39 A–C. *Spermacoce filituba*; D, E. *S. chaetocephala*. Photo by GWH.

5.5 cm × 1.2–5.5 mm, acute at the apex, narrowed at the base into the stipule-sheath; stipules often white, bearing 5–9 fimbriae 6.5 mm long. Flowers in very dense terminal heads consisting of several condensed nodes, the leaves of which form radiating bracts. Calyx-tube narrowly oblong, 2 mm long, hairy above; lobes 4, filiform, 1.5 mm long, hairy. Corolla greenish or white; tube funnel-shaped, ca. 1 mm long; lobes ovate-triangular, 0.6 mm long. Capsules compressed-cylindrical, 3 mm long.

Distribution: Western Kenya. [Senegal to Cameroon and Sudan, also extend to southern Africa].

Habitat: Roadsides or grasslands; ca. 1300 m.

West Pokot: Kacheliba, *Leippert 5071a* (K?).

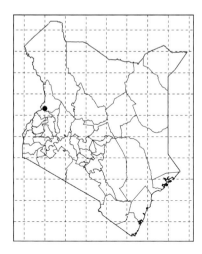

5. **Spermacoce pusilla** Wall., Fl. Ind. 1: 379. 1820; F.T.E.A. Rubiac. 1: 356. 1976; U.K.W.F.F.: 228. 2013. —Type: unknown (Plate 40)

Annual herb, up to 60 cm tall. Leaf-blades linear-lanceolate to narrowly lanceolate, 10–53 cm × 2–5.5 mm, acute at the apex, narrowed to the base, stipule-sheath 1.5–2 mm long, with 5–7 setae 2–4 mm long. Flowers congested in dense spherical clusters; bracteoles filiform, numerous, ca. 2 mm long. Calyx-tube ovoid, ca. 1 mm long, 4-toothed. Corolla white or pink; tube narrowly funnel-shaped, ca. 1.3 mm long; lobes 0.8–1.1 mm long. Capsules ellipsoid, ca. 1.5 mm long.

Distribution: Western, central, and coastal Kenya. [Native in tropical and subtropical Asia and India, exotic in Africa].

Habitat: Forests, hillsides, woodlands, bushlands, or grasslands with scattered trees and shrubs, or in old cultivations; up to 2050 m.

Baringo: road from Eldama Ravine to Nakuru, *Gilbert & Mesfin 6418* (EA). Kakamega: Kakamega Forest, *Gilbert & Mesfin 6639A* (EA). Kisumu: Songhor, *Kokwaro 3740* (EA). Kitui: Ngutwa Village, *Malombe et al. 967* (EA). Kwale: Shimba

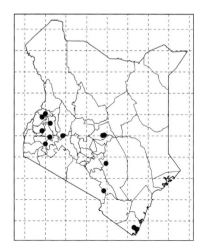

Hills, *SAJIT V0126* (HIB). Makueni: Chyulu Hills, *Gilbert 6244B* (EA). Meru: ca. 21 km southeast of Meru on road to Ishiara and Embu, *Gilbert et al. 5695* (EA). Trans Nzoia: Kitale, *Bogdan 4274* (EA). Uasin Gishu: ca. 13 km Eldoret-Kitale, *Gilbert & Mesfin 6512* (EA).

6. **Spermacoce laevis** Lam., Tabl. Encycl.

Plate 40 *Spermacoce pusilla*. Photo by GWH.

1: 273. 1792; F.T.E.A. Rubiac. 1: 357. 1976. —Type: Dominicana, Santo Domingo, J. Martin 102 [holotype: P-LAM (P00308647)]

Annual, erect or more often scrambling herb, up to 1.2 m tall. Leaf-blades narrowly elliptic or elliptic-lanceolate to ovate, 0.8–5.5 cm × 0.3–2.5 cm, acute at the apex, cuneate at the base, glabrous; stipules rather bright reddish brown when dry; sheath slightly pubescent, 2–3 mm long, with 5–7 fine fimbriae 2.5 mm long. Flowers congested into terminal and axillary clusters. Calyx-tube narrowly obconic, 2–2.5 mm long; lobes triangular, ca. 0.6 mm long. Corolla white; tube ca. 1.2 mm long; lobes lanceolate, ca. 1.3 mm long. Capsules ellipsoid or obovoid to fusiform, 2.5–4 mm long.

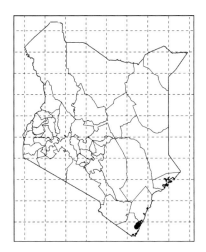

Distribution: Coastal Kenya. [Exotic from Mexico, Caribbean and South Tropical America].

Habitat: Lowland forests, woodlands, and open grasslands; 100–400 m.

Kwale: Shimba Hills, *Magogo & Glover 661* (EA).

7. **Spermacoce minutiflora** (K. Schum.) Verdc., Kew Bull. 30(2): 306. 1975; F.T.E.A. Rubiac. 1: 360. 1976; U.K.W.F.F.: 227. 2013. ≡ *Borreria minutiflora* K. Schum., Bot. Jahrb. Syst. 33: 373. 1903. —Type: Kenya, Nandi, *G.F. Scott-Elliot 6972* [holotype: B (destroyed); isotype: BM, **designated here**]

Perennial herb, with many decumbent to suberect stems from a woody rootstock. Leaf-blades narrowly to broadly elliptic or oblong-elliptic, 0.6–2 cm × 1–8 mm, rounded or obscurely acute at the apex, narrowed to the base, glabrous; petioles up to 1 mm long; stipule-sheath 1–2 mm long, with 3–5 fimbriae 0.5–2 mm long. Flowers congested into small terminal and axillary clusters. Calyx-tube obconic, ca. 1 mm long; lobes 4, oblong, 0.8–0.9 mm long. Corolla white; tube ca. 0.5 mm

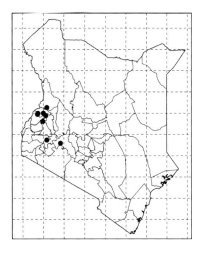

long; lobes oblong-triangular, ca. 1 mm long. Capsules oblong, ca. 2 mm long.

Distribution: Western Kenya. [Sudan, South Sudan, and Uganda].

Habitat: Woody grasslands; 1500–2350 m.

Kericho: Kericho-Kisumu Road, *Gillett 19341* (EA). Nakuru: Mau, *Baker 342* (EA). Trans Nzoia: east of Mount Elgon, *Tweedie 39* (EA). Uasin Gishu: Kipkarren, *Brodhurst-Hill*

468 (EA). West Pokot: Cherangani, *Symes 38* (EA).

8. Spermacoce princeae (K. Schum.) Verdc., Kew Bull. 30: 307. 1975; F.T.E.A. Rubiac. 1: 362. 1976; U.K.W.F.F.: 227. 2013. ≡ *Borreria princeae* K. Schum., Bot. Jahrb. Syst. 34: 341. 1904. —Type: Tanzania, Kilimanjaro, 27 Oct. 1901, *C. Uhlig 252* [lectotype: EA (EA000001595), designated by B. Verdcourt in Kew Bull. 17(3): 500. 1964] (Plate 41A–C)

= *Spermacoce princeae* (K. Schum.) Verdc. var. *pubescens* (Hepper) Verdc. —Type: Equatorial Guinea, Moka, 6 Dec. 1951, *A.S. Boughey 32* [holotype: K (K000422885)]

Perennial scrambling or decumbent herb, up to 1.5 m long; stems square, sparsely to densely hairy. Leaf-blades elliptic to ovate, 1.2–7 cm × 0.3–3 cm, acute at the apex, cuneate at the base, glabrous to pubescent above, sparsely hairy to pubescent beneath; petioles 0–1 mm long; stipule-sheath 4–6 mm long, with 5–9 fimbriae 7–8.5 mm long. Flowers congested into dense axillary clusters. Calyx-tube fusiform or obconic, 1.2–3 mm long; lobes (2–)4(–5), foliaceous, 3.5–5(–9) mm long. Corolla white; tube cylindrical or narrowly funnel-shaped, 5–10.5 mm long; lobes oblong or elliptic, 3–4(–6) mm long. Capsules oblong-ellipsoid, 5–6 mm long.

Distribution: Western Kenya, central, and southern Kenya. [Tropical Africa].

Habitat: Forests, bamboo zones, bushlands, riverine grasslands, or swamp-sides; 1400–2650 m.

Baringo: Katimok Forest, *Dale 2445* (EA). Bomet: Chemasingi Tea Estate, *Perdue & Kibuwa 9296* (EA). Elgeyo-Marakwet: Marakwet, *Brodhurst-Hill 257* (EA). Kajiado: Ngong Hills, *Fukuoka K68* (EA). Kericho: Tinderet Forest, *Perdue & Kibuwa 9191* (EA). Kiambu: Tigoni, *Luke & Luke 8371* (EA); near Limuru, *Fukuoka K115* (EA); Thika River Fishing Camp, *Bally B6452* (EA). Kirinyaga: south Mount Kenya, *SAJIT 003034* (HIB); near Castle Forest Station, *Perdue et al. 8374* (EA). Kisii: near Kisii, *Law 66* (EA). Makueni: ca. 1 km south of Nunguni Trading Center, *Mwangangi 2058* (EA). Meru: near Ontulili Forest Station, *SAJIT 003157* (HIB). Murang'a: ca. 10 km west from Gakoe, *Kuchar 8347* (EA). Nakuru: Enadabala, *Bally 4868* (EA). Narok: Endama, *Glover et al. 1947* (EA). Nandi: Kimondi, *Smith et al. s.n.* (EA). Nyeri: Kiandongoro Forest, *Mathenge 265* (EA). Teita Taveta: Taita Hills, Iyale Forest, *SAJIT 006410* (HIB). Trans Nzoia: Kitale, *Verdcourt 2465* (EA). Uasin Gishu: Hoey's Bridge, *Smith & Paulo 839* (EA). West Pokot: Cherangani Hills, *Thulin & Tidigs 274* (EA).

9. Spermacoce senensis (Klotzsch) Hiern, Fl. Trop. Afr. 3: 236. 1877; F.T.E.A. Rubiac. 1: 365. 1976; U.K.W.F.F.: 228. 2013. ≡ *Diodia senensis* Klotzsch, Naturw. Reise Mossambique 1: 289. 1861. —Type: Mozambique, Tete, Rios de Sena, *W. Peters s.n.* [holotype: B (destroyed)] (Plate

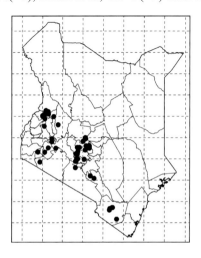

41D)

Annual herb up to 60 cm tall, with 4-angled densely hairy stems. Leaf-blades narrowly elliptic to elliptic-lanceolate, 2–7 cm × 0.3–2.2 cm, acute at the apex, narrowly cuneate at the base, hairy on both surfaces and scabrid at the margins; stipule-sheath 3 mm long, densely hairy, with 7 setae 0.2–1.1 cm long. Flowers congested into clusters at the nodes. Calyx-tube subcylindric, ca. 2.5 mm long; lobes 4, lanceolate to narrowly oblong, 2–4 mm long. Corolla white; tube funnel-shaped, 3–8 mm long; lobes triangular, 1.5–4.5 mm long. Capsules ellipsoid, ca. 3 mm long, densely hairy.

Distribution: Central and southern Kenya. [East Africa, from Ethiopia to Namibia and Mozambique].

Habitat: Bushlands, roadsides, or slopes of dams; 550–1250 m.

Embu: Mwea National Reserve, *Mwangangi et al. 5018* (EA). Kitui: Mutha Matikoni Village, *Kirika et al. NMK268* (EA). Machakos: Kindaruma Dam, *Gillett & Faden 18120* (EA). Makueni: Wote, *Muasya et al. NMK681* (EA). Teita Taveta: Ndara Hill, *Luke & Luke 4553* (EA). Tharaka-Nithi: ca. 24 km north of Ishiara, *Gilbert et al. 5712* (EA).

10. **Spermacoce sphaerostigma** (A. Rich.) Oliv., Trans. Linn. Soc. London 29: 88. 1873; F.T.E.A. Rubiac. 1: 367. 1976; U.K.W.F.F.: 227. 2013. ≡ *Hypodematium sphaerostigma* A. Rich., Tent. Fl. Abyss. 1: 348. 1847. —Types: Ethiopia, near Adua (Adowa), 30 Sept. 1837, *G.H.W. Schimper 100* [syntype: P (P00462457); isosyntypes: BM, BR (BR0000008358745), FI (FI000750), HBG (HBG521453), HOH (HOH 009714), JE (JE00000360), K (K000422866), LG (LG0000090029875 & LG0000090029776), M (M0186955 & M0186956), MPU (MPU 022590, MPU022591 & MPU022592), P (P0046 2456 & P00462458), REG (REG000626), TUB (TUB004618 & TUB004619), WAG (WAG000 3068)]; Ethiopia, Shire (Chiré), R. *Quartin-Dillon s.n.* [syntype: P]

Annual erect herb, up to 90 cm tall; stems covered with spreading hairs. Leaf-blades lanceolate to elliptic-lanceolate or narrowly ovate, 1–6 cm × 0.3–2 cm, acute at the apex, cuneate at the base, sparsely to distinctly hairy on both surfaces; stipule-sheath 1–4.5 mm long, hairy, with 5–7 setae 2–8 cm long. Flowers mostly congested into usually quite large terminal and

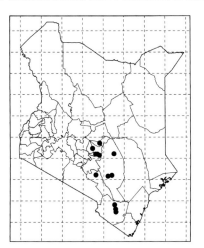
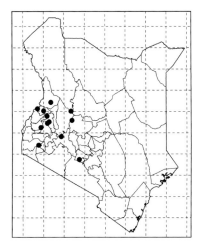

Plate 41 A–C. *Spermacoce princeae*; D. *S. senensis*. Photo by GWH (A–C) and NW (D).

axillary heads. Calyx-tube obconic, 2.5–3.5 mm long; lobes 4, ovate to lanceolate, 2.5–5 mm long. Corolla white or pale mauve; tube funnel-shaped, 5–8.5 mm long; lobes triangular, 3–4 mm long. Capsules ovoid, 3–4 mm long.

Distribution: Western Kenya. [Arabian Peninsula and tropical Africa].

Habitat: Forests, bushlands, roadsides, or in old cultivations; 1650–1900 m.

Kakamega: Kakamega Forest, Kokwaro 3761 (EA). Kisii: *Napier s.n.* (EA). Laikipia: Lakipia Ranch, *Brett 28* (EA). Nairobi: Kabete, *Hudson 298* (EA). Nakuru: Rongai, *van Someren s.n.* (EA). Trans Nzoia: Kitale, *Bogdan 5557* (EA). Turkana: Amaya, *Powys 221* (EA). Uasin Gishu: ca. 4 miles from Eldoret, *Agnew et al. 10608* (EA). West Pokot: Ortum, *Agnew et al. 10243* (EA).

32. **Richardia** L.

Annual or perennial erect or prostrate herbs. Leaves opposite, sessile or shortly petiolate, mostly ovate, elliptic or oblong; stipule-sheath connate with the petioles, bearing several fimbriae. Inflorescences terminal, capitate, several- to many-flowered, enclosed by 2–4 leaflike bracts. Calyx-tube turbinate or subglobose; lobes 4–8, lanceolate, ovate or subulate, persistent. Corolla shortly funnel-shaped; lobes 3–6, ovate to lanceolate, valvate. Stamens 3–6, inserted in corolla throat, exserted. Ovary 3–6-locular; ovule solitary in each locule; style filiform; stigmas 3–4, linear or spathulate, exserted. Capsules 3–4-coccous; cocci obovoid, indehiscent, subcrustaceous. Seeds oblong-ellipsoid or obovoid, dorsally convex, ventrally with 2 grooves; endosperm corneous.

About 16 species widespread in the Antilles, north and south America, several species naturalized in the Old World tropics; two species in Kenya.

1a. Corolla-tube 5.8–6.3 mm long, lobes ca. 2.5 mm long; cocci inner face dorsiventrally flattened, with 2 broad parallel depressions .... 1. *R. scabra*
1b. Corolla-tube 2.7–3.2 mm long, lobes 1–1.4 mm long; cocci inner face triangular to rounded, with 1 narrow groove ....2. *R. brasiliensis*

1. **Richardia scabra** L., Sp. Pl. 1: 330. 1753; F.T.E.A. Rubiac. 1: 380. 1976. —Type: Mexico, Vera Cruz, *Herb. Linn. No. 451.1* [lectotype: LINN, designated by Lewis & Oliver in Brittonia 26: 282. 1974] (Figure 30; Plate 42A–C)

Annual herb, with erect or ascending stems up to 80 cm long. Leaf-blades lanceolate, elliptic or ovate, 1–5 cm × 0.5–2.5 cm, base acute to cuneate, apex acute to obtuse, stipule-sheath 2–4 mm long, with 3–7 fimbriae 1–4 mm long. Inflorescences terminal, capitate,

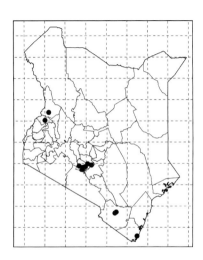

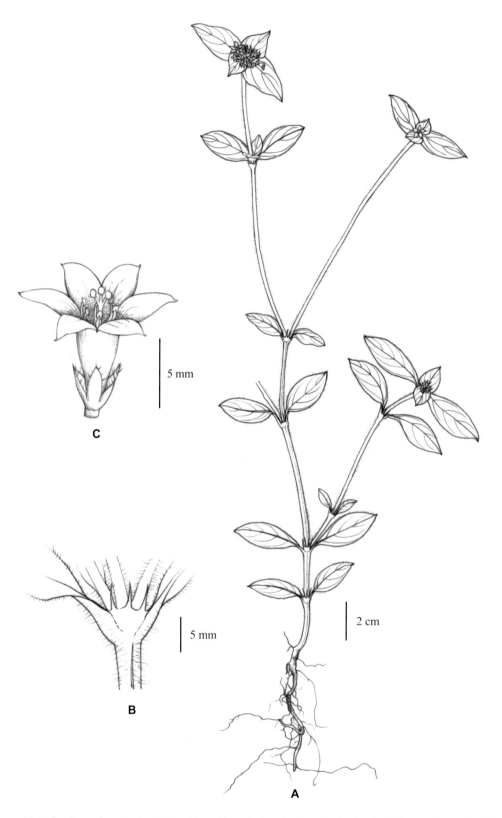

Figure 30 *Richardia scabra*. A. plant; B. portion of branch showing the stipule-sheath; C. flower. Drawn by NJ.

0.6–1.6 cm in diameter; bracts ciliate, lanceolate, 0.5–2.7 cm long. Calyx-tube ovate to obovate, up to 1 mm long; lobes triangular to oblong-lanceolate, 1.8–2.3 mm long. Corolla white; tube ca. 6 mm long; lobes ovate-triangular, ca. 2.5 mm long. Ovary 3-locular; style filiform, ca. 7.5 mm long; stigmas 3, ca. 0.7 mm long. Capsules 3-coccous; cocci 2.5–2.8 mm long, inner face somewhat dorsiventrally flattened, with 2 broad parallel depressions.

Distribution: Western, central, southern, and coastal Kenya. [Tropics].

Habitat: Grasslands, bushlands, or old cultivations and other disturbed places; up to 2200 m.

Kiambu: Tigoni, *Luke et al. 8368* (EA); Thika, *Faden 67/214* (EA). Kwale: Gongoni Forest, *SAJIT 006169* (HIB). Nairobi: Muthaiga, *Gillett 17316* (EA, K). Teita Taveta: Mbololo Hill Forest, *Kamau et al. 321* (EA). Trans-Nzoia: Kitale, *Tweedie 2875* (EA). West Pokot: north of Chepareria, *Hansson & Thomasson 28* (EA).

2. **Richardia brasiliensis** Gomes, Mem. Ipecacuanha Bras. 31. 1801; F.T.E.A. Rubiac. 1: 378. 1976; U.K.W.F.F.: 227. 2013; C.P.K.: 386. 2016. —Type: Brazil, Rio de Janeiro, *Gomes s.n.* [holotype: LISU] (Plate 42D–E)

Annual or perennial prostrate herb from a thick rootstock, with stems up to 40 cm long. Leaf-blades elliptic to ovate, 1–6.5 cm × 0.4–2.7 cm, acute at the apex, cuneate at the base; petioles up to 1.5 cm long; stipules sheathing at the base, with 3–5 fimbriae 1–4 mm long. Inflorescences terminal, capitate, 0.7–1.2 cm in diameter; bracts ciliate, ovate-elliptic, 1–3.5 cm long. Calyx-tube 1.2–1.7 mm long; lobes ovate-triangular, 1–1.5 mm long. Corolla white; tube 2.7–3.2 mm long; lobes 4–6, 1–1.4 mm long. Ovary 3-locular; style filiform, 3–4 mm long; stigmas 3, minute. Capsules 3-coccous; cocci oblong-obovoid, 2–2.6 mm long, inner face triangular to somewhat rounded, with 1 narrow grove.

Distribution: Western, central and southern Kenya. [Tropics].

Habitat: Grasslands, forest edges, or roadsides and other disturbed places; up to 2000 m.

Kiambu: Tigoni, *Granville 15* (EA). Meru: Ngaia Forest, *Luke et al. 11721* (EA). Murang'a: Fort Hall, *Mberi 191* (EA). Nairobi: near Kabete, *Patel in E.A.H. 14287* (EA). Teita Taveta: Taita Hills, *SAJIT 004557* (HIB). **Trans-Nzoia,** Saiwa Swamp National Park, *Kirika 122* (EA).

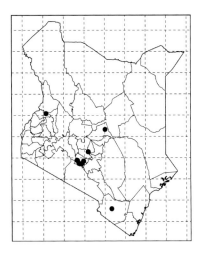

Plate 42 A–C. *Richardia scabra*; D, E. *R. brasiliensis*. Photo by GWH.

7. Trib. **Anthospermeae** Cham. & Schlecht.

Herbs or shrubs. Leaves opposite or whorled; stipules free or often connate with petioles to form sheaths. Flowers often unisexual. Corolla valvate in bud; stamens inserted at the throat, base or near the base of corolla. Ovary 1–4-locular; ovule solitary in each locule, erect. Fruits dry, dividing into cocci or capsular.

Only one genus occurs in Kenya.

33. **Anthospermum** L.

Herbs or small shrubs. Leaves opposite or ternately verticillate; leaf-blades linear or lanceolate, rarely ovate or oblong; stipule-sheath adnate to the petioles at the base with single tooth or rarely 2–3 teeth. Flowers small, dioecious, polygamous or hermaphrodite, axillary, sessile or rarely in panicles; bracts 2–3, minute. Calyx-tube ovoid, ellipsoid or obovoid; limb minute, 4–5-toothed. Hermaphrodite. Male flowers: corolla-tube campanulate, funnel-shaped or cylindrical, lobes 4–5, linear to elliptic-lanceolate; stamens exserted. Female flowers: corolla-tube very minute, lobes 2–4; styles 2, or joined at the base; stigmas 2, long exserted, hairy; ovary 2-locular; ovule single in each locule. Fruits 2-coccous; cocci compressed ellipsoid, ventrally plane or grooved, dorsally convex, indehiscent or sometimes dehiscing ventrally. Seeds erect; radicle inferior.

About 40 species mostly distributed in South Africa, several in east and south-central Africa, and a few in Madagascar; four species in Kenya.

1a. Straggling, trailing or ascending annual or perennial herbs; leaf-blades lanceolate to lanceolate-ovate, up to 1.6 cm wide 1
1b. Erect annual or perennial herbs, subshrubs or shrubs; leaf-blades mostly linear or narrowly lanceolate, less than 0.5 cm wide 3
2a. Fruits glabrous or minutely muriculate .... 1. *A. herbaceum*
2b. Fruits densely covered with spreading white hairs .... 2. *A. villosicarpum*
3a. Leaf-blades mostly less than 1.5 cm long, with inrolled margins; always found in typical heath zone of high mountains .... 3. *A. usambarense*
3b. Leaf-blades up to 4 cm long, flat; always found in rocky ground of lower altitude grassland or thicket .... 4. *A. welwitschii*

1. **Anthospermum herbaceum** L. f., Suppl. Pl. 440. 1781 [1782]; F.T.E.A. Rubiac. 1: 325. 1976; U.K.W.F.F.: 227. 2013. —Type: South Africa, Cape of Good Hope, *C.P. Thunberg 317* [holotype: (LINN HL1233.5); isotype: UPS] (Figure 31; Plate 43A, B)

Straggling, trailing or ascending annual or perennial herb, up to 2 m long/tall. Leaves opposite; blades lanceolate to lanceolate-ovate,

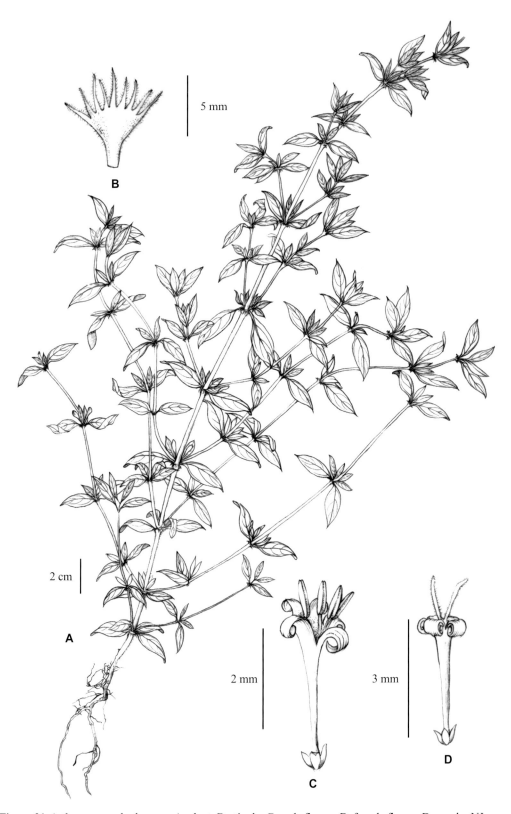

Figure 31 *Anthospermum herbaceum*. A. plant; B. stipule; C. male flower; D. female flower. Drawn by NJ.

0.5–5.5 cm × 0.1–2.5 cm, acute at the apex, cuneate to rounded at the base; stipules with 3–5 linear fimbriae 1.5–5.5 mm long. Flowers several in axillary, sessile or somewhat elongated clusters. Hermaphrodite. Male flowers: calyx-tube oblong, 1.5–4 mm long, lobes 1.5–4 mm long, anthers 1–2.5 mm long; calyx-tube and lobes as in male, stigmas 2, 2–6.5 mm long, ovary 0.5–1.2 mm long. Female flowers: calyx-tube 0.3–1.2 mm long, lobes 0.3–1 mm long, stigmas 2, 2–10 mm long, ovary as in hermaphrodite. Fruits yellowish-brown or reddish-brown, oblong to ovoid, 2-coccous, 2–3 mm long. Seeds ellipsoid, 1.8–2.8 mm long.

Distribution: Western, central and southern Kenya. [Arabian Peninsula, Eritrea to southern Africa].

Habitat: Forest edges, woodlands, scrubs, thickets, heathlands, grasslands, roadsides, and old cultivations; 900–2650 m.

Baringo: Eldama Ravine, *Whyte 1898* (K). Elgeyo-Marakwet: Kipkunurr Forest Reserve, *Hopper & Field 5001* (K). Kajiado: Chyulu Hills, *Gilbert 6198* (EA, K). Kericho: near Malagat Forest station, *Perdue & Kibuwa 9194* (EA, K). Kiambu: Muguga, *Verdcourt 728* (EA, K). Kisii: Marongo Ridge, *Vuyk 212* (EA). Makueni: Chyulu Hills, *Luke 2979* (EA, K). Murang'a: Kimakia Forest Reserve, *Kerfoot 477* (EA). Nakuru: west of Mau Forest, *Gillett 19030* (EA). Narok: Ol'Pusimoru Sawmill, *Glover et al. 1395* (EA, K). Nyamira: Magombo Valley, *Plaizier 1205* (WAG). Nyandarua: west edge of Kinangop Plateau, *Kuchar 12313* (EA). Samburu: Mathew's Range, *Luke 14267* (EA). Teita Taveta: Yale Forest, *SAJIT 004548* (HIB). Tharaka-Nithi: Chogoria, *Kahurananga 869* (EA). Trans-Nzoia: Saiwa Swamp National Park, *Pearce 711* (EA). West Pokot: Sekerr, *Agnew et al. 10345* (EA).

2. Anthospermum villosicarpum (Verdc.) Puff, Biosyst. Study African & Madagascan Rubiac.-Anthosperm. (Pl. Syst. Evol., Supp. 3): 292. 1986; F.T.E.A. Rubiac. 3: 927. 1976. ≡ *Anthospermum herbaceum* L. f. var. *villosicarpum* Verdc., Kew Bull. 30(2): 299. 1975; F.T.E.A. Rubiac. 1: 328. 1976. —Type: Kenya, Marsabit, Furroli, 20 Sept. 1952, *J.B. Gillett 13946* [holotype: K (K000318618)]

Shrub, subshrub or woody herb, up to 0.6 m tall, with densely spreading pubescent stems. Leaves 3-whorled; blades narrowly elliptic-lanceolate, 0.5–1.3 cm × 0.1–0.4 cm,

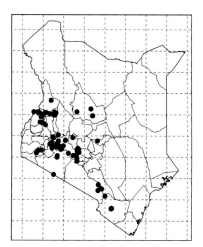
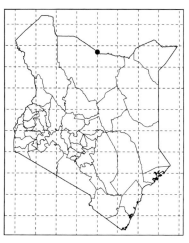

acuminate to mucronate at the apex, narrowed to the base, with bristly hairs on both sides. Stipule-sheath with a small seta. Flowers 4-merous, bisexual, several in axillary, sessile or somewhat elongated clusters. Calyx-lobes subobsolete. Corolla greenish to yellow-greenish, tube narrowly funnel-shaped, 1–2 mm long. Fruits densely covered with spreading white hairs.

Distribution: Northern Kenya. [Ethiopia].

Habitat: Plains, open deserts, grass-deserts, scrubs, or mountain slopes; 950–1900 m.

Marsabit: Furroli, *Gillett 13946* (K).

3. **Anthospermum usambarense** K. Schum., Bot. Jahrb. Syst. 28: 112. 1899; F.T.E.A. Rubiac. 1: 331. 1976; K.T.S.L.: 501. 1994; U.K.W.F.F.: 227. 2013. —Types: Tanzania, western Usambara Mts., Mar. 1892, *C. Holst 420* [syntype: B (destroyed)]; Tanzania, Kwai, *Eich s.n.* [syntype: B (destroyed)]; Tanzania, Usambara Mts., Matondwe Hill, head of Kwai Valley, 2000 m, 28 Feb. 1953, *R.B. Drummond & J.H. Hemsley 1351* [neotypes: B, BR, K, S, designated by C. Puff in Biosyst. Study African & Madagascan Rubiac.-Anthosperm. (Pl. Syst. Evol., Supp. 3): 196. 1986; second-step neotype: K, **designated here**; isoneotypes: B, BR (BR0000017843881), S] (Plate 43C–E)

= *Anthospermum aberdaricum* K. Krause, Notizbl. Bot. Gart. Berlin-Dahlem 10: 609. 1929. —Type: Kenya, Aberdares, Kinangop summit, 4000 m, 21 Mar. 1922, *R.E. & T.C.E. Fries 2574* [lectotype: S (S-G-504), designated by Puff in Biosyst. Study African & Madagascan Rubiac.-Anthosperm. (Pl. Syst. Evol., Supp. 3): 196. 1986; isolectotypes: BR (BR0000008826800), K (K000422836), S]

Shrub, up to 4.5 m tall. Leaves opposite or 3–6(–many)-whorled; blades narrowly obovate, oblanceolate to linear-lanceolate, 0.2–3.5 cm × 2–5 mm, margins often revolute, apex apiculate, base narrowed, glabrous, sparsely ciliate or hairy on mid-rib beneath; stipules with 1–7 setae 0.6–5 mm long. Flowers in sessile axillary clusters, unisexual and dioecious. Male flowers: calyx-lobes 4, very small; corolla-tube 0.7–1.6 mm long, funnel-shaped, lobes 1.4–2.7 mm long; anthers 0.7–1.7 mm long. Female flowers: calyx-lobes 4, subequal or 2 small and 2 larger; corolla-tube 0.3–0.7 mm long, lobes 0.2–0.7 mm long, style 0–0.5 mm long; stigmas 3–6 mm long; ovary 0.7–0.8 mm long. Fruits brown, oblong-ellipsoid, 1.5–2 mm long, crowned with the small calyx-lobes.

Distribution: Northwestern, western, central, and southern Kenya. [South Sudan to south tropical Africa].

Habitat: Mostly in heath zones of east Africa mountains, rarely extending to grasslands and forests; 1300–3800 m.

Bungoma: Mount Elgon, *Dale 3206* (EA, K). Elgeyo-Marakwet: Marakwet Hills, *Dale 896* (EA). Meru: Mount Kenya, *SAJIT 001615* (HIB). Kericho: southwest Mau Forest, *Whittall 214* (K). Nakuru: Mount Longonot, *Gilbert 6292* (K). Nandi: Nandi Forest, *Johnston 1901* (K). Narok: Entasekera, *Glover et al. 2119* (EA, K). Nyandarua: Aberdare

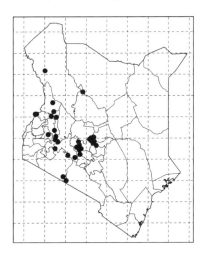

Plate 43 A, B. *Anthospermum herbaceum*; C–E. *A. usambarense*. Photo by GWH (A, B), BL (C, D) and YDZ (E).

Mountains, *SAJIT 006497* (HIB). Samburu: Mount Nyiru, *Kerfoot 1999* (EA, K). Turkana: Moruassigar, *Newbould 7192* (EA). Uasin Gishu: Timboroa Summit, *Bally B13898* (EA). West Pokot: Mount Sekerr, *Agnew et al. 10515* (EA).

4. **Anthospermum welwitschii** Hiern, Cat. Afr. Pl. 1(2): 500. 1898; F.T.E.A. Rubiac. 1: 330. 1976; K.T.S.L.: 502. 1994; U.K.W.F.F.: 227. 2013. —Type: Angola, Huila, Panda Forest, near Eme, Apr. 1860, *F.M.J. Welwitsch 5335* [holotype: LISU (LISU208673); isotypes: BM (BM000903647), G (G00014355), K (K000422833), P (P00462493)]

Perennial herb, subshrub or shrub, up to 1.8 m tall. Leaves opposite or pseudo-verticillate; blades oblanceolate elliptic to linear-lanceolate, 0.7–4 cm × 0.8–5 mm, acute or very shortly acuminate at the apex, narrowed to the base; petioles obsolete; stipule-sheath 1.5 mm long, hairy, with 1–5 setae, 0.7–4.5 mm long. Flowers in clusters of many at node. Male flowers: calyx-tube obconic, 0.5–1.2 mm long; lobes unequal, 0.5–1.5 mm long; corolla whitish or yellowish green, sometimes dotted with purple; tube funnel-shaped, 0.5–1 mm long; lobes oblong-elliptic, 1.2–3.5 mm long. Female flowers: calyx-tube oblong or ellipsoid, 0.5–2 mm long; lobes triangular or ovate to elliptic-lanceolate, 0.3–1.8 mm long; corolla-tube ellipsoid, 0.5–1 mm long, lobes oblong or elliptic-lanceolate, 0.5 mm long; ovary 0.6–1 mm long; style 0.5–1 cm long. Fruits reddish-brown, glabrous, oblong, elliptic to obovate, 2 mm long, covered with the persistent calyx-teeth.

Distribution: Western, central, and southern Kenya. [East Africa, from Kenya to southern Africa].

Habitat: Grasslands, thickets, or bushlands; 1650–2750 m.

Kericho: Mount Blackett, *Blunt 182* (EA). Narok: Mount Suswa, *Glover & Samuel 3363* (EA). Nakuru: Menengai Crater, *Kokwaro & Mathenge 3394* (EA, K); Mount Longonot, *Gilbert & Hedberg 6292* (EA). Samburu: Isiolo to Mathew's Range and Mount Nyiru, *Newbould s.n.* (K).

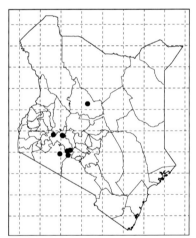

8. Trib. **Paederieae** DC.

Lianas or trees and shrubs, rarely herbs, usually fetid-smelling when bruised. Leaves opposite; stipules between petioles, free. Ovary 2–5-locular; ovules solitary in each locule, erect; style 2–5-lobed, branchlets often filiform, stigma often papillate. Fruits 2-coccous, with thin epicarp, or capsular.

Only one genus occurs in Kenya.

34. Paederia L.

Lianas, shrubs, subshrubs, or rarely suberect herbs, usually fetid-smelling when bruised. Leaves opposite or rarely 3–4-whorled; stipules interpetiolar, ovate to triangular, or sometimes bilobed. Inflorescences axillary and/or terminal, thyrsiform, paniculate, cymose or spiciform, several–many-flowered. Flowers bisexual, monomorphic. Calyx-lobes 4–6, triangular or subulate. Corolla-tube cylindrical, campanulate or narrowly funnel-shaped; lobes 4–6, induplicate-valvate in bud, with margins often crisped to irregular. Stamens 4–6, included; filaments reduced. Ovary 2–3-locular; ovule single in each locule; stigma-lobes 2–3, filiform, included or exserted. Fruits globose or compressed globose to compressed ellipsoid, indehiscent. Seeds with thin testa; cotyledons broadly cordate; radicle short, hypogeous.

About 30 species distributed in tropical regions of both hemispheres; only one species in Kenya.

1. **Paederia pospischilii** K. Schum., Bot. Jahrb. Syst. 23: 469. 1897; F.T.E.A. Rubiac. 1: 177. 1976; K.T.S.L.: 525. 1994; U.K.W.F.F.: 226. 2013. —Types: Kenya, Teita Taveta, on plain at foot of Kilimanjaro, *A. Pospischil s.n.* [holotype: B (destroyed)]; Kenya, Teita Taveta, Tsavo West, 615 m, 6 May 2003, *P.A. & W.R.Q. Luke 9456* [neotype: EA, **designated here**; isoneotype: UPS] (Figure 32; Plate 44)

Woody fetid-smelling climber or more rarely suberect herb up to 9 m long. Leaves opposite; blades round, ovate, elliptic or lanceolate, 1–6.5 cm × 0.2–2.7 cm, acute to rounded at the apex, narrowly cuneate to rounded at the base; petioles up to 1 cm long; stipules ovate or triangular, 1–2.5 mm long. Flowers solitary or several in axillary clusters; pedicels 0.3–1 cm long. Calyx-tube ovoid, 1–2 mm long, glabrous; lobes 5, narrowly triangular, 1.5–4 mm long, densely pubescent. Corolla white or greenish white with a reddish purple center; tube funnel-shaped, 0.6–1.6 cm long; lobes ovate, 5–8 mm long, densely hairy. Styles joined at the very base or free, 1–1.2 cm long. Fruits compressed-ellipsoid, 0.8–1.2 cm long. Seeds black, ca. 8 mm long.

Distribution: Northern, central, eastern, and southern Kenya. [Ethiopia and Somalia].

Habitat: Grasslands, woodlands, bushlands, or thickets; 300–1200 m.

Isiolo: ca. 10 km north of Garba Tula-Merti Road, *Bally & Smith B14695* (EA, K). Kitui: Endau, Malalani Village, *SAJIT MU0119* (HIB); Endau next to Chiefs camp, *Gatheri et al. 79/53* (EA, K); Mutha, *Kirika*

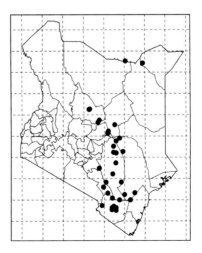

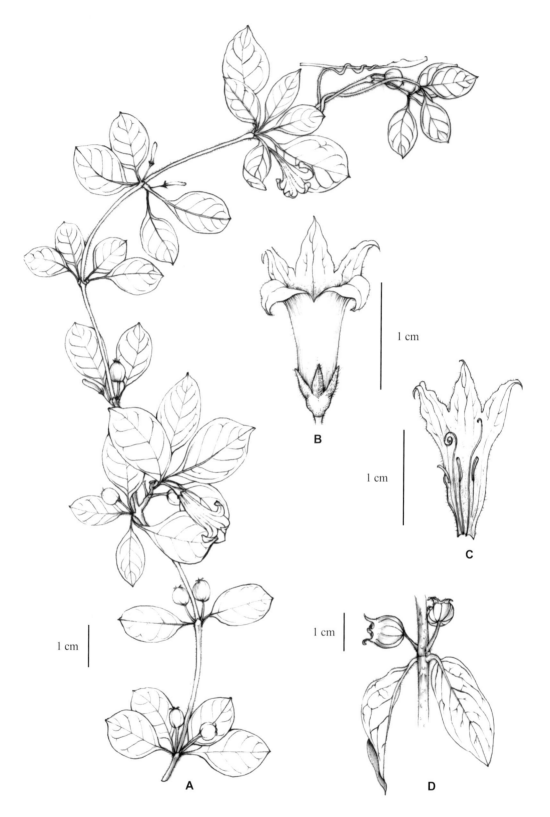

Figure 32 *Paederia pospischilii*. A. a fruiting branch; B. flower; C. longitudinal section of corolla, showing the stamens and styles; D. fruit. Drawn by NJ.

Plate 44 A, B. *Paederia pospischilii*. Photo by NW.

et al. NMK 274 (EA). Makueni: Kibwezi, *Battiscombe 883* (K). Mandera: Dandu, *Gillett 12756* (EA, K). Marsabit: Moyale, *Gillet 14124* (EA, K). Meru: Meru National Park, *Ament & Magogo 380* (EA). Samburu: Isiolo to Mathew's Range and Mount Nyiru, *Newbould 3497* (K). Tcita Taveta: Tsavo National Park East, *Gillett 17213* (EA, K); Between Voi and Sagala Hill, *Polhill & Paulo 951* (EA). Tana River: Nairobi-Garissa Road, *Gillett & Gachathi 20517* (EA, K).

9. Trib. **Rubieae** Baill.

Herbs or rarely subshrubs. Leaves always whorled with leaf-like stipules. Flowers heterostylous. Calyx rudimentary. Ovary 2-locular; ovule single in each locule, attached to the septum. Fruits dry or fleshy, didymous, seeds usually adhering to the pericarp.

Two genera occur in Kenya.

1a. Leaf-blades large, ovate to lanceolate, with petiole very well developed; flowers 5-merous ....35. *Rubia*
1b. Leaf-blades small, linear to lanceolate, or rarely ovate, sessile to shortly petiolate; flowers 4-merous .... 36. *Galium*

35. Rubia L.

Erect or climbing herbs, or rarely subshrubs or lianas; stems often prickly and/or longitudinally ribbed or winged. Leaves with leaflike stipules in whorls of 4–6(–8); blades linear, lanceolate, obovate or cordate, often prickly, with palmate veins. Inflorescences thyrsoid, with axillary and/or terminal cymes. Flowers hermaphrodite. Calyx-tube ovoid or globose; limb reduced and obsolete. Corolla rotate or sub-campanulate; lobes (4–)5(–6), valvate in bud, often long acuminate. Stamens usually 5, inserted in the corolla-tube, exserted. Ovary 1–2-locular, with single ovule in each locule. Styles 2-fid, included or exserted; stigmas capitate. Fruits baccate or berrylike; mericarps 1–2, subglobose. Seeds 2, suberect, ellipsoid, subglobose, or plano-convex, with membranous testa.

About 60–80 species, extending from tropical and temperate Asia to southwest Asia, Europe, Mediterranean and North, tropical and South Africa, also locally introduced and persisting from cultivation in America; only one species in Kenya.

1. **Rubia cordifolia** L., Syst. Nat., ed. 12. 3: 229. 1768; F.T.E.A. Rubiac. 1: 381. 1976; U.K.W.F.F.: 229. 2013. —Type: unknown

Climbing or scrambling herbs; stems up to 6 m long, quadrangular, with curved prickles on the ribs. Leaves 4–8(–12)-whorled; blades 5–7-nerved from the base, lanceolate, oblong-lanceolate, ovate or oblong-ovate, 0.7–8.5 cm × 0.2–4.5 cm, acuminate at the apex, rounded to cordate at the base, margins and veins below with recurved prickles; petioles 1–6 cm long, with recurved prickles. Inflorescences several- to many-flowered; bracts linear-lanceolate to elliptic, 1.2–1.5 mm long; pedicels 1–6 mm long. Flowers 5–6-merous, hermaphroditic. Corolla pale yellow or greenish yellow; tube minute; lobes triangular, 1.5–3 mm long, apiculate. Fruits brownish black; lobes globose, 2.5–6 mm in diameter; pyrenes globose. Seeds globose, 1–3 mm in diameter.

subsp. **conotricha** (Gand.) Verdc., Kew Bull. 30: 323. 1975; F.T.E.A. Rubiac. 1: 381. 1976; C.P.K.: 388. 2016. ≡ *Rubia conotricha* Gand., Bull. Soc. Bot. France 65: 35. 1918. —Type: South Africa, E. Griqualand, near R. Umziuklowa, 1500 m, 31 Jan. 1895, *F.R.R. Schlechter 6550* [holotype: LY; isotypes: K, S (S-G-5366)] (Figure 33)

Leaves always 4-whorled; blades often hairy beneath. Fruits with cocci less than 3 mm in diameter.

Distribution: Northern, Northwestern, western, central, and southern Kenya. [South Sudan to southern Africa].

Habitat: Sparse forests, forest margins, or grasslands; up to 2800 m.

Note: *Rubia cordifolia* subsp. *cordifolia* occurs in Asia, also reported in Greece.

Bomet: western Mau Forest, *Mass Geesteranus 5754* (K). Bungoma: Mount Elgon, *Dugard 204* (K). Elgeyo-Marakwet: Marakwet, *Lindsey 135A* (EA). Homa Bay: Mfangano Island, *Kirika 342* (EA). Kajiado: Loitokitok, *Rauh 215* (EA). Kericho: Sambret Catchment, *Kerfoot 4320* (EA). Kiambu: Katamayu Forest,

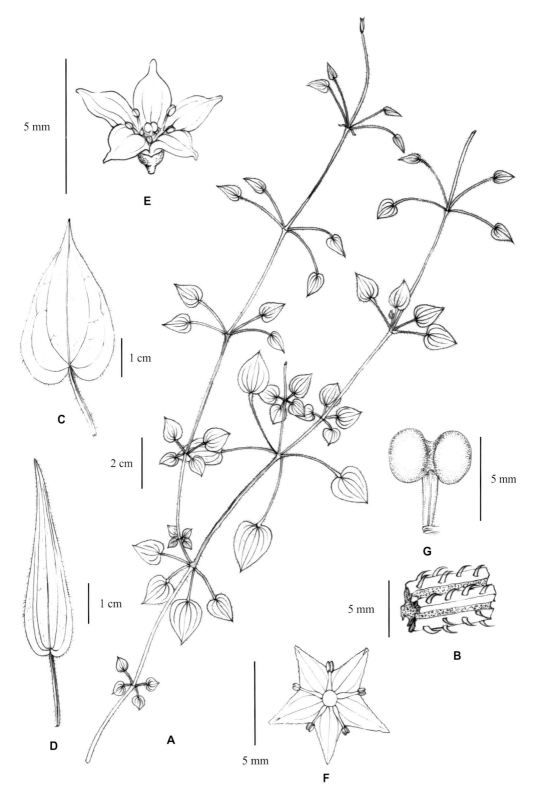

Figure 33 *Rubia cordifolia* subsp. *conotricha*. A. plant; B. stem; C, D. leaves; E, F. flowers; G. fruit. Drawn by NJ.

Verdcourt 620 (EA, K); near Upland, *Fukuoka 165* (EA). Kisii: ca. 17 km southwest of Kisii, *Vuyk & Breteler 116* (EA, K). Kitui: Boma, *Bally B1540* (K). Laikipia: Ol-Ari-Nyiro Ranch, *Muasya 1478* (EA). Machakos: Summit Mua Hills, *Gillett 16220* (EA). Makueni: north of Nunguni at western side of mountain slopes in Kithembe Village, *Mwangangi 863* (EA, K). Marsabit: Mount Kulal, *Bally B5506* (K). Meru: *Hancook 65* (K). Migori: Outskirts of Migori Town, *Kurimoto 20* (EA). Murang'a: Blue Posts Hotel, *Faden 6734* (EA). Nairobi: Eastleigh, *Mwangangi & Kasyoki 1787* (K). Nakuru: Morendat Station, *Polhill 144* (EA). Narok: Orengitok, *Glover et al. 1232* (EA); Lolgorrien, *Kuchar et al. 5539* (EA). Nyandarua: Aberdare Range, *SAJIT 006503* (HIB). Nyeri: west Mount Kenya, *Battiscombe K827* (EA, K). Samburu: Karisia Hills, *Ichikawa 738* (EA). Trans-Nzoia: on hedge of Kitale Hotel, *Verdcourt 731* (EA, K). Turkana: Mooruassigar, *Newbould 7214* (EA). Uasin Gishu: Lesso Reservoir, *Gilbert & Tadessa 6781* (EA).

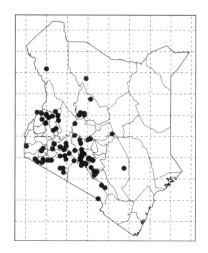

36. Galium L.

Annual or perennial herbs. Stems often 4-angled, erect, decumbent or climbing, glabrous, hispid or prickly. Leaves often with leaflike stipules, sessile to shortly petiolate, 3 to several-whorled. Inflorescences of terminal and axillary cymes or rarely solitary. Flowers hermaphroditic or sometimes unisexual, pedicellate to sessile, usually quite small. Calyx-tube ovoid or globose; limb reduced. Corolla white, yellow or reddish; tube reduced; lobes (3–)4(–5). Stamens (3–)4(–5), inserted on the corolla-tube near base, exserted. Ovary 2-locular; ovule solitary in each locule; styles 2, short; stigmas capitate. Fruits 2-lobed, dry or fleshy. Seeds small; testa membranous; endosperm corneous.

A very large genus with more than 600 species and widespread in temperate regions, also in subtropical and tropical zones at high elevations; 12 species in Kenya.

1a. Leaves always 4-whorled at each node, ovate-elliptic or rounded ....1. *G. thunbergianum*
1b. Leaves more than 4 at each node, linear, lanceolate, oblanceolate or narrowly elliptic............2
2a. Stems and the margins of leaf-blades without prickles 3
2b. Stems and the margins of leaf-blades usually with somewhat recurved prickles 7
3a. Stems erect; inflorescences dense; flowers crowded towards the stem apex ....
.... 12. *G. ossirwaense*
3b. Stems decumbent; inflorescences lax; flowers scattered on the stem 4

4a. Inflorescences many-flowered .... 11. *G. scioanum*
4b. Inflorescences 1–3(–4)-flowered 5
5a. Leaves less than 0.8 mm wide, blunt and without acumen at the apex .... 10. *G. kenyanum*
5b. Leaves more than 1 mm wide, often acute or acuminate and with acumen at the apex 6
6a. Leaves oblanceolate or spathulate; corolla greenish cream or pink, with tube 0.1–0.3 mm long .... 8. *G. glaciale*
6b. Leaves linear to obovate-oblanceolate; corolla white, with tube ca. 4 mm long .... 9. *G. acrophyum*
7a. Leaves elliptic, narrowly elliptic or oblanceolate, up to 12 mm wide 8
7b. Leaves usually linear or linear-lanceolate, less than 3 mm wide 10
8a. Leaves elliptic; flowers solitary; fruits glabrous .... 7. *G. brenanii*
8b. Leaves narrowly elliptic to oblanceolate; inflorescences 1–several-flowered; fruits sparsely to densely covered with hooked hairs 9
9a. Inflorescences, 1–3-flowered; fruits densely covered with glossy yellow hooked hairs .... 2. *G. chloroionanthum*
9b. Inflorescence several-flowered; fruits sparsely covered with white hooked hairs .... 3. *G. aparinoides*
10a. Leaves less than 1.5 mm wide, acuminate at the apex; fruits glabrous .... 6. *G. ruwenzoriense*
10b. Leaves up to 3 mm wide, with a distinct acumen at the apex; fruits glabrous or hairy 11
11a. Leaves with a tuft of minute pale hairs at the base; fruits fleshy .... 4. *G. simense*
11b. Leaves without such tuft of hairs at the base; fruits dry .... 5. *G. spurium*

1. **Galium thunbergianum** Eckl. & Zeyh., Enum. Pl. Afr. Austral. 369. 1837; F.T.E.A. Rubiac. 1: 387. 1976; U.K.W.F.F.: 228. 2013. —Type: South Africa, Cape Province, Katriviersberg, *C.F. Ecklon & C.L.P. Zeyher 2321* [lectotype: S (S-G-2770), designated by C. Puff in Fl. S. Afr. 31 (1–2): 66. 1986; isolectotypes: M (M0106196), NU, SAM, WU]

Straggling, climbing or procumbent herb, up to 75 cm long, glabrous or with short spreading white hairs. Leaves 4-whorled; blades 3-nerved, elliptic to ovate, 7–18 mm × 2–10 mm, acute or rounded and shortly mucronate at the apex, cuneate at the base. Inflorescences of several–many-flowered cymes. Calyx-tube densely covered with spreading bristly hairs. Corolla greenish, white to yellowish; lobes oblong, 0.7–1.5 mm long. Fruits densely covered with white tuberculate hooked hairs; cocci (1–)2, subglobose, 0.8–1.5 mm in diameter.

var. **hirsutum** (Sond.) Verdc., Kew Bull. 30: 326. 1975; F.T.E.A. Rubiac. 1: 388. 1976; C.P.K.: 379. 2016. ≡ *Galium rotundifolium* L. var. *hirsutum* Sond., Fl. Cap. 3: 39. 1865. —Type: South Africa, *Masson s.n.* [holotype: UPS-Thunb] (Plate 45A–C)

Stems and leaves sparsely to densely covered with spreading bristly white hairs.

Distribution: Western and central Kenya. [East Africa from Ethiopia to South Africa].

Habitat: Montane forests; 2000–3800 m.

Note: *Galium thunbergianum* var. *thunbergianum* has glabrous stems and foliage, and is recorded in Ethiopia, Sudan, South Sudan, Kenya and South Africa in FTEA. After the inspection of the specimen [Kerfoot 2947 (EA)] which was collected from Kenya, we speculated it should be treated as *G. thunbergianum* var. *hirsutum*, for it obviously has spreading bristly hairs on its stems.

Bungoma: Mount Elgon, *Knox 3787* (EA). Elgeyo-Marakwet: Cherangani Hills, *Knox 3376* (EA). Kericho: southwest Mau Forest, *Kerfoot 2947* (EA). Kirinyaga: south Mount Kenya, *Faden et al. 71/897* (EA). Meru: near Mount Kenya Lodge, *SAJIT 002007* (HIB). Nakuru: Eburru Forest Reserve, *Luke et al. 8876* (EA). Nyandarua: Aberdares Mountains, *Knox 3091* (EA). Nyeri: west Mount Kenya, *Fries & Fries 748* (K). Trans-Nzoia: northeast Mount Elgon, *Tweedie 929* (K). West Pokot: Cherangani Hills, *SAJIT 006857* (HIB).

(destroyed); lectotype: K (K000422934), **designated here**; isolectotypes: BM, BR (BR0000008847003)] (Plate 45D, E)

Climbing or procumbent perennial herb, up to 1.5 m long. Leaves 6-whorled; blades obovate, narrowly elliptic to elliptic-oblanceolate, 0.5–4.5 cm × 0.2–1.5 cm, acuminate at the apex, cuneate at the base, margins and midnerve above and beneath with recurved prickles. Inflorescences 1–3-flowered; peduncles 0.8–2.6 cm long; pedicels 0.5–1 mm long. Calyx-tube greenish, purplish or brownish, hairy. Corolla greenish-white to greenish-yellow; lobes 1.2–2.1 mm long. Fruits covered with brownish hooked hairs; cocci 1–3 mm in diameter.

Distribution: Western and central Kenya. [Tanzania].

Habitat: Montane forests; 2200–2800 m.

Kericho: ca. 6.5 miles east of Kericho, *Perdue & Kibuwa 9283* (EA, K). Kiambu:

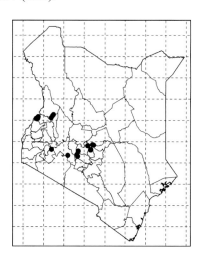

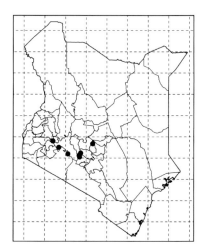

2. **Galium chloroionanthum** K. Schum., Bot. Jahrb. Syst. 30: 417. 1901; F.T.E.A. Rubiac. 1: 388. 1976; U.K.W.F.F.: 228. 2013. —Type: Tanzania, Rungwe Mt., 1898, *W. Goetze 1162* [holotype: B

Gatamayu Forest, *Beentje 2713* (EA). Kirinyaga: south Mount Kenya, *Townsend 2199* (K). Murang'a: Kimakia Forest, *Kerfoot 66* (EA). Nakuru: Mau Forest, *Bally B4891* (EA). Narok: Enesambulai Valley, *Greenway*

& *Kanuri 13636* (EA, K).

3. **Galium aparinoides** Forssk., Fl. Aegypt.-Arab.: 30. 1775; F.T.E.A. Rubiac. 1: 389. 1976; W.F.E.A.: 151. 1987; U.K.W.FF: 229. 2013. —Type: Yemen, Hadie Mts., Mokhaja, Mar. 1763, *P. Forsskål 1291* [lectotype: C (C10018166), designated by B. Verdcourt in Kew Bull. 28: 60. 1973] (Plate 45F–H)

Scrambling herb, up to 3 m long; stems usually prickly, glabrous to hairy. Leaves (4–)6(–7)-whorled; blades narrowly elliptic to oblanceolate, 0.4–3 cm × (0.5–)3–8 mm, with an acumen at the apex, cuneate at the base, sparsely hairy on both sides. Cymes lateral, several-flowered. Calyx-tube subglobose, 0.3–0.9 mm long. Corolla greenish-white, tinged with purple or reddish brown; tube 0.5–0.7 mm long; lobes narrowly triangular, 1.4–1.7 mm long. Fruits with cocci 1.5–3 mm in diameter, black, sparsely clothed with short white hooked hairs.

Distribution: Central, western and southern Kenya. [Tropical East Africa, also in Ethiopia and Arabia].

Habitat: Montane forests, bamboo zones, or roadside rocks; 1900–3300 m.

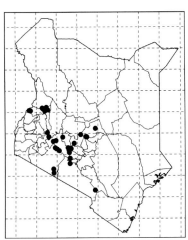

Bungoma: Mount Elgon, *Knox 3777* (EA). Elgeyo-Marakwet: Kipkunurr Forest Reserve, *Hepper & Field 5000* (EA). Kajiado: Chyulu Hills, *Luke & Luke 3896* (EA). Kiambu: Muguga Forest, *Kokwaro & Kabuye 357* (EA). Meru: near Mount Kenya Lodge, *SAJIT 001952* (HIB); Nyambeni Forest, *Knox 4199* (EA). Nakuru: Mau Escarpment, *Knox 2739* (EA). Narok: Loita Hills, *Kuchar & Msafiri 6965* (EA). Nyandarua: Bahati Forest, *Knox 3048* (EA). Nyeri: Kiandongoro Forest, *Mathenge 215* (EA). Trans-Nzoia: Mount Elgon, *Knox 2641* (EA). Uasin Gishu: Lake Narasha, *Knox 2672* (EA). West Pokot: Cherangani Hills, *Knox 3378* (EA).

4. **Galium simense** Fresen., Mus. Senckenberg. 2: 165. 1837; F.T.E.A. Rubiac. 1: 391. 1976; U.K.W.F.F.: 228. 2013. —Type: Ethiopia, Temben to Simien, 1832, *E. Rüeppell s.n.* [holotype: FR (FR0030048)]

Scrambling herb, up to 35 cm long. Leaves 6-whorled; blades linear-oblanceolate to obovate-oblanceolate, (3–)5.5–9 mm × 1–1.4 mm, acuminate at the apex, cuneate at the base; margins with short forwardly directed hairs or weak prickles, glabrous to covered with short forwardly directed hairs. Cymes 1–3-flowered. Calyx-tube half-ellipsoid, 0.5–0.7 mm long, glabrous. Corolla white; tube ca. 0.4 mm long; lobes triangular to ovate, 0.5–1 mm long. Fruits with rounded or ellipsoid cocci 1.5–3 mm in diameter.

Distribution: Central, western and southern Kenya. [Tropical Africa].

Habitat: Forest edges, woodlands, lakesides, or swamps; 1500–2700 m.

Kiambu: below Mzee Mungai's Farm, *Knox 3113* (EA). Murang'a: Kimakia, *Taiti 2027* (EA). Nakuru: Naivasha, *Someren et al. 1200* (K); Eburru Forest, *Luke & Ndeche 9009*

Plate 45 A–C. *Galium thunbergianum* var. *hirsutum*; D, E. *G. chloroionanthum*; F–H. *G. aparinoides*. Photo by GWH.

(K); Nakuru National Park, *Hingley 38* (EA). Trans-Nzoia: Saiwa Swamp National Park, *Kirika PK57* (EA).

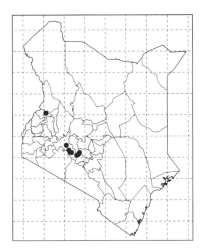

5. Galium spurium L., Sp. Pl. 106. 1753; F.T.E.A. Rubiac. 1: 390. 1976; U.K.W.F.F.: 229. 2013. —Type: "Habitat in Europae cultis", ex herb. Alstroemerii [lectotype: S-Linn (No. 55.17), designated by Natali & Jeanmonod in Jeanmonod, Compl. Prodr. Fl. Corse, Rubiac.: 53. 2000]

Scrambling herb, up to 2 m long, with recurved prickles on stems. Leaves 6–8-whorled; blades linear-lanceolate or linear-oblanceolate, 1–4(–4.5) cm × 1–3(–7) mm, with a filiform acumen at the apex, cuneate at the base; margins and midvein with coarse recurved prickles. Cymes 1–9-flowered. Calyx-tube 0.5–0.7 mm long, hairy or glabrous. Corolla greenish white, cream white, green or yellowish; tube 0.3–0.4 mm long; lobes triangular, 0.7–1.1 mm long. Fruits covered with white hooked hairs or glabrous; cocci globose, 2–4 mm in diameter.

var. **africanum** Verdc., Kew Bull. 30: 324. 1975; F.T.E.A. Rubiac. 1: 390. 1976. —Type: Kenya, Kiambu, Muguga, Oct. 1955, *E. Milne-Redhead & P. Taylor 7147* [holotype: K; isotype: BR (BR0000008846907)]

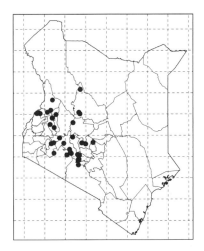

Fruits covered with white hooked hairs.

Distribution: Western and central Kenya. [East Africa, from Sudan and Ethiopia to South Africa].

Habitat: Forest edges, bushlands, swamps, roadsides, or other disturbed places; 1600–2900 m.

Note: *Galium spurium* var. *spurium* is widely distributed in Europe, western Asia and northern Africa and introduced into other places. B. Verdcourt mentioned it occurs in western Kenya in FTEA, but we did not find the specimens.

Baringo: Londiani, *Graham 877* (EA). Bomet: Tinderet Forest Reserve, *Geesteranus 5024* (K). Bungoma: Mount Elgon, *Mumiukha 227* (EA). Elgeyo-Marakwet: Marakwet, *Lindsay 135B* (EA). Kajiado: Ngong Hills, *Beentje 1762* (EA). Kericho: southwest Mau Forest Reserve, *Geesteranus 5681* (K). Kiambu: Brackenridge Farm, *Luke & Luke 8367* (EA); Gatamayu Forest, *Knox 3213* (EA). Kirinyaga: Thiba River, *Bally 397* (K). Laikipia: Ol Ari Nyiro Ranch, *Muasya 1503* (EA). Nakuru: Eburru Forest Reserve, *Luke & Mwatsuma 9009* (EA). Narok:

Melili, *Galaty 69* (EA). Nyandarua: Ol Joro Orok, *Peers c12* (EA). Nyeri: Zawadi Estate, *Faden et al. 74/839* (K). Samburu: ca. 3 km north of Maralal on road to Baragoi, *Gilbert et al. 5119* (EA). Trans-Nzoia: Mount Elgon, *Knox 2634* (EA). Uasin Gishu: Moiben, *Hill 230* (EA). West Pokot: Cherangani Hills, *Knox 3354* (EA).

6. **Galium ruwenzoriense** (Cortesi) Ehrend., Ann. Bot. (Rome) 9: 69. 1911; F.T.E.A. Rubiac. 1: 391. 1976; U.K.W.F.F.: 228. 2013. ≡ *Rubia ruwenzoriensis* Cortesi, Ann. Bot. (Rome) 6: 152. 1907. —Types: Uganda, Ruwenzori, Valley of the Lakes, *Abruzzi Exped. s.n.* [syntype: TO]; Uganda, Ruwenzori, Bujuku Valley, Bujongolo, *Abruzzi Exped. s.n.* [syntype: TO] (Plate 46A, B)

= *Galium afroalpinum* Bullock, Bull. Misc. Inform. Kew 1932: 498. 1932. —Type: Kenya, Mt. Elgon, Dec. 1930, *E.J. & C. Lugard 365* [holotype: K (K000422933); isotype: EA]

Scrambling or climbing herb, up to 2 m or more long. Leaves 6–8-whorled; blades linear to narrowly elliptic, 0.2–1.9 cm × 0.3–1.5(–3.5) mm, acuminate at the apex, cuneate at the base; margins covered with very coarse curved prickles. Cymes 1–3-flowered. Calyx-tube glabrous. Corolla green to yellow; tube 0.2–0.8 mm long; lobes ovate, 1.4–2.2 mm long, acute. Fruits purplish black, black or blue, becoming distinctly fleshy, 3.5–7 mm in diameter, glabrous.

Distribution: Western and central Kenya. [Tropical East Africa].

Habitat: Montane grasslands, forest edges, heath zones, or moorlands; 2900–4500 m.

Bungoma: southeast of Mount Elgon, *Major et al. 365* (EA). Elgeyo-Marakwet: Cherangani Hills, *Thulin & Tidigs 101* (EA). Meru: Mount Kenya, Sirimon Track, *SAJIT 001401* (HIB); Mount Kenya, Timau Track, *Knox 3951* (EA); Meru Mount Kenya Lodge, *Beentje 2699* (EA). Nakuru: Kari Naivasha, *Nyamongo et al. GBK09/04/02* (EA). Narok: Nsampolai Valley, *Greenway & Kanuri 15046* (EA). Nyandarua: Aberdare National Park, *Beentje 3229* (EA). Nyeri: Mount Kenya, Naro Moru Track, *Knox 3823* (EA). Trans-Nzoia: Mount Elgon, *Gillett 18450* (EA); Mweiga, Aberdare National Park, *Mbale et al. NMK856* (EA).

7. **Galium brenanii** Ehrend. & Verdc., Kew Bull. 28: 485. 1974; F.T.E.A. Rubiac. 1: 393. 1976. —Type: Tanzania, Uluguru Mts., Salaza Forest, 2000 m, 15 Mar. 1953, *R.B. Drummond & J.H. Hemsley 1625* [holotype: K (K000353093); isotype: B (B_10_0279686)] (Plate 46C–F)

Scrambling herb, up to 4 m or more long. Leaves 4–6-whorled; blades elliptic, 0.9–3.6 cm × 0.3–1.2 cm, acute and apiculate at the apex, cuneate at the base, glabrous except the petioles and the bases of the whorls; petioles up to 5 mm long. Inflorescences

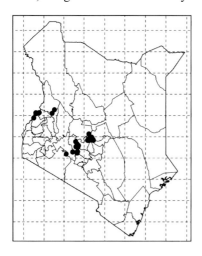

Plate 46 A, B. *Galium ruwenzoriense*; C–F. *G. brenanii*. Photo by BL (A, B) and GWH (C–F).

mostly 1(–3)-flowered; peduncles 0.3–1.3 cm long; pedicels 0–9 mm long. Calyx-tube 0.3–0.6 mm in diameter, glabrous. Corolla green, white, greenish white or yellowish white; tube 0.5–1.1 mm long; lobes ovate, 1–1.5 mm long, shortly acuminate. Fruits blackish purple; cocci 3–5.5 mm in diameter, glabrous.

Distribution: Southern Kenya. [Tanzania].
Habitat: Forests; 1900–2200 m.

Teita Taveta: Vuria Hill, *Faden & Faden 72/248* (EA).

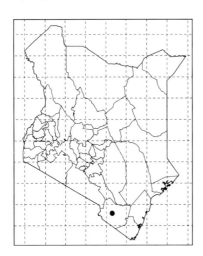

8. **Galium glaciale** K. Krause, Bot. Jahrb. Syst. 43: 159. 1909; F.T.E.A. Rubiac. 1: 394. 1976; U.K.W.F.F.: 229. 2013. —Types: Tanzania, Kilimanjaro, Kibo, *Jeager 150* [holotype: B (destroyed)]; Tanzania, Kilimanjaro, saddle between Kibo and Mawenzi, 23 June 1948, *O. Hedberg 1345* [neotype: UPS, designated by Hedberg in Symb. Bot. Ups. 15(1): 178. 1957; isoneotypes: BR (BR0000008847096), EA (EA000001590), K (K000319829), LD, S]

Dwarf herb with prostrate or ascending stems, up to 15(–20) cm long, glabrous, pubescent or slightly prickly. Leaves 3–6-whorled; blades oblanceolate or spathulate, 2.5–8 mm × 1–2.5 mm, acute and with an acumen at the apex, cuneate at the base. Inflorescences 1–2(–4)-flowered, on 3–4 mm long lateral shoots. Calyx-tube ovoid, 0.6–0.9 mm long. Corolla greenish cream or pink; tube 0.1–0.3 mm long; lobes ovate, 0.6–1.2 mm in diameter, acute. Fruits black, 1.5–2.3 mm in diameter, glabrous; cocci black, ovoid, ca. 1.7 mm in diameter.

1a. Leaf-blades glabrous .... a. var. *glaciale*
1b. Leaf-blades covered with short curved hairs .... b. var. *satimmae*

a. var. **glaciale** (Plate 47A, B)

Leaf-blades glabrous above.
Distribution: Western and central Kenya. [Tanzania].
Habitat: Moorlands; 3500–4700 m.

Meru: east Mount Keya, Chogoria Track, *Knox 3185* (EA). Nyeri: west Mount Kenya, Teleki Valley, *Hedberg 1733* (EA). Trans-Nzoia: Mount Elgon, *Herdberg 892* (EA).

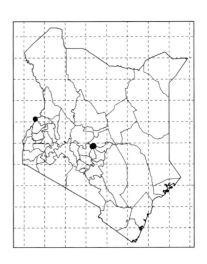

b. var. **satimmae** Verdc., Kew Bull. 28: 493. 1974; F.T.E.A. Rubiac. 1: 394. 1976. —Type: Kenya, Aberdare Mts., Satima, 19 Mar. 1922, *R.E. & T.C.E. Fries 2680* [holotype: UPS (099616)]

Leaf-blades covered with short curved hairs.

Distribution: Central Kenya. [Tanzania].
Habitat: Heath zones or moorlands; 3600 m.
Nyeri: Aberdare Mountains, *Fries & Fries 1276B* (K, S, BR).

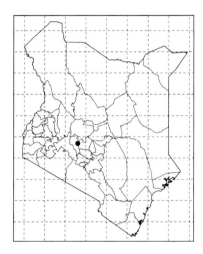

9. **Galium acrophyum** Hochst. ex Chiov., Ann. Bot. (Rome) 9(1): 69. 1911. ≡ *Galium hochstetteri* Pic. Serm., Webbia 7: 339. 1950, *nom. illeg.*; F.T.E.A. Rubiac. 1: 394. 1976; U.K.W.F.F.: 229. 2013. —Type: Ethiopia, Semien, Bachit, 3 Aug. 1842, *G.H.W. Schimper 548* [Holotype: FT; isotypes: BM, K (K000422936 & K000422938), LG (LG0000090029714), MPU (MPU024397), P, TUB (TUB004643)]

= *Galium simense* A. Rich. var. *hysophilum* R.E. Fr., Kongl. Svenska Vetensk. Acad. Handl. III, 25(5): 79. 1948. —Type: Kenya, W. slope of Mt. Kenya, 2 May 1922, *R.E. & T.C.E. Fries 1276b* [holotype: UPS; isotypes: BR, K (K000319827), S (S-G-2767)]

= *Galium simense* A. Rich. var. *keniense* R.E. Fr., Kongl. Svenska Vetensk. Acad. Handl. 25: 79. 1948. —Type: Kenya, Mt. Kenya, *R.E. & T.C.E. Fries 1368* [holotype: UPS]

Scrambling herb, up to 35 cm long. Leaves 6-whorled; blades linear-oblanceolate to obovate-oblanceolate, 3–9 mm × 1–1.4 mm, acuminate and with an acumen at the apex, cuneate at the base, glabrous to slightly

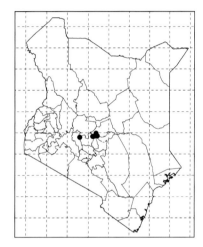

hairy. Inflorescences 1–3-flowered. Calyx-tube 0.5–0.7 mm long, glabrous. Corolla white; tube ca. 0.4 mm long; lobes triangular to ovate, 0.5–1 mm long. Fruits black, smooth; cocci rounded or ellipsoid, 1.5–3 mm in diameter.

Distribution: Central Kenya. [Ethiopia and Tanzania].
Habitat: Heath zones or moorlands; 2700–3100 m.
Meru: east Mount Kenya, Chogoria Track, *Knox 3893* (EA). Nyandarua: north of Aberdares Mountains, *Kokwaro 1947* (EA, K). Nyeri: west Mount Kenya, Naro Moru Track, *Kirika et al. NMK1191* (EA, K).

10. Galium kenyanum Verdc., Kew Bull. 28: 493. 1974; F.T.E.A. Rubiac. 1: 395. 1976; U.K.W.F.F.: 229. 2013. —Type: Kenya, Mt. Aberdares National Park, near Fort Jerusalem, 15 Mar. 1964, *B. Verdcourt 3998* [holotype: K (K000319835); isotypes: BR (BR0000008847225), EA (EA000001589)]

Procumbent mat-forming herb, up to 60 cm long, glabrous or with a few scattered spreading hairs. Leaves 6-whorled; blades linear-lanceolate, 1.5–4 mm × 0.3–0.8 mm, obtuse or subacute at the apex, cuneate at the base. Flowers 1–2 at the end of short lateral shoots. Calyx-tube glabrous. Corolla white or tinged pink, broadly campanulate-rotate; tube 0.5–0.7 mm long; lobes ovate, 0.5–0.9 mm long. Fruits dry, 3–4 mm in diameter, glabrous; cocci ellipsoid, 2.3 mm long.

Distribution: Western and central Kenya. [Tanzania].

Habitat: Heath zones or moorlands; 2800–3400 m.

Elgeyo-Marakwet: Cherangani Hills, *Mabberley 518* (EA, K). Meru: near Meru Mount Kenya Lodge, *Beentje 2624* (EA). Nyeri: Aberdare National Park, *Polhill 239* (K). West Pokot: Cherangani Hills, *Thulin & Tidigs 231* (EA).

11. Galium scioanum Chiov., Ann. Bot. (Rome) 9(3): 322. 1911; F.T.E.A. Rubiac. 1: 398. 1976; U.K.W.F.F.: 229. 2013. —Type: Ethiopia, Addis Ababa, 2400 m, 8 June 1909, *G. Negri 1410* [holotype: FT (FT003483)]

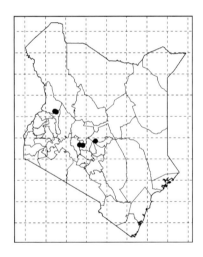

Creeping, scrambling or climbing herb, up to 90 cm long, glabrous or quite densely covered with spreading hairs. Leaves 6-whorled; blades linear-lanceolate to narrowly elliptic, 0.4–1.7(–2.5) × 1–2.8(–10) mm, with a short acute apical acumen, cuneate at the base, glabrous to hairy; margins revolute or flat, ciliate. Inflorescences of terminal and axillary panicles, many-flowered. Corolla white; tube 0.2–1 mm long; lobes narrowly to broadly ovate-triangular, 0.8–2.1 mm long, subacute. Fruits reddish purple; cocci 1.5–2 mm in diameter, glabrous.

1a. Stems and leaves distinctly hairy .... a. var. *scioanum*
1b. Stems and leaves glabrous or glabrescent ....b. var. *glabrum*

a. var. **scioanum** (Plate 47C, D)

Stems and leaves distinctly hairy.

Distribution: Western and southern Kenya. [D.R. Congo, Ethiopia, and Tanzania].

Habitat: Swamp edges, riversides or wet savannas; 1800–2500 m.

Kericho: southwest Mau Forest, *Kerfoot 2917* (EA). Kiambu: below Mzee Mungai's Farm, *Knox 3110* (EA). Nandi: Nandi Hills, *Bickford 21* (EA). Narok: Loita Hills, *Fayad 120* (EA). Trans-Nzoia: Sandum's Bridge,

Knox 3847 (EA). West Pokot: Cherangani Hills, *Thulin & Tidigs 175* (EA).

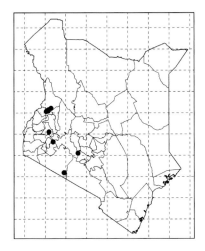

b. var. **glabrum** Brenan, Kew Bull. 5: 372. 1951; F.T.E.A. Rubiac. 1: 398. 1976. —Type: Zambia, Mwinilunga, R. Lunga, just below R. Mudjanyama junction, 25 Nov. 1937, *E. Milne-Redhead 3401* [holotype: K (K000422921); isotypes: BM (BM000903786), BR (BR0000008846921), K (K000422922), PRE (PRE 0594451-0)]

Stems and leaves glabrous or glabrescent.

Distribution: Western and central Kenya. [Angola, Ethiopia, Tanzania, and Zambia].

Habitat: Swampy grassy places, riversides, or streamside; 1800–2100 m.

Busia: ca. 25 km south of Eldoret on road to Nakuru, *Gilbert & Mesfin 6772* (EA, K). Nandi: ca. 5 km east of Kapsabet, *Hooper & Townsend 1524* (EA, K). Nyandarua: Ol Joro Orok, *Nattrass 607* (EA). Uasin Gishu: Eldoret, *Williams 231* (EA). West Pokot: south Cherangani Hills, *Symes 107* (EA).

12. **Galium ossirwaense** K. Krause, Bot. Jahrb. Syst. 43: 159. 1909; F.T.E.A. Rubiac. 1: 400. 1976; U.K.W.F.F.: 229. 2013. —Types: Tanzania, Masai, western slope of Ossirwa, *F. Jaeger 506* [holotype: B (destroyed)]; Tanzania, Arusha Ngorongoro, 25 May 1989, *S. Chuwa 2758* [neotype: K (K000319880), **designated here**] (Plate 47E)

= *Galium mollicomum* Bullock, Bull. Misc. Inform. Kew 1932: 498. 1932. —Type: Kenya, Mt. Elgon, Jan. 1931, *E.J. & C. Lugard 400a* [holotype: K (K000422930)]

= *Galium mollicomum* Bullock var. *friesiorum* Bullock, Bull. Misc. Inform. Kew 1932: 498. 1932. —Type: Kenya, Mt. Kenya, 23 Dec. 1921, *R.E. & T.C.E. Fries 479* [holotype: K (K000319866); isotypes: S, UPS]

Straggling herb, up to 35 cm tall, with erect flowering shoots, sparsely to densely covered with white hairs or less often glabrous. Leaves (7–)8(–9)-whorled; blades linear, 0.3–1.5 cm × 0.5–1.5(–2.5) mm, attenuate at the apex into a filiform acumen, glabrous or with a few stiff hairs near the inrolled margins. Inflorescences of dense cymes crowded towards the stem apex. Calyx-tube glabrous. Corolla pale to bright yellow; tube 0.3–0.6 mm long; lobes ovate-triangular, 1.3–2 mm long, acute. Fruits glabrous.

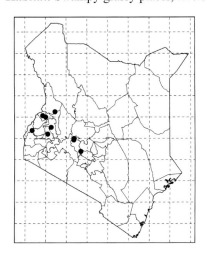

Plate 47 A, B. *Galium glaciale* var. *glaciale*; C, D. *G. scioanum* var. s*cioanum*; E. *G. ossirwaense*. Photo by YDZ (A, B) and SDZ (C, D) and GWH (E).

Distribution: Western, central, and southern Kenya. [Malawi, Mozambique, Tanznaia, and Uganda].

Habitat: Grasslands or montane forest edges; 2100–3800 m.

Bungoma: Mount Elgon, *Knox 3784* (EA). Kajiado: Amboseli, *Hindorf 808* (K). Laikipia: *Dowson 587* (EA). Meru: north Mount Kenya, *SAJIT 002735* (HIB). Nakuru: Eburru Forest Reserve, *Luke et al. 8877* (EA). Narok: Enesambulai Valley, *Greenway & Kanuri 14860* (EA). Nyandarua: north Kinangop, *Kuchar*

12353 (EA). Nyeri: west Mount Kenya, *Clarke 72M* (EA). Trans-Nzoia: Mount Elgon, *Mwangangi 321* (EA). Uasin Gishu: Tinderet Forest Reserve, *Geesteranus 5505* (K). West Pokot: Cherangani Hills, *Tweedie 3907* (K).

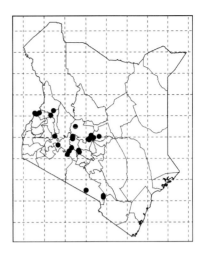

2. Subfam. **Cinchonoideae** Raf.

Small trees, shrubs, or rarely herbs. Raphides absent. Stipules usually entire or rarely bifid. Corolla-lobes valvate, contorted or imbricate. Ovary 1–many-locular, each with numerous or rarely solitary ovules. Fruits dry or succulent, dehiscent or indehiscent. Seeds with or without albumen.

Three tribes and five genera occur in Kenya.

Key to tribes

1a. Ovary 2–many-locular, with solitary pendulous ovule in each locule....... Trib. 12. Guettardeae
1b. Ovary 2-locular, with single to numerous ovules in each locule 2
2a. Inflorescences spicate to racemose or racemose to corymbiform; fruits large, dry, dehiscent from the top, 2-valved, covered with distinct lenticels...................... Trib. 10. Hymenodictyeae
2b. Inflorescences congested into spherical heads; fruits mostly free or sometimes fused into multiple, fleshy syncarps or pseudosyncarps... Trib. 11. Naucleeae

10. Trib. **Hymenodictyeae** Razafim. & B. Bremer

Trees, shrubs, epiphytic, or rarely lianas. Raphides absent. Leaves opposite; stipules interpetiolar, triangular or narrowly to deeply bifid, deciduous. Inflorescences terminal or/ and axillary, elongated, spicate to racemose or racemose to corymbiform, with or rarely without a pair of long-petiolate, scarious, caducous or persistent bracts at the base. Calyx-tube ovoid to subglobose, 5(–6)-lobed. Corolla funnel-shaped or narrowly campanulate; lobes 5, valvate. Stamens 5, inserted at the throat, included; filaments short; anthers basifixed, 2-thecous. Ovary 2-locular, with several to numerous ovules per locule; style slender, long-exserted. Fruits large, dry, dehiscent from the top, 2-valved, covered with distinct lenticels. Seeds winged.

Only one genus occurs in Kenya.

37. **Hymenodictyon** Wall.

Trees, shrubs, epiphytic, or rarely lianas. Leaves opposite, usually with domatia in the nerve-axils; stipules interpetiolar, deciduous, triangular to ligulate, entire or glandular-serrate. Inflorescences elongated, spicate to racemose, often with a pair of foliaceous reticulated marcescent long-petiolate bracts at the base. Calyx-tube subglobose; limb deeply 5(–6)-lobed, lobes linear to lanceolate, entire, deciduous. Corolla funnel-shaped or narrowly campanulate; lobes 5, valvate. Stamens 5, inserted below the throat, included; filaments short; anthers basifixed, 2-thecous. Ovary 2-locular; ovules numerous in each locule. Style filiform, exserted; stigma oblong-capitate. Capsules oblong or fusiform, 2-locular, loculicidal. Seeds numerous.

About 23 species, occurring in Africa, tropical Asia, and Madagascar; two species in Kenya.

1a. Inflorescences 6–22 cm long, with foliaceous reticulated marcescent long-petiolate bracts at the base; capsules 1–1.5 cm long..1. *H. floribundum*
1b. Inflorescences 1–8 cm long, without bracts at the base; capsules 1.5–4 cm long ....
.... 2. *H. parvifolium*

1. **Hymenodictyon floribundum** (Hochst. & Steud.) B.L. Rob., Proc. Amer. Acad. Arts 45: 404. 1910; F.T.E.A. Rubiac. 2: 452. 1988; K.T.S.L.: 516. 1994. ≡ *Kurria floribunda* Hochst. & Steud., Flora 25: 234. 1842. ≡ *Hymenodictyon kurria* Hochst., Flora 26: 71. 1843; Fl. Trop. Afr. 3: 42. 1877. —Type: Ethiopia, Mt. Scholoda, 11 June 1837, *G.H.W. Schimper 277* [holotype: B (destroyed); lectotype: P (P00214371), designated by Heine in Fl. Gabon 12(1): 58. 1966; isolectotypes: BM, BR (BR0000008359285 & BR0000008847966), G (G00436259, G00436260 & G00436261), GH (GH00096629), K (K000173360 & K000173361), M (M0106427 & M0106428), P (P00214373)] (Figure 34; Plate 48A–D)

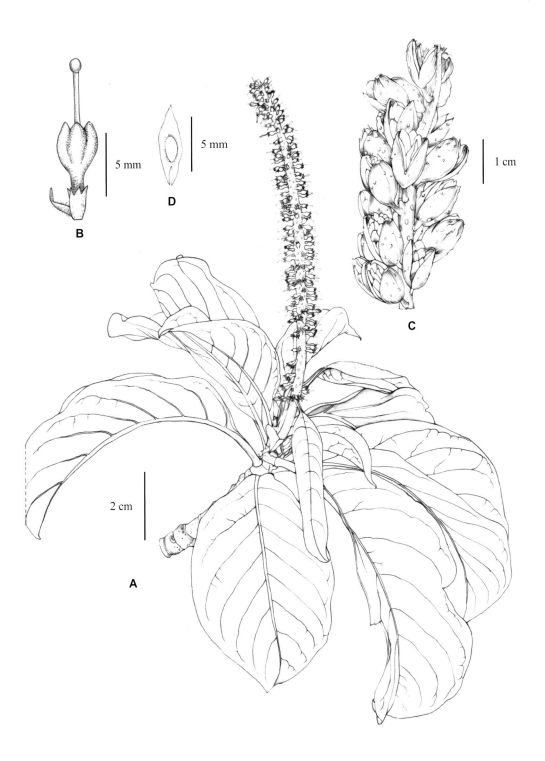

Figure 34 *Hymenodictyon floribundum*. A. flowering branch; B. flower; C. infructescence; D. seeds. Drawn by NJ.

Shrub or small tree, up to 12 m tall, with grey-black, reticulate bark. Leaves turning red before falling; blades elliptic to obovate, 5–25 cm × 2–12 cm, glabrous to densely tomentellous beneath, apex acuminate, base cuneate; stipules triangular to lanceolate or strap-shaped, 0.8–1.2 cm long; petiole up to 4 cm long. Inflorescences terminal, cylindrical, dense, raceme-like, up to 22 cm long, with peduncles up to 6 cm long; bracts paired, yellowish green to red, lanceolate to elliptic, 3.5–10 cm long, with petioles up to 6 cm long. Flowers sweet-scented. Calyx-tube minute, 5–6-lobed. Corolla yellowish red; tube 4–7 mm long; lobes 5–6. Ovary 2-locular; style up to 1 cm long, exserted; stigma 0.8–1 mm long. Capsules lenticellate, ellipsoid, 1–1.5 cm long. Seeds very compressed, 0.7–1 cm long, winged.

Distribution: Western Kenya. [Tropical Africa to north of southern Africa].

Habitat: Rocky hills and pavements in bushlands, woodlands, or grasslands; 540–3000 m.

Baringo: Kabarnet, *Bally 15240* (EA). Bungoma: Mount Elgon, *SAJIT PR0093* (EA). Kakamega: Kakamega Forest, *FOKP 814* (EA). Kericho: west of Kedowa, *Birch 60/306* (EA). Nandi: North Nandi Forest, *SAJIT 006608* (EA).

2. **Hymenodictyon parvifolium** Oliv., Hooker's Icon. Pl. 15: 69, t. 1488. 1885; F.T.E.A. Rubiac. 2: 454. 1976; K.T.S.L.: 517. 1994. —Type: Kenya, Mombasa, Nov. 1884, *R.T. Wakefield s.n.* [holotype: K (K000173362)] (Plate 48E, F)

Shrub, small tree or occasionally a liana, with grey and smooth bark. Leaves elliptic to, obtuse to acute at the apex, narrowly cuneate at the base; stipules deltoid or oblong-triangular, 2–7 mm long, entire or sometimes bifid; petioles up to 5.5 cm long. Inflorescences terminal, up to 8 cm long. Flowers very sweet-scented. Calyx-tube subglobose to oblong-ellipsoid, 1–1.5 mm long; lobes lanceolate, up to 2.5 mm long. Corolla white, greenish-white or yellow, tube 2.5–5 mm long; lobes 5–6, ovate, ca. 1.5 mm long. Ovary 2-locular; style up to 1.5 cm long, exserted; stigma ellipsoid. Capsules brown, ellipsoid, prominently lenticellate, splitting into 2 valves. Seeds straw-coloured, broadly elliptic.

Distribution: Western, central, southern, and coastal Kenya. [D.R. Congo and east Tropical Africa to east southern Africa].

Habitat: Dry rocky places in bushlands or woodlands; up to 1800 m.

Embu: Gwa-chengecha, along road to Ishiara, *Kirika et al. GBK045239/NMK 147* (K). Isiolo: Lengishu Hills, *Adamson*

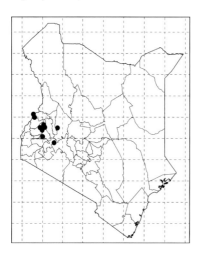

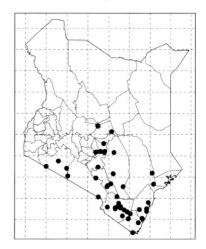

Plate 48 A–D. *Hymenodictyon floribundum*; E, F. *H. parvifolium*. Photo by GWH (A, C, E), NW (B), SDZ (D) and BL (F).

EAH12686 (EA). Kajiado: Loitokitok, *Ibrahim 706* (K). Kilifi: Kilifi Creek, *Robertson 6929* (EA, K). Kitui: Mutomo Hill, *SAJIT MU0050* (HIB). Kwale: Lungalunga-Ramisi Road, *Faden et al. 77/730* (K). Machakos: west slopes of Kindaruma Dam, *Gillett & Faden 18186* (EA, K). Makueni: Kibwezi, *Kokwaro 3831* (EA, K). Meru: Meru National Park, *Ament & Magogo 153* (EA). Mombasa: Mowesa, *Graham 1787* (EA, K). Narok: Nguruman Escarpment, *Kuchar 13804* (EA). Teita Taveta: Sagala Forest, *SAJIT 005365* (HIB); Buchuma, *Polhill & Paulo 477* (EA, K). Tana River: near Dalu, *Luke & Robertson 1299* (EA).

11. Trib. **Naucleeae** Kostel.

Trees, shurbs or rarely lianas. Raphides absent. Leaves opposite or rarely whorled; stipules interpetiolar, entire, caducous. Inflorescences terminal or/and axillary, congested into spherical heads. Flower 4–5(–6)-merous. Corolla aestivation imbricate or valvate. Stamens inserted in the tube or throat; filaments short; anthers mostly basifixed or rarely medifixed. Ovary 2-locular; ovules single to numerous in each locule, on predominantly imbricately pendulous or ascendent placentas attached to the base, one third, middle, upper third or apex of the septa. Fruits mostly free or sometimes fused into multiple, succulent syncarps or pseudosyncarps. Seeds winged or not.

Four genera occur in Kenya.

1a. Woody lianas or scrambling shrubs, armed with recurved spines .... 40. *Uncaria*
1b. Shrubs or trees, unarmed 2
2a. Ovaries and fruitlets connate; fruit a syncarp .... 41. *Nauclea*
2b. Ovaries and fruitlets free; fruit not a syncarp 3
3a. Leaves opposite, broadly elliptic to obovate; stipules entire, membranous, broadly elliptic ......
.... 38. *Mitragyna*
3b. Leaves 3–4-whorled, lanceolate or oblong-lanceolate; stipules bifid, triangular .... 39. *Breonadia*

38. **Mitragyna** Korth.

Trees with flattened buds. Leaves opposite, sometimes with domatia; stipules interpetiolar, caducous, large, generally ovate to obovate, entire. Inflorescences terminal and/or axillary, of 3–many globose heads in fascicles, cymes, umbels, thyrses or panicles. Flowers sessile, bisexual, monomorphic. Calyx tubular, with limb truncate to 5-lobed. Corolla-tube funnel-shaped; lobes 5, valvate. Stamens 5, inserted near corolla throat, exserted or included; filaments short; anthers basifixed. Ovary free, 2-locular; ovules numerous in each locule on fleshy, pendulous, axile placentas attached in upper third of septum; style distinctly exserted; stigma clavate to mitriform.

Fruit capsular, obovoid to ellipsoid, septicidally then loculicidally dehiscent. Seeds numerous, small, with winged margins.

About 12 species, discontinuouly distributed in Asia and Africa, nine of which in South Asia and China, three found in Africa; only one species in Kenya.

1. **Mitragyna rubrostipulata** (K. Schum.) Havil., J. Linn. Soc., Bot. 33: 73. 1897; K.T.S.L.: 521. 1994. ≡ *Adina rubrostipulata* K. Schum., Pflanzenw. Ost-Afrikas, C: 378. 1895. ≡ *Hallea rubrostipulata* (K. Schum.) J.-F. Leroy, Adansonia, n.s. 15: 66. 1975; F.T.E.A. Rubiac. 2: 449. 1988. ≡ *Fleroya rubrostipulata* (K. Schum.) Y. F. Deng, Taxon 56(1): 247. 2007. —Type: Congo, Landschaft Kiboscho Sinas Boma, Steppe am Ouare, *G. Volkens 1583* [lectotype: K (K000394921), designated by G. D. Haviland in J. Linn. Soc. Bot. 33: 73. 1897; isotypes: BM, E (E00130749)] (Figure 35; Plate 49)

= *Adina rubrostipulata* K. Schum. var. *discolor* Chiov., Racc. Bot. 51. 1935. —Type: Kenya, Meru, Nyambeni, *G. Balbo 462* [holotype: FT]

Large tree, up to 30 m tall, with greyish-brown, fairly smooth to rough bark. Leaves opposite, broadly elliptic, oblong-elliptic or obovate, 4.5–35 cm × 2.5–25 cm, acuminate at the apex, rounded to subcordate at the base; petioles up to 5.5 cm long; stipules often bright red, membranous, broadly elliptic, 3.5–7.5(–15) cm × 2.5–5(–10) cm. Inflorescences terminal and axillary, paniculate, of several to many heads to 3 cm in diameter. Flowers

Plate 49 *Mitragyna rubrostipulata*. Photo by YDZ.

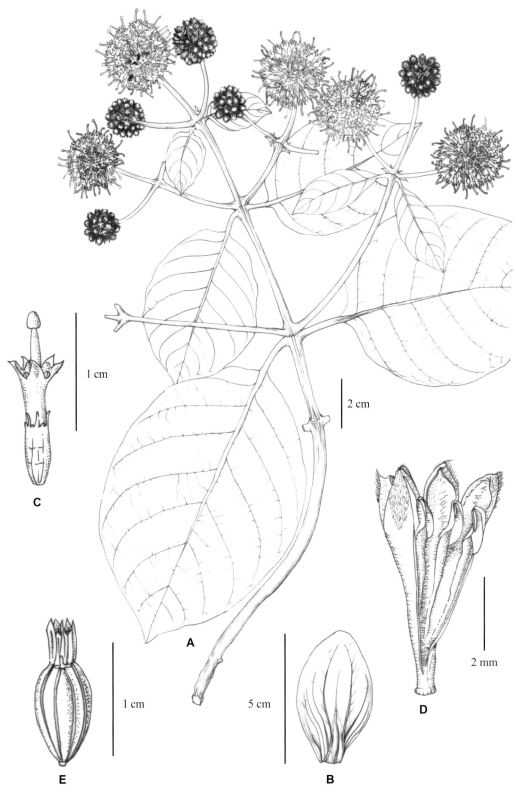

Figure 35 *Mitragyna rubrostipulata*. A. flowering branch; B. stipule; C. flower; D. dissected corolla, showing the stamens; E. fruit. Drawn by NJ.

sweet-scented. Calyx-tube 2–3 mm long, ribbed; lobes elongate-triangular, 1–2 mm long. Corolla green or yellowish white; tube narrowly funnel-shaped, 4–6 mm long; lobes ovate-triangular, 1.5–3 mm long. Stamens 5, inserted near corolla throat. Style 1–1.5 cm long, exserted; stigma oblong-ellipsoid, ca. 1 mm long. Fruits ellipsoid, loculicidally ribbed, 0.9–1.3 cm long, crowned with the persistent calyx. Seeds winged at both ends.

Distribution: Central and southern Kenya. [Tropical Africa].

Habitat: Moist forests along the rivers; 900–2190 m.

Embu: southeast Mount Kenya, *Battiscombe 690* (EA, K). Kirinyaga: streamside near Kerugoya, *Brunt 1477* (K). Meru: near Maua, *SAJIT Z0139* (HIB); Kangeta Forest Station, *Mabberley 419* (EA, K). Teita Taveta: Mbololo Forest, *Faden et al. 770* (EA, K). Tharaka-Nithi: Meru National Park, *Luke & Luke 10290* (EA, K).

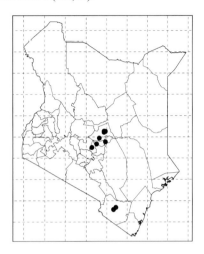

39. **Breonadia** Ridsdale

Small to medium-sized trees. Leaves 3–4-whorled, lanceolate or oblong-lanceolate; stipules bifid, triangular, deciduous. Inflorescences lateral, solitary; peduncles long; bracts membranous, often connated and calyptra-like; bracteoles spathulate. Flowers 5-merous. Calyx-tubes free from each other; lobes oblong. Corolla-tube hypocrateriform; lobes oblong, imbricate. Stamens inserted in the throat, exserted; anthers basifixed; filaments short. Ovary 2-locular; placentas shortly obovoid, attached to the upper third of the septum; ovules 2–5 per locule, pendulous; style exserted. Fruit a head of free capsules. Seeds ovoid, bilaterally compressed, not winged.

Only one species occur in Yemen, tropical and sub-tropical Africa including Kenya and Madagascar.

1. **Breonadia salicina** (Vahl) Hepper & J.R.I. Wood, Kew Bull. 36: 860. 1982; F.T.E.A. Rubiac. 2: 445. 1988. ≡ *Nerium salicinum* Vahl, Symb. Bot. 2: 45. 1791. —Types: Yemen, Hadie, Mar. 1763, *P. Forsskål 236* [lectotype: C (C10002639, only the right vegetative branch), **designated here**)]; ibid., *P. Forsskål 237* [syntype: C (C10002640, only the left vegetative branch)] (Figure 36; Plate 50)

= *Nauclea microcephala* Delile, Cent. Pl. Afr. Voy. Méroé: 67. 1826. ≡ *Adina microcephala* (Delile) Hiern, Fl. Trop. Afr. 3: 40. 1877. ≡ *Breonadia microcephala* (Delile) Ridsdale, Blumea 22: 549. 1975; K.T.S.L.: 502. 1994. —Type: Sudan, Sennaar, Singue, *F. Cailliaud s.n.* [holotype: MPU (MPU007021)]

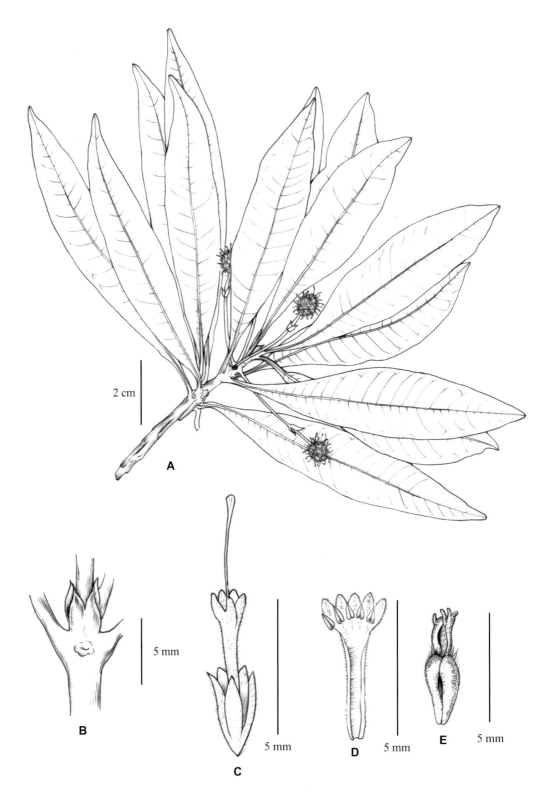

Figure 36 *Breonadia salicina*. A. a flowering branch; B. stipules; C. flower; D. dissected corolla, showing the stamens; E. fruit. Drawn by NJ.

Plate 50 *Breonadia salicina*. Photo by YDZ.

Medium to large tree up to 20 m tall, with grey, fissured and scaly bark. Leaves 3–4-whorled; blades leathery, lanceolate to narrowly elliptic, 8–33 cm × 2–9 cm, subacute to slightly acuminate at the apex, narrowly cuneate at the base, petioles up to 2.5 cm long; stipules interpetiolar, triangular, bifid. Flowers in small axillary spherical heads up to 2.5 cm in diameter; peduncles up to 10 cm long; bracts ovate, membranous, often connate and calyptra-like, 0.4–1 cm long; bracteoles filiform-spathulate, up to 5 mm long. Calyx-tube 1–2 mm long; lobes oblong, linear or triangular, up to 4.5 mm long. Corolla white or yellowish brown; tube up to 6.5 mm long; lobes imbricate, linear, 1–2.5 mm long. Stamens inserted in the throat, exserted. Ovary 2-locular; ovules 2–5 per locule, pendulous; style distinctly exserted. Fruits septicidal capsules. Seeds narrowly compressed, ovoid, 2.2–2.8 mm long, not winged.

Distribution: Central, southern, and coastal Kenya. [Arabian Peninsula, tropical and southern Africa, and Madagascar].

Habitat: Lowland riverine forests; up to 1500 m.

Embu: bank of Thiba River, *Robertson 2033* (EA, K). Kiambu: Fourteen Falls, *SAJIT Z002-19* (HIB). Kilifi: Pangani Rocks, *Luke & Robertson 1920* (EA, K). Kwale: Shimba Hills, *Luke & Luke 4576* (EA). Makueni: Kibwezi, *Davidson 38* (K). Meru: Rojewero River, *Hamilton 717* (EA). Murang'a: Blue Post Hotel, *Faden 66119* (EA). Teita Taveta: Taita Hills, *Mwachala et al. 3336* (EA). Tana River: Tana Falls, *Sampson 8* (EA, K). Tharaka-Nithi: Ura River between Kampi ya Elsa and Teziwa, *Hamilton 743* (EA).

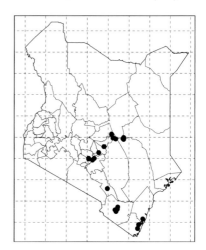

40. **Uncaria** Schreb.

Woody lianas or scrambling shrubs, armed with recurved spines. Raphides absent. Leaves opposite; stipules interpetiolar, entire to bifid, usually reflexed. Inflorescences axillary or terminal, usually of single globose head, pedunculate, bracteate. Flowers 5-merous, bisexual, monomorphic. Calyx-tube variable; limb 5-lobed. Corolla funnel-form; lobes 5, imbricate. Stamens 5, inserted in corolla-tube near throat, exserted; filaments short; anthers dorsifixed. Ovary 2-locular; ovules numerous in each locule on axile placentas attached in upper third of septum; stigma globose or clavate, exserted. Fruits capsular, fusiform to obovoid, loculicidally dehiscent into 2 valves. Seeds numerous, fusiform, winged.

About 50 species mainly distributed in tropical Asia, two in tropical America, one in Madagascar, and three in Africa; only one species with two subspecies in Kenya.

1. **Uncaria africana** G. Don, Gen. Hist. 3: 471. 1834; F.T.E.A. Rubiac. 2: 450. 1988; K.T.S.L.: 550. 1994. —Type: Congo, Orientale, Yangumbi, 13 Nov. 1937, *J. Louis 6593* [holotype: BR (BR0000008859044); isotypes: MO (MO-391836), U (U0006332), YBI (YBI191981860)]

Climbing shrub or liana, up to 15 m long, armed with opposite hooks, equal or unequal; young stems often quadrangular. Leaves opposite; blades elliptic, 3.5–15 cm × 1–

6 cm, acuminate at the apex, rounded to cuneate at the base; petioles up to 1 cm long; stipules interpetiolar, bifid, often joined at the base. Inflorescences terminal, of solitary single globose head, 4–6 cm in diameter. Calyx-tube ellipsoid, campulate or funnel-shaped, 1–4 mm long; lobes 5, short to distinctly triangular, up to 3 mm long. Corolla yellowish white, bristly pubescence outside; tube up to 1.5 cm long; lobes oblong, 2.5–4 mm long. Stamens 5, inserted in corolla-tube near throat, exserted; filaments very short; anthers dorsifixed. Ovary 2-locular. Style 1.5–2 cm long. Capsules fusiform, 1–2.5 cm long, 10-ribbed, with persistent calyx.

1a. Calyx with small deltoid teeth, ca. 0.5 mm long .... a. subsp. *africana*
1b. Calyx with distinct narrowly triangular teeth, 1.5–3 mm long .... b. subsp. *lacus-victoriae*

a. subsp. **africana**

Calyx with small deltoid teeth, ca. 0.5 mm long.

Distribution: Coastal Kenya. [Tropical Africa, also in Comoros Islands and Madagascar].

Habitat: Coastal dry evergreen forests; up to 270 m.

Kilifi: Pangani Rocks, *Luke 1841* (EA, K).

Plate 51 *Uncaria africana* subsp. *lacus-victoriae*. Photo by TD.

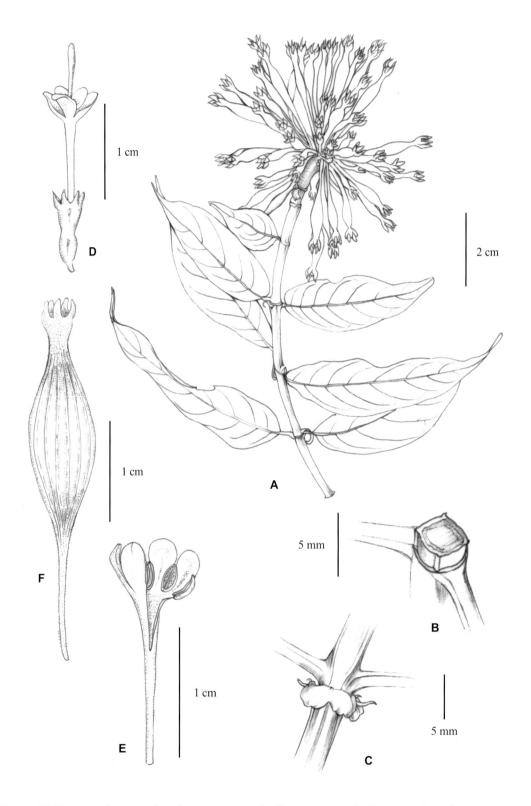

Figure 37 *Uncaria africana* subsp. *lacus-victoriae*. A. flowering branch; B. transection of young stem; C. node showing the stipule; D. flower; E. dissected corolla, showing the stamens; F. fruit. Drawn by NJ.

Kwale: Shimba Hills, *Luke & Robertson 2752* (EA, K). Tana River: near Hewani, *Robertson & Obara 6428* (EA, K); Tana River Primate Reserve, *Luke & Robertson 1163* (EA, K).

Calyx with distinct narrowly triangular teeth, 1.5–3 mm long.

Distribution: Western Kenya. [Ethiopia, Uganda, and Tanzania].

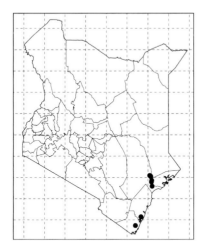

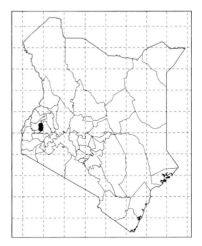

b. subsp. **lacus-victoriae** Verdc., Kew Bull. 31: 181. 1976; F.T.E.A. Rubiac. 2: 450. 1976; K.T.S.L.: 550. 1994. —Type: Uganda, Bunyoro, Bugoma Forest, 28 Nov. 1905, *M.T. Dawe 704* [holotype: K (K000394913)] (Figure 37; Plate 51)

Habitat: Inland wet evergreen forests; 1500–1650 m.

Kakamega: Kakamega Forest, *SAJIT 006748* (HIB).

41. Nauclea L.

Shrubs or small trees. Raphides absent. Leaves opposite; stipules interpetiolar, elliptic, obovate or deltoid, entire or slightly bifid. Inflorescences terminal and/or lateral, of solitary or few globose heads. Flowers 4–5-merous, bisexual, monomorphic. Calyx-tubes fused to each other; lobes obtuse to triangular. Corolla-tube funnel-form to salver-form; lobes oblong, imbricate. Stamens 4–5, inserted in the upper part of corolla-tube, exserted or not; filaments short; anthers basifixed. Ovary 2-locular; placentas attached to the middle or upper third of the septum; ovules numerous in each locule. Style exserted; stigma fusiform, exserted. fruits a fleshy syncarp. Seeds ovoid or ellipsoid, sometimes slightly compressed, not winged.

About 16 species widely distributed in tropical Africa, Asia, and Australia; six species in Africa; only one species at the border of Kenya and Uganda.

1. **Nauclea latifolia** Sm., Cycl. 24: 5. 1813. ≡ *Sarcocephalus latifolius* (Sm.) E.A. Bruce, Kew Bull. 2: 31. 1947; F.T.E.A. Rubiac. 2: 439. 1988; K.T.S.L.: 523. 1994. —Type: Sierra Leone, *H. Smeathman s.n.* [holotype: BM (BM000902821)] (Figure 38)

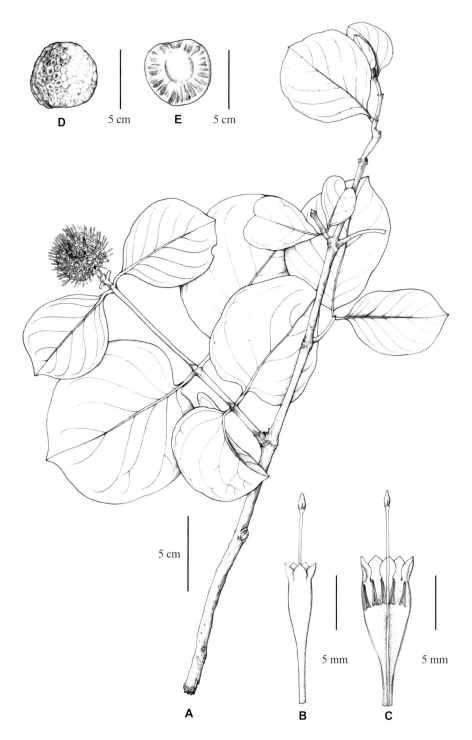

Figure 38 *Nauclea latifolia*. A. flowering branch; B. flower; C. dissected corolla, showing the stamens and style; D. fruit; E. longitudinal section of fruit. Drawn by NJ.

Small tree or shrub, up to 9 m tall, with greyish brown, deeply fissured bark. Leaves opposite; blades broadly elliptic to rounded-ovate, 8–22 cm × 5–14 cm, shortly acuminate at the apex, cuneate to rounded or subcordate at the base; petioles up to 2 cm long; stipules deltoid, up to 5 mm long, entire or slightly bifid. Inflorescences terminal, of solitary globose heads 4–5 cm in diameter; peduncles short, 1.5–2.5 cm long. Calyx-tubes fused to each other; lobes triangular, 0.5–1 mm long. Corolla yellowish white, tube narrowly funnel-shaped; lobes 2–2.5 mm long. Stamens 5, inserted in the upper part of corolla-tube. Ovary 2-locular; style exserted; stigma 2.5–4 mm long. Fruit a syncarp, ovoid or globose, up to 8 cm in diameter. Seeds numerous, subglobose or ellipsoidal, 1–1.2 mm long.

Distribution: Western Kenya. [West tropical Africa to Ethiopia and northwest Angola].

Habitat: Wet grasslands or savannas; 1100–1200 m.

Busia: near Busia, *Brunt 1412* (EA, K).

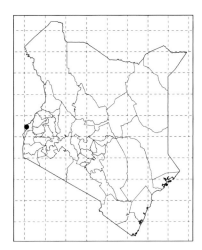

12. Trib. **Guettardeae** DC.

Trees or shrubs. Raphides absent. Leaves opposite or rarely verticillate; stipules interpetiolar or intrapetiolar, free or connate, persistent or caducous. Flowers 4–10-merous. Corolla-lobes valvate or imbricate. Stamens inserted in corolla throat, included or exserted; filaments short; anthers dorsifixed. Ovary 2–many-locular, with solitary pendulous ovule in each locule; style sometimes unequal; stigma included or exserted. Fruits drupaceous or with a woody putamen, with calyx limb persistent. Seeds with little or no albumen.

Only one genus occurs in Kenya.

42. **Guettarda** L.

Trees or shrubs. Raphides absent. Leaves opposite or rarely ternate; stipules caducous, interpetiolar or intrapetiolar, simple, generally triangular. Inflorescences axillary, cymose, single- to many-flowered. Flowers 4–9-merous, sessile or shortly pedicellate, unisexual or bisexual, monomorphic. Calyx-tube globose or ovoid; limb truncate or 4–9-toothed. Corolla salver-shaped; lobes 4–9, imbricate. Stamens 4–9, inserted in corolla-tube, included; anthers dorsifixed. Ovary 4–9-locular, ovules single in each locule, pendulous. Style slender; stigma capitate, included.

Fruits drupaceous, ellipsoid, or subglobose, with 2–9 pyrenes. Seeds with little or no albumen.

About 80 species, mostly confined in tropical America and Pacific region; only one species widespread on the coasts of Indian Ocean and east Pacific Ocean, also on the coast of Kenya.

1. **Guettarda speciosa** L., Sp. Pl. 2: 991. 1753; F.T.E.A. Rubiac. 3: 924. 1991; K.T.S.L.: 514. 1994. —Type: India or Java, *Herb. Linn. No. 1121.1*, excl. fruit [lectotype: LINN, designated by K.M. Wong & B. Verdcourt in Kew Bull. 43: 496. 1988] (Figure 39; Plate 52)

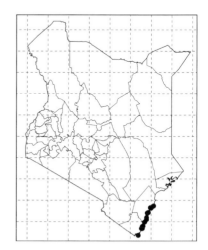

Shrub or small tree, up to 8 m tall, with brown, smooth or often lenticellate bark. Leaves often crowded at the ends of the branchlets, opposite; blade elliptic to broadly obovate, 5–30 cm × 3.5–18 cm, rounded to obtuse at the apex, rounded to cordate at the base; petioles stout, 0.5–5 cm long; stipules ovate, up to 2 cm long, caducous. Inflorescences cymose, single- to several-flowered. Flowers fragrant, sessile or shortly

Plate 52 *Guettarda speciosa*. Photo by BL.

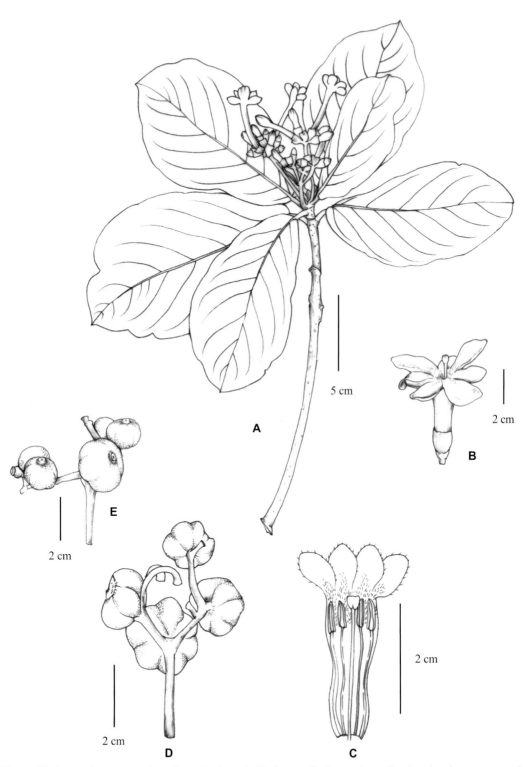

Figure 39 *Guettarda speciosa*. A. a flowering branch; B. flower; C. dissected corolla, showing the stamens and style; D, E. inflorescence. Drawn by NJ.

pedicelate. Calyx-tube campanulate, 1–3 mm long, limb tubular, 2–5 mm long, truncate. Corolla yellowish white, slightly pink tinged; tube narrowly cylindrical, 2–5 cm long; lobes 4–9, oblong or obovate, 2–15 mm long, obtuse to rounded. Fruits drupaceous, ellipsoid or subglobose, 1.5–3 cm in diameter.

Distribution: Coastal Kenya. [Widespread throughout Indian and Pacific Ocean].

Habitat: Sandy and limestone coasts; up to 50 m.

Kilifi: Mida Creek, *Kuchar 12026* (EA); Kikambala, *Bock 71/4* (EA). Kwale: Diani Forest, *Gillett & Kibuwa 19896* (EA); ca. 1 mile south of Jadini, *Greenway 9797* (EA, K).

3. Subfam. **Ixoroideae** Raf.

Trees, shrubs, subshrubs or rarely herbs. Raphides absent. Stipules entire, bifid or rarely fimbricate, interpetiolar or rarely intrapetiolar. Flowers dioecious, monoecious or protogynous, usually actinomorphic or rarely zygomorphic, heterostylous, and often with secondary pollen presentation. Calyx mostly persistent. Corolla aestivation contorted to left, or rarely to right, imbricate or valvate. Ovary 2-locular, with 1–many ovules. Fruits fleshy or dry, indehiscent or capsular.

Nine tribes and 41 genera occur in Kenya.

Key to tribes

1a. Some calyx-lobes enlarged, foliaceous or developed into membranous calycophylls; ovary 2–many-locular; ovules numerous in each locule .... Trib. 13. Mussaendeae
1b. Calyx-lobes never enlarged; ovary 2- to few-locular; ovules single to few, or numerous in each locule 2
2a. Fruit a capsule, loculicidally splitting into 2 valves; seeds winged .... Trib. 14. Crossopterygeae
2b. Fruit an indehiscent drupe; seeds unwinged 3
3a. Ovules pendulous from near the apex of locule .... Trib. 16. Vanguerieae
3b. Ovules erect or attached by middle to the septum 4
4a. Petioles articulate; flowers 4(–5)-merous .... Trib. 15. Ixoreae
4b. Petioles not articulate; Flowers 4–8(–12)-merous 5
5a. Inflorescences mostly axillary; ovary 2-locular with 1–several ovules in each locule; style-arms divergent; seeds few, never held together in a matrix 6
5b. Inflorescences terminal, laterial or occasionally axillary; ovary 1–2(–4)-locular with 1–numerous ovules in each locule; style-arms seldom divergent; seeds many, sometimes held together in a matrix 7
6a. Flowers 4–8(–12)-merous; ovary 2-locular; seed coat more or less entire, or with a distinct longitudinal ventral invagination .... Trib. 17. Coffeeae
6b. Flowers (4–)5(–6)-merous, rarely 6–7(–8)-merous; ovary 1–2-locular; seed coat with a very distinct fingerprint-like or sometimes reticulate pattern invagination .... Trib. 18. Octotropideae
7a. Ovary 2-locular, with axile placentas; seeds with adaxial excavation; pollen grains in monads, 3–4(–5)-colporate .... Trib. 20. Pavetteae
7b. Ovary 1–2-locular, with axile or parietal placentas; seeds with or without adaxial excavation; pollen grains in monads, tetrads, or massulae, 3(–4)-aperturate, porate or pororate, 3–4(–5)-colporate, or rarely pantoporate 8
8a. Seeds with secondary thickenings and folded testa .... Trib. 19. Sherbournieae
8b. Seeds testa unfolded .... Trib. 21. Gardenieae

13. Trib. Mussaendeae Benth. & Hook. f.

Trees, erect or scrambling shrubs, or lianas. Leaves opposite, decussate; stipules usually bifid, persistent or caducous. Flowers often heterodistylous, without secondary pollen presentation. Some calyx-lobes always enlarged, foliaceous or developed into membranous calycophylls. Corolla valvate or imbricate. Ovary 2–many-locular; ovules numerous in each locule. Fruits fleshy or dry, indehiscent or capsular.

Three genera occur in Kenya.

1a. Some calyx-lobes enlarged, subfoliaceous ....43. *Heinsia*
1b. Some calyx-lobes developed into a stalked white to colored, membranous, stipitate calycophylls 3
2a. Flower buds bearing 5 apical filiform appendages; fruits dehiscing at the apex ....44. *Pseudomussaenda*
2b. Flower buds without appendages; fruits indehiscent..45. *Mussaenda*

43. Heinsia DC.

Shrubs, small trees, or rarely subshrubs or climbers. Leaves opposite, shortly petiolate; stipules bifid, persistent or caducous. Flowers solitary or few to many in lax to dense cymes, often large and sweet-scented, 4–6-merous, heterostylous. Calyx-tube campanulate or oblong; lobes enlarged, subfoliaceous. Corolla salver-shaped; tube slender, appressed hairy outside and densely hairy inside throat; lobes spreading, unequally 5–6-lobed, large, imbricated. Stamens 5–6, inserted on the throat of the corolla, included; filaments short; anthers linear. Ovary 2-locular; ovules numerous in each locule. Style rather slender, with 2 shortly linear stigma-arms. Fruits dry, indehiscent, mostly crowned by the persistent calyx-lobes, many-seeded.

A small genus of 5 species restricted to tropical and subtropical Africa; two species in Kenya.

1a. Leaves less than 4.5 cm long, almost glabrous except for venation; flowers always solitary or few in lax terminal cymes..1. *H. crinita*
1b. Leaves up to 15 cm long, densely pubescent beneath; flowers mostly many in dense terminal cymes..2. *H. zanzibarica*

1. **Heinsia crinita** (Widmark) G. Taylor, Cat. Vasc. Pl. S. Tome 209. 1944; F.T.E.A. Rubiac. 2: 476. 1988; K.T.S.L.: 516. 1994. ≡ *Gardenia crinita* Widmark, Strip Guinea Med. Sp. Nov. 2: 13. 1829. —Type: Sierra Leone, *A. Afzelius s.n.* [holotype: BM (BM000902993)]

Shrub or small tree, up to 7.5 m tall.

Leaves opposite; blades lanceolate, oblong or narrowly to broadly elliptic, 1–14 cm × 0.5–7 cm, slightly to markedly acuminate at the apex, cuneate at the base; petioles very short; stipules bifid, up to 5 mm long, caducous. Flowers solitary or in lax terminal cymes, with peduncles up to 4 cm long. Calyx-tube turbinate; lobes 5–6, subfoliaceous, oblong, elliptic or lanceolate, up to 2 cm long. Corolla white, sweet-scented; tube slender, up to 3 cm long; lobes 5–6, linear-oblong to broadly elliptic, 1–3 cm long, shortly apiculate at the apex. Fruits oblong-ellipsoid to subglobose, up to 2 cm long, crowned with persistent calyx-lobes. Seeds strongly compressed.

subsp. **parviflora** (K. Schum. & K. Krause) Verdc., Kew Bull. 31: 184.1976; W.F.E.A.: 152. 1987; F.T.E.A. Rubiac. 2: 477. 1988; K.TS.L: 516. 1994; C.P.K.: 381. 2016. ≡ *Heinsia parviflora* K. Schum. & K. Krause, Bot. Jahrb. Syst. 39: 530. 1907. ——Types: Tanzania, Makonde Plateau, Mkomadatchi, 27 Jan. 1901, *W. Busse 1083* [syntype: B (destroyed); lectotype: EA, **designated here**]; Tanzania, Uzaramo, Mogo Forest (Sachsenwald), 28 Nov. 1901, *C. Holtz 341* [syntype: B (destroyed)] (Figure 40; Plate 53A, B)

Shrub or rarely small tree, up to 4.5 m tall. Leaves 1.5–4.5 cm × 0.7–2 cm, glabrous except for venation. Calyx-lobes 3.5–6.5 cm. Corolla-tube 1.8–2.5 cm long; lobes 1–1.8 cm long. Fruits ellipsoid, 0.8–1.2 cm long.

Distribution: Coastal Kenya. [Tanzania, Somalia, Mozambique, Malawi, Zimbabwe, South Africa].

Habitat: Coastal bushlands or forest edges; up to 300 m.

Note: *Heinsia crinita* subsp. *crinita* usually has much larger leaves and calyx-lobes as well as bigger fruits, and occurs from west tropical Africa to northwest Zambia.

Kilifi: Arabuko-Sokoke Forest, *SAJIT 006454* (HIB); Kaya Kivara, *Robertson & Luke 4769* (EA). Kwale: Shimba Hills, *De Block et al. 465* (EA, K); Dzombo Hill, *Robertson et al. MDE330* (EA, K). Lamu: ca. 2 km north of Hindi, *Gillett 20338* (EA, K); Witu, *Rawlins 11255* (EA, K). Mombasa: Mtwapa Creek, *Napier 3559* (EA, K). Tana River: Nairobi Ranch, *Festo & Luke 2514* (EA, K).

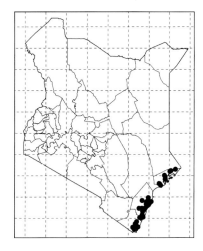

2. **Heinsia zanzibarica** (Bojer) Verdc., Kew Bull. 35: 422. 1980; F.T.E.A. Rubiac. 2: 477. 1988; K.T.S.L.: 516. 1994. ≡ *Mussaenda zanzibarica* Bojer, Ann. Sci. Nat., Bot., sér. 2, 4: 264. 1835. —Type: Tanzania, Zanzibar, *W. Bojer s.n.* [holotype: P (P00546198); isotype: MPU (MPU021495)] (Plate 53C, D)

Shrub or small tree, up to 4.5 m tall. Leaves opposite, shortly petiolate; blades lanceolate, elleptic or oblong-elliptic, 5–15 cm × 2.5–6 cm, acuminate at the apex, cuneate to almost rounded at the base; stipules bifid, up to 1.3 cm long. Flowers sweet-scented, mostly many in dense terminal cymes. Calyx-tube turbinate, hairy; lobes subfoliaceous, up to 1.2 cm long. Corolla white, tube slender, 2–

Figure 40 *Heinsia crinita* subsp. *parviflora*. A. fruiting branch; B. flower; C. fruit. Drawn by NJ.

Plate 53 A, B. *Heinsia crinita* subsp. *parviflora*; C, D. *H. zanzibarica*. Photo by GWH (A), BL (B) and VMN (C, D).

3 cm long; lobes elliptic to oblong-elliptic, 2–3.5 cm long, caudate-acuminate at the apex. Fruits dry, indehiscent, ellipsoid, 6–9 mm long, bristly pilose.

Distribution: Coastal Kenya. [Mozambique and Tanzania].

Habitat: Coastal forest edges; up to 400 m.

Kilifi: Cha Simba Rocks, *Robertson et al. 6980* (EA, K); Kaya Jibana, *Robertson & Luke 4493* (EA). Kwale: Shimba Hills, *SAJIT V0165 & V0177* (HIB); Gongoni Forest Reserve, *Robertson & Luke 6352* (EA).

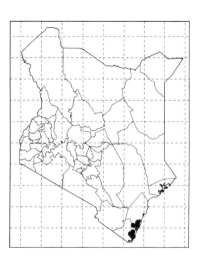

44. Pseudomussaenda Wernham

Shrubs or subshrubs. Raphides absent. Leaves opposite, or rarely 3-whorled in lower part, petiolate; stipules broad, with 1–2 subulate lobes. Inflorescences of terminal cymes, several- to many-flowered, with some calyx-lobes enlarged into stalked, cream white to coloured, membranous calycophylls. Flowers 5-merous, sessile, heterostylous. Calyx-lobes 5, subulate to linear or enlarged. Corolla-tube narrowly cylindrical; lobes ovate, apiculate. Stamens included. Ovary 2-locular, each with many ovules; style slender; stigmas 2, slightly exserted. Capsules oblong, dehiscent or not. Seeds small, reticulate.

A small genus of six species restricted in tropical Africa; one species in Kenya.

1. **Pseudomussaenda flava** Verdc., Kew Bull. 6: 378. 1952; F.T.E.A. Rubiac. 2: 467. 1988; K.T.S.L.: 533. 1994. —Type: Uganda, Toro, Bwamba, Kabango, 29 Spet. 1932, *A.S. Thomas 725* [holotype: EA; isotypes: K (K000414482), KAW (KAW000084)] (Figure 41)

Shrub, up to 3.5 m tall; stems pubescent to densely hairy. Leaves opposite, or rarely 3-whorled in lower part; blades elliptic, 2–11 cm × 1–5 cm, acute to acuminate at the apex, cuneate at the base, glabrescent to sparsely or densely pubescent; petioles up to 8 mm long; stipules with 2 filiform lobes up to 4.5 mm long. Cymes terminal, lax, few to several-flowered. Calyx-tube narrowly turbinate, ca. 3 mm long; lobes filiform, 2–6 mm long, or with one enlarged into a long-stalked calycophylls; blades white to yellow, membranous, oblong, ovate to rounded or subcordate, 1.5–5.5 cm × 1–4.5 cm; stipe up to 3.5 cm long. Corolla yellow; tube 2.5–3.5 cm long; lobes ovate to oblong, 4–8 mm long, apiculate. Capsules oblong, ca. 7 mm long, loculicidally 2-valved. Seeds many, small, angular.

Distribution: Northwestern Kenya [D.R. Congo, Ethiopia, Nigeria, and Uganda].

Habitat: Glasslands or bushlands; 600–1400 m.

Turkana: Songot Hills, *Champion 180* (K?); Lake Turkana, *Wellby s.n.* (K?).

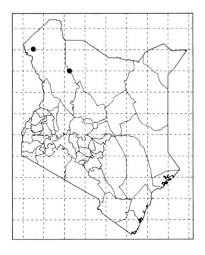

45. Mussaenda L.

Trees, shrubs or lianas. Raphides absent. Leaves petiolate, opposite or occasionally 3-whorled, stipules persistent or caducous, interpetiolar, entire or 2-lobed. Inflorescences terminal

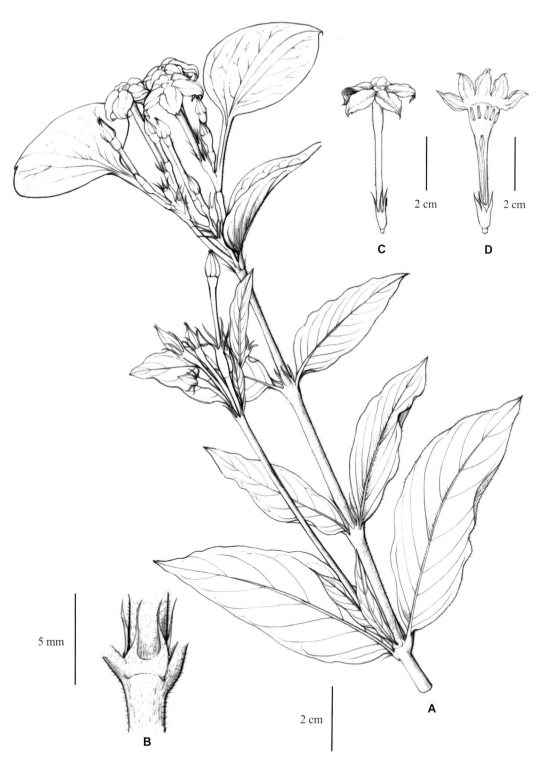

Figure 41 *Pseudomussaenda flava*. A. flowering branch; B. portion of branch showing the stipule; C. flowers; D. longitudinal section of flower, showing stamens and style. Drawn by NJ.

or lateral, cymose, paniculate or thyrsiform, several to many flowered. Flowers occasionally sweet-scented, isostylous or heterostylous. Calyx-lobes 5, persistent or caducous, frequently several lobes on each inflorescence developed into a stalked white to colored, membranous, stipitate calycophylls. Corolla white or colored; tube slender; lobes 5, valvate-reduplicate in bud. Stamens 5, inserted in the middle to upper part of corolla-tube, included; anthers basifixed. Ovary 2-locular; ovules numerous in each locule. Style slender; stigmas 2-lobed, lobes linear, included or exserted. Fruits globose, ellipsoid or oblong, indehiscent, sometimes with persistent calyx-lobes. Seeds numerous, small, angled to flattened.

About 190 species widespread in tropical Africa, Asia, Madagascar and Pacific islands; four species in Kenya.

1a. Leaves coriaceous, glabrous; foliaceous calyx-lobes absent or not fully developed..................
....1. *M. arcuata*
1b. Leaves herbaceous, always hairy at least on the nerves beneath; foliaceous calyx-lobes usually well developed, white or colored 2
2a. Foliaceous calyx-lobes bright red .... 2. *M. erythrophylla*
2b. Foliaceous calyx-lobes cream white or yellow..3
3a. Coastal species; fruits narrowly ellipsoid .... 3. *M. monticola*
3b. Inland species; fruits narrowly globose...4. *M. microdonta*

1. **Mussaenda arcuata** Lam. ex Poir., Encycl. 4: 392. 1797; F.T.E.A. Rubiac. 2: 461. 1988; K.T.S.L.: 522. 1994. —Type: Mauritius, *P. Commerson 304* [holotype: P-LAM] (Plate 54)

Erect or scrambling shrub, up to 7 m tall. Leaves opposite, coriaceous, oblong, elliptic or rounded, 3–18 cm × 1–8 cm, acuminate to caudate at the apex, cuneate, acute or rounded at the base; petioles up to 3 cm long. Stipules caducous, 3–12 mm long, entire or 2-lobed, reflexing. Inflorescences terminal, paniculate, dense or lax, few- to many-flowered. Flowers sweet-scented, isostylous or heterostylous. Calyx-tube turbinate to ellipsoid, 2–4 mm long; lobes linear or slightly spathulate, 0.7–1.5 mm long; foliaceous calyx-lobes absent or not fully developed. Corolla pale yellow with a bright orange star of hairs in the center; tube up to 2.5 cm long; lobes lanceolate to broadly ovate, up to 1.5 cm long. Fruits ellipsoid or subglobose, with a round whitish scar at the apex, edible.

Distribution: Western Kenya. [Tropical Africa to southern Africa].

Habitat: Bushlands, woodlands, wooded grasslands, or forest margins; 1000–2200 m.

Bungoma: ca. 4 km from Webuye on road to Kitale, *Gilbert & Mesfin 6591* (EA). Kakamega: Kakamega Forest, *SAJIT 006774* (HIB). Kericho: Mau Forest, *Mutangah &*

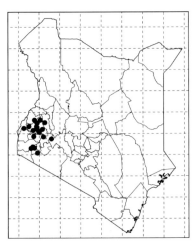

Plate 54 A–F. *Mussaenda arcuata*. Photo by GWH.

Kamau M149 (EA). Kisii: near Nyamarambe, *Vuyk 149* (EA). Kisumu: Kisumu, *Brown 1226* (EA). Migori: ca. 2 km south of Ogumo, *Vuyk 428* (EA). Nandi: Nandi Forest, *Makin 307* (EA, K). Siaya: Jera, East Ugenya, *Kokwaro 4389* (EA). Uasin Gishu: Turbo, *Brodhurst-Hill 374* (EA).

2. **Mussaenda erythrophylla** Schumach. & Thonn., Beskr. Guin. Pl.: 116. 1827; F.T.E.A. Rubiac. 2: 463. 1988; K.T.S.L.: 522. 1994. —Type: Ghana, Aquapim Mt., *P. Thonning 93* [holotype: C (C10004192); isotype: G-DC]

Erect or scrambling shrub, up to 8 m tall. Leaves opposite, elliptic to round, 2.5–18 cm × 1–8 cm, acute to acuminate at the apex, rounded, cuneate or cordate at the base, hairy on both surfaces; petioles up to 5 cm long; stipules bifid, up to 1.2 cm long, usually reflexed, persistent. Inflorescences terminal, of dense panicles; foliaceous calyx-lobes 2–6, elliptic to round, 3–10 cm × 2–9 cm, bright red. Flowers heterostylous. Calyx-tube obovoid; lobes lanceolate, caducous. Corolla-tube up to 3 cm long; lobes round, up to 1.2 cm long. Fruits yellowish, ellipsoid, 1–2.4 cm long, hairy.

Distribution: Western Kenya. [Tropical Africa].

Habitat: Wet forests; 1500–1600 m.

Kakamega: Kakamega Forest, *Dale K3110* (EA).

3. **Mussaenda monticola** K. Krause, Bot. Jahrb. Syst. 48: 406. 1912; F.T.E.A. Rubiac. 2: 466. 1988; K.T.S.L.: 523. 1994. —Types: Tanzania, Morogoro, Uluguru Mts., Apr. 1907, *C. Holtz 1720* [holotype: B (destroyed)]; Tanzania, Kilombero, 250 m, 24 May 1976, *K.B. Vollesen MRC3679* [neotype: K (K 000319540), **designated here**] (Plate 55A–D)

Shrub or small tree, up to 10 m tall. Leaves opposite, elliptic to ovate, 10–23 cm × 5–11 cm, sparsely hairy on both surfaces, acuminate at the apex, acute to obtuse or rounded at the base; petioles up to 5.5 cm long; stipules ovate, up to 1.2 cm long, usually reflexed. Inflorescences terminal, of lax panicles; foliaceous calyx-lobes several, elliptic, 5.5–7 cm × 3.5–5.5 cm, cream to yellow. Calyx-tube narrowly turbinate; lobes linear, 3–4 mm long. Corolla-tube 1.5–2 cm long; lobes broadly ovate, 3–4 mm long. Fruits narrowly ellipsoid to cylindrical, 2–3 cm long, crowned by persistent calyx-lobes.

Distribution: Coastal Kenya. [Tanzania].

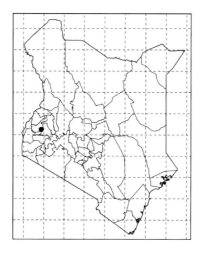
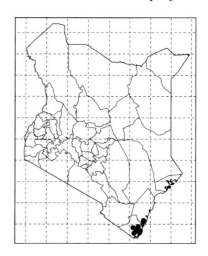

Habitat: Lowland moist forests; up to 400 m.

Kilifi: Pangani Rocks, *Luke & Robertson 1915* (EA); Chonyi-Ribe Road, *Faden & Faden 74/1264* (EA, K). Kwale: Shimba Hills, *SAJIT 006080* (HIB); Gongoni Forest Reserve, *Robertson & Luke 5952* (EA).

4. **Mussaenda microdonta** Wernham, J. Bot. 51: 239. 1913; F.T.E.A. Rubiac. 2: 464. 1988; K.T.S.L.: 523. 1994. —Type: Tanzania, Usambara Mts., June 1900, *J. Buchwald 431* [holotype: BM (BM000902915); isotypes: BR (BR0000008848352 & BR0000008848345), K (K000319549)]

Shrub or small tree, up to 9 m tall. Leaves opposite, elliptic, 8–22 cm × 3–11 cm, acuminate at the apex, cuneate at the base, sparsely hairy on both surfaces; petioles up to 3.5 cm long; stipules narrowly triangular, 3–13 mm long. Inflorescences terminal, of lax to dense panicles; foliaceous calyx-lobes several, ovate to broadly ovate, 3–9 cm × 2.5–8 cm, white to cream. Flowers sweet-scented, heterostylous or isostylous. Calyx-tube turbinate, 2.5–5 mm long; lobes linear, up to 1.2 cm long, caducous. Corolla-tube up to 4 cm long; lobes ovate, 5–12 mm long. Fruits globose, 6–10 mm in diameter, dotted with brown lenticels.

subsp. **odorata** (Hutch.) Bridson, Kew Bull. 30: 696. 1976; F.T.E.A. Rubiac. 2: 466. 1988; K.T.S.L.: 523. 1994. ≡ *Mussaenda odorata* Hutch., Bull. Misc. Inform. Kew 1914: 247. 1914. —Type: Kenya, Embu, S.E. Mt. Kenya, *E. Battiscombe 708* [holotype: K (K000319584); isotypes: BM, EA (EA 000002958)] (Figure 42; Plate 55E–G)

= *Mussaenda keniensis* K. Krause, Notizbl. Bot. Gart. Berlin-Dahlem 10: 603. 1929. —Types: Kenya, Meru, Kaseri R., 22 Feb. 1922, *R.E. & T.C.E. Fries 1818* [lectotype: UPS, **designated here**]; Kenya, Nyeri, Mukengeria R., 5 Mar. 1922, *R.E. & T.C.E. Fries 2060* [syntype: UPS]

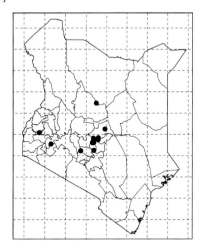

Leaves 13–22 cm long. Calyx-lobes 5–12 mm long. Corolla-lobes with long acumen 2–4 mm long. Fruits ca. 10 mm in diameter.

Distribution: Western and central Kenya. [Tanzania].

Habitat: Upland moist forests; 1500–2100 m.

Note: *Mussaenda microdonta* subsp. *microdonta* only distributed in northeast Tanzania, which has heterostylous flowers, subulate to narrowly triangular calyx-lobes, glabrescent foliaceous lobes, and smaller fruits.

Embu: Thiba River, *Kabuye 68* (EA, K). Kericho: Changena Tea Estate, *Perdue & Kibuwa 9237* (EA). Kirinyaga: Mount Kenya, *Faden et al. 71/902* (EA, K). Meru: Nyambeni Hills, *Verdcourt in Polhill & Verdcourt 282* (K). Murang'a: Kimakia, *Greenway 9680* (EA, K). Samburu: Mathew's Range, *Luke 14289* (EA). Tharaka-Nithi: Thungura Hill Forest, *SAJIT 003956* (HIB). Vihiga: western Vihiga Mbale, *Thomas 2593* (K).

Plate 55 A–D. *Mussaenda monticola*; E–G. *M. microdonta* subsp. *odorata*. Photo by GWH (A–D) and YDZ (E–G).

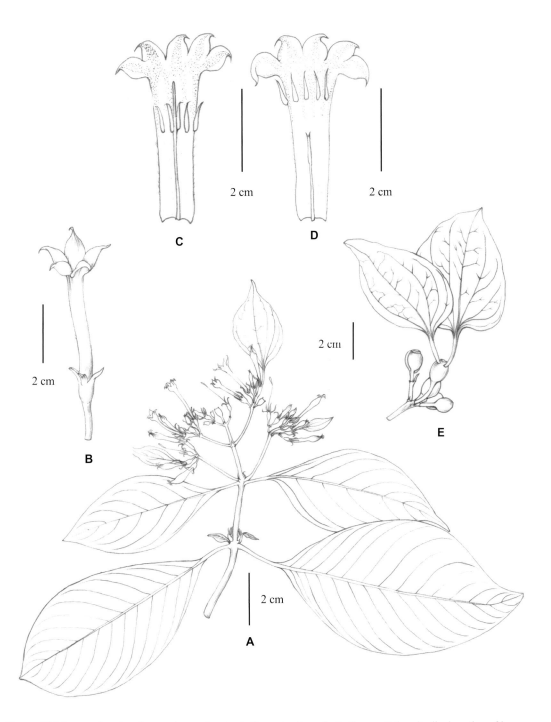

Figure 42 *Mussaenda microdonta* subsp. *odorata*. A. flowering branch; B. flower; C. longitudinal section of long-styled flower, showing the stamens and style; D. longitudinal section of short-styled flower, showing the stamens and style; E. fruit with foliaceous calyx-lobes. Drawn by NJ.

14. Trib. **Crossopterygeae** F. White ex Bridson

Small trees or shrubs. Leaves opposite; stipules interpetiolar, subpersistent. Flowers 4–6-merous, small, numerous in dense terminal corymbose panicles. Calyx-tube obovoid to globose. Corolla-lobes contorted, spreading. Ovary 2-locular; ovules few in each locule. Style filiform, long-exserted; stigma 2-lobed. Capsule loculicidally splitting into 2 valves. Seeds peltate, winged.

Only one genus occurs in Kenya.

46. **Crossopteryx** Fenzl

Small trees or shrubs. Leaves opposite; stipules interpetiolar, triangular, subpersistent. Flowers 4–6-merous, small, strongly scented, numerous in dense branched terminal corymbose panicles. Calyx-tube short, broadly oblong; lobes obtuse, deciduous. Corolla salver-shaped; tube slender, straight; lobes 4–6, contorted, spreading. Stamens 4–6, subsessile, inserted at the throat of the corolla; anthers lanceolate, oblong, apiculate. Ovary 2-locular; ovules few in each locule. Style filiform, exserted; stigma clavate, bilobed. Capsules crustaceous, globose, loculicidally splitting into 2 valves. Seeds several, peltate, compressed, without membranous wings.

Only one species and confined to the savanna areas of tropical Africa, also in Kenya.

1. **Crossopteryx febrifuga** (Afzel. ex G. Don) Benth., Niger Fl.: 381. 1849; F.T.E.A. Rubiac. 2: 457. 1988; K.T.S.L.: 512. 1994. ≡ *Rondeletia febrifuga* Afzel. ex G. Don, Gen. Hist. 3: 516. 1834. —Type: Sierra Leone, near Freetown, *G. Don s.n.* [holotype: BM (BM000902895); isotype: K (K000394956)] (Figure 43; Plate 56)

Small tree or shrub, up to 5(–15) m tall. Leaves opposite, elliptic, elliptic-oblong, ovate, obovate or almost round, 1.5–13.5 cm × 1.2–7.5 cm, rounded to shortly acuminate at the apex, broadly cuneate to rounded at the base; petioles up to 1.8 cm long; stipules interpetiolar, triangular, 2–3 mm long. Inflorescences dense, branched, terminal, corymbose panicles. Flowers strongly scented, 4–6-merous. Calyx-tube broadly oblong, ca. 1 mm long; lobes elliptic to linear, 0.5–1.5 mm long. Corolla cream-white or pale yellow, densely pubescent outside, tube 5–10 mm long; lobes round, ca. 1.5 mm long and wide. Style filiform, exserted for 3–7.5 mm; stigma

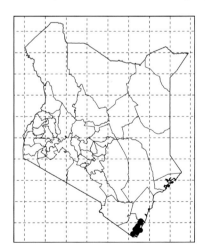

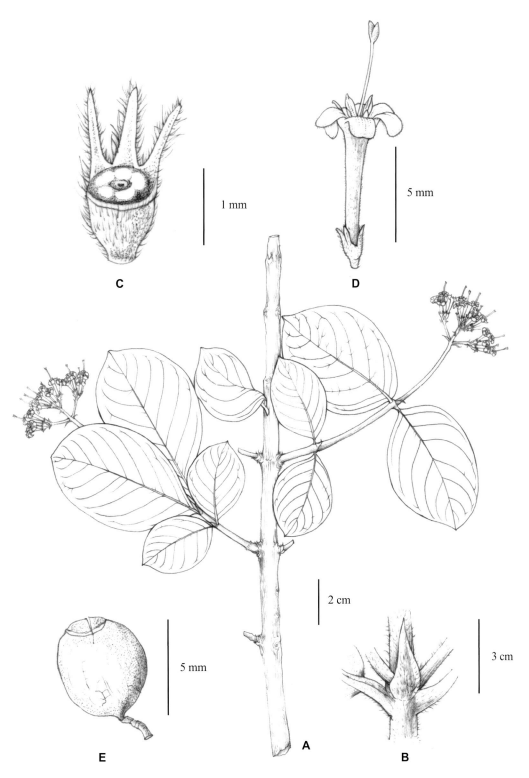

Figure 43 *Crossopteryx febrifuga*. A. flowering branch; B. a portion of branch showing the stipule; C. calyx; D. flower; E. fruit. Drawn by NJ.

Plate 56 A, B. *Crossopteryx febrifuga*. Photo by GWH.

clavate, bilobed. Capsule ellipsoid or globose, crustaceous, 6–10 mm long. Seeds thin, flat, 3.2–5 mm × 2.5–3.5 mm.

Distribution: Coastal Kenya. [Tropical and southern Africa].

Habitat: Lowland woodlands or grasslands; up to 400 m.

Kilifi: Kaya Rabai, *Luke & Robertson 2283* (EA). Kwale: Shimba Hills, *SAJIT 005482* (HIB); Godoni Forest Area, *Spjut 4586* (EA); Mwachi Forest Reserve, *Robertson & Luke 6169* (EA, K).

15. Trib. **Ixoreae** Benth. & Hook. f.

Shrubs or small trees. Raphides absent. Leaves opposite or rarely ternate, chartaceous to coriaceous; petioles articulate at the base; stipules persistent to caducous. Inflorescences terminal or rarely axillary, cymose to corymbiform or paniculiform. Flowers 4(–5)-merous, usually fragrant. Corolla hypocrateriform; lobes contorted in bud. Stamens inserted at corolla throat, partially to fully exserted. Ovary 2(–7)-locular, ovule single in each locule. Style slender; stigmas 2-lobed, exserted. Fruits drupaceous, spherical or 2(–7)-lobed; pyrenes 1–2, 1-seeded, leathery to crustaceous.

Only one genus occurs in Kenya.

47. Ixora L.

Shrubs or small trees. Leaves opposite or rarely ternate, chartaceous to coriaceous, entirely glabrous; petioles articulate at the base; domatia and bacterial nodules absent; stipules persistent to caducous, interpetiolar, usually connated into a sheath, with an elongated or pointed tip. Inflorescences terminal or rarely axillary, cymose to corymbiform or paniculiform, few- to many-flowered, bracts or bracteate reduced; axes often articulate; inflorescence-supporting leaves often present at the base of peduncles. Flowers 4(–5)-merous, bisexual, monomorphic, often fragrant. Calyx-tube ovoid; limb truncate or 4(–5)-lobed. Corolla-tube cylindrical, usually slender; lobes contorted in bud, spreading or reflexing. Stamens 4, inserted at corolla throat, partially to fully exserted; filaments short; anthers linear, dorsifixed near the base. Ovary 2-locular, with single ovule in each locule; style slender; stigmas 2-lobed, exserted. Drupes spherical or 2-lobed, slightly fleshy or coriaceous; pyrenes 1–2, 1-seeded. Seeds globose, ellipsoid or oblanceolate; embryo dorsal, curved.

A large genus of about 550 species widespread in tropical Africa, Madagascar, America, Asia, and Pacific islands; only two species in Kenya.

1a. Inland species; inflorescence-supporting leaves often present; pedicels absent or less than 4 mm long; corolla-lobes 4–7 mm long .... 1. *I. scheffleri*
1b. Coastal species; inflorescence-supporting leaves often absent; pedicels up to 12 mm long; corolla-lobes 6–10 mm long .... 2. *I. narcissodora*

1. Ixora scheffleri K. Schum. & K. Krause, Bot. Jahrb. Syst. 39: 553. 1907; F.T.E.A. Rubiac. 2: 612. 1988; K.T.S.L.: 517. 1994. —Type: Tanzania, E. Usambara Mts., Derema, 1899, *G. Scheffler 218* [holotype: B (destroyed); lectotype: Z (Z-000021794), designated by P. de Block in Rev. Ixora Afr.: 150. 2005; isolectotype: E (E00279366)]

Small tree or shrub, up to 8 m tall. Leaves opposite; blades glabrous, chartaceous, elliptic to oblong-elliptic, 3–18 cm × 2–6 cm, acute to shortly acuminate at the apex, cuneate at the base; petioles up to 2 cm long; stipule-limbs truncate, 2–5 mm long, with a subulate to filiform apical lobe up to 5 mm long, caducous. Inflorescences terminal, cymose; supporting leaves oblong-elliptic, ovate or rarely rounded, 2–10 cm × 1–6 cm; bracteoles triangular to subulate. Calyx-tube 1–1.5 mm long; lobes shortly triangular, minute. Corolla white; tube up to 2.5 cm long; lobes oblong-elliptic or narrowly obovate, 4–7 mm long. Drupes ellipsoid to rounded, 8–10 mm long.

subsp. **keniensis** Bridson, Kew Bull. 32: 603. 1978; F.T.E.A. Rubiac. 2: 613. 1988; K.T.S.L.: 517. 1994. —Type: Kenya, Meru, Marania, 30 Apr. 1944, *J. Bally 30* in *P.R.O. Bally 3530* [holotype: K (K000311727); isotype: EA]

Calyx-lobes rounded, truncate or obtuse at the apex; stipule-lobes less than 1 mm long.
Distribution: Central Kenya. [Endemic].

Habitat: Mountane forests; 1900–2700 m.

Note: *Ixora scheffleri* subsp. *scheffleri* occurs in Malawi, Mozambique and Tanzania, with calyx-lobes acute or less often obtuse-acute at the apex and longer stipule-lobes.

Embu: *Abraham* in *Bally 6514* (EA, K). Kirinyaga: near Castle Forest Station, *Perdue & Kibuwa 8358* (EA, K, WAG). Meru: Marania, *Bally 3530* (K). Nyeri: Ragati Forest, *Höft 4117* (EA).

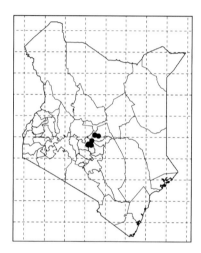

2. **Ixora narcissodora** K. Schum., Bot. Jahrb. Syst. 33: 356 1903; F.T.E.A. Rubiac. 2: 616. 1988; K.T.S.L.: 517. 1994. —Types: Tanzania, Pangani, Pangani R., Makinyumbi, 300 m, 4 Aug. 1900, *G. Scheffler 263* [holotype: B (destroyed)]; Tanzania, Tuliani, Luale stream, 360 m, 6 July 1933, *H.J. Schilieben 4053* [neotype: B, designated by P. de Block in Rev. Ixora Afr.: 136. 2005; isoneotypes: BM, BR (BR0000006268374), HBG, LISC, M (M0106278), P, S (S05-9472), Z] (Figure 44; Plate 57)

Shrub or small tree, up to 10 m tall. Leaf-blades lanceolate, oblanceolate or elliptic to oblong-elliptic, 3–25 cm × 1–6 cm, acute to shortly acuminate at the apex, cuneate at the base; petioles up to 1 cm long; stipule-limbs truncate to triangular, 1.5–4 mm long, with an aristate lobe up to 3.5 mm long. Inflorescences lax, cymose; supporting leaves usually absent; bracteoles triangular to filiform. Calyx-tube 1–2 mm long; lobes minute. Corolla white to pink; tube slender, up to 8 cm long; lobes narrowly oblong to oblong, 6–10 mm long. Drupes reddish, bilobed or rounded, 6–9 mm in diameter.

Distribution: Coastal Kenya. [Malawi, Mozambique, and Zimbabwe].

Habitat: Lowland forests; up to 450 m.

Kilifi: Pangani Rocks, *Luke & Robertson 1902* (EA). Kwale: Shimba Hills, *SAJIT 006040* (HIB); Kaya Kinondo, *Malombe et al. 1648* (EA). Lamu: Witu Forest, *Robertson & Luke 5509* (EA). Tana River: Tana River Primate National Reserve, *Kirika et al. 568* (EA); Hewani Forest, *Robertson & Luke 5307* (EA, K).

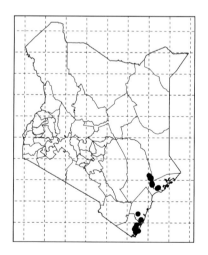

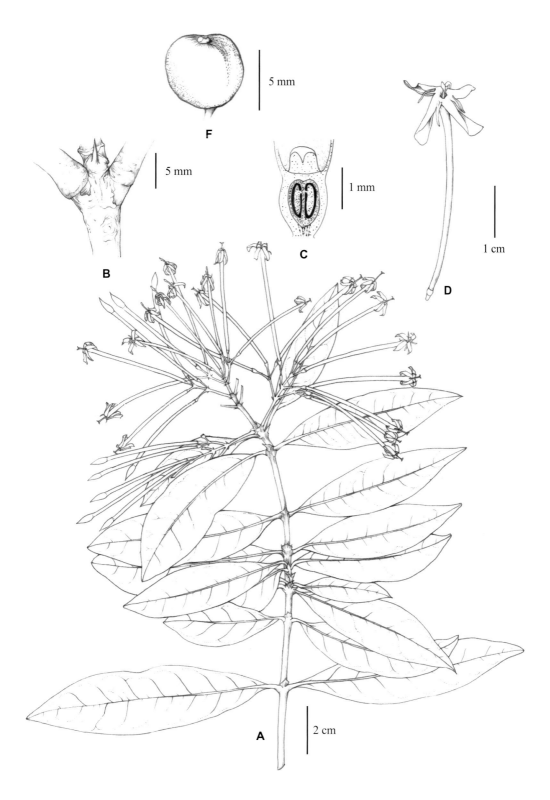

Figure 44 *Ixora narcissodora*. A. flowering branch; B. portion of branch showing the stipule; C. ovary; D. flower; E. fruit. Drawn by NJ.

Plate 57 A, B. *Ixora narcissodora*. Photo by GWH.

16. Trib. **Vanguerieae** A. Rich. ex Dumort.

Trees, erect or scandent shrubs, or rarely woody herbs. Spines present or absent, paired or rarely whorled. Stipules always connate. Flowers hermaphroditic or rarely dioecious. Corolla-lobes (4–)5(–6), valvate. Ovary 2–5(–20)-locular; ovule solitary and pendulous in each locule. Fruits fleshy drupe with 1–10 pyrenes.

11 genera occur in Kenya.

1a. Calyx-lobes conspicuous, leafy, 8–16 mm long, and much exceeding the corolla-tube .... 58. *Vangueria* (*V. pallidiflora*)
1b. Calyx-lobes not conspicuous, or if conspicuous not exceeding the corolla-tube 2
2a. Scandent shrubs or lianas 3
2b. Erect woody herbs, shrubs or trees, rarely scandent 4
3a. Leaves chartaceous to subcoriaceous, rarely coriaceous; calyx-limb dentate or slightly lobed .... 48. *Keetia*
3b. Leaves usually subcoriaceous to coriaceous; calyx-limb a dentate to repand rim, usually much smaller .... 50. *Psydrax*
4a. Ovule solitary in each locule .... 55. *Fadogia*
4b. Ovules 2–many in each locule 5
5a. Inflorescences umbellate or 1-flowered, entirely enclosed in bud by paired connate bracts .... 52. *Pyrostria*
5b. Inflorescences various, never enclosed by paired bracts 6
6a. Corolla-lobes linear-lanceolate, 2–4 cm long, much exceeding the tube ...... 57. *Vangueriopsis*
6b. Corolla-lobes not as above 7
7a. Subshrubby herbs or single-stemmed shrubs from a woody rootstock, up to 2 m tall .... 56. *Multidentia*
7b. Shrubs or small trees, up to 20 m tall 8
8a. Stipules glabrous within; style usually at least twice as long as corolla-tube; stigmatic knob cylindric, about twice as long as wide .... 50. *Psydrax*
8b. Stipules hairy or glabrous within; style usually much less than twice of the length of corolla-tube; stigmatic knob mostly as broad as long 9
9a. Ovary 2-locular; stigmatic knob 2-lobed; fruit 2-seeded 10
9b. Ovary 2–5(–6)-locular; stigmatic knob 2–5-lobed; fruit 1–5-seeded 12
10a. Calyx with a well-developed tubular part; pyrenes very thickly wooded, strongly irregularly ridged with lines of dehiscence apparent .... 56. *Multidentia*
10b. Calyx without such conspicuous tubular part; pyrenes not as above 11
11a. Flowers unisexual, (4–)5–6-merous, functionally male ones 1–20 in fascicules, functionally females usually solitary .... 51. *Bullockia*

11b. Flowers bisexual, 4–5-merous, few to many in pedunculate cymes .... 49. *Afrocanthium*
12a. Leaves usually with domatia; corolla-tube glabrous or hairy, but not bearded at the throat; fruits usually less than 30 mm in diameter .... 54. *Rytigynia*
12b. Leaves without domatia; corolla-tube bearded at the throat; fruits large, up to 50 mm in diameter .... 58. *Vangueria*

48. Keetia E.P. Phillips

Scandent shrubs. Leaves opposite, chartaceous or occasionally coriaceous; stipules lanceolate, triangular or ovate. Flowers 4–6-merous, several in pedunculate cymes, axillary. Calyx-tube ellipsoid to ovoid, dentate or slightly lobed. Corolla white, cream or yellow; tube cylindrical, short; lobes reflexed. Stamens insert at the throat of the corolla; filaments well-developed; anthers narrowly ovate or oblong, exserted. Ovary 2-locular, with single ovule in each locule; style long, slender, distinctly exserted; stigmatic knob cylindrical. Fruit a 2-seeded drupe, slightly to strongly bilobed; pyrenes woody or less often cartilaginous. Seeds ovoid, convoluted.

A genus about 35 species, confined to tropical and south Africa; four species in Kenya.

1a. Young branches sparsely to densely covered with golden to rust-coloured hairs; leaf-blades with finely reticulate tertiary veins 2
1b. Young branches glabrous to sparsely pubescent; leaf-blades with rather spaced or obscure tertiary veins 3
2a. Stipules lanceolate to ovate; leaf-blades frequently subcordate to cordate at the base ....
.... 1. *K. gueinzii*
2b. Stipules linear from a triangular base; leaf-blades obtuse to rounded or rarely subcordate at the base .... 2. *K. venosa*
3a. Leaf-blades with secondary nerves in 3–4 main pairs; flowers 15–20 in pedunculate compact cymes .... 3. *K. lukei*
3b. Leaf-blades with secondary nerves in 7–10 main pairs; flowers 30–60 in pedunculate cymes .
.... 4. *K. zanzibarica*

1. **Keetia gueinzii** (Sond.) Bridson, Kew Bull. 49: 803. 1994; F.T.E.A. Rubiac. 3: 911. 1991; K.T.S.L.: 517. 1994. ≡ *Canthium gueinzii* Sond., Linnaea 23: 54. 1850. —Type: South Africa, Port Natal., *W. Gueinzius s.n.* [holotype: S; isotype: K (K000422579)] (Figure 45; Plate 58)

Scandent shrub or liana, up to 9(–25) m long; young branches sparsely to densely covered with crisped or spreading golden to rust-coloured hairs. Leaf-blades oblong-lanceolate to ovate, 5–13 cm × 2–6.5 cm, acuminate at the apex, obtuse to rounded or more frequently subcordate to cordate at the base, glabrous to rarely glabrescent above, glabrescent to densely pubescent beneath; secondary nerves in 6–9 main pairs; tertiary nerves finely reticulate; domatia present as tufts of hairs; petioles up to 7 mm long; stipules

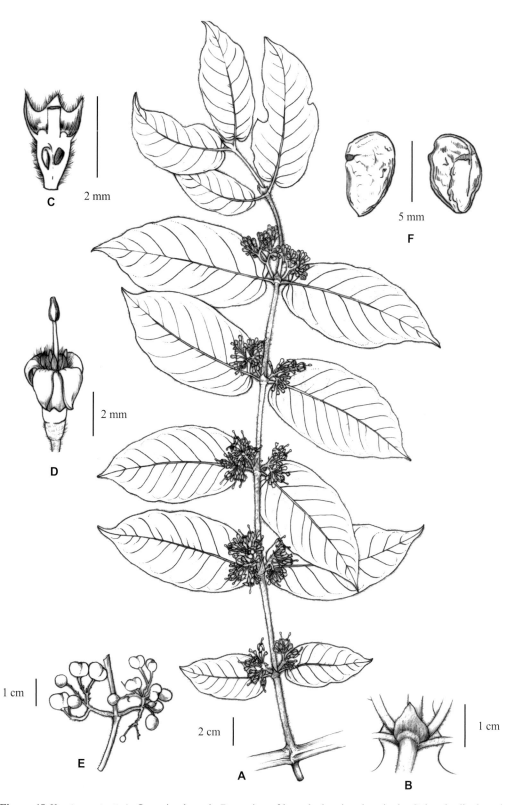

Figure 45 *Keetia gueinzii*. A. flowering branch; B. portion of branch showing the stipule; C. longitudinal section of ovary and calyx; D. flower; E. infructescences; F. pyrenes. Drawn by NJ.

Plate 58 A–F. *Keetia gueinzii*. Photo by CL (A), GWH (B–F).

lanceolate to ovate, up to 1.3 cm × 0.6 cm, pubescent. Flowers (4–)5(–6)-merous, 20–50 in pedunculate cymes; peduncles to 15 mm long. Calyx-tube up to 1 mm long; limb to 1.5 mm long. Corolla cream-white; tube up to 4 mm long; lobes oblong-lanceolate to ovate, 2.5–4 mm long. Style up to 1 cm long, glabrous; stigmatic knob 1.2–2.3 mm long. Fruits oblong, 7–9 mm long; pyrenes 6–7 mm long, obovoid to the mispherical.

Distribution: Western, central, southern, and coastal Kenya. [Cameroon, Ethiopia to Angola and southern Africa].

Habitat: Moist forests, secondary bushlands, or woodlands; 50–2500 m.

Baringo: Katimok Forest, *Dale 2423* (EA, K). Bomet: Tinderet Forest, *Geesteranus 4999* (K). Bungoma: Webuye Falls, *Tweedie 3078* (K). Elgeyo-Marakwet: Iten, Kibukuimet, *SAJIT 007060* (HIB). Embu: Ena Bridge, *Kayu 531* (EA, K). Kakamega: Kakamega Forest, *SAJIT 006775* (HIB). Kericho: Sambret Catchment, *Kerfoot 3728* (EA). Kiambu: Limuru, *Napier s.n.* (EA). Kilifi: Kaya Jibana, *Luke & Robertson 2635* (EA). Kisii: on Marongo Ridge, *Vuyk & Breteler 216* (EA, K). Kwale: Shimba Hills, *SAJIT 005500* (HIB). Makueni: Kilungu, Kavatanzou School, *Mwangangi 1666* (EA, K). Meru: Meru Forest, *SAJIT 003902* (HIB). Nakuru: Endabarra, *Bally 4846* (K). Nandi: North Nandi Forest, *SAJIT 006603* (HIB). Nithi: Chuka Forest, *SAJIT VK0067* (HIB). Nyamira: 16 km NEE of Kisii, *Vuyk & Breteler 175* (EA). Nyeri: southeast Aberdares, *Moon 752* (K). Samburu: Mathew's Range, *Luke 14194* (EA). Siaya: Jera, East Ugenya, *Kokwaro 4384* (EA). Teita Taveta: Mbololo Forest, *SAJIT 005321* (HIB). Trans-Nzoia: Cherangani Forest Station, *Lind et al. 5084* (EA). West Pokot: Cherangani Hills, *SAJIT Z0025* (HIB).

2. **Keetia venosa** (Oliv.) Bridson, Kew Bull. 41(4): 970. 1986; F.T.E.A. Rubiac. 3: 914. 1991; K.T.S.L.: 518. 1994. ≡ *Plectronia venosa* Oliv., Trans. Linn. Soc. London 29: 85, t. 49. 1873. —Type: Uganda, W. Nile, Madi, Dec. 1862, *J.A. Grant s.n.* [holotype: K] (Plate 59A–D)

Scandent shrub or liana, up to 7 m long; young branches sparsely to densely covered with rust-coloured hairs. Leaf-blades oblong-elliptic, elliptic or rarely round, 4.5–15 cm × 2.5–8 cm, acuminate at the apex, cuneate to rounded or rarely subcordate at the base, glabrous or rarely glabrescent above, glabrous or occasionally sparsely pubescent beneath, lateral nerves in 5–9 main pairs; tertiary nerves finely reticulate; domatia present as inconspicuous tufts of hairs; petioles up to 1.5 cm long; stipules linear from a triangular base, up to 1.7 cm long. Flowers 4–6-merous, 20–70 in pedunculate cymes, peduncles up to 1.7 cm long, bracteoles linear-lanceolate, up to 6 mm long. Calyx-tube up to 1 mm long; limb up to 1.3 mm long. Corolla cream-white; tube 2–3 mm long; lobes lanceolate to ovate, 1.5–2.5 mm long. Style up to 6.5 mm long. Fruits 8–11 mm in diameter; pyrenes 6–7 mm long,

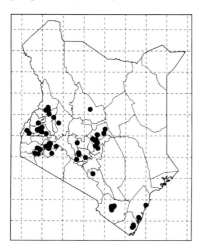

hemispherical to suborbicular.

Distribution: Coastal Kenya. [Tropical Africa].

Habitat: Moist forests or thickets; up to 500 m.

Kwale: Shimba Hills, *SAJIT 005506* (HIB); Gongoni Forest, *Luke 2425* (EA, K); Buda Mafisini Forest Reserve, *Luke & Robertson 1706* (EA, K).

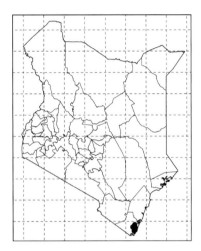

3. **Keetia lukei** Bridson, Kew Bull. 49: 803. 1994. —Type: Kenya, Kwale, Kaya Muhaka, 2 Dec. 1992, *W.R.Q. Luke 3397* [holotype: K (K000319787); isotypes: EA, MO (MO-716825), US]

Scandent shrub or liana, up to 4 m tall; young stems glabrous, becoming lenticellate when older. Leaf-blades glabrous and glossy, broadly elliptic, 4–10 cm × 2–6 cm, acuminate to acute at the apex, acute to obtuse at the base; secondary nerves in 3–4 main pairs; tertiary nerves obscure; domatia present as inconspicuous tufts of hairs; petioles up to 8 mm long; stipules linear from a triangular base, up to 7 mm long. Flowers 5-merous, 15–20 in pedunculate cymes; peduncles up to 2 cm long; bracts and bracteoles linear. Calyx-tube up to 1.2 mm long; limb ca. 0.5 mm long; lobes up to 0.8 mm long. Corolla pale red-brown; tube ca. 3.5 mm long; lobes triangular-ovate, ca. 3 mm long. Style up to 7.5 mm long. Fruits circular-cordate, 1.1–1.4 cm in diameter. Pyrenes broadly ellipsoid, ca. 1.2 cm long.

Distribution: Southern and coastal Kenya. [Endemic].

Habitat: Moist forests or thickets; 40–

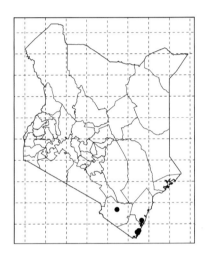

1600 m.

Kilifi: Kaya Jibana, *Luke & Robertson 2634* (EA). Kwale: Gongoni Forest, *Luke & Luke 3950* (EA, K); Kaya Muhaka, *Luke 3080* (EA, K). **Teita Taveta**, Taita Hills, Mbololo Forest, *De Block et al. 468* (EA).

4. **Keetia zanzibarica** (Klotzsch) Bridson, Kew Bull. 41: 979. 1986; F.T.E.A. Rubiac. 3: 917. 1991; K.T.S.L.: 518. 1994. ≡ *Canthium zanzibaricum* Klotzsch, Naturw. Reise Mossambique 1: 291. 1861. —Types: Tanzania, Zanzibar I., *W. Peters s.n.* [holotype: B (destroyed)]; Tanzania, Zanzibar I., near Chuini, 31 Jan. 1929, *P.J. Greenway 1266* [neotype: K (K000319422), **designated here**] (Plate 59E–H)

Plate 59 A–D. *Keetia venosa*; E–H. *K. zanzibarica*. Photo by GWH (A, B, D, H), BL (C), YDZ (E, G) and SWW (F).

Scandent shrub, small tree or liana, up to 9 m tall; young stems glabrous to pubescent. Leaf-blades glabrous, elliptic, ovate or rounded, 5–15 cm × 2–7.5 cm, acuminate, acute or apiculate at the apex, truncate, obtuse or rarely subcordate at the base; lateral nerves in 7–10 main pairs; tertiary nerves rather spaced; petioles up to 1.5 cm long; stipules subulate or linear from a triangular base. Flowers 4–5-merous, 30–60 in pedunculate cymes; peduncles up to 1.5 cm long. Calyx-tube up to 1 mm long; limb 0.5–1 mm long. Corolla white; tube up to 3.3 mm long; lobes oblong-lanceolate, 1.5–2.8 mm long. Style up to 8 mm long. Fruits oblong or obcordate, 8–13 mm long. Pyrenes 8–11 mm long.

Distribution: Northern and coastal Kenya. [Tropical Africa].

Habitat: Forest margins, woodlands, or thickets; up to 1300 m.

Garissa: Bura, *Homby 3100* (EA, K). Kilifi: Chasimba, *Faden et al. 77/408* (EA); Kaloleni, *Waaijenberg 37* (EA). Kwale: Shimba Hills, *Robertson 5303* (K); Tiwi Beach, *Kuchar 10123* (EA). Lamu: Pangani, *Hooper & Townsend 1210* (K). Marsabit: Marsabit Forest, near Lake Paradise, *SAJIT Z0274* (HIB). Mombasa: *Wakefield s.n.* (K). Teita Taveta: Mbololo Forest, *De Block et al. 468* (EA). Tana River: Mchelelo Forest, *Kimberly 205* (EA). Wajir: Mchelelo West Forest, *Medley 205* (K).

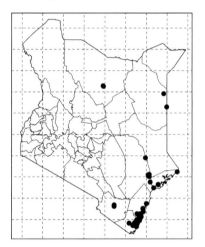

49. **Afrocanthium** (Bridson) Lantz & B. Bremer

Shrubs or small trees without spines. Leaves opposite, petiolate, papery to subcoriaceous; domatia present or absent; stipules sheathing, always caducous, with tufts of white or rust-coloured hairs within. Flowers 4–5-merous, few to many in pedunculate cymes. Calyx-tube broadly ellipsoid to ovoid; limb reduced to a rim or shortly dentate. Corolla white or greenish yellow; tube broadly cylindrical; lobes always reflexed, obtuse to acute or subacuminate. Ovary 2-locular; style slender; stigmatic knob rounded, 2-lobed. Fruit a 2-seeded drupe; pyrenes ellipsoid. Seeds obovoid, flattened.

About 17 species restricted to eastern and southern tropical Africa and South Africa; five species in Kenya.

1a. Lateral branches always reduced, appearing to have four leaves on each node; leaf-blades with lateral nerves in 2–4 main pairs .... 1. *A. pseudoverticillatum*
1b. Lateral branches clearly apparent; leaf-blades with lateral nerves in 3–10 main pairs 2
2a. Leaf-blades elliptic to round, with lateral nerves in 8–10 main pairs; inflorescences with 20–50 flowers .... 2. *A. lactescens*

2b. Leaf-blades occasionally round, with lateral nerves in 4–8 main pairs; inflorescences with less than 25 flowers 3

3a. Inland species; domatia present; petioles less than 5 mm long .... 4. *A. keniense*

3b. Coastal species; domatia absent; petioles up to 10 mm long 4

4a. Leaf-blades papery, with tertiary nerves obscure; stipules less than 5 mm long .... 3. *A. peteri*

4b. Leaf-blades subcoriaceous, with tertiary nerves rather coarsely reticulate; stipules 4–9 mm long .... 5. *A. kilifiense*

1. **Afrocanthium pseudoverticillatum** (S. Moore) Lantz, Bot. J. Linn. Soc. 146: 278. 2004. ≡ *Canthium pseudoverticillatum* S. Moore, J. Bot. 43: 352. 1905; F.T.E.A. Rubiac. 3: 886. 1991; K.T.S.L.: 506. 1994. —Type: Kenya, Kwale, Shimba Hills, 19 Mar. 1902, *T. Kässner 383* [holotype: BM (BM000903285); isotype: K] (Figure 46; Plate 60A)

= *Plectronia microterantha* K. Schum. & K. Krause, Bot. Jahrb. Syst. 39: 541. 1907. —Type: Kenya, Kwale, Shimba Hills, 19 Mar. 1902, *T. Kässner 383* [holotype: BM (BM000903285); isotype: K]

= *Canthium robynsianum* Bullock, Bull. Misc. Inform. Kew 1932: 377, f. 2. 1932. —Type: Kenya, Kilifi, Malindi, Mida, Mar. 1930, *R.M. Graham* in *F.D. 2341* [holotype: K (K000318895); isotypes: BM, EA (EA000001539), K (K000318896)]

Shrub or small tree, up to 6 m tall. Lateral branches always reduced, each node appears to have four leaves. Leaves elliptic to ovate, 2.5–8 cm × 1–5 cm, acute to subacuminate at the apex, acute to cuneate at the base, glabrous or pubescent at least beneath; lateral nerves in 2–4 main pairs; tertiary nerves obscure or apparent; petioles up to 9 mm long; stipules with a linear lobe from a triangular base, caducous. Flowers (4–)5-merous, few to several in shortly pedunculate cymes. Calyx-tube 1–2 mm long; limb truncate or shortly dentate. Corolla yellow or cream green; tube ca. 1 mm long; lobes 1.5–2.3 mm long. Fruits oblong, 2-lobed, ca. 9 mm in diameter; pyrenes narrowly oblong-ovoid, ca. 9 mm long.

Distribution: Coastal Kenya. [Mozambique, Tanzania].

Habitat: Coastal bushlands; up to 1000 m.

Kilifi: Arabuko-Sokoke Forest, *Robertson et al. 5252* (EA, K). Kwale: Shimba Hills, *SAJIT 006037* (HIB). **Lamu**, Boni Forest Reserve, *Robertson & Luke 5616* (EA, K). Mombasa: Bamburi Quarry North Edge, *Robertson 5917* (EA, K). Tana River: Kurawa-Shekiko, *Luke & Luke 9448* (EA, K).

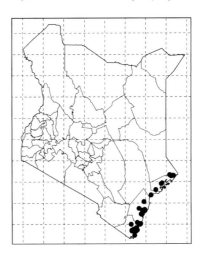

2. **Afrocanthium lactescens** (Hiern) Lantz, Bot. J. Linn. Soc. 146: 278. 2004. ≡ *Canthium lactescens* Hiern, Cat. Afr. Pl. 1: 511. 1898; F.T.E.A. Rubiac. 3: 871. 1991; K.T.S.L.: 505. 1994. —Type: Angola, Huila, Dec. 1859,

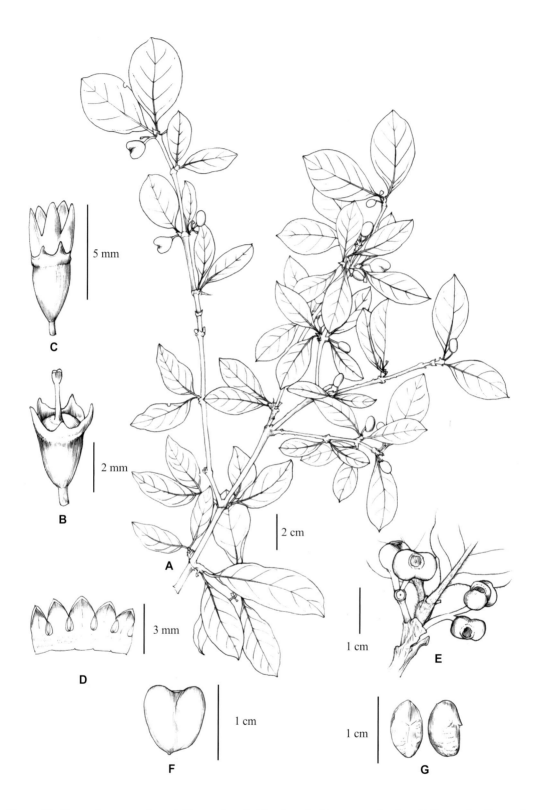

Figure 46 *Afrocanthium pseudoverticillatum*. A. a fruiting branch; B. ovary and calyx; C. flower; D. dissected corolla, showing the stamens; E. infructescence; F. fruit; G. pyrene. Drawn by NJ.

F.M.J. Welwitsch 3157 [holotype: LISU (LISU208593); isotypes: BM (BM000903286), COI, K (K000412401), P (P00546336)] (Plate 60B, C)

Shrub or small tree up to 12 m tall; young stems glabrous or occasionally pubescent. Leaves usually paired at branch tips, blades broadly elliptic to round, 7–18 cm × 5–13 cm, acute to subacuminate at the apex, acute to obtuse or truncate at the base, glabrous or occasionally pubescent; lateral nerves in 8–10 main pairs; petioles up to 2 cm long; stipules broadly triangular to ovate, up to 12 mm long. Flowers 4–5-merous, 20–50 in pedunculate cymes. Calyx-tube ca. 1.5 mm long; limb reduced to a rim. Corolla cream to yellowish; tube 2–2.5 mm long; lobes triangular-ovate, 1.5–3 mm long. Fruits edible, 2-blobed, 1.0–1.5 cm in diameter; pyrenes ellipsoid to ovoid, 0.8–1 cm long.

Distribution: Northwestern and western Kenya. [East tropical Africa to southern Africa].

Habitat: Bushlands, thickets, or woodlands; 1000–2400 m.

Baringo: near Chepkesin, *Bonnefille & Riollet 75/33* (EA). Laikipia: Kisima Ranch, *Beentje & Powys 4099* (EA). Narok: ca. 15 km NE of Maji Moto, *Muchiri 600* (EA). Samburu: ca. 2 km south of Marti on the Baragoi Road, *Briscoe 215* (EA). Turkana: Loima Mounts, *Newbould 7071* (EA, K). Uasin Gishu: Ol Dane Sapuk-Kaposoret, *Williams 211* (EA). West Pokot: Cherangani Hills, *SAJIT 006879* (HIB).

3. **Afrocanthium peteri** (Bridson) Lantz, Bot. J. Linn. Soc. 146: 278. 2004. ≡ *Canthium peteri* Bridson, F.T.E.A. Rubiac. 3: 873. 1991. —Type: Tanzania, Uzaramo, Msua to Bagala, 5 Nov. 1925, *A. Peter 56193* [holotype: K (K000318811); isotype: B (B100160723)]

Shrub, up to 4 m tall. Leaf-blades glabrous, papery, elliptic to round, 4.5–8.5 cm × 2.5–5 cm, rounded, obtuse or acute at the apex, acute to attenuated at the base; lateral nerves in 4–5 main pairs; tertiary nerves obscure; domatia absent; petioles 5–8 mm long; stipules triangular, acuminate, 2–5 mm long. Flowers 5-merous, several in pedunculate cymes. Calyx-tube ca. 1 mm long; limb reduced. Corolla yellow; tube up to 2 mm long; lobes triangular-ovate, ca. 2 mm long. Fruits heart-shaped, 2-lobed, ca. 1 cm wide; pyrenes

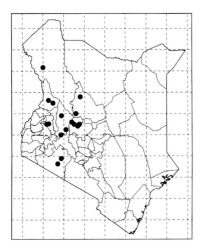

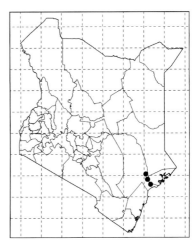

oblong-ellipsoid, ca. 1 cm long.

Distribution: Coastal Kenya. [Tanzania].

Habitat: Lowland forests; up to 450 m.

Tana River: Tana River National Primate Reserve, *Luke et al. 659* (EA); Wema Forest, *Robertson & Gafo 6596* (EA, K).

4. **Afrocanthium keniense** (Bullock) Lantz, Bot. J. Linn. Soc. 146: 278. 2004. ≡ *Canthium keniense* Bullock, Kew Bull. 1932: 377. 1932; F.T.E.A. Rubiac. 3: 874. 1991; K.T.S.L.: 504. 1994. —Type: Kenya, Nairobi-Kikuyu, 28 Apr. 1919, *E. Battiscombe* in *F.D. 872* [holotype: K (K000311655); isotype: EA (EA000001535)]

Shrub or small tree, up to 9 m tall. Leaf-blades glabrous, elliptic, 6–14 cm × 2.5–6.5 cm, acute to acuminate at the apex, acute to obtuse at the base; lateral nerves in 5–6 main pairs; domatia present as conspicuous tufts of whitish hair; petioles 3–5 mm long; stipules triangular, up to 1 cm long. Flowers 5-merous, several in pedunculate cymes. Calyx-tube ca. 1.5 mm long; limb reduced. Corolla yellowish green; tube up to 2 mm long; lobes ovate, 1.7–2 mm long. Fruits heart-shaped, 2-lobed, 1.3–1.5 cm wide. pyrenes obovoid, ca. 1.4 cm long.

Distribution: Western and central Kenya. [Endemic].

Habitat: Forests; 1400–2100 m.

Kajiado: Ololua Forest, *Bytebier 798* (EA). Kiambu: near Chania Falls, *Faden 66223* (EA); Karura Forest, *Agnew & E.A. Nat. Hist. Soc. 7241* (EA). Machakos: Mumandu Forest Reserve, *KSCP/PGRWG 003/19/2000* (K). Nairobi: Ngong Road Forest, *Perdue & Kibuwa 8117* (K); Nairobi City Park, *Mwangangi & Kamau 3962* (EA). Nakuru: Mau Forest, *Mutanga & Kamau 215* (EA). Samburu: Mathew's Range, *Luke 14133* (EA, K).

5. **Afrocanthium kilifiense** (Bridson) Lantz, Bot. J. Linn. Soc. 146: 278. 2004. ≡ *Canthium kilifiense* Bridson, F.T.E.A. Rubiac. 3: 874. 1991. —Type: Kilifi, Roka, June 1937, *I.R. Dale 3840* [holotype: K (K000311656); isotype: EA (EA000001536)] (Plate 60D)

Shrub or small tree, up to 7 m tall. Leaf-blades glabrous, subcoriaceous, elliptic, 6–10.5 cm × 3–7 cm, rounded, obtuse to acute at the apex, rounded, obtuse or occasionally acute at the base; lateral nerves in 5–7 main pairs; tertiary nerves coarsely reticulate; domatia absent; petioles up to 1 cm long; stipules

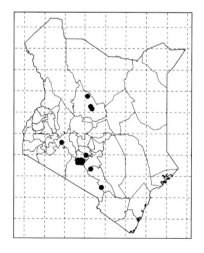
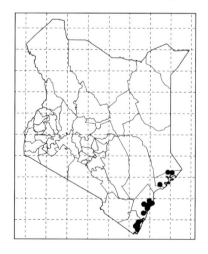

Plate 60 A. *Afrocanthium pseudoverticillatum*; B, C. *A. kilifiense*; D. *A. kilifiense*. Photo by GWH (A, D) and VMN (B, C).

triangular with a subulate lobe, 4–9 mm long. Flowers 5-merous, few in pedunculate cymes. Calyx-tube ca. 1.3 mm long; limb reduced. Corolla-tube ca. 3 mm long; lobes ca. 2 mm long. Fruits slightly square, 2-lobed, 0.9–1.2 cm long and wide; pyrenes oblong-ellipsoid, ca. 1 cm long.

Distribution: Coastal Kenya. [Endemic].
Habitat: Coastal or lowland forests; up to 400 m.

Kilifi: Arabuko-Sokoke Forest, *Musyoki & Hansen 1012* (EA, K); Kaya Kivara, *Robertson 4651* (EA, K). Kwale: Shimba Hills, *SAJIT V0154* (HIB); Mwaluganje Elephant Sanctuary, *Mwadime et al. 136* (EA). Lamu: Lunghi Forest Reserve, *Luke & Robertson 1545* (EA); Boni Forest Reserve, *Luke & Robertson 1516A* (EA).

50. **Psydrax** Gaertn.

Small trees, shrubs or lianas, unarmed. Raphides absent. Leaves opposite or sometimes ternate; leaf-blades coriaceous or less often chartaceous, with or without domatia; stipules persistent, usually with truncate to triangular base and a keeled lobe. Inflorescences axillary, cymose, few- to several-flowered, or rarely solitary, sessile to pedunculate; bracteate or bracts reduced. Flowers 4–5-merous, bisexual, monomorphic. Calyx-tube broadly ellipsoid to hemispherical; limb truncate or dentate. Corolla white to yellow; tube broadly cylindrical, hairy at the throat; lobes markedly reflexed at anthesis. Stamens inserted in corolla throat; filaments developed, reflexed at anthesis. Ovary 2-locular, each with single ovule; style long, slender; stigmatic knob cylindrical. Fruits drupaceous, fleshy, subglobose to ellipsoid or sometimes dicoccous; pyrenes 2, woody or cartilaginous. Seeds with fleshy endosperm.

A large genus of about 80 species throughout the Old World tropics. There are eight species known to occur in Kenya.

We treat *Psydrax* sp. A of FTEA and *P. robertsoniae* as one species, after we had carefully compared the specimens of *P.* sp. A with the type specimens [*Robertson 6152* (EA, K)] of the later one. *Luke & Luke 9446* (EA), *Robertson & Luke 5880* (EA) and *Robertson & Luke 5937* (EA) were identified as *P. robertsoniae*, which have very small and narrowly oblong leaves, should be treated as a new species, while lacking flowers and fruits. *Luke 3142* (EA, K) and *Hawthorne 463* (K) maybe also wrongly identified as *P. robertsoniae*, but its leaf-blades have an acuminate base, and should be treated as distinct.

1a. Inflorescences clearly pedunculate 2
1b. Inflorescences sessile to shortly pedunculate 4
2a. Deciduous plant; leaf-blades chartaceous to subcoriaceous .... 2. *P. lividus*
2b. Evergreen plant; leaf-blades subcoriaeous to coriaeous 3
3a. Leaves 5.5–15 cm long; fruits 0.8–1.4 cm in diameter .... 1. *P. parviflorus*
3b. Leaves 4.5–7.5 cm long; fruits 0.7–0.8 cm in diameter .... 3. *P. faulknerae*
4a. Inflorescences 1–2-flowered .... 8. *P. recurvifolius*
4b. Inflorescences few- to many-flowered 5

5a. Leaves mostly restricted to apices of reduced lateral spurs .... 6. *P. polhillii*
5b. Leaves well spaced along the branches 6
6a. Leaf-blades acute or more often gradually acuminate at the apex; corolla-tube 2–2.5 mm long .... 4. *P. schimperianus*
6b. Leaf-blades rounded or obtuse at the apex; corolla-tube 3–5 mm long 7
7a. Corolla-tube 3–4 mm long; lobes 2–3 mm long .... 5. *P. robertsoniae*
7b. Corolla-tube 4–5 mm long; lobes 5–6 mm long .... 7. *P. kaessneri*

1. **Psydrax parviflorum** (Afzel.) Bridson, Kew Bull. 40: 700. 1985, '*parviflora*'; F.T.E.A. Rubiac. 3: 897. 1991; K.T.S.L.: 539. 1994. ≡ *Pavetta parviflora* Afzel., Remed. Guin. 47. 1815. —Type: Sierra Leone, *K. Afzelius 3441a* [holotype: UPS (V-028020); isotype: BM (BM000903242)]

Shrub or tall timber tree, sometimes scandent, up to 27 m tall; young stems usually square. Leaves opposite; blades elliptic, ovate or oblong-elliptic, 5.5–15.5 cm × 2–8 cm, acuminate at the apex, acute to obtuse or rounded at the base, subcoriaceous to coriaceous, glabrous; midribs whitish or red; secondary nerves in 4–8 main pairs; domatia present; petioles 3–10 mm long; stipules 2–7 mm long. Flowers 4-merous, many in pedunculate corymbs; peduncles 3–20 mm long; pedicels 2–9 mm long. Calyx-tube 0.7–1.3 mm long; limb truncate to repand. Corolla whitish; tube 2–3 mm long, hairy at throat; lobes oblong ovate, 1.5–2.5 mm long. Style up to 8 mm long. Fruits black, globose or bispherical, 5–8 mm in diameter.

1a. Leaf-blades with pale mid-ribs and glabrous domatia................................a. subsp. *parviflorus*
1b. Leaf-blades with red mid-ribs and hairy domatia .... b. subsp. *rubrocostatus*

a. subsp. **parviflorus**

Leaf-blades with pale mid-ribs and glabrous domatia.

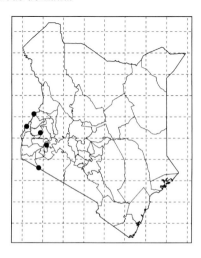

Distribution: Western Kenya. [West and east Africa, also in north part of southern Africa].

Habitat: Forest edges, wooded bushlands, roadsides, or clumps in grasslands; 1200–2000 m.

Busia: Nyaranga Village, *Omondi & Obunyali KEFRI312* (K). Bungoma: Mount Elgon, *Jackson 322* (EA). Kakamega: Kakamega Forest, *Bridson 37* (EA). Kericho: Belgut, *Kerfoot 2191* (EA). Narok: Masai Mara Game Reserve, *Msafiri et al. 411* (EA).

b. subsp. **rubrocostatus** (Robyns) Bridson, Kew Bull. 40: 702. 1985, '*rubrocostata*'; F.T.E.A. Rubiac. 3: 898. 1991; K.T.S.L.: 539. 1994. ≡ *Canthium rubrocostatum*

Robyns, Notizbl. Bot. Gart. Berlin-Dahlem 10: 616. 1929. —Type: Kenya, between rivers Kiringa and Mekengeria, 5 Mar. 1922, *R.E. & T.C.E. Fries 2107* [holotype: UPS(V-087564); isotypes: BR (BR0000008854759 & BR0000008854797), K] (Plate 61A)

Leaf-blades with red mid-ribs and hairy domatia.

Distribution: Western, central, and southern Kenya. [West and east Africa, also in north part of southern Africa].

Habitat: Forests, bushlands, or rocky hillsides; 1500–2250 m.

Baringo: above Bartabwa, *Luke 219* (EA). Bomet: Mau Forest, *Mutangah & Kamau M40* (EA). Embu: Kiangombe Hill, *Beentje 3704* (EA). Kajiado: Chyulu Hills, *Luke & Luke 3911* (EA). Kericho: Kipsonoi, *Perdue & Kibuwa 9313* (EA). Kiambu: Katamayu, *van Someren 6720* (K). Murang'a: Tusu, *Rammell 1054* (K). Nairobi: *Dale 2901* (K). Nakuru: Koibatek Forest, *Kirika et al. 42* (EA). Nandi: Esero, *Brunt 1338* (K). Samburu: Mathew's Range, *Luke et al. 14130* (EA). Teita Taveta: Chawia Forest, *Beentje et al. 844* (EA); Mbololo Hill, *Faden & Faden 72/271* (EA).

Tharaka-Nithi: Chogoria Track, *Kirika et al. KMK5* (EA). Trans Nzoia: Saiwa Swamp National Park, *Mwangangi & Mutanga 3908* (EA). Uasin Gishu: Kapseret, *Gardner OX544* (EA).

2. **Psydrax lividus** (Hiern) Bridson, Kew Bull. 40: 705. 1985, '*livida*'; F.T,E.A. Rubiac. 3: 898. 1991; K.T.S.L.: 538. 1994. ≡ *Canthium lividum* Hiern, Fl. Trop. Afr. 3: 144. 1877. —Type: Mozambique, Moramballa, 18 Jan. 1863, *J. Kirk s.n.* [holotype: K (K000422507)] (Figure 47)

Deciduous shrub or small tree, up to 8 m tall. Leaves opposite, usually on the young branches; blades 1.5–13 cm × 1–7 cm, ovate to oblong, acute to acuminate at the apex, obtuse to rounded at the base, herbaceous, glabrous to pubescent on both sides; secondary nerves in 4–5 main pairs; domatia present; petioles 2–8 mm long; stipules 4–7 mm long. Flowers 4-merous, 6–70 on pedunculate compact corymbs; peduncle 3–15 mm long, pubescent; pedicels 3–8 mm long, pubescent. Calyx-tube 0.5–1 mm long; limb irregularly dentate. Corolla pale cream; tube 1.2–3 mm long, pubescent at throat; lobes oblong, 1.2–

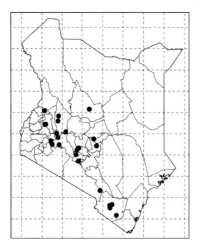

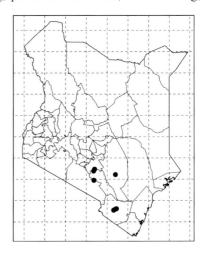

Figure 47 *Psydrax lividus*. A. fruiting branch; B. flower; C. fruit. Drawn by NJ.

2 mm long. Style up to 6 mm long. Fruits black, broadly oblong, 5–6 mm long.

Distribution: Central and southern Kenya. [Eastern and southern Africa].

Habitat: Bushlands, thickets, or other rocky hills; 750–1700 m.

Kajiado: Emali Hill, *Someren 23* (K). Kitui: Mutha Hill, *Bally B1659* (EA). Makueni: near Tulimani, *Muasya 1136* (EA). Teita Taveta: Mwawache Area, *Mwachala et al. in EW 3341* (EA).

3. **Psydrax faulknerae** Bridson, Kew Bull. 40: 707. 1985; F.T.E.A. Rubiac. 3: 901. 1991; K.T.S.L.: 538. 1994. —Type: Tanzania, Tanga, Mwarongo, 26 Nov. 1957, *H.G. Faulkner 2105* [holotype: K (K000352692)] (Plate 61B)

Shrub or small tree, sometimes scandent, up to 7 m tall. Leaves opposite; blades elliptic, 1–7.5 cm × 0.5–5 cm, acuminate at the apex, acute at the base, coriaceous, glabrous; lateral nerves in 3–4 main pairs; domatia present; petioles 2–10 mm long; stipules 3.5–6 mm long. Flowers 4-merous, many in pedunculate compact cymes; peduncles 2–8 mm long; pedicels 2.5–6 mm long. Calyx-tube ca. 1 mm long; limb shortly toothed. Corolla whitish; tube 1.5–2.8 mm long, hairy at the throat; lobes oblong, 1.7–2 mm long.

Style 5.7–6.5 mm long. Fruits black, oblong, 5–5.5 mm long.

Distribution: Coastal Kenya. [Tanzania].

Habitat: Coastal forests, thicket edges, woodlands, or open grasslands; up to 500 m.

Kilifi: Mangea Hill, *Luke & Robertson 625* (EA); Arabuko-Sokoke *Robertson 3880* (EA); Kaya Kivara, *Robertson 4653* (EA). Kwale: Shimba Hills, *SAJIT V0267* (HIB); Jombo Hill, *Gilbert et al. 4959* (EA). Lamu: Lunghi Forest Reserve, *Luke & Robertson 1543* (EA).

4. **Psydrax schimperianus** (A. Rich.) Bridson, Kew Bull. 40: 714. 1985, '*schimperiana*'; F.T.E.A. Rubiac. 3: 901. 1991; K.T.S.L.: 539. 1994. ≡ *Canthium schimperianum* A. Rich., Tent. Fl. Abyss. 1: 350. 1848. —Types: Ethiopia, Tigre, Mt. Seleuda [Scholoda], 3 Oct. 1837, *G.H.W. Schimper 328* [lectotype: P (P00546360), **designated here**; isosyntypes: BR (BR0000008358417, BR0000008854698), HAL (HAL 0114300), HBG (HBG 521272), JE (JE0000 4259), L (L0057927), LG (LG0000090029806), M (M0106336), MPU (MPU021757), P (P00546361, P00546362, P 00546363), TUB (TUB004572)]; Ethiopia, near Maye-Gouagoua, Nov. 1839, *R. Quartin Dillon & A. Petit s.n.* [syntypes: P (P00546356, P00546357, P00546358, P00546359)]; Ethiopia, Maye-Gouagoua, s.d., *R. Quartin Dillon 20* [syntype: P

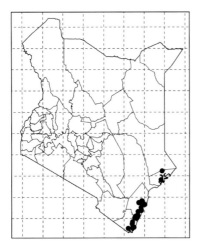

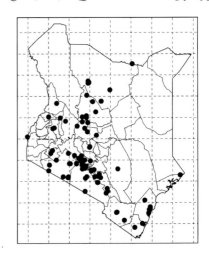

(P00551138)] (Plate 61C)

Shrub or tree, up to 10 m high, with young branches square or slightly winged. Leaves opposite, narrowly to broadly elliptic or obovate, 3–10.5 cm × 1.3–5 cm, acute or acuminate at the apex, acute or obtuse at the base; petioles 2–3 mm long; stipules 3–10 mm long. Inflorescences of axillary umbellate cymes, sessile to subsessile. Pedicels 2–7 mm long, pubescent. Calyx-tube 1–2 mm long; limb reduced to a shortly toothed rim. Corolla whitish; tube 2–2.5 mm long; lobes oblong-lanceolate, 3–3.5 mm long. Fruits black, 5–6.5 mm in diameter.

Distribution: Northern, western, central, southern, and coastal Kenya. [East Africa from Ethiopia to Zambia, also in Yemen].

Habitat: Forest edges, thickets, woodlands, rocky bushlands, bushed grasslands, or scattered tree grasslands; up to 2250 m.

Baringo: above Bartabwa, *Luke 221* (EA). Busia: Port Victoria, *Glasgow 45/39* (EA). Elgeyo-Marakwet: *Lindom 150* (EA). Kajiado: Ilpartimaro, *Kuchar & Msafiri 8017* (EA); Ngong Escarpment, *van Someren B818* (EA). Kericho: Cheptuiyet, *Kerfoot 2143* (EA). Kiambu: Karimenu, *Malombe & Kirika 79* (EA). Kilifi: Dakabuko, *Luke & Robertson 2549* (EA); near Jilore Forest Station, *Perdue & Kibuwa 10009* (EA). Kitui: Makongo Forest Reserve, *Mwangangi et al. 4480* (EA). Kwale: ca. 3 miles west of Kwale, *Moomaw 1077* (EA). Laikipia: Ol Ari-Nyiro Ranch, *Muasya 1112* (EA). Lamu: Mundane Range, *Robertson & Luke 5587* (EA). Machakos: Lukenya Hill, *Ng'weno 24* (K). Makueni: Thikuni Hill, *Trapnell & Birch 60/75* (EA). Marsabit: Mount Kulal, *Kamau et al. GBK18* (K); Moyale, *Gillett 13523* (EA); West of Marsabit, *Faden 68/435* (EA). Meru: Ndare Ngare Forest, *Beentje 2103* (EA). Nairobi: Ngong Road Forest, *Perdue & Kibuwa 8128* (EA). Nakuru: Ol Longonot Estate, *Kerfoot 3517* (EA). Narok: Chepkorobotik Forest, *Glover et al. 21* (EA); Lebetero Hills, *van Someren in EA 12262* (EA). Nyandarua: Karati, *Dale 420* (EA). Nyeri: near Bantu Lodge, *SAJIT 002215* (HIB). Samburu: Karisia Hill, *Ichikawa 753* (EA). Teita Taveta: Mount Kasigau, *Luke & Luke 4115* (EA); Bura, *Mwachala in EW 68* (EA). Trans Nzoia: Kitale, *Evans et al. 1479* (K). Uasin Gishu: Eldoret, *Tweedie 4068* (K). West Pokot: Mount Sekerr, *Agnew et al. 10417* (EA).

5. **Psydrax robertsoniae** Bridson, F.T.E.A. Rubiac. 3: 903. 1991. —Type: Kenya, Kilifi, Watamu, Knocker Plot 34, 29 Apr. 1990, *S.A. Robertson 6152* [holotype: K (K000311721); isotypes: EA (EA000001550), MO, US]

Shrub to small tree, up to 4 m tall. Leaves opposite; blades elliptic to rounded, 1–8 cm × 0.5–3.5 cm, obtuse to rounded at the apex, acute, obtuse to truncate at the base, coriaceous, glabrous; lateral nerves in 2–5 main pairs; domatia present; petioles 0.5–1 cm long; stipules 4.5–5.5 mm long. Flowers 4-merous,

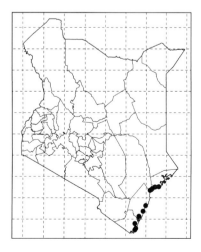

Plate 61 A. *Psydrax parviflora* subsp. *rubrocostata*; B. *P. faulknerae*; C. *P. schimperiana*. Photo by YDZ (A, C) and SDZ (B).

10–30 in sessile subumbellate cymes; pedicels 0.3–1 cm long; bracteoles inconspicuous. Calyx-tube 0.5–1 mm long; limb reduced to a repand rim. Corolla-tube 3–4 mm long, hairy at the throat; lobes narrowly oblong, 2–3 mm long. Style 8–9 mm long. Fruits didymous or broadly oblong, 5–6 mm long.

Distribution: Coastal Kenya. [Endemic].

Habitat: Forests, coast thickets, or edges of mangroves; up to 15 m.

Note: We have compared the specimens of *Psydrax* sp. A [*Festo & Luke 2402* (EA), *Luke & Luke 9447* (EA), *Luke & Robertson 1361* (EA), *Luke & Robertson 1368* (EA), *Luke et al. 9031* (EA), *Luke et al. 9449A* (EA), *Luke et al. 9449B* (EA, K), *Nyange et al. 0561* (K) and *Nyange M0561* (EA)] with the specimens of *P. robertsoniae* Bridson [*Moggridge 334* (EA, K) and *Robertson 6152* (EA, K)], and found these two groups have no significant differences. We treat *Psydrax* sp. A of FTEA. and *P. robertsoniae* as one species, which is easily separated from others with elliptic and coriaceous leaves.

Kilifi: Watamu, *Robertson 6152* (EA). Kwale: Funzi Peninsula, *Luke et al. 9031* (K). Tana River: Shekiko, *Luke & Robertson 1368* (EA, K); Kurawa, *Luke et al. 9449B* (EA, K).

6. **Psydrax polhillii** Bridson, Kew Bull. 40: 716. 1985; F.T.E.A. Rubiac. 3: 903. 1991; K.T.S.L.: 539. 1994. —Type: Kenya, Tana River, Kurawa, 24 Sept. 1961, *R.M. Polhill & S. Paulo 556* [holotype: K (K000311720); isotypes: BR (BR0000008854728), EA (EA 000001549), PRE (PRE0593544-0)]

Shrub, up to 6 m. Leaves always on lateral short spurs; blades broadly elliptic to round, 0.8–2.8 cm × 0.5–2.3 cm, rounded at the apex, obtuse to rounded at the base, coriaceous, lateral nerves in 2–3 main pairs; domatia present; petioles 1.5–4 mm long; stipules 0.5–3 mm long. Flowers 4-merous, 5–20 in subsessile cymes; peduncles up to 2.5 mm long; pedicels 2–5 mm long. Calyx-tube 1 mm long; limb reduced to a repand rim. Corolla greenish cream; tube 2–3 mm long; lobes up to 4 mm long. Style up to 6 mm long. Fruits black, broadly oblong, 5–6.5 mm long.

Distribution: Eastern and coastal Kenya. [Endemic].

Habitat: Forests, thickets near rocks, dry coastal bushlands, or scattered tree grasslands; up to 450 m.

Kilifi: Mangea Hill, *Luke & Robertson 1785* (EA); Arabuko-Sokoke Forest, *Robertson 4075* (EA). Kwale: Shimba Hills, *Luke & Luke 9467* (EA); Maluganji Forest Reserve, *Robertson & Luke 6031* (EA). Lamu: Boni Forest Reserve, *Kuchar 13542* (EA). Tana River: near Dalu, *Luke & Robertson 1291* (EA); Tana River National Primate Reserve, *Luke et al. 267* (EA).

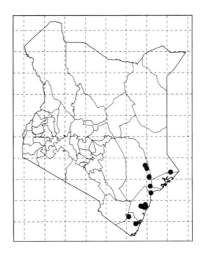

7. **Psydrax kaessneri** (S. Moore) Bridson, Kew Bull. 40: 719. 1985; F.T.E.A. Rubiac. 3: 904. 1991; K.T.S.L.: 538. 1994. ≡ *Canthium kaessneri* S. Moore, J. Bot. 43: 351. 1905. —Type: Kenya, Kwale, Gadu,

Mar. 1902, *T. Käassner 418* [holotype: BM (BM000903250); isotype: K (K000422547)]

Scandent shrub, up to 3 m tall, usually with short perpendicular lateral branches. Leaves opposite; blades elliptic to round or obovate, 3–8 cm × 1.5–6 cm, rounded to obtuse at the apex, acute to obtuse at the base, subcoriaceous to coriaceous; lateral nerves in 3–4 main pairs; domatia present; petioles 2–4 mm long; stipules 4–5 mm long. Flowers 5-merous, 5–30 in subumbelliform cymes; peduncles up to 2 mm long; pedicels 2–8 mm long; bracteoles inconspicuous. Calyx-tube 1.3 mm long; limb reduced to a dentate rim. Corolla whitish; tube 4–5 mm long, hairy inside; lobes oblong, 5–6 mm long. Style 7–9 mm long. Fruits black, subglobose, 7–9 mm in diameter.

Distribution: Coastal Kenya. [Mozambique, Somalia, and Tanzania].

Habitat: Riverine forests, floodplain forests, thickets, or wooded grasslands; up to 200 m.

Kwale: Kaya Puma, *Luke et al. 6328* (EA). Lamu: Lunghi Forest Reserve, *Luke & Robertson 1532* (EA). Tana River: Tana River National Primate Reserve, *Luke et al. 568* (EA); ca. 3 km north of Wema, *Gillett & Kibuwa 19927* (EA).

8. **Psydrax recurvifolius** (Bullock) Bridson, Kew Bull. 40: 722. 1985; F.T.E.A. Rubiac. 3: 906. 1991; K.T.S.L.: 539. 1994. ≡ *Canthium recurvifolium* Bullock, Bull. Misc. Inform. Kew 1932: 385. 1932. —Type: Tanzania, Pemba I., 24 Dec. 1930, *P.J. Greenway 2781* [holotype: K (K000352732); isotype: EA (EA000001548)]

Scandent shrub, up to 5 m tall. Leaves opposite; blades elliptic to ovate, 1–6 cm × 0.7–3.5 cm, obtuse to rounded at the apex, obtuse to rounded at the base, coriaceous, lateral nerves in 3–4 main pairs; petioles 1–3 mm long; stipules 3–5 mm long. Flowers 5-merous, single or 2 in axillary fascicles, with pedicels 3–12 mm long. Calyx-tube 1–1.5 mm long, glabrous; limb dentate. Corolla white; tube 3–5 mm long, hairy inside; lobes oblong, 3–4.5 mm long. Style 6–8 mm long. Fruits obovate, 7–8 mm long.

Distribution: Coastal Kenya. [Tanzania].

Habitat: Forests, dry bushlands, or hillsides; 5–500 m.

Kilifi: Arabuko-Sokoke Forest, *Robertson et al. 5246* (EA); Mangea Hill, *Luke & Robertson 1037* (EA). Kwale: Maluganji Forest Reserve, *Robertson & Luke 5993* (EA).

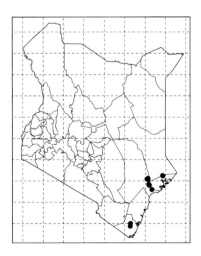

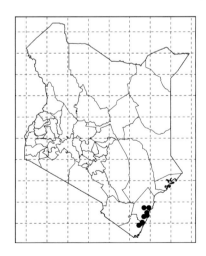

51. **Bullockia** (Bridson) Razafim., Lantz & B. Bremer

Shrubs or small trees, sometimes scandent. Leaves opposite, petiolate or subsessile; blades chartaceous, glabrous to pubescent; domatia absent or present; stipules narrowly ovate to ovate or triangular at the base, sometimes produced into a lobe, with silky hairs and colleter inside. Flowers unisexual, (4–)5–6-merous; functionally male ones 1–20 in fascicules, shortly pedunculate umbels or subumbellate cymes; functionally female ones usually solitary. Calyx-tube very reduced in functionally male flowers and ovoid in functionally female ones; lobes irregular. Corolla-tube with a well-defined ring of deflected hairs; lobes erect, thickened towards the apex, not apiculate. Stamens inserted at the throat of corolla-tube; anthers ovate, with a dark connective tissue on the central area of the dorsal face or rarely entirely without dark connective tissue. Ovary 2-locular, with single ovule in each locule; style slender; stigmatic knob 2-lobed. Fruits fleshy, obovate to broadly obovate or heart-shaped; pyrenes thinly woody, narrowly obovoid, flattened. Seeds narrowly obovoid, flattened; embryo slightly curved.

A small genus with eight species in Africa and Madagascar; five species in Kenya.

1a. Leaf-blades glabrous above; pedicels and calyx-tube usually glabrous 2
1b. Leaf-blades glabrescent to pubescent; pedicels and calyx-tube sparsely to densely pubescent 4
2a. Petioles 2–20 mm long; stipules ovate to narrowly ovate or sometimes lanceolate .... 1. *B. mombazensis*
2b. Petioles 1–3 mm long; stipules triangular at the base, linear above 3
3a. Leaf-blades broadly oblong-elliptic, 1–3 cm long, obtuse to rounded at the apex; stipules 3–5 mm long .... 2. *B. dyscritos*
3b. Leaf-blades narrowly elliptic to elliptic, 2–6 cm long, acute to obtuse at the apex; stipules 4–10 mm long .... 3. *B. fadenii*
4a. Plant covered with straw- to rust-coloured, upwardly directed hairs; leaf-blades 1.2–6.5 cm long .... 4. *B. setiflora*
4b. Plant covered with pale, soft, crisped or patent hairs; leaf-blades 0.6–3.3 cm long .... 5. *B. pseudosetiflora*

1. **Bullockia mombazensis** (Baill.) Razafim., Lantz & B. Bremer, Ann. Missouri Bot. Gard. 96: 175. 2009. ≡ *Canthium mombazense* Baill., Adansonia 12: 188. 1878; F.T.E.A. Rubiac. 3: 881. 1991; K.T.S.L.: 505. 1994. —Type: Kenya, Mombasa, Nov. 1848, *L.H. Boivin s.n.* [holotype: P (P00546338); isotype: K (K000172760)]

(Figure 48; Plate 62A–C)

= *Canthium inopinatum* Bullock, Bull. Misc. Inform. Kew 1932: 389. 1932. —Type: Kenya, Machakos, Ukambani, 1893, *G.F. Scott-Elliot 6380* [holotype: K (K000412411)]

Shrub or small tree, up to 7.5 m tall. Leaf-blades elliptic, 2–13 cm × 1–7.5 cm, obtuse to

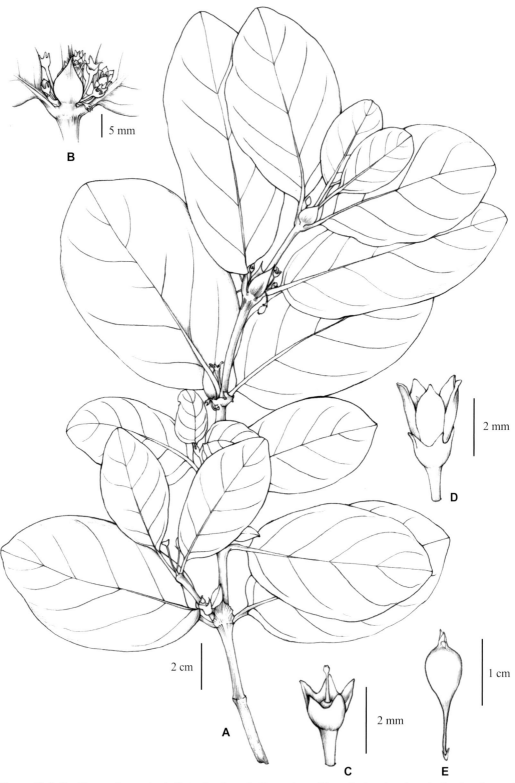

Figure 48 *Bullockia mombazensis*. A. flowering branch; B. portion of branch showing the stipule; C. calyx and style; D. flower; E. fruit. Drawn by NJ.

rounded at the apex, cuneate or rounded at the base, glabrous above, glabrous or rarely pubescent beneath; domatia absent or inconspicuous; petioles up to 1 cm long; stipules ovate, 0.4–1.6 cm long, acuminate. Flowers unisexual, 5–6-merous. Functionally male flowers 3–20 in subsessile umbels; pedicels 3–7 mm long; calyx-tube reduced, lobes unequal, linear-triangular, up to 2 mm long; corolla-tube ca. 1.5 mm long, lobes triangular-ovate. Functionally female flowers solitary; pedicels 0.4–1 cm long; calyx-tube 1.7–2.5 mm long, lobes spathulate, 2–3 mm long. Fruits obovoid, compressed, 0.7–1.2 cm long; pyrenes obovoid, ca. 7 mm long.

Distribution: Central and coastal Kenya. [Mozambique, Somalia, and Tanzania].

Habitat: Forests, bushlands, or wooded grasslands; up to 1200 m.

Kilifi: Arabuko-Sokoke Forest, *SAJIT 004642* (HIB); Pumwani, *Beentje 2339* (EA, K); Mnarani, *Faden & Faden 71/820* (EA). Kitui: Mutha Hill, *Luke & Luke 8202* (EA, K). Kwale: Shimba Hills, *SAJIT V0137* (HIB); Mwache Forest Reserve, *Luke & Luke 8355* (EA); Chuma Forest, *Luke & Rorbertson 567* (EA). Lamu: Utwani Forest, *Rawlins 281* (EA, K); Boni Forest, *Robertson & Luke 5618* (EA, K); Lunghi Forest Reserve, *Luke & Rorbertson 1466* (EA). Mombasa: Mazeras, *Reitsma 223* (K). Teita Taveta: Maungu Hills, *Faden et al. 70/176* (EA, K); Mount Kasigau, *Medley 804* (EA). Tana River: south of Bfunbe, *Robertson & Luke 5335* (EA); Ozi, *Luke & Robertson 1362* (EA).

2. **Bullockia dyscritos** (Bullock) Razafim., Lantz & B. Bremer, Ann. Missouri Bot. Gard. 96: 175. 2009. ≡ *Canthium dyscriton* Bullock, Bull. Misc. Inform. Kew 1936: 478. 1936; F.T.E.A. Rubiac. 3: 882. 1991; K.T.S.L.: 504. 1994. —Type: Kenya, Taita Hills, Nyatchi, Sept. 1932, *H.M. Gardner in F.D. 3000* [holotype: K (K000311652); isotype: EA (EA000001545)] (Plate 62D, E)

Shrub, up to 3 m tall. Leaf-blades glabrous, broadly oblong-elliptic, 1–2.8 cm × 0.5–1.8 cm, rounded to obtuse at the apex, rounded at the base; domatia present as small tufts; petioles 1–2 mm long; stipules triangular at the base, tapering to a linear lobe 3–5 mm long. Flowers unisexual, 5-merous. Functionally male flowers 2–8 in sessile umbels; pedicels 4–6 mm long; calyx-tube reduced, lobes minute; corolla-tube ca. 1.5 mm long, lobes erect, triangular-ovate, ca. 1.2 mm long, acute, thickened at the apex. Functionally females flowers solitary; pedicels 0.5–1.5 cm long; calyx-tube ca. 0.7 mm long. Fruits ellipsoid to obovoid, ca. 9 mm long.

Distribution: Central and southern Kenya. [Tanzania].

Habitat: Rocky outcrops or roadsides; up to 1600 m.

Kajiado: Chyulu Plains, *Luke & Luke 11173* (EA, K). Machakos: Donyo Sabuk National Reserve, *Mbare et al. NMK828* (EA).

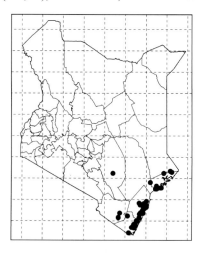

Makueni: Tututha Hill, *Muasya 1112* (EA). Teita Taveta: Choke Forest, *Mwachala et al. 3252* (EA); Sagala Hill, *Gilbert & Gilbert 6102* (EA, K).

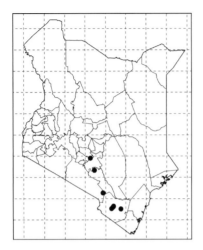

3. **Bullockia fadenii** (Bridson) Razafim., Lantz & B. Bremer, Ann. Missouri Bot. Gard. 96: 175. 2009. ≡ *Canthium fadenii* Bridson, Kew Bull. 42: 632. 1987; F.T.E.A. Rubiac. 3: 882. 1991; K.T.S.L.: 504. 1994. —Type: Kenya, Kiambu, Thika, behind Blue Posts Hotel, 23 Mar. 1968, *R.B. Faden 68/012* [holotype: K (K000311651); isotype: EA]

Shrub, up to 4.5 m tall. Leaf-blades elliptic to narrowly elliptic, 1.7–6 cm × 0.7–2.5 cm, acute to obtuse at the apex, acute and often unequal at the base, glabrous or sometimes pubescent beneath; domatia absent or obscure; petioles 2–3 mm long; stipules triangular at the base, linear above, up to 1 cm long. Flowers unisexual, 5-merous. Functionally male flowers 2–4 in sessile umbels; pedicels 2–5 mm long; calyx-tube reduced, lobes minute; corolla-tube ca. 1 mm long, lobes triangular-ovate, ca. 1.3 mm long, acute. Functionally female flowers solitary; pedicels 4–6 mm long; calyx-tube 1.5 mm long, lobes linear to spathulate, 0.7–1.5 mm long. Fruits broadly obovate, ca. 7 mm long.

Distribution: Central Kenya. [Endemic].
Habitat: Bushlands or forest edges; 1450–

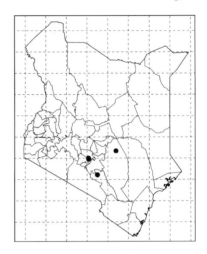

1900 m.

Kiambu: Blue Post Hotel, *Luke & Luke 9014* (EA). Kitui: Igomena (Ngomeni) Hills, *Rauh KE818* (EA). Makueni: Makuli Forest, *MuasyaNMK 688* (EA).

4. **Bullockia setiflora** (Hiern) Razafim., Lantz & B. Bremer, Ann. Missouri Bot. Gard. 96: 175. 2009. ≡ *Canthium setiflorum* Hiern, Fl. Trop. Afr. 3: 134. 1877; F.T.E.A. Rubiac. 3: 883. 1991; K.T.S.L.: 506. 1994. —Type: Mozambique, near Tete, Jule 1860, *J. Kirk s.n.* [holotype: K (K000412395)] (Plate 62F, G)

≡ *Plectronia telidosma* K. Schum., Bot. Jahrb. Syst. 23: 460. 1897. ≡ *Canthium setiflorum* Hiern subsp. *telidosma* (K. Schum.) Bridson, Kew Bull. 42: 634. 1987; F.T.E.A. Rubiac. 3: 883. 1991. —Type: Tanzania, Uzaramo, near Madimola, *F.L. Stuhlmann 6690* [holotype: B (destroyed); lectotype: K (K000318805), **designated here**]

Plate 62 A–C. *Bullockia mombazensis*; D, E. *B. dyscritos*; F, G. *B. setiflora*. Photo by GWH (A, B), VMN (C, F, G) and BL (D, E).

Scandent shrub, up to 4 m tall, covered with straw- to rust-coloured, upwardly directed hairs. Leaf-blades elliptic, 1.2–6.5 cm × 0.6–3.5 cm, acute to obtuse and apiculate at the apex, obtuse to rounded at the base, sparsely pubescent to glabrescent above, glabrescent to pubescent beneath; domatia absent or inconspicuous; petioles 1–2 mm long; stipules triangular at the base, with a linear lobe 2–5 mm long. Flowers unisexual, (4–)5(–6)-merous. Functionally male flowers 2–12 in subsessile to pedunculate cymes; pedicels 1–4 mm long; calyx-tube reduced, lobes linear to spathulate, up to 3 mm long; corolla-tube 1.5–2 mm long, lobes triangular-ovate, ca. 1 mm long. Functionally female flowers 1(–2); pedicels 1.5–8 mm long; calyx-tube 1.5–2 mm long, lobes 1.5–3 mm long; corolla-tube 1–1.3 mm long, lobes up to 1–3 mm long. Fruits broadly obovoid, 7–12 mm long.

Distribution: Coastal Kenya. [East Africa, from Kenya to northeast South Africa].

Habitat: Coastal forests; up to 800 m.

Kwale: Shimba Hills, *SAJIT 006125* (HIB); Mount Jombo, *Polhill & Robertson 4842* (K); Marenji Forest Reserve, *Luke & Robertson 1734* (EA, K).

5. **Bullockia pseudosetiflora** (Bridson) Razafim., Lantz & B. Bremer, Ann. Missouri Bot. Gard. 96: 175. 2009. ≡ *Canthium pseudosetiflorum* Bridson, Kew Bull. 42: 635. 1987; F.T.E.A. Rubiac. 3: 884. 1991; K.T.S.L.: 505. 1994. —Type: Kenya, Marsabit, Moyale, 20 Oct. 1952, *J.B. Gillett 14096* [holotype: K (K000311649); isotypes: BR (BR0000008827920), EA]

Shrub, up to 3 m tall, covered with pale, soft, crisped or patent hairs. Leaf-blades elliptic or oblong-elliptic, acute or rarely obtuse at the apex, cuneate or obtuse at the base, glabrescent or pubescent above, pubescent beneath; domatia absent or inconspicuous; petioles 1–3 mm long; stipules triangular at the base, with a linear lobe 2–4 mm long. Flowers unisexual, 5-merous. Functionally male flowers few in subsessile to pedunculate cymes; pedicels 1–5 mm long; calyx-tube reduced; lobes linear to spathulate, 0.5–1.5 mm long; corolla-tube 1.2–1.8 mm long, lobes triangular-ovate 1–1.3 mm long. Functionally female flowers solitary; pedicels 3–9 mm long; calyx-tube 1.5–2 mm long, lobes 1–2 mm long. Fruits broadly obovoid, 8–11 mm long.

Distribution: Northern, northwestern, western, and central Kenya. [Ethiopia, Uganda, Tanzania].

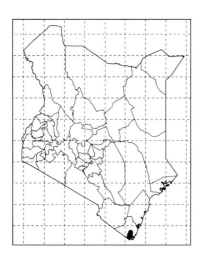
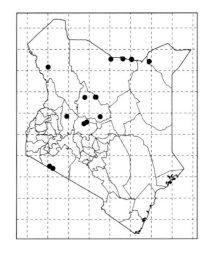

Habitat: Dry bushlands or rocky places; 700–1800 m.

Baringo: Vatya to Chepliesin, *Luke BFFP207* (EA, K). Laikipia: Kisima Farm, *Barry 1540* (EA); Mpala Reaserch Center, *Wasike et al. 3302* (EA). Mandera: Dandu, *Gilbert 12680* (EA, K). Marsabit: Moyale, *Gillett 14069* (EA, K); Mount Burole, *Barry & Smith 14875* (EA, K). Narok: Masai Mara Game Reserve, *Kokwaro & Mathenge 2699* (EA). Samburu: Lolokwe Mount, *Gilbert 20207* (EA, K). Turkana: Moruassigar, *Newbould 7294* (EA).

52. **Pyrostria** Comm. ex Juss.

Shrubs to small trees. Leaves sessile to petiolate; blades subcoriaceous, usually glabrous to infrequently pubescent; tertiary nerves obscure; stipules triangular at the base with an ovate to linear lobe. Flowers unisexual or bisexual, 4–5-merous, few to many in pedunculate umbels; bracts paired, persistent. Calyx-tube ovoid; limb reduced. Corolla rather fleshy; tube shorter than or longer than the lobes; lobes spreading, reflexed or sometimes erect. Stamens inserted at the throat of the corolla, with very short filaments. Style slender, slightly longer than corolla-tube. Fruits subspherical to heart-shaped, 2-lobed; pyrenes woody, obovoid.

A genus containing about 45 species in Africa, Madagascar and the Mascarenes; two species in Kenya.

1a. Lateral branches always reduced to spurs; leaf-blades linear to oblong, 0.5–4.5 cm × 0.3–1.7 cm .... 1. *P. phyllanthoidea*
1b. Lateral branches well developed; leaf-blades elliptic, 4–12.5 cm × 1–7 cm ...... 2. *P. bibracteata*

1. **Pyrostria phyllanthoidea** (Baill.) Bridson, Kew Bull. 42: 629. 1987; F.T.E.A. Rubiac. 3: 891. 1991; K.T.S.L.: 540. 1994. ≡ *Canthium phyllanthoideum* Baill., Adansonia 12: 220. 1876. —Type: Kenya, Mombasa, *L.H. Boivin s.n.* [holotype: P (P00553443)] (Plate 63A, B)

Shrub or rarely small tree, up to 6 m tall, with lateral branches often reduced to spurs. Leaf-blades subcoriaceous, linear to oblong, 0.5–4.5 cm × 0.3–1.7 cm, rounded at the apex, acute to cuneate at the base; domatia absent or present. Flowers 4-merous, few in sessile or subsessile umbels; bracts ovate, 2–2.5 mm long. Calyx-tube reduced; limb-lobes triangular, 1–1.2 mm long. Corolla-tube 2–5 mm long; lobes ovate, 2–4 mm long. Fruits heart-shaped, 9–11 mm wide; pyrenes obovoid, ca. 7 mm long.

Distribution: Northern, western, central, eastern, southern, and coastal Kenya. [Arabian Peninsula, and northeast Africa to Tanzania].

Habitat: Bushlands or woodlands; up to 2100 m.

Busia: Busia Hills, *Makin 154* (EA, K). Homa Bay: south of Nyakweri, *Kirika 296* (EA). Kajiado: Mount Suswa, *SAJIT 002361* (HIB); Ngong Hills, *Gillett 16970* (EA, K). Kilifi: Arabuko-Sokoke Forest, *Faden & Evans 71/712* (K); Gede Forest, *Gerhardt & Steiner 40* (EA). Kwale: Kaya Lunguma, *Luke 3350* (EA). Laikipia: Ol Ari Nyiro Ranch, *Muasya 1868* (EA). Lamu: Ras Tenewi, *Luke*

& *Robertson 1435* (EA, K). Machakos: Katumani Farm, *Thomas 1007* (EA). Mandera: Dandu, *Gillett 12764* (EA, K). Marsabit: Mount Kulal, *Hepper & Jaeger 6950* (EA, K). Meru: between Timau and Isiolo, *Blackwell 6* (EA). Nairobi: Nairobi National Park, *Gillett 21654* (K). Nakuru: Margaret Hill, *Bally 967* (K). Narok: ca. 9 km north of Sekenani Hill, *Kuchar 10980* (EA). Samburu: north side of Mount Lolokwe, *Gillett 20200* (EA). Teita Taveta: Kasigau Forest, Bungule Village, *SAJIT 005402* (HIB); east of Bura river, *Mwachala 182* (EA). Tana River: Kora Game Reserve, *Mungai et al. 417/83* (EA).

elliptic, 4–12.5 cm × 1–7 cm; tertiary nerves obscure; petioles up to 8 mm long. Flowers 4-merous, few to many in shortly pedunculate umbels; bracts acuminate, 2–7 mm long. Calyx-tube almost globose; limb reduced. Corolla yellowish cream; tube 2–3.5 mm long; lobes ovate-triangular, 2–3.5 mm long. Fruits almost globose, 5–8 mm in diameter, edible; pyrenes obovoid, 5–6 mm long.

Distribution: Central, southern, and coastal Kenya. [East Africa from Kenya to Zimbabwe, also in Madagascar].

Habitat: Bushlands or forest edges; up to 870 m.

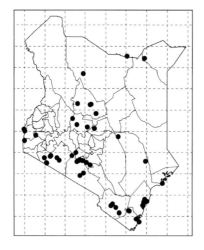

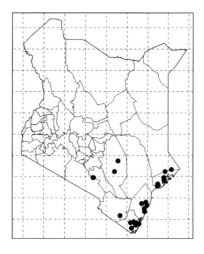

2. **Pyrostria bibracteata** (Baker) Cavaco, Bull. Mus. Natl. Hist. Nat., sér. 2, 39: 1015. 1968; F.T.E.A. Rubiac. 3: 887. 1991; K.T.S.L.: 540. 1994. ≡ *Plectronia bibracteata* Baker, Fl. Mauritius 146. 1877. —Type: Seychelles, 1841, *M. Pervillé 82* [holotype: K (K000412467); isotype: P] (Figure 49; Plate 63C, D)

Shrub or small tree, up to 10 m tall. Leaf-blades papery to subcoriaceous, glabrous,

Kajiado: Emali Hill, *Luke & Robertson 109* (K). Kilifi: Mangea Hills, *Luke & Robertson 622* (EA). Kitui: Mutha Hill, *Luke & Stone 8228* (K). Kwale: Dzombo Hill, *Robertson et al. MDE387* (EA); Shimba Hills, *Magogo & Glover 294* (EA, K). Lamu: Witu Forest, *Robertson & Luke 5523* (EA, K); Boni Forest Reserve, *Luke & Robertson 1523* (EA). Teita Taveta: Kasigau, *Bally 9403* (K). Tana River: Kanwe Mayi Forest, *Luke & Robertson 5384* (K).

Plate 63 A, B. *Pyrostria phyllanthoidea*; C, D. *P. bibracteata*. Photo by GWH (A, B) and VMN (C, D).

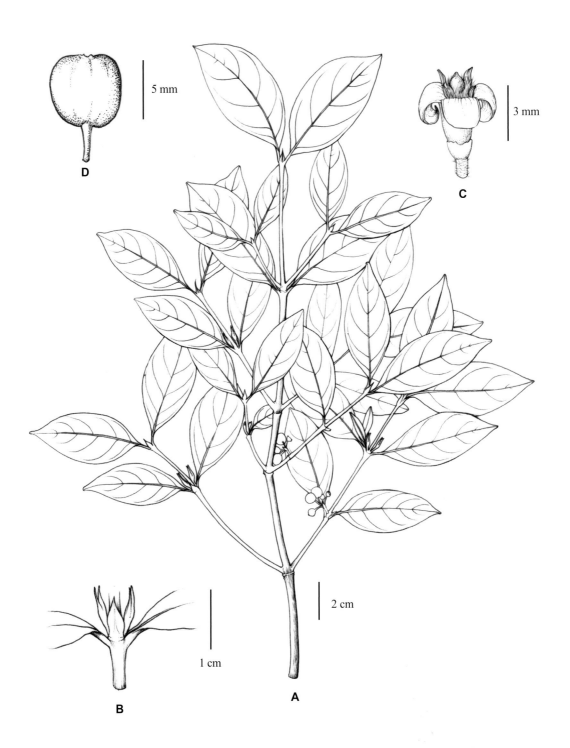

Figure 49 *Pyrostria bibracteata*. A. a fruiting branch; B. a portion of branch showing the stipule; C. flower; D. fruit. Drawn by NJ.

53. **Canthium** Lam.

Shrubs, small trees or rarely lianas, usually with paired supra-axillary spines. Leaves petiolate, opposite and spaced along the branches or constructed on very short lateral spurs, with or without domatia; stipules triangular, apiculate or aristate, pubescent within. Flowers bisexual, 4–5-merous, few to several in subsessile to pedunculate cymes. Calyx-tube broadly ellipsoid to ovoid; limb reduced to dentate or 4–5-lobed. Corolla white to green; tube broadly cylindrical, hairy inside; lobes reflexed, acute or shortly apiculate. Stamens inserted at corolla throat, partially to fully exserted; anthers dorsifixed near the base. Ovary 2–5-locular; ovule 1 in each locule. Style slender; stigmatic knob almost spherical, 2-lobed. Fruit a fleshy drupe; pyrenes 1–5, ellipsoid.

A genus containing about 50 species and widely distributed in tropical and subtropical Africa and Asia; three species in Kenya.

1a. Leaves always opposite and spaced along the branches; corolla-tube 2–4.3 mm long, fruits 1.3–2.5 cm in diameter.. 1. *C. oligocarpum*
1b. Leaves always constructed on short lateral spurs, appearing to be 4-whorls in each node; corolla-tube less than 2 mm long; fruit less than 1.7 cm long 2
2a. Fruits square in outline, 0.9–1.2 cm long and wide; pyrencs 2, 0.7–1.2 cm long ...... 2. *C. glaucum*
2b. Fruits subglobose, ca.1.7 cm × 1.9–2 cm; pyrenes 4–5, ca.1.6 cm long .... 3. *C. tetraphylla*

1. **Canthium oligocarpum** Hiern, Fl. Trop. Afr. 3: 138. 1877; F.T.E.A. Rubiac. 3: 881. 1991; K.T.S.L.: 505. 1994. —Type: Ethiopia, Begemder, Aug. 1863, *G.H.W. Schimper 1127* [holotype: K; isotypes: BR, E (E00193680), PRE (PRE0593385-0)]

Shrub or small tree, up to 7.5(–20) m tall. Leaf-blades elliptic, 3–14.5 cm × 1–6 cm, obtuse to acute or acuminate at the apex, rounded or cuneate at the base; secondary nerves in 3–7 main pairs; tertiary nerves obscure; domatia pit-like, ciliate to densely hairy; petioles up to 1.5 cm long; stipules broadly triangular, slightly connate at the base, up to 6 mm long. Flowers 5-merous, few to many in pedunculate cymes. Calyx-tube up to 2 mm long; lobes linear to narrowly triangular, up to 1.5 mm long. Corolla-tube 2–4.2 mm long; lobes triangular, 1.5–4 mm long. Fruits obovoid to somewhat square, 1.3–2.5 cm long; pyrenes ellipsoid, 1.2–1.4 cm long.

1a. Leaf-blades obtuse to acute or rarely subacuminate at the apex; domatia of conspicuous densely hairy tufts...b. subsp. *friesiorum*
1b. Leaf-blades subacuminate to acuminate at the apex; domatia with ciliate to pubescent pits 2
2a. Domatia ciliate to pubescent; flowers (2–)7–23(–40) in cymes............... a. subsp. *oligocarpum*
2b. Domatia densely pubescent; flowers 3–12 in cymes .... c. subsp. *intermedium*

a. subsp. **oligocarpum**

Leaf-blades distinctly acuminate at the apex, secondary nerves in 3–5 main pairs; domatia ciliate to pubescent. Flowers (2–)7–23(–40) in pedunculate cymes. Fruits 1.3–2.4 cm in diameter.

Distribution: Western Kenya. [Burundi, D.R. Congo, Ethiopia, Rwanda, Sudan, Tanzania, and Uganda].

Habitat: Upland forests; 2100–2200 m.

Kericho: southwest of Mau Escarpment, *Kerfoot 4551* (K).

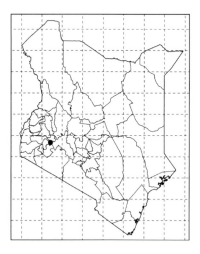

b. subsp. **friesiorum** (Robyns) Bridson, F.T.E.A. Rubiac. 3: 878. 1991; K.T.S.L.: 505. 1994. ≡ *Rytigynia friesiorum* Robyns, Notizbl. Bot. Gart. Berlin-Dahlem 10: 614. 1929. —Type: Kenya, Nyeri, West Mount Kenya Forest Station, 16 Jan. 1922, *R.E. & T.C.E. Fries 956* [lectotype: K (K000311654), designated by B. Verdcourt & D. Bridson in F.T.E.A. Rubiac. 3: 878. 1991; isolectotypes: BR (BR0000008827968), S (S05-10406)] (Figure 50; Plate 64A, B)

Leaf-blades always small, 2.5–5 cm × 1.3–2.8 cm, obtuse to acute at the apex; lateral nerves in 4–5 main pairs; domatia densely pubescent. Flowers 5–15 in pedunculate cymes. Fruits 1–1.5 cm in diameter.

Distribution: Central Kenya. [Endemic].

Habitat: Upland forests or thickets; 2000–2500 m.

Kajiado: Ngong Hills, *Beentje 2251* (EA, K). Kiambu: Limuru, *Dummer 1536* (K). Kirinyaga: Irangi Forest, *SAJIT 004021* (HIB). Meru: Nyambeni Hills, *Luke et al. 10281B* (EA, K). Nyandarua: Aberdare Mountains, *Fey EA11703* (EA, K). Narok: Loita Hills, *Luke et al. 9272* (EA, K). Samburu: Mathew's Range,

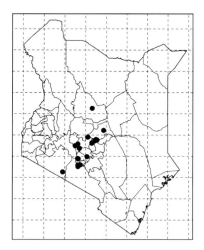

Luke 14237 (EA).

c. subsp. **intermedium** Bridson, F.T.E.A. Rubiac. 3: 878. 1991. —Type: Tanzania, Pare, Tona, 13 Jule 1915, *A. Peter 55763* [holotype: K (K000318845); isotypes: B (B100160725), BR (BR0000008827111)]

Leaf-blades subacuminate or acuminate at the apex; lateral nerves in 5–7 main pairs; domatia densely pubescent. Flowers 3–12 in cymes. Fruits 1.4–1.7 cm in diameter.

Distribution: Southern Kenya. [Tanzania].
Habitat: Lowland forests; 1350–2000 m.

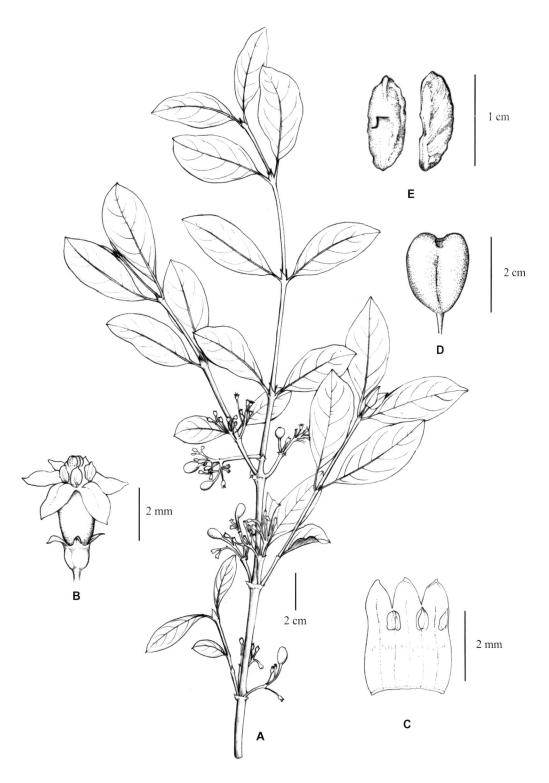

Figure 50 *Canthium oligocarpum* subsp. *friesiorum*. A. a fruiting branch; B. flower; C. longitudinal section of corolla, showing the stamens; D. fruit; E. pyrenes. Drawn by NJ.

Teita Taveta: Mbololo Forest, *De Block et al. 479* (K); Mount Kasigau, *Faden et al. 71/81* (K).

Kilifi: *Langridge 30* (K). Kwale: Shimba Hills, *Luke 3159* (EA, K); Mwaluganje Elephant Sanctuary, *Mwadime et al. 100* (EA).

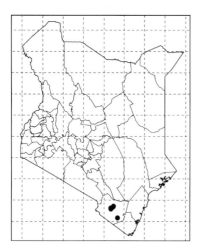

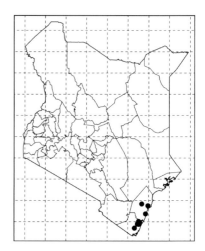

2. **Canthium glaucum** Hiern, Fl. Trop. Afr. 3: 134. 1877; F.T.E.A. Rubiac. 3: 862. 1991; K.T.S.L.: 504. 1994. —Type: Somalia, Tula (Tola) R., Nov. 1858, *J. Kirk s.n.* [lectotype: K (K000412406), designated by A.A. Bullock in Bull. Misc. Inform. 8: 359. 1932] (Plate 64C)

Shrub or small tree, sometimes scandent, up to 5 m tall. Leaves always constructed on short lateral spurs, shortly petiolate, blades elliptic, 1.5–6 cm × 1–3 cm, acute to obtuse at the apex, acute to cuneate at the base, lateral nerves in 3–5 main pairs; tertiary nerves apparent; domatia present as small tufts of hair; stipules triangular, apiculate, up to 4 mm long. Flowers 4–5-merous, 2–8 in pedunculate cymes. Calyx-tube ca. 1 mm long; limb lobes narrowly triangular, 0.5–1 mm long. Corolla-tube up to 1.3 mm long; lobes 2–2.5 mm long. Fruits edible, square, 0.9–1.2 cm long; pyrenes ellipsoid, 7–12 mm long.

Distribution: Coastal Kenya. [East Africa, from Somalia to Botswana].

Habitat: Coastal forests; up to 200 m.

3. **Canthium tetraphyllum** (Schweinf. ex Hiern) Baill., Adansonia 12: 192. 1878. ≡ *Vangueria tetraphylla* Schweinf. ex Hiern, Fl. Trop. Afr. 3: 152. 1877. ≡ *Meyna tetraphylla* (Schweinf. ex Hiern) Robyns, Bull. Jard. Bot. État Bruxelles 11: 232. 1928; F.T.E.A. Rubiac. 3: 859. 1991; K.T.S.L.: 521. 1994; U.T.S.K.: 308. 2005. —Type: Sudan, Jur, Abu Qurun, 10 May 1869, *G. Schweinfurth 1856* [holotype: K (K000412114); isotypes: BM, M (M0106356)] (Plate 64D, E)

= *Meyna tetraphylla* (Schweinf. ex Hiern) Robyns subsp. *comorensis* (Robyns) Verdc., Kew Bull. 36: 540. 1981. ≡ *Meyna comorensis* Robyns, Bull. Jard. Bot. État 11: 233. 1928. —Type: Coromor Is., Moely, *L.H. Boivin s.n.* [holotype: P(P00209608)]

Shrub or small tree, up to 9 m tall, or sometimes scandent, up to 20 m long. Leaves always constructed on short lateral spurs, shortly petiolate; blades elliptic, 2–7(–12) cm × 1.5–3.5(–6) cm, obtuse to acuminate at the apex, cuneate or rarely rounded at the base, glabrescent

Plate 64 A, B. *Canthium oligocarpum* subsp. *friesiorum*; C. *C. glaucum*; D, E. *C. tetraphyllum*. Photo by SWW (A, B), BL (C), and GWH (D, E).

to pubescent; domatia absent or feebly developed; stipules with subulate tips 1–2 mm long. Flowers 2–10 in subsessile to shortly pedunculate cymes. Calyx-tube 1.5–2.5 mm long; lobes triangular, less than 0.7 mm long. Corolla-tube 1.2–2(–3) mm long; lobes narrowly triangular, 2–3 mm long. Ovary 4–5-locular; style slender. Fruits subglobose, bluntly 5-angled, 1.3–1.7 cm × 1.6–2 cm; pyrenes 4–5, 1.2–1.6 cm long.

Distribution: Widespread in Kenya. [Northeast Africa, from Sudan to Tanzania].

Habitat: Thickets, woodlands, bushlands, or coastal forests; up to 1700 m.

Baringo: Chemolingot, *Timberlake 676* (EA). Kilifi: Arabuko-Sokoke Forest, *SAJIT 006452* (HIB). Kitui: Kora Game Reserve, *Gillett 21104* (EA). Kwale: Dzombo Hill, *Robertson et al. 165* (EA). Makueni: Kibwezi, *Muasya et al. GBK02/31/07* (EA, K). Marsabit: Gulni, *Sato 430* (EA). Meru: Kinadani, *Hamilton 770* (EA). Teita Taveta: Kasigau Forest, *SAJIT 005408* (HIB). Tana River: Tana River National Primate Reserve, *Luke et al. TPR 773* (K). Turkana: Oropoi, *Newbould 6999* (EA). West Pokot: Amolem, *Lyles 56* (EA).

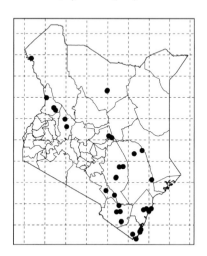

54. **Rytigynia** Blume

Shrubs or small trees, sometimes spiny. Leaves opposite or occasionally in 3-whorled; stipules united at the base, oblong or triangular, hairy within, ending in a mostly linear or subulate lobe. Flowers mostly 5-merous, axillary, solitary or in few- to several-flowered cymes; peduncles and pedicels usually developed. Calyx-tube subglobose; limb-tube short, mostly truncate or denticulate only, or few distinctly lobed. Corolla-tube cylindrical, glabrous or hairy within; lobes acute to long apiculate or with a filiform appendage. Ovary 2–5(–6)-locular; ovule single in each locule; stigmatic club coroniform, subglobose or cylindrical, distinctly 2–5-lobed at the apex. Fruits mostly subglobose, with 1–5 pyrenes.

A genus with about 82 species restricted in tropical Africa, south Africa and Madagascar; nine species in Kenya. *Homewood 62* (EA) and *Homewood 64* (EA) were recorded as *Rytigynia* sp. I in FTEA. and some other specimens can also be classified into this group, such as *Festo & Luke 2398* (EA), *Kimberly 267* (EA, K), *Kimberly 289* (EA), *Kimberly 320* (EA, K) and *Luke & Robertson 1431* (K). These specimens seems to be the intermediate forms between *R. parvifolia* Verdc. and *R. celastroides* (Baill.) Verdc. Alternative approach is to combine *R.* sp. I, *R. parvifolia* and *R. celastroides* as one speices. *Luke & Robertson 6309* (EA) was recorded as *R.* sp. L in FTEA, and *Luke 3040* (EA) and *Luke et al. 9039* (EA) also could be classified into this group. Because of the lack of fertile materials, especially flowers, we could not deal with them.

1a. Plants with spines 2
1b. Plants unarmed 5
2a. Calyx-lobes linear-lanceolate; corolla-lobes much shorter than corolla-tube ....
 .... 9. *R. bugoyensis*
2b. Calyx-lobes not as above; corolla-lobes as long as or longer than corolla-tube 3
3a. Calyx-limb distinctly toothed .... 6. *R. mrimaensis*
3b. Calyx-limb truncate or slightly toothed 4
4a. Leaves 1–2.5 cm × 0.5–1.5 cm, rounded at the apex; ovary 3-locular .... 7. *R. parvifolia*
4b. Leaves 2–5.5 cm × 0.8–2.5 cm, acuminate at the apex; ovary 2-locular .... 8. *R. celastroides*
5a. Leaves densely pubescent to velvety .... 5. *R. decussata*
5b. Leaves glabrous or sparsely pubescent, or sometimes pubescent on the veins beneath 6
6a. Inflorescences of (1–)2-flowered cymes with usually distinct pedicels and peduncles 7
6b. Inflorescences with several flowers in sessile fascicles or shortly pedunculate umbel-like
 cymes 8
7a. Ovary 3–5-locular; corolla-tube usually longer than lobes .... 3. *R. uhligii*
7b. Ovary 2-locular; corolla-tube much shorter than lobes .... 4. *R. eickii*
8a. Peduncles up to 2 mm long; ovary 2-locular .... 2. *R. neglecta*
8b. Peduncles 2–4 mm long; ovary 3–5-locular .... 1. *R. acuminatissima*

1. **Rytigynia acuminatissima** (K. Schum.) Robyns, Bull. Jard. Bot. État Bruxelles 11: 169. 1928; F.T.E.A. Rubiac. 3: 811. 1991; K.T.S.L.: 543. 1994. ≡ *Vangueria acuminatissima* K. Schum., Pflanzenw. Ost-Afrikas, C: 385. 1895. —Types: Tanzania, Bukoba, Nov. 1890, *F.L. Stuhlmann 1014* [holotype: B (destroyed)]; Bukoba, Aug. 1931, *A.E. Haarer 2128* [neotype: K, **designated here**; isoneotype: EA] (Plate 65)

Shrub or tree, up to 9 m tall; spines absent. Leaf-blades oblong-elliptic, elliptic or obovate, 3–15 cm × 0.5–8 cm, narrowly and distinctly acuminate at the apex, cuneate, rounded or truncate at the base; petioles 3–6 mm long; stipules connate at the base, 2–4 mm long, with a subulate lobe up to 7 mm long. Inflorescences (1–)2–7(–13)-flowered; peduncles 2–4 mm long; pedicels 1.5–8 mm long, elongated in fruit. Calyx-tube 1–1.8 mm long; limb short, truncate or slightly toothed. Corolla-tube 2–3.5 mm long; lobes ovate-triangular, 2–3 mm long. Ovary 3–4-locular. Fruits subglobose, 7–13 mm in diameter.

Distribution: Western Kenya. [Tropical East Africa].

Habitat: Forests; 1600–2200 m.

Bomet: Tinderet Forest Reserve, *Geesteranus 4981* (K). Kakamega: Kakamega Forest, *SAJIT 006787* (HIB). Narok: Kilgoris, *Birch 60/403* (EA). Trans Nzoia: Mount Elgon, *Hamilton 76/770* (EA). Uasin Gishu: west Eldoret, *Birch 61/144* (EA).

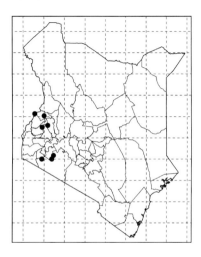

Plate 65 A–F. *Rytigynia acuminatissima*. Photo by GWH.

2. **Rytigynia neglecta** (Hiern) Robyns, Bull. Jard. Bot. État Bruxelles 11: 183. 1928; F.T.E.A. Rubiac. 3: 813. 1991; K.T.S.L.: 544. 1994. ≡ *Canthium neglectum* Hiern, Fl. Trop. Afr. 3: 135. 1877. —Type: Ethiopia, Begemder, Jule 1863, *G.H.W. Schimper 1106* [holotype: K (K000412423); isotypes: BM (BM000903274), E (E00193708), PRE (PRE0593545-0)] (Plate 66A)

Shrub or small tree, up to 9 m tall; spines absent. Leaf-blades elliptic, 2–14.5 cm × 1.5–6.5 cm, acuminate at the apex, cuneate to rounded at the base; petioles 5–8 mm long; stipules connate, 3–4 mm long, with a subulate lobe 2–7 mm long. Inflorescences of 3–10-flowered umbel-like cymes or fascicles; peduncles 1–2 mm long; pedicels 1–3 mm long, elongated in fruit. Calyx-tube ca. 1.5 mm long; limb very short, minutely denticulate. Corolla-tube ca. 2 mm long, hairy inside; lobes 3–4 mm long. Ovary 2-locular. Fruits black, 2-lobed, 7–11 mm in diameter.

Distribution: Northern, western, and central Kenya. [Ethiopia and Sudan to Cameroon and D.R. Congo].

Habitat: Forests or hillsides; 1350–2300 m.

Marsabit: Mount Kulal, *Herlocker H-408* (EA); Marsabit Forest, *Faden 68/536* (EA). Meru: Ngaia Forest, *Luke et al. 9594* (EA). Samburu: Mathew's Range, *Luke 14134* (EA). West Pokot: Sekerr Mountain, *Agnew et al. 10461* (EA); Kapenguria, *Okwaro OW57* (EA).

3. **Rytigynia uhligii** (K. Schum. & K. Krause) Verdc., Kew Bull. 42: 165. 1987; F.T.E.A. Rubiac. 3: 818. 1991; K.T.S.L.: 544. 1994. ≡ *Vangueria uhligii* K. Schum. & K. Krause, Bot. Jahrb. Syst. 39: 534. 1907. —Type: Tanzania, Kilimanjaro, *C. Uhlig 521* [holotype: B (destroyed); isotype: EA, **designated here**] (Plate 66B–E)

= *Canthium sarogliae* Chiov., Racc. Bot. Miss. Consol. Kenya: 55. 1935. —Types: Kenya, Meru, 5 May 1929, *P.G. Balbo 17* [lectotype: FT (FT003430), **designated here**]; Kenya, Nyeri, 2 Apr. 1911, *P.G. Balbo 826* [syntype: FT (FT003430)]

Shrub or tree, up to 9 m tall; spines absent. Leaf-blades elliptic to ovate, 1.5–11 cm × 0.5–5 cm, narrowly or abruptly acuminate at the apex, cuneate, rounded or truncate at the base; petioles 2–6 mm long; stipule-sheath connate, triangular or ovate, 3–5 mm long, with a subulate lobe up to 4.5 mm long.

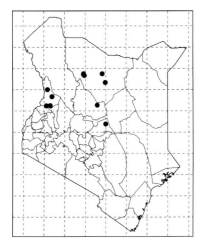

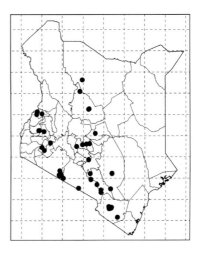

Plate 66 A. *Rytigynia neglecta*; B–E. *R. uhligii*. Photo by BL (A), YDZ (B) and GWH (C–E).

Inflorescences of 1–2-flowered cymes, or rarely more; peduncles up to 6(–10) mm long; pedicels (2)6–12 mm long, elongate in fruit. Calyx-tube 1.2–1.5 mm long; limb very short, truncate or slightly toothed. Corolla-tube 4.5–5.5 mm long; lobes ovate, ca. 2 mm long. Ovary 2–5-locular. Fruits blue-black, globose, 7–10 mm in diameter.

Distribution: Northern, western, cental, and southern Kenya. [East Africa, from Kenya to Zimbabwe and Mozambique].

Habitat: Montane forests, forest edges, hills, wooded grasslands, bushed woodlands, or footpaths; 1150–2450 m.

Bomet: Sotik, *Bally B7449* (EA). Kajiado: Emali Hill, *van Someren 135* (K). Kakamega: Kakamega Forest, *Friis & Hansen 2577* (EA). Kericho: southwest Mau Forest, *Kerfoot 3887* (EA). Kitui: Mutha Hill, *Luke & Stone 8217* (K). Makueni: Kilungu Forest, *Mwangangi 1942* (EA). Marsabit: Mount Kulal, *Nyamongo et al. GBK21* (K). Meru: Nkunga Crater Lake, *Luke et al. 7258* (EA). Nandi: Kapsabet, *Williams & Piers 607* (EA). Narok: Keshumoruo Forest, *Morris 47* (EA). Nyamira: Aitibu, *Dale K1032* (EA). Nyeri: Kagochi, *Kerfoot 1492* (EA). Samburu: Mount Nyiru, *Bytebier et al. 277* (EA). Teita Taveta: Iyale Forest, *SAJIT 006408* (HIB); Ndiwenyi Forest, *Block et al. 330* (K); Mgange Nyika, *Mwachala et al. A3067* (EA). Trans Nzoia: Kitale Forest, *Thorold 2730* (EA).

4. **Rytigynia eickii** (K. Schum. & K. Krause) Bullock, Bull. Misc. Inform. Kew 1932(8): 389. 1932; F.T.E.A. Rubiac. 3: 817. 1991; K.T.S.L.: 544. 1994. ≡ *Plectronia eickii* K. Schum. & K. Krause, Bot. Jahrb. Syst. 39: 538. 1907. —Types: Tanzania, Lushoto, near Muafa, *J. Buchwald 611* [syntype: BM (destroyed); lectotype: K (K000284340), **designated here**]; Tanzania, Lushoto, near Kwai, *Erick 86* [syntype: B (destroyed); isosyntype (frag.): K (K000284340 lower pack)] (Plate 67A)

Shrub or small tree, up to 5 m tall; spines absent. Leaf-blades oblong to ovate-oblong, 4–14.5 cm × 1.7–6.8 cm, shortly obtusely acuminate at the apex, cuneate to rounded at the base; petioles 4–8 mm long; stipules connate at the base, 1.5–3 mm long, with a subulate lobe 2–5 mm long. Inflorescences 1–2-flowered; peduncles up to 5 mm long; pedicels 2–4 mm long. Calyx-tube 1–1.2 mm long; limb-tube very short, truncate or shortly toothed. Corolla-tube ca. 3 mm long; lobes 2.5–3.5 mm long. Ovary 2-locular. Fruits black, 1–2-lobed, 1–1.1 cm in diameter.

Distribution: Central and southern Kenya. [Tanzania].

Habitat: Montane forests; 1200–1850 m.

Kitui: Muumoni Hill Forest, *Mwachala et al. 588* (EA). Meru: Ngaia Forest, *Luke et al. 11719* (EA). Teita Taveta: Ngangao Forest, *Faden et al. 535* (K).

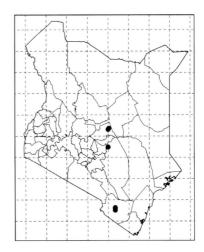

5. **Rytigynia decussata** (K. Schum.) Robyns, Bull. Jard. Bot. État Bruxelles 11: 195. 1928; F.T.E.A. Rubiac. 3: 835. 1991. ≡ *Pachystigma*

decussatum K. Schum., Pflanzenw. Ost-Afrikas C: 387. 1895. —Types: Tanzania, Uzaramo, 1894, *F.L. Stuhlmann 6426* [syntype, B (destroyed); lectotype, Z (Z-000023129), **designated here**]; ibid., *F.L. Stuhlmann 6637* [syntype: B (destroyed), former lectotype designated by W. Robyns in Bull. Jard. Bot. État Bruxelles 11: 196. 1928]; ibid., *F.L. Stuhlmann 6836* [syntype, B (destroyed); isosyntype, K(K000319916)]; ibid., *F.L. Stuhlmann 6788* [syntype, B (destroyed)]; ibid., *F.L. Stuhlmann 7022* [syntype, B (destroyed)]; ibid., *F.L. Stuhlmann 7045* [syntype, B (destroyed)]; ibid., *F.L. Stuhlmann 7118* [syntype, B (destroyed)] (Plate 67B–D)

Shrub, up to 1.5 m tall; spines absent. Leaves paired or rarely 3-whorled, elliptic, oblong or ovate, 1.7–11 cm × 0.8–6 cm, shortly pubescent above, velvety-woolly beneath, obtuse to acute at the apex, cuneate or rarely rounded at the base; petioles 4–7 mm long; stipules connate at the base, 1.5–2.5 mm long, with a subulate lobe 2–3 mm long. Inflorescences of 1–5-flowered cymes; peduncles very short; pedicels up to 6 mm long. Calyx-tube campanulate or globose, 1–1.5 mm long; limb-tube very short, lobes linear-lanceolate, 1.5–4 mm long. Corolla-tube 2–4.5 mm long; lobes triangular, 2.5–4 mm long, apiculate. Ovary 2-locular. Fruits black, ca. 1 cm in diameter.

Distribution: Coastal Kenya. [East Africa, from Kenya to Mozambique].

Habitat: Coastal forests or bushlands; 200–400 m.

Kwale: Shimba Hills, *SAJIT V0131* (HIB).

6. **Rytigynia mrimaensis** Verdc., Kew Bull. 42(1): 170. 1987; F.T.E.A. Rubiac. 3: 823. 1991; K.T.S.L.: 544. 1994. —Type: Kenya, Kwale, Mrima Hill, *J.P.M. Brenan et al. 14616* [holotype: K; isotype: EA (EA 000001726)]

Shrub, up to 4 m tall; spines supra-axillary, up to 3.8 cm long. Leaves ovate, 2.5–7 cm × 1.3–4.5 cm,, acuminate at the apex, rounded at the base; petioles up to 4 mm long; stipules ca. 3.5 mm long, with a subulate lobe up to 2 mm long. Flowers solitary; pedicels 0.8–1.2 mm long. Calyx-tube obconic, ca. 2 mm long; limb-tube very short, with narrow acute triangular teeth. Corolla-tube ca. 4.5 mm long; lobes narrowly triangular, ca. 5 mm long. Ovary 2-locular. Fruits black, 2-lobed, up to 1.4 cm × 1.8 cm.

Distribution: Coastal Kenya. [Endemic].

Habitat: Forest patches, hillsides, bushlands or coastal thickets; up to 400 m.

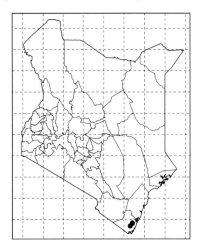
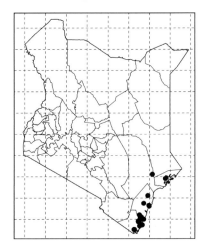

Plate 67 A. *Rytigynia eickii*; B–D. *R. decussata*. Photo by YDZ (A) and VMN (B–D).

Kilifi: Mangea Hill, *Luke 1618* (EA); Kaya Fungo, *Robertson & Luke 5747* (EA). Kwale: Mrima Hill, *Robertson et al. MDE14* (EA); Shimba Hills, *Luke et al. 4386* (EA); Muluganji Forest, *Robertson & Luke 5994* (EA). Lamu: Hindei to Bargoni, *Festo & Luke 2599* (EA). Mombasa: Bamburi Nature Trail, *Robertson 6527* (K). Tana River: Tana River National Primate Reserve, *Luke et al. TPR61* (EA).

7. **Rytigynia parvifolia** Verdc., Kew Bull. 42(1): 172. 1987; F.T.E.A. Rubiac. 3: 825. 1991; K.T.S.L.: 544. 1994. —Type: Kenya, Kilifi, 52 km west of Malindi, Kakoneni, 8 Dec. 1973, *R.W. Spjut 3956* [holotype: K (K000311659); isotypes: BR (BR0000008856463), EA (EA000001725)] (Plate 68A–C)

Shrub, up to 6 m tall; spines very short, up to 5 mm long, or sometimes absent. Leaf-blades elliptic, 1–5 cm × 0.6–2 cm, rounded at the apex, cuneate at the base; petioles 1–2 mm long; stipules ca. 1 mm long, with a reflexed lobe. Inflorescences of 2–3-flowered cymes; pedicels up to 14 mm long. Calyx-tube 0.8–1 mm long; limb very short, truncate or denticulate. Corolla-tube 1.8–3 mm long, hairy inside; lobes 2–3.5 mm long, acuminate. Ovary 3-locular. Fruits ellipsoid, 7–7.5 mm long.

Distribution: Coastal Kenya. [Somalia].

Habitat: Forest edges, bushlands, wooded grasslands, or roadsides; up to 400 m.

Kilifi: Arabuko-Sokoke Forest, *Simpson 191* (EA); Kakoneni, *Spjut 3956* (EA). Kwale: Shimba Hills, *SAJIT 005467* (HIB). Lamu: Boni Forest Reserve, *Muchiri 429* (EA); Lunghi Forest Reserve, *Luke & Robertson 1533* (EA). Tana River: Tana River National Primate Reserve, *Kimberly 267* (EA); Nairobi Ranch, *Festo & Luke 2398* (EA); Ozi, *Green et al. 1360* (EA).

8. **Rytigynia celastroides** (Baill.) Verdc., Kew Bull. 42: 171. 1987; F.T.E.A. Rubiac. 3: 823. 1991; K.T.S.L.: 543. 1994. ≡ *Canthium celastroides* Baill., Adansonia 12: 190. 1878. —Type: Kenya, Mombasa, *L.H. Boivin s.n.* [lectotype, P (P03789458), **designated here**; isolectotypes: P (P03789457), K (photo)]

= *Vangueria microphylla* K. Schum., Pflanzenw. Ost-Afrikas, C: 385. 1895. ≡ *Rytigynia microphylla* (K.Schum.) Robyns, Bull. Jard. Bot. État Bruxelles 11: 171. 1928. — Types: Kenya, Kitui, Ukambani, Malemba, June 1877, *J.M. Hildebrandt 2836* [syntype: B (destroyed); lectotype: K (K000284376),

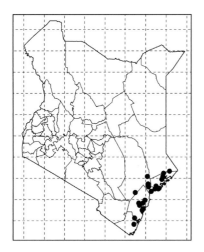
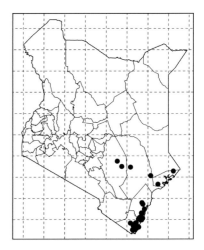

designated here]; Tanzania, Usaramo, 200 m, Jan. 1894, *F.L. Stuhlmann 6242* [syntype: B (destroyed), former lectotype designated by Robyns in Bull. Jard. Bot. État Bruxelles 11: 172. 1928]; Tanzania, Usaramo, *F.L. Stuhlmann 6842* [syntype: B (destroyed)]; Tanzania, Usaramo, 200 m, Jan. 1894, *F.L. Stuhlmann 6868* [syntype: B (destroyed)]; Kenya, Fimboni at Rabai, Jan. 1877, *J.M. Hildebrandt 2303* [syntype: B (destroyed)]

Shrub or small tree, up to 7.5 m tall; spines frequently present, solitary or paired, 7–13 mm long. Leaves usually paired or rarely 3-whorled; blades narrowly elliptic to ovate, 2–5.5(–9) cm × 0.8–2.5(–4) cm, obtuse to narrowly acuminate at the apex, cuneate or rarely rounded at the base; petioles 0.5–2 mm long; stipules connate at the base, 1–1.5 mm long, truncate, with a reflexed lobe. Inflorescences of 2–4(7)-flowered cymes; peduncles obsolete or up to 3(–7) mm long; pedicels 2–9 mm long. Calyx-tube 1–1.5 mm long; limb truncate to toothed or rarly shortly lobed. Corolla-tube 1.5–2 mm long; lobes oblong, 1.5–3.5 mm long. Ovary 2(–5)-locular. Fruits black, subglobose, 6–9 mm in diameter.

Distribution: Cental and coastal Kenya. [East Africa, from Kenya to KwaZulu-Natal of South Africa].

Habitat: Wet or dry forests, thicket edges, woodlands, open grasslands, hillsides, or disturbed roadsides; up to 400 m.

Kilifi: Mangea Hill, *Luke & Robertson 620A* (EA); Kaya Kivara, *Luke & Robertson 4770* (EA); Arabuko-Sokoke Forest, *Langridge 29* (EA). Kitui: Endau Hill, *Mwachala 482* (EA). Kwale: Godoni Forest, *Spjut 4580* (EA); Mrima Hill, *Robertson et al. MDE107* (EA). Lamu: Boni Forest Reserve, *Robertson & Luke 5648* (K). Mombasa: *Boivin s.n.* (K). Tana River: Baomo Lodge Forest, *Luke et al. TPR519* (K).

9. **Rytigynia bugoyensis** (K. Krause) Verdc., Bull. Jard. Bot. Natl. Belg. 50: 515. 1980; F.T.E.A. Rubiac. 3: 833. 1991; K.T.S.L.: 543. 1994. ≡ *Plectronia bugoyensis* K. Krause, Wiss. Ergebn. Deut. Zentr.-Afr. Exped., Bot. 2: 327. 1911. —Type: Rwanda, Gisenyi (Kissenye), Bugoie (Bugoye), 4 Nov. 1907, *J. Mildbraed 1484* [holotype: B (destroyed); lectotype: BM (BM000903261), **designated here**; isolectotyes: BR (BR0000008855671), K (K000412433)] (Plate 68D–G)

= *Canthium urophyllum* Chiov., Racc. Bot. Miss. Consol. Kenya: 54. 1935. —Type: Kenya, S.W. Mt. Kenya, Mukarara, 17 Dec. 1910, *G. Balbo 24* [holotype: FT (FT003429)]

Shrub or small tree, up to 6 m tall; spines supra-axillary, paired, up to 2.5 cm long. Leaves open paired on shortened lateral opposite shoots, so each node appears to have four leaves; blades ovate to elliptic, 3–10.5 cm × 1.2–7 cm, long-acuminate at the apex, rounded to cuneate at the base; petioles 3–10 mm long; stipules connate at the base, ca. 4 mm long with a filiform lobe. Inflorescences of 1–3-flowered

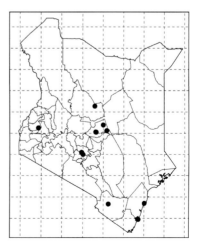

Plate 68 A–C. *Rytigynia parvifolia*; D–G. *R. bugoyensis*. Photo by BL (A), VMN (B, C), GWH (D–F) and YDZ (G).

axillary cymes; peduncles 4 mm long; pedicels 3–10 mm long. Calyx-tube hemispherical, 1.2–1.5 mm long; lobes linear-lanceolate, 1–3.5 mm long. Corolla-tube 3–6 mm long, hairy inside; lobes 1.5–3.5 mm long, ovate-oblong. Ovary 2-locular. Style 4.5–5 mm long. Fruits black, 2-lobed, 12–13 mm in diameter.

Distribution: Western, central, southern, and coastal Kenya. [Malawi, Tanzania, and Zimbabwe].

Habitat: Forests; up to 1950 m.

Kakamega: Kakamega Forest, *Mutangah et al. 111* (EA). Kiambu: Thika, *Faden 67/142* (EA). Kilifi: Gede Forest, *Gerhardt & Steiner 316* (EA). Meru: Meru Forest, *SAJIT 003891* (HIB); Ngaia Forest, *Luke et al. 10269* (EA). Mombasa: Nyali, *Birch 6278* (EA). Murang'a: Mokkarara, *Balbo 24/373* (K). Samburu: Mathew's Range, *Luke 14135* (EA). Teita Taveta: Fururu Forest, *Block et al. 338* (K).

55. **Fadogia** Schweinf.

Woody herbs, subshrubs or rarely shrubs or small trees. Leaves paired or 3–6-whorled, subsessile to shortly petiolate; stipules connate at the base, hairy inside at the base. Flowers small, solitary or few in shortly pedunculate cymes. Calyx-tube campanulate or subglobose; limb truncate, toothed or lobed. Corolla-tube cylindrical; lobes 5(–9). Ovary 3–4(–9)-locular; ovule solitary in each locule; style slender or rather thick. Fruits subglobose or oblique; pyrenes 1–3(–9).

A genus of about 50 species restricted to eastern and southern Africa; only one species in Kenya.

1. **Fadogia cienkowskii** Schweinf., Reliq. Kotschy: 47. 1868; F.T.E.A. Rubiac. 3: 788. 1991. —Type: Burundi, Kitega, 5 Dec. 1922, *O.A.J. Elskens 237* [holotype: BR (BR0000008829986)] (Figure 51)

Woody herb with several stems from a woody rootstock, up to 1.2 m tall. Leaves 3–4-whorled, shortly petiolate; blades narrowly lanceolate to elliptic, 2–8.5 cm × 0.5–4.5 cm, acute or shortly acuminate at the apex, cuneate or rounded at the base, sparsely to densely hairy above, very densely tomentose beneath with ferruginous matted hairs; stipules shortly connate at the base. Flowers solitary or 2–6 in pedunculate cymes. Calyx-tube 1–1.5 mm long; limb 7–10-toothed. Corolla-tube cylindrical 2.5–3.5 mm long; lobes narrowly lanceolate 3.5–5.5 mm long. Ovary 3–4-locular. Fruits subglobose, up to 1 cm in diameter; pyrenes 1–3.

Distribution: Western Kenya. [Tropical Africa].

Habitat: Grasslands or woodlands; 1000–2100 m.

Trans-Nzoia: Saiwa Swamp National Park, *Pearce 766* (EA). West Pokot: Kacheliba, *Napier 2005* (EA, K).

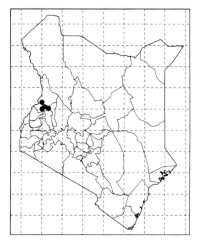

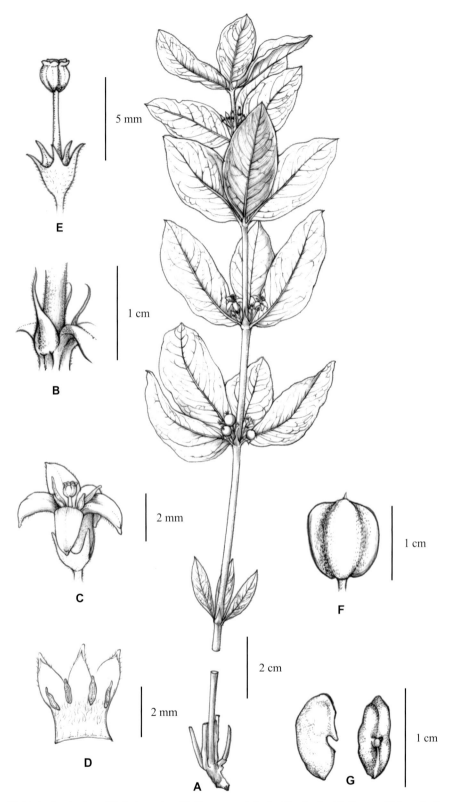

Figure 51 *Fadogia cienkowskii*. A. branch with flowers and fruits; B. stipule; C. flower; D. longitudinal section of corolla, showing the stamens; E. style; F. fruit; G. pyrenes. Drawn by NJ.

56. **Multidentia** Gilli

Subshrubs, shrubs or small trees, up to 12 m tall. Leaves petiolate, paired or rarely ternate; blades chartaceous to coriaceous; domatia present as tufts of hair or absent; stipules connate at the base, pubescent within. Flowers (4–)5(–6)-merous, few to many in pedunculate cymes. Calyx-tube very short; limb slightly to distinctly wider than tube. Corolla-tube cylindrical; lobes reflexed. Ovary 2-locular with single ovule in each locule. Fruit a 2-seeded drupe, subglobose, laterally compressed; pyrenes broadly ellipsoid.

A genus of about ten species restricted to tropical Africa; two species in Kenya.

1a. Leaf-blades papery or subcoriaceous, narrowly elliptic to elliptic, less than 11 cm × 4.5 cm; fruits 1–2 cm in diameter .... 1. *M. sclerocarpa*
1b. Leaf-blades coriaceous, oblong-elliptic, oblong or rounded, up to 27.5 cm × 15.5 cm; fruits 2–3 cm in diameter .... 2. *M. crassa*

1. **Multidentia sclerocarpa** (K. Schum.) Bridson, Kew Bull. 42: 650. 1987; F.T.E.A. Rubiac. 3: 844. 1991. ≡ *Plectronia sclerocarpa* K. Schum., Bot. Jahrb. Syst. 34: 334. 1904. —Types: Tanzania, East Usambara, riparian forest of the Sigitales between Muhesa and Lungusa, 12 Sept. 1902, *A. Engler 389 & 394* [syntypes: B (destroyed)]; Lushoto, E. Usambara Mts., Sigi to Longuza, 1 May 1917, *A. Peter 56025* [neotype: B (B100160704), designated by D. M. Bridson in Kew Bull. 42: 650. 1987] (Plate 69)

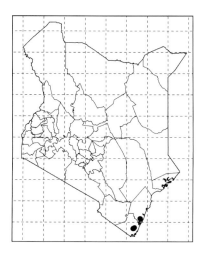

Shrub or small tree, up to 13 m tall. Leaf-blades papery or subcoriaceous, narrowly elliptic to elliptic, 6–11 cm × 2.5–4.5 cm, distinctly acuminate at the apex, acute at the base; secondary nerves in 6–7 main pairs; domatia present as tufts of hairs; petioles up to 1.2 cm long; stipules connate at the base, up to 5 mm long. Flowers 5-merous, 15–40 in pedunculate cymes. Calyx-tube ca. 1.5 mm long; limb 1.5–2 mm long, distinctly wider than tube. Corolla-tube ca. 3 mm long; lobes triangular ca. 4 mm long. Fruits subglobose, laterally compressed, 1–2 cm × 1.3–2.5 cm.

Distribution: Coastal Kenya. [Tanzania].
Habitat: Coastal forests; up to 400 m.
Kilifi: Kaya Jibana, *Luke & Luke 4327* (EA).
Kwale: Shimba Hills, *SAJIT 006058* (HIB).

2. **Multidentia crassa** (Hiern) Bridson & Verdc., Kew Bull. 42: 652. 1987; F.T.E.A. Rubiac. 3: 845. 1991; K.T.S.L.: 522. 1994. ≡ *Canthium crassum* Hiern, Fl. Trop. Afr. 3: 145. 1877. —Type: Sudan, Jur, Kurschuk Ali, 4 May 1869, *G.A. Schweinfurth 1707*

Plate 69 *Multidentia sclerocarpa*. Photo by GWH.

[lectotype: K (P00553470), designated by A.A. Bullock in Bull. Misc. Inform. Kew 1932: 380. 1932; isolectotypes: BM (BM000903294), P (P00553470)] (Figure 52)

Shrub or small tree, up to 6 tall. Leaf-blades oblong-elliptic to oblong or round, 3–27.5 cm × 1.7–15.5 cm, obtuse to acute at the apex, cuneate to rounded at the base; lateral nerves in 5–7 main pairs; domatia present as tufts of hairy; petioles up to 2.5 cm long; stipules connate at the base, up to 5 mm long. Flowers 5-merous, many in pedunculate cymes. Calyx-tube obconic, 1.5–2 mm long; limb 1.5–4 mm long, truncate or shortly toothed. Corolla-tube 2.5–4.5 mm long; lobes narrowly triangular, ca. 3 mm long. Fruits depressed globose, 2.2–3 cm × 2.6–3.7 cm.

Distribution: Western Kenya. [East Africa from Ethiopia to Zimbabwe].

Habitat: Woodlands, thickets, grasslands, or other hillsides and rocky places; 900–2100 m.

Bungoma: Lugari, *Bogdan in E.A.H. 10414* (EA).

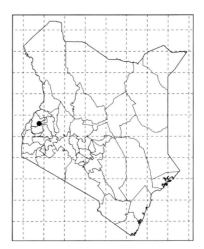

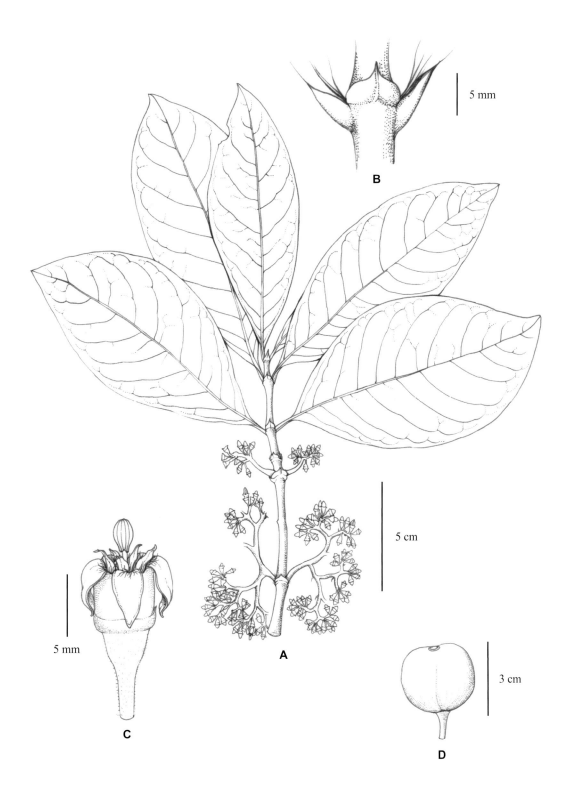

Figure 52 *Multidentia sclerocarpa*. A. flowering branch; B. portion of branch showing the stipule; C. flower; D. fruit. Drawn by NJ.

57. **Vangueriopsis** Robyns

Shrubs or small trees. Leaves opposite, somewhat coriaceous, hairy or velvety tomentose; stipules thick, triangular, joined into a sheath at the base, long-caudate at the apex. Flowers many in axillary much-branched cymes. Calyx-lobes narrowly triangular to linear-lanceolate. Corolla-tube cylindrical; lobes linear-lanceolate. Ovary 2(–5)-locular, with single pendulous ovule in each locule; style slender, long-exserted; stigmatic club cylindrical. Fruits obovoid or ellipsoid, with 2 well developed pyrenes or rather unilateral and oblique with single pyrene by abortion, smooth or irregularly ribbed.

A small genus containing only five species restricted to eastern and southern Africa; only one species in Kenya.

1. **Vangueriopsis shimbaensis** A.P. Davis & Q. Luke, Nordic J. Bot. 28: 513. 2010. —Type: Kenya, Kwale, Shimba Hills, Mwele Forest, 360 m, 22 Jan. 2005, *W.R.Q. Luke & I. Lehman 10894* [holotype: EA; isotypes: BR, K (K000843162), NHT, MO, UPS] (Figure 53)

Small tree up to 8 m tall. Leaf-blades oblong-ovate, 13–17 cm × 7.5–13 cm, acuminate at the apex, rounded to subcordate at the base, ferruginous-hirsute above, tomentose beneath; domatia absent; petioles up to 1.7 cm long; stipules broadly triangular, 6–9 mm long, with a long upper appendage 2–2.7 cm long. Flowers 5-merous, 2–5 in pedunculate cymes. Calyx-tube campanulate-obconical, 2.5–4 mm long; lobes linear to narrowly triangular, 1.5–3 cm long. Corolla-tube broadly funnel-shaped, 5–9 mm long; lobes linear to very narrowly triangular, reflexed to twisted, 3.5–6 cm long. Ovary ellipsoid, 2-locular; style well exserted, 4.2–5.3 cm long; stigmatic club cylindrical, 3–4 mm long. Fruits asymmetrical-obovoid, slightly laterally compressed, 2.4–2.6 cm long, containing a single pyrene (by abortion).

Distribution: Coastal Kenya. [Endemic].
Habitat: Coastal forests; up to 400 m.
Kwale: Shimba Hills, *Luke 8316* (EA, K).

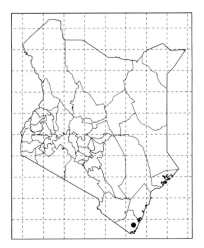

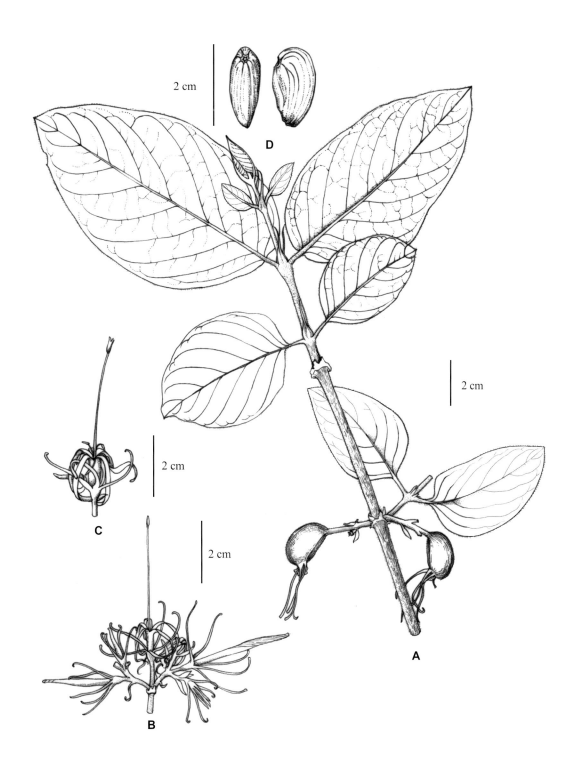

Figure 53 *Vangueriopsis shimbaensis*. A. fruiting branch; B. inflorescence; C. flower; D. pyrenes. Drawn by NJ.

58. **Vangueria** Juss.

Shrubs or small trees, unarmed, or rarely with paired spines. Leaves opposite or rarely 3-wholred; stipules usually broad at the base, produced into a narrow appendage at the apex, hairy inside. Inflorescences of axillary fascicles or dichasial cymes on opposite sides of the nodes, few- to many-flowered; bracts and bracteoles inconspicuous. Flowers 5(–6)-merous. Calyx-tube hemispherical or depressed campanulate; lobes 5, always well developed, triangular, oblong, lanceolate, linear, or spathulate. Corolla usually white, yellow or greenish yellow, often distinctly apiculate in bud; corolla-tube shortly cylindrical or campanulate, hairy at the throat; lobes 5, usually reflexed, often with distinct narrow appendage at the apex. Ovary 2–5(–6)-locular; ovule single in each locule; style slender, exserted; stigmatic knob cylindrical, obscurely 5-lobed. Fruits subglobose or globose, glabrous or pubescent, indehiscent and fleshy, sometimes crowned by the persistent clayx-limb, with 2–5 woody pyrenes.

A genus of about 56 species in tropical Africa and Madagascar; ten species in Kenya.

1a. Calyx-lobes 8–16 mm long, oblong to spathulate .... 5. *V. pallidiflora*
1b. Calyx-lobes usually shorter, triangular, oblong, lanceolate or linear 2
2a. Inflorescences of lax dichasial cymes, with flowers scattered along the arms; fruits often large, over 1.5 cm in diameter 3
2b. Inflorescences of fascicles or condensed dichasial cymes; fruits often less than 1.5 cm in diameter 7
3a. Calyx-lobes triangular, triangular-oblong or narrowly oblong, 0.5–1.5 mm long 4
3b. Calyx-lobes narrowly triangular, linear, oblong or slightly spathulate, more than 1.5 mm long 5
4a. Young branches, leaf-blades, petioles and corolla always glabrous ..... 6. *V. madagascariensis*
4b. Young branches, leaf-blades and petioles always pubescent; corolla pubescent or rarely glabrous .... 7. *V. infausta*
5a. Calyx-lobes linear or oblong-linear; corolla-tube 2–3 mm long .... 8. *V. randii*
5b. Calyx-lobes oblong to slightly spathulate; corolla-tube more than 3.5 mm long 6
6a. Leaf-blades glabrous .... 9. *V. apiculata*
6b. Leaf-blades densely pubescent to velvety .... 10. *V. volkensii*
7a. Leaf-blades entirely glabrous .... 1. *V. loranthifolium*
7b. Leaf-blades sparsely pubescent to tomentose 8
8a. Leaf-blades subacute to obtusely acuminate at the apex; fruits 1.4–1.5 cm in diameter .... 2. *V. induta*
8b. Leaf-blades rounded to obtuse at the apex; fruits less than 1.4 cm in diameter 9
9a. Leaf-blades pubescent; calyx-lobes triangular-lanceolate, 3.2–4.5 mm long .... 3. *V. gillettii*
9b. Leaf-blades tomentose; calyx-lobes ovate to oblong-triangular, 1–2.5 mm long .... 4. *V. schumanniana*

1. **Vangueria loranthifolia** K. Schum., Pflanzenw. Ost-Afrikas, C: 385. 1895. ≡ *Pachystigma loranthifolium* (K. Schum.) Verdc., Kew Bull. 42: 140. 1987; F.T.E.A. Rubiac. 3: 769. 1991; K.T.S.L.: 525. 1994. —Type: Tanzania, Tanga, Doda Creek, June 1893, *C. Holst 2935* [holotype: B (destroyed); lectotype: K (K000316467), **designated here**]

Shrub or small tree, up to 7 m tall. Leaf-blades elliptic to obovate, 0.7–4 cm × 0.4–2 cm, rounded at the apex, cuneate at the base; petioles up to 3 mm long; stipules rounded, ca. 1.5 mm long, apical. Inflorescences umbel-like, 1–4-flowered; peduncles up to 3 mm long; pedicels 7–9 mm long. Calyx-tube campanulate, 1.5–2 mm long; lobes broadly triangular, 1–2.5 mm long. Corolla green to yellow or orange; tube 3–4 mm long; lobes oblong-lanceolate, 3–5 mm long, with apical appendage. Ovary 5-locular. Fruit globose, ca. 1.5 cm in diameter, crowned by the persistent calxy-limb, with 4–5 pyrenes.

1a. Leaf-blades 1.4–4 cm long; pedicels 7–9 mm long .... a. subsp. *loranthifolia*
1b. Leaf-blades 0.5–0.7 cm long; pedicels 1–2 mm long .... b. subsp. *salaensis*

a. subsp. **loranthifolia** (Plate 70A–C)

Leaf-blades 1.4–4 cm long. Pedicels 7–9 mm long.
Distribution: Coastal Kenya. [Tanzania].
Habitat: Lowland forest, bushland or thickets; up to 800 m.
Kilifi: Mangea Hill, *Luke & Robertson 1806* (EA); below Jilore Forest Station, *Spjut & Ensor 2627* (EA). Kwale: Shimba Hills, *SAJIT 005974* (HIB); Maluganji Forest Reserve, *Robertson & Luke 6032* (EA); Mwaluganje Elephant Sanctuary, *Mwadime et al. 160* (EA). Lamu: Bodhei to Basuba, *Festo et al. 2657* (EA); near Mkunumbi, *Gillett 20431* (EA). Mombasa: Nyali Beach, *Napier 3305* (EA). Tana River: Kurawa, *Polhill & Paulo 652* (EA).

b. subsp. **salaensis** (Verdc.) Lantz, Pl. Syst. Evol. 253: 180. 2005. ≡ *Pachystigma loranthifolium* (K. Schum.) Verdc. subsp. *salaense* Verdc., Kew Bull. 42: 140. 1987; F.T.E.A. Rubiac. 3: 770. 1991; K.T.S.L.: 525. 1994. —Type: Kenya, Teita Taveta, Sala, 24

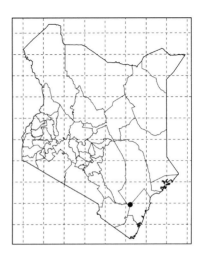

Dec. 1966, *M. Hucks 932* [holotype: EA (EA000002931)]

Leaf-blades 0.5–0.7 cm long; pedicels 1–2 mm long.
Distribution: Southern Kenya. [Tanzania].
Habitat: Bushlands; 200–300 m.
Teita Taveta: Sala area, *Hucks 932* (EA, K).

2. **Vangueria induta** (Bullock) Lantz, Pl. Syst. Evol. 253: 180. 2005. ≡ *Canthium indutum* Bullock, Bull. Misc. Inform. Kew 1932: 366. 1932. ≡ *Rytigynia induta* (Bullock) Verdc. & Bridson, Kew Bull. 42: 169. 1987; F.T.E.A. Rubiac. 3: 821. 1991; K.T.S.L.: 554. 1994. —Type: Tanzania, Kondoa, Kolo, Jan. 1928, *B.D. Burtt 1294* [holotype: K (K000319912); isotypes: BR (BR0000008856425), EA, MO (MO-391815)] (Plate 70D–F)

Shrub or small tree, up to 11 m tall. Leaves opposite, shortly petiolate; blades elliptic to ovate-oblong, 2.5–7.5 cm × 1.5–5 cm, subacute to obtusely acuminate or rarely rounded at the apex, cuneate to rounded at the base, softly pubescent on both sides; stipules broadly triangular, 3–5 mm long, acumintae. Cymes 1–4-flowered, with peduncles 0–3 mm long; pedicels 2–3.5 mm long. Calyx-tube 1–2 mm long; lobes triangular-oblong, 0.5–1 mm long. Corolla-tube 3–6 mm long; lobes triangular-oblong, 3–5 mm long. Ovary 2–3-locular. Fruits subglobose, 1.4–1.5 cm in diameter.
Distribution: Central and southern Kenya. [Tanzania].
Habitat: Forests or woodlands; 1400–2400 m.
Kajiado: Olmuntus Hill, *Morris 001* (EA). Makueni: Nzaui Hill, *Agnew et al. 8410* (EA). Narok: Loita Hills, *Luke et al. 9273* (EA); Entasekera, *Morris 051* (EA); Legeri Hills, *Kuchar 10692* (EA).

3. **Vangueria gillettii** (Tennant) Lantz, Pl. Syst. Evol. 253: 180. 2005. ≡ *Rytigynia gillettii* Tennant, Kew Bull. 19: 279. 1965. ≡ *Pachystigma gillettii* (Tennant) Verdc., Kew Bull. 42: 140. 1987; F.T.E.A. Rubiac. 3: 767. 1991; K.T.S.L.: 524. 1994. —Type: Kenya, Marsabit, Dandu, 750 m, 1 May 1952, *J.B. Gillett 12765* [holotype: K (K000311607); isotypes: BR (BR0000008849885), EA]

= *Pachystigma kenyense* Verdc., Kew Bull. 36: 542, f. 10. 1981. —Type: Kenya, Meru, Meru National Park, Kiolu R., 28 Nov. 1979, *P.H. Hamilton 593* [holotype: EA (EA000001740); isotype: K (K000311606)]

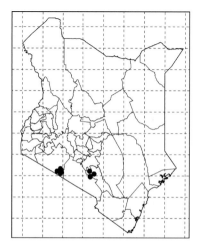
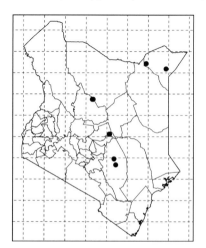

Plate 70 A–C. *Vangueria loranthifolia* subsp. *loranthifolia*; D–F. *V. induta*. Photo by GWH (A, C), VMN (B) and BL (D–F).

Shrub, up to 3 m tall. Leaves on lateral short shoots, shortly petiolate; blades broadly elliptic to obovate, 1–3 cm × 0.7–1.8 cm, rounded at the apex, cuneate at the base, pubescent on both surfaces; stipules narrowly triangular. Flowers in few-flowered fascicles on lateral short shoots; pedicels up to 5 mm long. Calyx-tube 1–2 mm long; lobes triangular-lanceolate, 2–4.5 mm long, pubescent. Corolla yellow or white; tube 3–3.7 mm long; lobes ovate-triangular, ca. 3 mm long, with apical appendage. Ovary 3–4-locular. Fruits ellipsoid, 8–9.5 mm long, sparsely pubescent.

Distribution: Central and eastern Kenya. [Somalia].

Habitat: Bushlands or other rocky places; 400–900 m.

Kitui: Endau, *Kirika et al. NMK325* (EA). Mandera: War Gedud, *Gilbert & Thulin 1281* (EA). Marsabit: Gulni, *Sato 369* (EA). Meru: Meru National Park, *Hamilton 805* (EA). Samburu: near Ndoto Mountains, *Gilbert & Gachathi 5266* (EA).

4. **Vangueria schumanniana** (Robyns) Lantz, Pl. Syst. Evol. 253: 181. 2005. ≡ *Tapiphyllum schumannianum* Robyns, Bull. Jard. Bot. État Bruxelles 11: 109. 1928. ≡ *Pachystigma schumannianum* (Robyns) Bridson & Verdc., F.T.E.A. Rubiac. 3: 768. 1991. —Type: Tanzania, Moshi, Himo, Jan. 1894, *G. Volkens 1748* [holotype: B (destroyed); lectotype: BR (BR0000008849571), **designated here**; isolectotype: HBG (HBG520896)]

Shrub or small tree, up to 4.5(–10) m tall. Leaf-blades round to ovate, 0.9–4.5 cm × 0.8–4 cm, obtuse to rounded at the apex, rounded to subcordate or rarely cuneate at the base, densely grey velvety hairy on both sides; petioles up to 3 mm long; stipules triangular, 2.5–4 mm long, acuminate. Flowers few to several in fascicles; pedicels up to 5 mm long. Calyx-tube 1.5–2 mm long; lobes ovate to oblong-triangular, 1–2.5 mm long. Corolla-tube cylindrical, 3–4 mm long; lobes elliptic-oblong, 3.5–5 mm long. Fruits globose, 0.8–1.4 cm in diameter.

1a. Leaves usually less than 1.7 cm × 1.8 cm, often grey on both sides ...... a. subsp. *schumanniana*
1b. Leaves up to 4.5 cm × 4 cm, often dark above and grey beneath .... b. subsp. *mucronulatum*

a. subsp. **schumanniana**

Leaves usually small, less than 1.7 cm × 1.8 cm, often grey on both sides.

Distribution: Southern Kenya. [Tanzania].

Habitat: Bushlands or woodlands; 1600–1700 m.

Kajiado: Ngong Hills, *Kokwaro et al. 302* (EA).

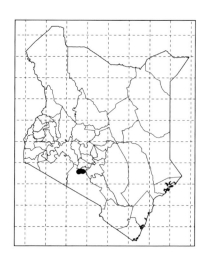

b. subsp. **mucronulata** (Robyns) Lantz, Pl. Syst. Evol. 253: 181. 2005. ≡ *Tapiphyllum mucronulatum* Robyns, Bull. Jard. Bot. État Bruxelles 32: 151. 1962. ≡ *Pachystigma schumannianum* (Robyns) Bridson & Verdc. subsp. *mucronulatum* (Robyns) Bridson & Verdc., F.T.E.A. Rubiac. 3: 768. 1991. —Type: Kenya, Kitui, Migwani Location, 5 May 1960, *Napper 1590* [holotype: BR; isotypes: EA (EA000002932), K (K000311605)]

= *Vangueria tomentosa* K. Schum. P.P.A.C.: 385. 1895, nom. illeg. —Type: Kenya, Kitui, Ikanga, *J.M. Hildebrandt 2835* [holotype: B (destroyed)]

Leaves sometimes very large, up to 4.5 cm × 4 cm, often dark above and grey beneath.

Distribution: Central and southern Kenya. [Endemic].

Habitat: Thickets, bushland, woodland or wooded grassland; 300–2000 m.

Embu: ca. 3.5 km towards Embu from turnoff to Siakago, *Faden & Faden 74/722* (EA); Kiritiri, *Smith et al. 293* (K); Mwea National Reserve, *Mwangangi et al. 4729* (EA). Kiambu: Thika, *Faden 67115* (EA). Kitui: Makongo Forest Reserve, *Mwangangi et al. 4535A* (EA); Tuvani, *Kuchar 14993a* (EA). Machakos: Yatta Plateau, *Gillett et al. 23948* (EA); Mbiuni, *Fliervoet 839* (EA). Makueni: beyond Kikoko Hill, *Mwangangi 1667* (EA); Kibwezi, *Elliot 6722* (K). Murang'a: Kakuzi Estate, *Luke et al. 17052* (EA). Teita Taveta: Tsavo National Park, *Gilbert 6009* (EA).

5. **Vangueria pallidiflora** (Bullock) Lantz, Pl. Syst. Evol. 253: 181. 2005. ≡ *Lagynias pallidiflora* Bullock, Kew Bull. 1931: 273. 1931; F.T.E.A. Rubiac. 3: 762. 1991; K.T.S.L.: 519. 1994. —Type: Tanzania, E. Usambara Mts., Sigi, 25 Oct. 1910, *Zimmermann in Herb. Amani 3219* [holotype: K (K000316809); isotypes: EA (EA000001743, EA000001744), PRE (PRE0594435-0)] (Figure 54; Plate 71A, B)

Shrub or small tree, up to 10(–20) m tall, less often with paired spines up to 3 cm long. Leaf-blades narrowly elliptic to elliptic or obovate, 1.5–14.5 cm × 0.7–5.5 cm, acuminate or rouded at the apex, cuneate at the base, dark above and pale beneath; petioles up to 10 mm long; stipules filiform, 4–10 mm long. Inflorescences umbel-like, (1–)2–6-flowered; peduncles 2–12 mm long; pedicels 9–30 mm long. Calyx-

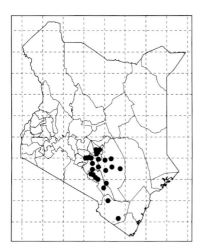
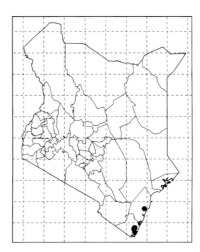

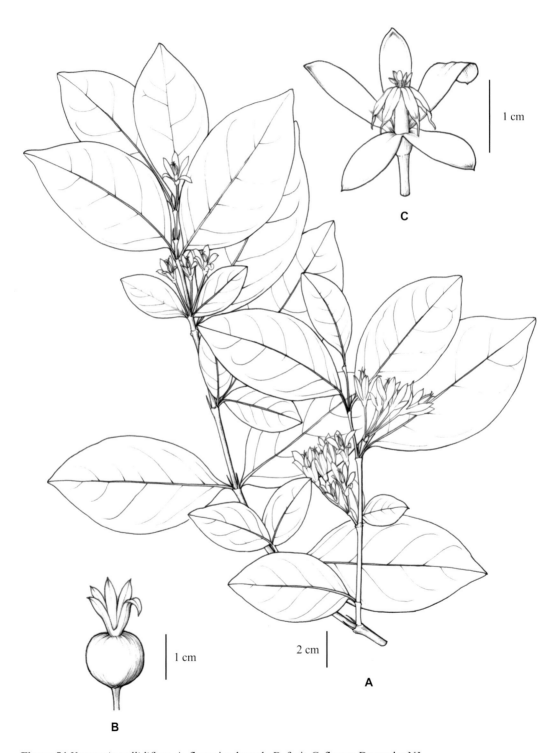

Figure 54 *Vangueria pallidiflora*. A. flowering branch; B. fruit; C. flower. Drawn by NJ.

tube subglobose, 1.5–2.5 mm long; lobes cream, green, greenish yellow or pale yellow, oblong to spathulate, 7–16 mm × 1.5–4 mm. Corolla cream, greenish white or greenish yellow; tube cylindrical, 9–12 mm long; lobes lanceolate, 6–8 mm long, with a tail-like appendage. Ovary 5-locular; style slightly exserted. Fruits ellipsoid to subglobose, 10–16 mm long.

Distribution: Coastal Kenya. [Tanzania].

Habitat: Lowland forests or thickets; up to 400 m.

Kilifi: Arabuko-Sokoke Forest, *Simpson 65* (EA). Kwale: Shimba Hills, *SAJIT 006057* (HIB); Buda Mafisini Forest, *Drummond & Hemsley 3940* (EA, K); Mwele Forest, *Luke & Robertson 232* (EA).

6. **Vangueria madagascariensis** J.F. Gmel., Syst. Nat.: 367. 1791; F.T.E.A. Rubiac. 3: 849. 1991; K.T.S.L.: 551. 1994; U.T.S.K.: 433. 2005. —Type: Mauritius, *P. Commerson s.n.* [holotype: P (P00553223)] (Plate 71C–E)

Shrub or small tree, up to 15 m tall. Leaf-blades elliptic to ovate, 8–28 cm × 3–15 cm, acute or shortly acuminate at the apex, cuneate or rounded at the base, glabrous or nearly so; petioles up to 1.8 cm long; stipules hairy, with base 3–5 mm long, apical part 0.4–1.8 cm long. Inflorescences of fairly dense cymes, with peduncles up to 1 cm long; pedicels ca. 2 mm long. Calyx-tube 1–3 mm long; lobes triangular to narrowly oblong, 0.5–1.5 mm long. Corolla-tube 3–4.5 mm long; lobes 3–4.5 mm long. Fruits green, subglobose, 2.5–5 cm in diameter.

Distribution: Northern, western, central, eastern, and southern Kenya. [Tropical and southern Africa].

Habitat: Forests, woodlands, bushlands, or wooded grasslands; 750–2450 m.

Homa Bay: Masisi, *Kirika 349* (EA). Isiolo: Nanyuki-Isiolo Road, *Kassam 35* (EA). Kajiado: Elangata Wuas, *Omondi et al. KEFRI 44* (EA); Namanga Hill Forest, *Kindeketa et al. 371* (K). Kakamega: Kakamega Forest, *Faden et al. 69/2084* (K). Kericho: Londiani, *Vorontsova et al. 22* (EA). Kiambu: Ruiru, *Mayerrnoff 45* (K). Kisii: Kebabe-Kisii, *Oxtoby 20* (EA). Kitui: Makongo, *Mwangangi et al. 4644* (EA); Mutha Plain, *Bally 7414* (K). Laikipia: Luoniek, *Muasya 2268* (EA). Makueni: Kibwezi, *Gillett 19398* (EA, K). Mandera: Dandu, *Gillet 12766* (EA, K). Marsabit: Marsabit Forest, *SAJIT Z0296* (HIB); Mount Kulal, *Adamson 110* (K). Meru: Ngaiya Forest, *Luke et al. 7135* (EA); Meru National Park, *Ament & Magogo 55* (EA). Nakuru: Gilgil, *van Someren 1688* (EA). Narok: Narosura, *Glover et al. 2414* (K); Masai Mara, *Melanie 4* (EA). Samburu: Mount Ndoto, *Gilbert et al. 5532* (K); Mathew's Range, *Newbould 3530.1* (K). Teita Taveta: Ngulia Hill, *Gilbert 2710* (EA); Mount Kasigau, *Kimberly 564* (EA). Trans Nzoia: Mount Elgon, *Jackson 331A* (K). West Pokot: Cherangani Hills, *SAJIT 006844* (HIB).

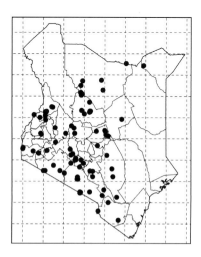

Plate 71 A, B. *Vangueria pallidiflora*; C–E. *V. madagascariensis*. Photo by GWH (A–D) and YDZ (E).

7. Vangueria infausta Burch., Trav. S. Africa 2: 258, f. p. 259. 1824; F.T.E.A. Rubiac. 3: 851. 1991; K.T.S.L.: 550. 1994; U.T.S.K.: 432. 2005. —Type: South Africa, Cape Province, Sensavan, 26 Dec. 1812, *W.J. Burchell 2629* [holotype: K (K000412117); isotype: M (M0106363)]

Shrub or small tree, up to 8 m tall. Leaf-blades elliptic or ovate, up to 30 cm × 18 cm, base cuneate or rounded, apex acute or shortly acuminate, densely pubescent or velvety; petioles 0.3–1 cm long, hairy; stipules hairy, with base 2–4 mm long, apical part 0.3–1.2 cm long. Inflorescences laxly cymose, with peduncles 6–8 mm long; pedicels 1–2.5 mm long. Calyx-tube 0.7–1.2 mm long; lobes triangular to narrowly oblong, 1–2 mm long. Corolla pubescent to densely hairy, green or yellow-green, with tube 3–4.5 mm long; lobes 3–4 mm long. Fruits subglobose, up to 4.7 cm in diameter.

subsp. **rotundata** (Robyns) Verdc., Kew Bull. 36: 549. 1981; F.T.E.A. Rubiac. 3: 852. 1991; K.T.S.L.: 550. 1994; C.P.K.: 389. 2016. ≡ *Vangueria rotundata* Robyns, Bull. Jard. Bot. État Bruxelles 11: 300, f. 27, 28. 1928. —Type: Tanzania, Rungwe, Kyimbila, 1910, *A. Stolz 432* [holotype: K (K000412132); isotypes: BR (BR0000008857422), EA, K (K000412133, K000412134, K000412135, K000412136), L (L.2967318), U (U.1585936)] (Plate 72)

= *Vangueria infausta* Burch. var. *campanulata* (Robyns) Verdc., Kew Bull. 36: 550. 1981.; F.T.E.A. Rubiac. 3: 852. 1991. ≡ *Vangueria campanulata* Robyns, Bull. Jard. Bot. État 11: 293. 1928. —Type: Kenya, Nairobi Government Farm, *Linton 21* [holotype: K]

Shrub or small tree, up to 8 m tall; young branches always pubescent. Leaf-blades elliptic to ovate, 4–30 cm × 2.5–18 cm, densely pubescent or velvety. Flowers white or yellow-green, in lax cymes. Corolla pubescent or rarely glabrous; tube 3–4.5 mm long; lobes 3–4 mm long. Fruits subglobose, 1.5–4.7 cm in diameter.

Distribution: Widespread in Kenya. [Tropical and southern Africa].

Habitat: Forest edges, riverine scrubs, woodland, or wooded grasslands; up to 1800 m.

Note: *Vangueria infausta* subsp. *infausta* widely distributed in central and southern Africa, from Rwanda to South Africa, with smaller leaves and hairy buds often rounded at the apex. Besides, *V. infausta* subsp. *rotundata* var. *campanulata* (Robyns) Verdc. was recorded as a variant under this subspecies by B. Verdcourt in Kew Bull. and FTEA. *V. infausta* subsp. *rotundata* var. *campanulata* could just be distinguished from *V. infausta* subsp. *rotundata* var. *rotundata* by its glabrous or slightly pubescent corolla, which is probably a continuous variation of this subspecies as a whole. We think there is no need to create new combinations, such as *V. infausta* subsp.

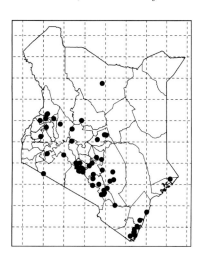

Plate 72 A–F. *Vangueria infausta* subsp. *rotundata*. Photo by GWH.

campanulata or *V. infausta* var. *rotundata*, so we just accepted *V. infausta* subsp. *rotundata* as a whole one, which have large leaves and apiculate buds compared with the typical subspecies.

Baringo: Bartabwa, *Luke BFFP220* (EA). Embu: Thiba River, *Robertson 2044* (EA, K). Kakamega: Kakamega Forest, *Mwangangi & Gliniars 7* (EA). Kericho: Cheptuiyet, *Kerfoot 2158* (EA, K). Kiambu: Kamuiti Forest, *Ossent 525* (EA). Kilifi: Gede, *Robertson & Ngonyo 6828* (EA, K). Kwale: Shimba Hills, *SAJIT V0200* (HIB); Dzombo, *Magogo & Glover 800* (EA, K); Gongoni Forest, *Luke et al. 8349* (EA). Laikipia: Olarinyiru, *Joshua 1349* (EA). Lamu: Boni Forest, *Luke & Robertson 1511* (EA). Machakos: Donyo Sabuk, *Gedye 3593* (EA). Makueni: Chyulu National Park, *SAJIT 006325* (HIB). Marsabit: *Sato 404* (EA). Meru: near Lake Nkuga, *SAJIT 003829* (HIB). Nairobi: Karura Forest, *Perdue & Kibuwa 8056* (EA); Bahati, *Gilbert 644* (EA). Nakuru: Kedong, *Bally B987* (EA). Narok: Masai Mara Game Reserve, *Kuchar 9700* (EA). Samburu: Leroghi Range, *Kerfoot 2108* (EA). Trans Nzoia: Saiwa, *Beentje 3424* (EA); Mount Elgon, *Padwa 3* (EA, K). West Pokot: Cherangani Hills, *SAJIT 006853* (HIB).

8. **Vangueria randii** S. Moore, J. Bot. 40: 252. 1902; F.T.E.A. Rubiac. 3: 855. 1991; K.T.S.L.: 551. 1994. —Type: Zimbabwe, Bulawayo, Jan. 1898, *R.F. Rand 123* [holotype: BM (BM000903240)]

Shrub or small tree, up to 7 m tall. Leaf-blades elliptic to oblong, 2–15.5 cm × 0.8–6.6 cm, acute to acuminate at the apex, cuneate to rounded at the base, glabrous to pubescent on both sides; stipules filiform, up to 11 mm long; petioles 3–7 mm long. Inflorescences of dichasial cymes, with ultimate branches 3–8-flowered; peduncles 4–15 mm long; pedicels 1–5 mm long. Calyx-tube subglobose, ca. 1 mm long; lobes linear to oblong, 2–7 mm long. Corolla white, green or golden green; tube 2–3 mm long; lobes narrowly triangular, 2.5–3 mm long. Fruits yellow, subglobose, 15–20 mm in diameter.

subsp. **acuminata** Verdc., Kew Bull. 36: 553. 1981; F.T.E.A. Rubiac. 3: 856. 1991; K.T.S.L.: 551. 1994. —Type: Tanzania, Tanga, Kange Gorge, 21 Mar. 1956, *H.G. Faulkner 1828* [holotype: K (K000352849); isotype: K (K000352850)] (Plate 73)

Shrub, up to 3 m tall. Leaf-blades 2–15 cm × 1.5–6 cm, glabrous. Inflorescences with peduncles up to 15 mm long; ultimate branches 3–5-flowered. Calyx-lobes 2–3 mm in diameter.

Distribution: Coastal Kenya. [Tanzania].
Habitat: Lowland forests; up to 500 m.
Note: There are four subspecies recorded in *Vangueria randii*: subsp. *randii* occurs in Zimbabwe and Zambia and has a pubescent calyx-tube, shorter calyx-lobes, glabrous

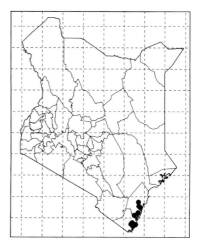

Plate 73 *Vangueria randii* subsp. *acuminata*. Photo by GWH.

leaves and short peduncles; subsp. *chartacea* (Robyns) Verdc. occurs in South Africa and has glabrous leaves and calyx-tube and short peduncles; subsp. *vollesenii* Verdc. occurs in Tanzania and has very narrow calyx-lobes and very short peduncles ca. 3 mm long.

Kilifi: Mangea Hill, *Luke & Robertson 63* (EA); Cha Shimba, *Robertson & Cunningham 6726* (EA); Chonyi, *Lavranos & Newton 12297* (K). Kwale: Shimba Hills, *SAJIT 005536* (HIB); Gongoni Forest Reserve, *Luke 8350* (EA); Diani Forest, *Robertson 4158* (EA).

9. **Vangueria apiculata** K. Schum., Pflanzenw. Ost-Afrikas, C: 385. 1895; F.T.E.A. Rubiac. 3: 853. 1991; K.T.S.L.: 550. 1994; U.T.S.K.: 431. 2005. —Type: Kenya, S. Kavirondo, Karachuonyo, Dec. 1891, *A. Fischer 294* [lectotype: K (K000412140)] (Plate 74)

Shrub or tree, up to 12 m tall. Leaf-blades ovate or elliptic, 3–15 cm × 1.5–6 cm, acuminate at the apex, cuneate, rounded or subcordate at the base, glabrous; petioles 0.7–1 cm long; stipules with filiform part 3–9.5 mm long and a short broad base. Inflorescences laxly or densely cymose, with peduncles 0.5–1 cm long; pedicels 2–4 mm long. Calyx-tube subglobose, 1–1.5 mm

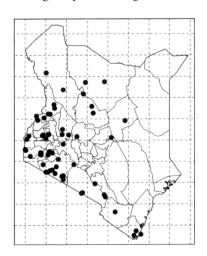

Plate 74 A–F. *Vangueria apiculata*. Photo by GWH (A–C, F) and SWW (D, E).

in diameter; lobes oblong to linear, 3–7 mm long. Corolla greenish-white, or green to yellow; tube 4–5 mm long, with whitish hairs on the throat; lobes 4–5 mm long. Fruits green to brown, subglobose, 1.7–2.2 cm in diameter.

Distribution: Northern, northwestern, western, central, southern, and coastal Kenya. [East Africa, from Ethiopia to Zimbabwe].

Habitat: Forest edges, dense or open woodlands, bushlands, thickets, scrublands, or grasslands with scattered trees; up to 2400 m.

Baringo: Morop Forest, *SAJIT 006541* (HIB). Bomet: southwest Mau Forest Reserve, *Geesteranus 5768* (K). Bungoma: Mount Elgon, *SAJIT PR0001* (HIB). Homa Bay: Sokli-Kitutu, *Kirika 341* (EA). Isiolo: *Adamson B3993* (EA). Kajiado: Ol Doinyo Orok, *Nyakundi 240* (EA). Kakamega: Kakamega Forest, *Paulo 554* (EA, K). Kericho: Kipsigis, *Kifikerich 164709* (EA). Kisii: Wanjare, *Vuyk 358* (EA). Kwale: Mwele Forest, *Luke & Lehman 10894* (K). Laikipia: Rumuruti, *Gerald 125* (K). Marsabit: Mount Kulal, *Hepper & Jaeger 6962* (EA, K). Nandi: Kosirai, *Brunt 1313* (K). Narok: Melelo Area, *Glover et al. 1758* (EA, K); Masai Mara Game Reserve, *Kuchar 11497* (EA). Samburu: Mount Ndoto, *Beentje 3983* (EA). Siaya: Yala, *Kokwaro 130* (EA). Teita Taveta: Taita Hills, *SAJIT 006402* (HIB). Turkana: Muruasigar, *Newbould 7059* (EA, K). West Pokot: Cherangani Hills, *SAJIT 005061* (HIB).

10. **Vangueria volkensii** K. Schum., Pflanzenw. Ost-Afrikas, C: 384. 1895; F.T.E.A. Rubiac. 3: 854. 1991; K.T.S.L.: 551. 1994. —Type: Tanzania, Kilimanjaro, Marangu, 1500 m, Apr. 1893, *G. Volkens 247* [lectotype **designated here**: BM (BM 000903235)] (Plate 75)

= *Vangueria volkensii* K. Schum. var. *fyffei* (Robyns) Verdc., Kew Bull. 36: 551. 1981. —Type: Uganda, Masaka, Malabigambo Forest Reserve, *R. Fyffe 53* [holotype: K (K000412138)]

= *Vangueria volkensii* K. Schum. var. *kyimbilensis* (Robyns) Verdc., Kew Bull. 36: 552. 1981. ≡ *Vangueria kyimbilensis* Robyns, Bull. Jard. Bot. État Bruxelles 11: 294. 1928. —Type: Tanzania, Rungwe, Mwakaleii, 13 Nov. 1913, *A.F. Stolz 2289* [holotype: K (K000352863); isotypes: BR (BR0000008859426), EA, K (K000352864)]

Shrub or small tree, up to 9 m tall. Leaf-blades elliptic to ovate, 3–17 cm × 1.5–10 cm, acuminate at the apex, cuneate, rounded or subcordate at the base, densely pubescent to velvety; petioles 0.5–1.3 cm long; stipules with filiform part 0.5–1.2 cm long and a short broad base 2 mm long. Flowers in dense cymes; peduncles 0.6–2 cm long; pedicels 0.7–3 mm long. Calyx-tube subglobose, hairy, 1–1.5 mm in diameter; lobes oblong to linear, 1.5–6 mm long. Corolla bright green or yellow-green; tube 3.5–5.5 mm long, with whitish hairs on the throat; lobes narrowly triangular, 4–4.5 mm long. Fruits subglobose, 2–2.5 cm in diameter.

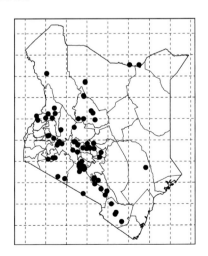

Plate 75 A–F. *Vangueria volkensii*. Photo by NW (A), GWH (B, C, E, F) and CL (D).

Distribution: Northern, northwestern, western, central, eastern, and southern Kenya. [Ethiopia to southern tropical Africa].

Habitat: Forest edges, thickets, bushlands, or bushed grasslands and roadsides; 1000–2550 m.

Bomet: Tinderet Forest Reserve, *Geesteranus 4983* (K). Bungoma: Mount Elgon, *SAJIT PR0024* (HIB). Kajiado: Ngong Hills, *Beentje 2255* (EA); Emali Hill, *Faden et al. 71/933* (EA, K). Kericho: southwest of Londiani, *Perdue & Kibuwa 9116* (EA, K). Kiambu: Muguga, *Kirika 490* (EA, K). Kirinyaga: Mount Kenya, *Bamps 6744* (K). Laikipia: Ol Ari Nyiro, *Joshua 2218* (EA). Makueni: Mukaa Secondary School, *Mwangangi 2045* (EA); Kibwezi, *Gillett 19398* (EA). Marsabit: Mount Kulal, *Herlocker H-147* (EA); Moyale, *Gillett 12949* (EA, K). Nairobi: Nairobi National Park, *Kokwaro 242* (EA, K). Nakuru: north Gilgil, *Gilbert 6036* (EA, K). Nandi: Kapsabet, *Elkens 139* (EA). Narok: Loita Hills, *Fayad 145* (EA). Nyandarua: Aberdare National Park, *Hepper & Field 4918* (EA, K). Nyeri: Kagochi, *Kerfoot 1495* (EA). Samburu: Mathew's Range, *Ichikawa 265* (EA). Teita Taveta: Chawia Forest, *Drummond & Hemsley 4376* (EA, K); Mount Kasigau, *Luke et al. 5378* (EA). Trans Nzoia: Kitale, *Mungai 139/84* (EA, K). West Pokot: Kapenguria, *Padwa 53* (EA, K).

17. Trib. **Coffeeae** DC.

Trees, shrubs, or less often lianas, unarmed. Raphides absent. Leaves opposite; stipules interpetiolar, paired, adnate or free, apex entire or apiculate. Inflorescences paired, axillary or axillary and then appearing terminal on very short shoots; bracts and bracteoles free or sometimes fused into calyculi. Flowers 4–8(–12)-merous. Corolla-lobes contorted; anthers exserted, rarely included. Ovary 2-locular, placentation axile; ovules 1–2(–10) in each locule. Fruit an indehiscent drupe, with 1–2(–10) seeds.

Four genera occur in Kenya.

1a. Ventral surface of seed with a distinct longitudinal ventral invagination (typical "coffee-beans") .... 59. *Coffea*
1b. Ventral surface of seed without such invagination 2
2a. Flowers (4–)5(–6)-merous; anthers with a conspicuous ribbon-like appendage........................
.... 61. *Empogona*
2b. Flowers 4–8(–12)-merous; anthers with or without an inconspicuous apical appendage 3
3a. Corolla-lobes longer than tube; ovules (seeds) 1–2 per locule; fruits 9–13 mm in diameter ......
.... 60. *Calycosiphonia*
3b. Corolla-lobes shorter than tube; ovules (seeds) 1–12 per locule; fruits usuall less than 7 mm in diameter .... 62. *Tricalysia*

59. Coffea L.

Shrubs or small trees. Leaves opposite, petiolate, glabrous; domatia usually present; stipules shortly united at the base, obtuse to aristate. Flowers 4–8(–12)-merous, bisexual, monomorphic, 1 to many in sessile to shortly pedunculate cymes. Calyx-tube campanulate to turbinate; limb obsolete or occasionally truncate or 4–6-toothed. Corolla white or pink; tube salver-form or funnel-form; lobes 4–9, convolute in bud. Stamens 4–8, inserted in corolla throat, exserted; anthers dorsifixed near the base. Ovary 2-locular; ovules solitary in each locule; style slender, glabrous; stigma 2-lobed, exserted. Fruit an ellipsoid drupe; pyrenes 2, 1-seeded. Seeds oblong-ellipsoid, longitudinally grooved on ventral face.

A genus of about 126 species native to tropical Africa, Madagascar and the Mascarenes Islands; six species in Kenya.

1a. Leaves and flowers on shortened lateral branches ....6. *C. rhamnifolia*
1b. Leaves spaced on branches; flowers always axillary 2
2a. Flowers (5–)6–8-merous; peduncules 2–6 mm long ....4. *C. pseudozanguebariae*
2b. Flowers 5(–6)-merous; peduncles less than 3 mm long 3
3a. Flowers 2–20 per axil .... 2. *C. arabica*
3b. Flowers 1–2(–3) per axil 4
4a. Leaf-blades with obtuse or rounded apex; corolla-tube ca. 3 mm long .... 1. *C. fadenii*
4b. Leaf-blades with acute to acuminate apex; corolla-tube more than 5.5 mm long 5
5a. Corolla-tube 5.5–10 mm long, lobes 5–12 mm long .... 3. *C. eugenioides*
5b. Corolla-tube 10–14 mm long, lobes 13–16 mm long .... 5. *C. sessiliflora*

1. **Coffea fadenii** Bridson, Kew Bull. 36: 827. 1982; F.T.E.A. Rubiac. 2: 709. 1988; K.T.S.L.: 510. 1994. —Type: Kenya, Teita Taveta, Taita Hills, Mbololo Hill, 2 Jan. 1971, *R.B. Faden et al. 71/56* [holotype: K (K000097517); isotype: EA]

Small tree, up to 15 m tall. Leaf-blades glabrous, broadly elliptic to rounded, 6–15 cm × 4–8 cm, obtuse to rounded at the apex, obtuse to acute at the base; domatia glabrous to sparsely hairy; petioles up to 1 cm long; stipules triangular-ovate, 3–4 mm long. Flowers 5-merous, 1–2(–3) per axil; peduncles up to 3 mm long. Calyx-limb reduced to a rim. Corolla-tube ca. 3 mm long; lobes ca. 9 mm long, rounded. Drupes obovoid, 16–22 mm long.

Distribution: Southern Kenya. [Endemic].

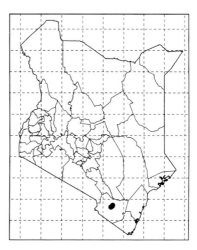

Habitat: Moist or montane forests; 1400–1800 m.

Teita Taveta: Mbololo Hill, *Faden & Faden 72/269* (EA, K); Ngangao Forest, *Goodrich EAH17159* (EA).

2. **Coffea arabica** L., Sp. Pl. 1: 172. 1753; F.T.E.A. Rubiac. 2: 712. 1988; K.T.S.L.: 509. 1994. —Type: Cultivated in Holland, *Hort. Cliff. 59* [lectotype: BM, designated by D. Bridson & B. Verdcourt in F.T.E.A. Rubiac. 2: 713. 1988] (Figure 55)

Shrub or small tree, up to 8 m tall. Leaf-blades glabrous, elliptic, elliptic-oblong, or occasionally ovate-lanceolate, (2–)7–18(–22) cm × 3–8.5 cm, distinctly acuminate at the apex, acute to obtuse at the base; domatia absent or rather inconspicuous; petioles up to 1.5 cm long; stipules broadly triangular, 3–8(–12) mm, aristate. Flowers (4–)5(–6)-merous, 2–20 per axil; peduncles up 2(–3) mm long. Calyx-tube 1–2 mm long; limb very short, truncate to undulate or denticulate. Corolla white; tube funnel-form, 5–11 mm long; lobes oblong, 9–16 mm long, rounded. Drupes red, ellipsoid to subglobose, 11–16 mm long.

Distribution: Northern Kenya. [Ethiopia, Sudan, and widely cultivated throughout the tropics].

Habitat: Forests; 1500–1600 m.

Marsabit: Mount Marsabit, *Kerfoot 3710* (EA, K).

3. **Coffea eugenioides** S. Moore, J. Bot. 45: 43. 1907; F.T.E.A. Rubiac. 2: 713. 1988; K.T.S.L.: 509. 1994. —Type: Uganda, Toro, near Mpanga, Jule 1906, *A.G. Bagshawe 1076* [holotype: BM (BM000903409)] (Plate 76)

= *Coffea arabica* L. var. *intermedia* A. Froehner, Bot. Jahrb. Syst. 25: 264. 1898. —Type: Kenya, Elgeyo, *A. Fischer 326* [holotype: B (destroyed)]

Shrub or small tree, up to 4.5 m tall. Leaf-blades glabrous, elliptic, 2–12 cm × 1–5.5 cm, acuminate at the apex, acute at the base; domatia absent or present, glabrous; petioles 2–7 mm long; stipules triangular, 1.5–2.5 mm long, aristate. Flowers 5-merous, 1–2(–5) per axil; peduncles up to 2.5 mm long. Calyx-limb reduced to a rim. Corolla-tube 5.5–10 mm long; lobes oblong-lanceolate, 5–12 mm long, acute. Drupes red, ellipsoid, 8–10 mm long.

Distribution: Western and central Kenya. [D.R. Congo, Rwanda, Sudan, South Sudan, Tanzania, and Uganda].

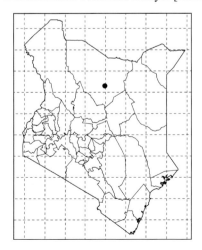

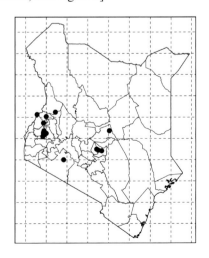

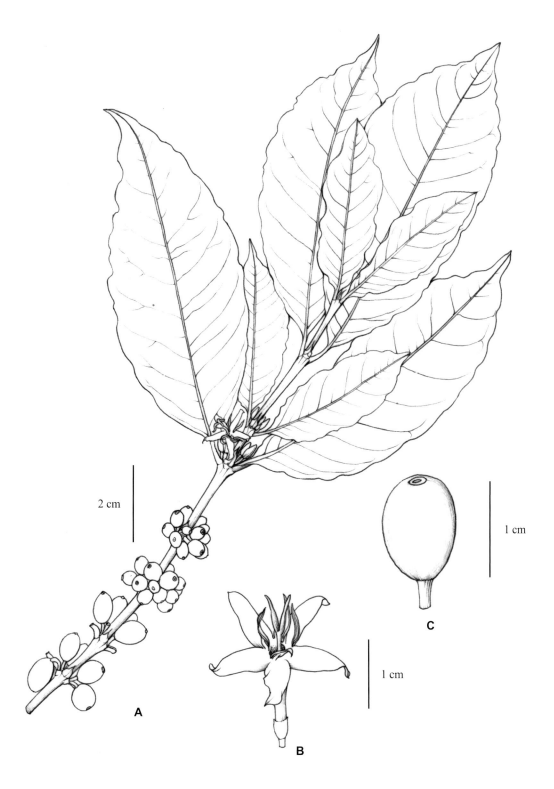

Figure 55 *Coffea arabica*. A. branch with flowers and fruits; B. flower; C. fruit. Drawn by NJ.

Plate 76 A–C. *Coffea eugenioides*. Photo by GWH.

Habitat: Forests; 1000–2100 m.

Bungoma: Malava, *Bamps 6521* (EA, K). Embu: Embu Forest, *Gillett A* (EA, K). Kakamega: Kakamega Forest, *Faden & Faden 77/854* (EA). Kirinyaga: east Mount Kenya, *Doughty 82(a)* (EA, K). Meru: Ngaia Forest, *Gillett s.n.* (EA). Nandi: Nandi Forest, *SAJIT 006718* (HIB). Narok: southern Uaso Nyiro, *Dowson 720* (K). Trans-Nzoia: Kitale, *Tweedie 2594* (K). Vihiga: Tiriki Forest, *Gardner K3817* (K). West Pokot: Cheptoket, *Kerfoot 2161* (EA, K).

4. **Coffea pseudozanguebariae** Bridson, Kew Bull. 36: 835. 1982; F.T.E.A. Rubiac. 2: 714. 1988; K.T.S.L.: 510. 1994. —Type: Tanzania, Tanga, Magunga Estate, 25 Feb. 1952, *H.G. Faulkner 1077* [holotype: K (K000097656); isotypes: BR (BR0000008380616), EA (EA000001753), LISC (LISC002656), P (P03826687), PRE (PRE0594502-0)]

Shrub or small tree, up to 4.5 m tall. Leaf-blades elliptic, 3–12.5 cm × 1.5–7 cm, acute to subacuminate or occasionally obtuse at the apex, acute to cuneate at the base; domatia present, ciliate to pubescent; petioles 2–6 mm long; stipules triangular, shortly aristate, 2–4 mm long. Flowers (5–)6–8-merous, 1–2(–5) per axil. Peduncles 2–6 mm long. Corolla-tube 7–12 mm long; lobes lanceolate to oblong, 10–14 mm long, rounded. Drupes ellipsoid, 8–11 mm long, distinctly beaked. Seeds 5–7.5 mm long.

Distribution: Southern and coastal Kenya. [Tanzania].

Habitat: Bushlands or forests; up to 800 m.

Kilifi: Kaya Kambe, *Robertson & Luke 4782* (EA, K); Mangea Hill, *Luke & Robertson 1821* (EA, K). Kwale: Diani Forest, *Robertson & Luke 5926* (EA, K); Shimba Hills, *Luke 3479* (EA, K). Lamu: Mambosasa Forest, *Goodrich EAH17167* (EA). Mombasa: *Wakefield s.n.* (K). Teita Taveta: Kighombo, *Mwachala et al.* in *EW3333* (EA).

5. **Coffea sessiliflora** Bridson, Kew Bull. 41: 307. 1986; F.T.E.A. Rubiac. 2: 716. 1988; K.T.S.L.: 510. 1994. —Type: Kenya, Kilifi, Rabai, 30 Aug. 1959, *B. Verdcourt 2402* [holotype: K (K000097710); isotype: EA (EA000001749)]

Shrub up to 4.5 m tall. Leaf-blades glabrous, narrowly to broadly elliptic, 4.5–12 cm × 1–6 cm, acute to shortly acuminate at the apex, acute or sometimes cuneate at the base;

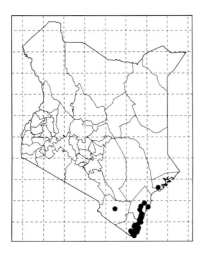
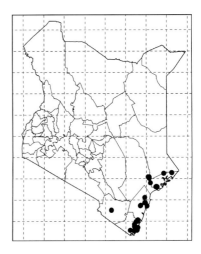

domatia glabrous to sparsely pubescent; petioles 2–6 mm long; stipules triangular, distinctly aristate, 3–5.5 mm long. Flowers 5–6-merous, 1–2(–3) per axil; peduncles less than 2 mm long. Corolla-tube 10–14 mm long; lobes oblong-obovate, 13–16 mm long. Drupes black, ellipsoid, 8–12 mm long.

Distribution: Southern and coastal Kenya. [Tanzania].

Habitat: Forests; up to 450 m.

Kilifi: Mangea Hill, *Luke & Robertson 1822* (EA, K); Kaya Dagamra, *Robertson & Luke 5723* (EA, K). Kwale: Shimba Hills, *Luke 3478* (EA, K); Kaya Gandini, *Gray et al. 451* (EA). Lamu: Boni Forest Reserve, *Robertson & Luke 5633* (EA, K). Mombasa: Rabai, *Verdcourt 2402* (EA, K). Teita Taveta: Ngurunyi Forest, *Mwachala et al.* in *EW3178* (EA, K). Tana River: Tana River National Primate Reserve, *Luke et al. 498* (EA).

6. **Coffea rhamnifolia** (Chiov.) Bridson, Kew Bull. 38: 320. 1983; F.T.E.A. Rubiac. 2: 722. 1988; K.T.S.L.: 510. 1994. ≡ *Plectronia rhamnifolia* Chiov., Res. Sci. Somal. Ital. 1: 93. 1916. —Type: Somalia, Fra Bur Eibi e Sahaieroi, 10 Nov. 1913, *G. Paoli 1163* [holotype: FT (FT003439)]

Small shrub, up to 3 m tall. Leaves always on lateral short spurs, appearing after flowers, narrowly obovate, 0.9–5 cm × 0.7–2.8 cm, obtuse to rounded or rarely apiculate at the apex, cuneate at the base; domatia absent; petioles 1–2 mm long; stipules triangular, acute, ca. 2 mm long. Flowers (5–)6–7-merous, terminal on short lateral spurs, solitary. Calyx-tube ca. 1.5 mm long; limb very short, truncate. Corolla-tube 0.8–1.2 cm long; lobes oblong to elliptic, 1.2–2.3 cm long, acute. Drupes yellow or red, ellipsoid, 8–9 mm long. Seeds 7–9 mm long.

Distribution: Eastern Kenya. [Somalia].

Habitat: Dry bushlands; up to 100 m.

Tana River: Garissa-Garsen, *Festo & Luke 2282* (EA, K); near Hola, *Luke & Luke 10350K* (EA).

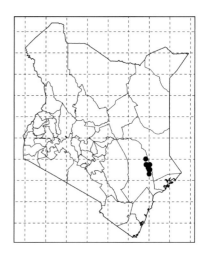

60. **Calycosiphonia** Pierre ex Robbr.

Shrubs or small trees. Leaves opposite, petiolate, papery to subcoriaceous, glabrous; domatia absent or inconspicuous and sparsely ciliate; stipules shortly sheathing, triangular, with subulate to linear lobe. Flowers bisexual, 7–8-merous, 1–4 per axil; bracts 2–3, cupular or saucer-shaped. Calyx-tube campanulate; limb truncate or slightly toothed. Corolla white; tube cylindrical; lobes spreading. Stamens 7–8, exserted; anthers narrowly linear-lanceolate. Ovary 2-locular; ovules 1–2; style cylindrical, slender; stigma exserted, 2-lobed. Fruits rather fleshy, 1–2-seeded, with persistent calyx-limb. Seeds hemispherical.

A small genus with two species in tropical Africa; one species in Kenya.

1. **Calycosiphonia spathicalyx** (K. Schum.) Robbr., Bull. Jard. Bot. État Bruxelles 51: 373. 1981; F.T.E.A. Rubiac. 2: 727. 1988; K.T.S.L.: 503. 1994. ≡ *Coffea spathicalyx* K. Schum., Bot. Jahrb. Syst. 23: 587. 1897. —Type: Cameroon, Yaúnde-Station, 1890, *G.A. Zenker & A. Staudt 79* [holotype: B (destroyed); lectotype: K (K000346922), **designated here**; isolectotypes: LE (LE00017393), P (P 00553346)] (Figure 56; Plate 77)

Shrub or small tree, up to 8.5 m tall. Leaf-blades glabrous, oblong-elliptic to elliptic or narrowly obovate, 6.5–22 cm × 2–8 cm, acute to acuminate at the apex, acute to obtuse or sometimes cuneate at the base; petioles up to 1.3 cm long; stipules broadly triangular, with subulate to linear lobe, 2–5 mm long. Flowers bisexual, 1–3 per axil; peduncules 2–4(–7) mm long. Calyx-tube 1.5–2 mm long; limb 2.5–7 mm long, truncate or slightly toothed. Corolla-tube 7–14 mm long; lobes oblong or oblong-lanceolate, 7–17 mm long, acute to acuminate.

Plate 77 *Calycosiphonia spathicalyx*. Photo by BL.

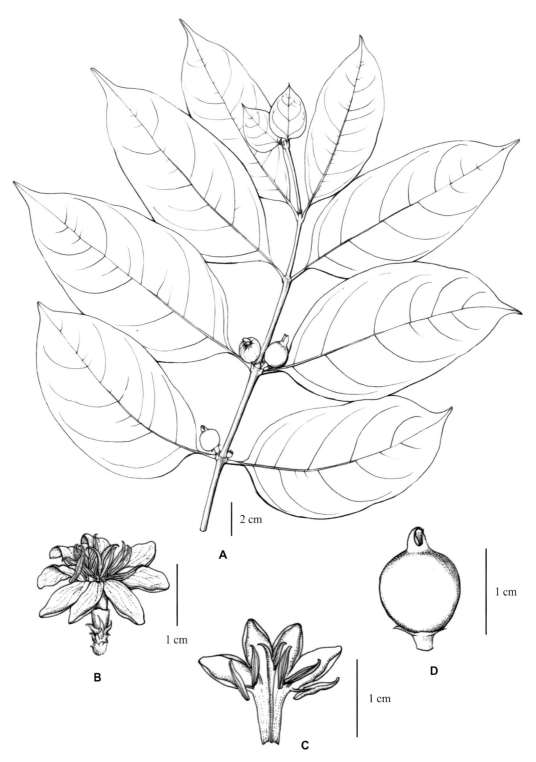

Figure 56 *Calycosiphonia spathicalyx*. A. a fruiting branch; B. flower; C. longitudinal section of corolla, showing the stamens and style; D. fruit. Drawn by NJ.

Fruits green-yellow, ellipsoid, 9–13 mm long. Seeds (1–)2, blackish, usually hemispherical.

Distribution: Coastal Kenya. [Tropical Africa].

Habitat: Lowland forests; 300–400 m.

Kwale: Shimba Hills, *Magogo & Glover 1100* (EA, K).

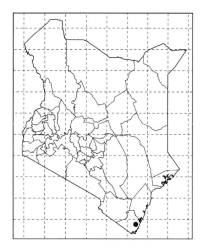

61. **Empogona** Hook. f.

Shrubs or small trees. Leaves opposite, often with domatia; stipules sheathing at the base, with needlelike awns. Flowers hermaphroditic, (4–)5(–6)-merous, sessile or rarely pedicellate; bracts and bracteoles free or fused into calyculi. Calyx-tube short; lobes well-developed and ofen overlapping. Corolla mostly white, salver-shaped, densely bearded at the throat; lobes (6–)8–17 mm long. Stamens attached to the throat; anthers with a conspicuous ribbon-like appendage. Ovary 2-locular, with 1–25 ovules on hemi-circular to hemi-ellipsoid placenta. Fruits drupaceous, 8–10 mm in diameter, mostly with persistent calyx.

A genus about 35 species widely distributed in sub-Sahara Africa and Madagascar; two species in Kenya.

1a. Leaf-blades glabrous to densely hairy; pedicels up to 16 mm long .... 1. *E. ovalifolia*
1b. Leaf-blades glabrous; pedicels very short, elongate to 7 mm in fruiting .... 2. *E. ruandensis*

1. **Empogona ovalifolia** (Hiern) Tosh & Robbr., Ann. Missouri Bot. Gard. 96(1): 208. 2009. ≡ *Tricalysia ovalifolia* Hiern, Fl. Trop. Afr. 3: 119. 1877; F.T.E.A. Rubiac. 2: 561. 1988; K.T.S.L.: 549. 1994. —Type: Tanzania, Zanzibar, *J. Kirk s.n.* [lectotype: K (K000346977), designated by E. Robbrecht in Bull. Jard. Bot. Natl. Belg. 49: 339. 1979; isolectotypes: K (K000346978, K000346979)]

Shrub or small tree, up to 6 m tall. Leaf-blades coriaceous, glabrous to densely hairy, elliptic to ovate or obovate, (3–)5–11 cm × (1–)2.5–6 cm, acute to shortly acuminate at the apex, cuneate at the base; domatia mostly absent; petioles up to 5 mm long, glabrous to hairy; stipules sheathing, with awns up to 6 mm long. Flowers 5-merous, sweet-scented, 3–6 per axil; pedicels up to 16 mm long; bracts cupular; bracteoles free, (1–)2, alternate. Calyx-tube campanulate; limb-lobes triangular. Corolla white or rarely pinkish; tube 3–6.5 mm long, densely bearded at the throat; lobes 3.5–

8 mm long, emarginate. Fruits purplish to black, globose, ca. 7 mm in diameter.

1a. Leaf-blades entirely glabrous; the upper part of petioles slightly hairy; bracteoles, pedicels and calyx-tube glabrous .... a. var. *ovalifolia*
1b. Leaf-blades puberulous on the nerves or densely hairy on all parts of the blades; bracteoles, pedicels and calyx-tube hairy 2
2a. Leaf-blades puberulous on the midnerves beneath .... b. var. *glabrata*
2b. Leaf-blades velutinous or softly hairy underneath .... c. var. *taylorii*

a. var. **ovalifolia** (Plate 78A, B)

Leaf-blades entirely glabrous. Petioles slightly hairy at the upper. Bracteoles, pedicels and calyx-tube glabrous.

Distribution: Central and coastal Kenya. [Aldabra, Assumption, Comoros, Madagascar, Somalia, Tanzania].

Habitat: Coastal forests, bushlands, or thickets; up to 400 m.

Kilifi: Arabuko-Sokoke Forest, *Robertson et al. 5238* (EA, K); Gede Forest, *Gerhardt & Steiner 136* (EA); Mangea Hill, *Luke 1616* (EA). Kwale: Kaya Waa, *SAJIT 005572* (HIB); Shaitani Forest, *Brenan et al. 14509* (EA, K); Diani Forest, *Robertson 4351* (EA). Lamu: Boni Forest Reserve, *Robertson & Luke 5607* (EA); Diani Forest, *Gillett & Kibuwa 19894* (EA). Machakos: Donyo Sabuk, *Ossent 687* (EA). Mombasa: Shanzu, *Robertson 4949* (EA). Tana River: Ozi, *Luke & Robertson 1357* (EA, K).

b. var. **glabrata** (Oliv.) Tosh & Robbr., Ann. Missouri Bot. Gard. 96(1): 208. 2009. ≡ *Empogona kirkii* Hook. f. var. *glabrata* Oliv., Trans. Linn. Soc. London, Bot. 2: 331. 1887. ≡ *Tricalysia ovalifolia* Hiern var. *glabrata* (Oliv.) Brenan, Kew Bull. 2: 59. 1949; F.T.E.A. Rubiac. 2: 563. 1988; K.T.S.L.: 549. 1994. —Type: Kenya or Tanzania, 40–60 miles from coast, 1884, *H.H. Johnston s.n.* [holotype: K (K000352901)] (Figure 57; Plate 78C)

Leaf-blades puberulous on the midnerves beneath. Bracteoles, pedicels and calyx-tube hairy.

Distribution: Cental, southern, and coastal Kenya. [Madagascar, Tanzania].

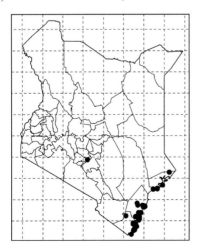

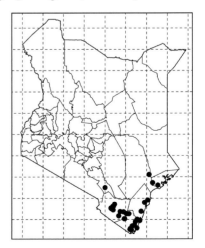

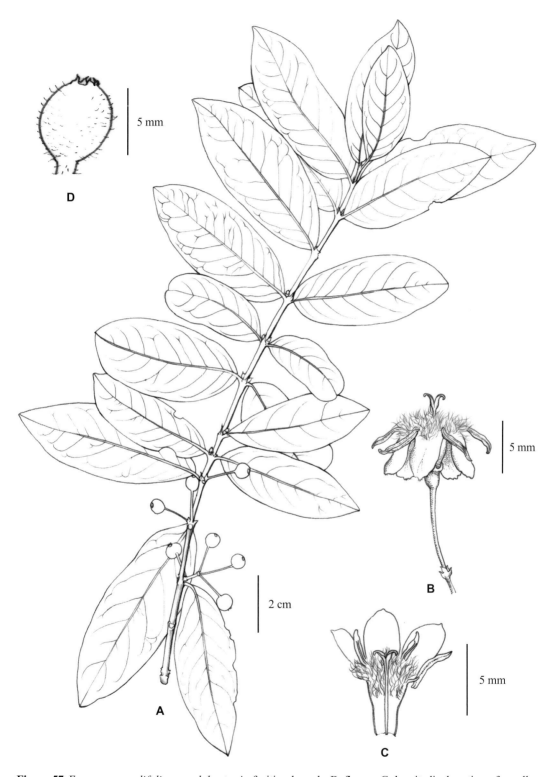

Figure 57 *Empogona ovalifolia* var. *glabrata*. A. fruiting branch; B. flower; C. longitudinal section of corolla, showing the stamens and style; D. fruit. Drawn by NJ.

Habitat: Dry thickets, wooded grasslands, or coastal forests; up to 1200 m.

Kilifi: ca. 7 km west of Marikebuni, *Beentje 2349* (EA). Kwale: Kilibasi Forest, *SAJIT 005437* (HIB); Shimba Hills, *SAJIT V0136* (HIB); Mwachi Forest, *Faden et al. 77/455* (EA). Lamu: ca. 2 km west of Nyangoro Bridge, *Robertson & Luke 5378* (EA, K). Makueni: Ukambani, *Scott 2304* (K). Mombasa: *Wakefield 1884* (K). Teita Taveta: Mwandongo Forest, *Mwachala et al.* in *EW 976* (EA); Teita Sisal Estate Zongoloni Hill, *Luke & Luke 6419* (EA); Mbololo Hill, *Wakanene & Mwangangi 522* (EA). Tana River: Tana River National Primate Reserve, *Luke et al. TPR428* (EA, K).

c. var. **taylorii** (S. Moore) Tosh & Robbr., Ann. Missouri Bot. Gard. 96(1): 208. 2009. ≡ *Empogona taylorii* S. Moore, J. Bot. 63: 145. 1925. ≡ *Tricalysia ovalifolia* Hiern var. *taylorii* (S. Moore) Brenan, Kew Bull. 2: 59. 1947; F.T.E.A. Rubiac. 2: 563. 1988; K.T.S.L.: 549. 1994. —Type: Kenya, Kilifi, Giriama, Oct. 1887, *W.E. Taylor s.n.* [holotype: BM (BM000903156)] (Plate 78D–G)

Leaf-blades velutinous or softly hairy underneath; bracteoles, pedicels and calyx-tube hairy.

Distribution: Southern and coastal Kenya. [Tanzania].

Habitat: Dry thickets, bushlands, or forests; up to 1000 m.

Kilifi: Arabuko-Sokoke Forest, *Robertson et al. 5237* (EA, K). Teita Taveta: Taita Hills, *SAJIT 004636* (HIB).

2. **Empogona ruandensis** (Bremek.) Tosh & Robbr., Ann. Missouri Bot. Gard. 96(1): 209. 2009. ≡ *Tricalysia ruandensis* Bremek., Bull. Jard. Bot. État Bruxelles 26: 253. 1956; F.T.E.A. Rubiac. 2: 564. 1988; K.T.S.L.: 549. 1994. —Type: Rwanda, Mayaga, Mutema, 19 May 1954, *L. Liben 1416* [holotype: U (U0006330); isotypes: BR (BR0000008381040, BR0000008381217), IUK (IUK07829), WAG, YBI (YBI192923470)]

Shrub or small tree, up to 5(–8) m tall. Leaf-blades glabrous, coriaceous, obovate to elliptic, 5–8.5 cm × 2–3.5 cm, acute to acuminate at the apex, cuneate at the base; domatia absent; petioles 2–7 mm long; stipules sheathing, with awns 1–3 mm long. Inflorescences 3–10-flowered; bracts and bracteoles cupular, 2-toothed. Flowers (4–)5-merous, sweet-scented,

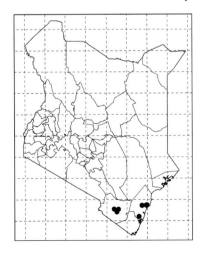
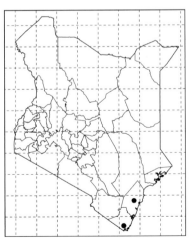

Plate 78 A, B. *Empogona ovalifolia* var. *ovalifolia*; C. *E. ovalifolia* var. *glabrata*; D–G. *E. ovalifolia* var. *taylorii*. Photo by GWH (A, B, D–G), VMN (C).

shortly pediculate, elongate to 7 mm in fruiting. Calyx-tube glabrous; limb-tube very short; lobes rounded or elliptic, 0.5–1 mm long. Corolla white; tube 2–5 mm long; lobes 2.7–4.5 mm long. Fruits white, purple to black, globose, 6–10 mm in diameter.

Distribution: Coastal Kenya. [East Africa].

Habitat: Coastal forests; 250–500 m.

Kilifi: Mangea Hill, *Luke & Robertson 648* (EA). Kwale: Dzombo Hill, *Robertson et al. 287* (EA).

62. **Tricalysia** A. Rich. ex DC.

Shrubs, small trees or rarely large trees. Leaves opposite, petiolate, often with domatia; stipules sheathing, crowned by two awns. Inflorescences axillary, sessile to subsessile, 1–many-flowered, contracted; bracts and bracteoles free or fused into calyculi. Flowers hermaphroditic or unisexual, 4–7(–12)-merous, sweet-scented. Calyx with well-developed limb-tube, truncate, toothed or with linear or triangular lobes. Corolla white, cream or rose, salver-shaped; glabrous to hairy at throat; lobes spreading or reflexed. Stamens inserted at the corolla-throat; filaments long; anthers medifixed, exserted, with or without an inconspicuous apical appendage. Ovary 2-locular, with 1–12 ovules on a semi-circular to hemi-ellipsoid placenta; style 2-lobed, exserted. Fruits drupaceous, red or rarely orange, with persistent calyx. Seeds 1–12 per locule; endosperm entire.

A genus of 82 species widespread in Africa and Madagascar; four species in Kenya.

1a. Flowers always solitary and sessile .... 4. *T. bridsoniana*
1b. Flowers in 3–many-flowered inflorescences 2
2a. Calyx-limb with subulate lobes; corolla-tube 8.5–10 mm long, lobes 8–10 mm long ....
.... 1. *T. microphylla*
2b. Calyx-limb shortly toothed; corolla-tube less than 8 mm long, lobes less than 6 mm long 3
3a. Western species; leaf-blades brownish above and green beneath .... 3. *T. niamniamensis*
3b. Central and coastal species; leaf-blades greenish or greyish green on both sides .. 2. *T. pallens*

1. **Tricalysia microphylla** Hiern, Fl. Trop. Afr. 3: 123. 1877; F.T.E.A. Rubiac. 2: 551. 1988; K. T.S.L.: 548. 1994. —Type: Tanzania, Zanzibar, June 1873, *J.M. Hildebrandt 1163* [holotype: BM (BM000903118)] (Plate 79A–C)

Shrub, up to 3 m tall. Leaves elliptical to obovate or ovate, 2–13 cm × 1–6.5 cm, acuminate at the apex, subcoriaceous, with 3–7 pairs of main lateral nerves; domatia hairy; petioles up to 1 cm long; stipules sheathing, short, apiculate. Flowers 6-merous, subsessile,

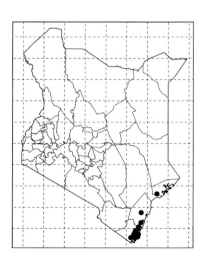

3–9 in axillary inflorescences, rarely solitary. Calyx-limb with subulate lobes up to 5 mm long. Corolla white, tube salver-shaped, 8.5–10 mm long, glabrous outside, slightly pubescent at throat; lobes ovate, 8–10 mm long, acute. Fruits red, ellipsoid, ca. 10 mm long.

Distribution: Coastal Kenya. [Tanzania].

Habitat: Lowland forests; up to 500 m.

Kilifi: Chasimba, *Faden et al. 77/413* (EA); Mangea Hill, *Luke & Robertson 657* (EA). Kwale: Shimba Hills, *SAJIT V0314* (HIB); Kinondo Forest, *SAJIT 005443* (HIB); Dzombo Hill, *Robertson et al. 352* (AE). Lamu: Witu Forest, *Robertson & Luke 5480* (EA).

2. **Tricalysia pallens** Hiern, Fl. Trop. Afr. 3: 121. 1877; F.T.E.A. Rubiac. 2: 548. 1988; K.T.S.L.: 549. 1994. —Type: Equatorial Guinea, Bioko, 1860, *G. Mann s.n.* [holotype: K (K000346910)] (Figure 58; Plate 79D, E)

Shrub or small tree, up to 10 m tall. Leaf-blades elliptic or obovate, 2–12 cm × 1.5–5 cm, acute to acuminate at the apex, cuneate at the base, papyraceous to subcoriaceous, with 4–5 pairs of main lateral nerves; domatia hairy; petioles up to 1 cm long; stiples sheathing, 1–2 mm long, awned. Flowers subsessile, few to many in axillary inflorescences; bracts and bracteoles cupular. Calyx-limb tubular, shortly toothed, 1–1.5 mm long. Corolla white; tube 2–8 mm long; lobes 3–6 mm long. Fruits red, globose, 3–6 mm in diameter.

Distribution: Central and coastal Kenya. [Tropical Africa].

Habitat: Lowland wet forests, dry bushlands, sandstone hills, or wooded grasslands; up to 1800 m.

Kilifi: Kaya Jibana, *Luke & Robertson 2644* (EA). Kitui: Mutha Hill, *Luke & Stone 8213* (EA). Kwale: Shimba Hills, *SAJIT V0160* (HIB); Gongoni Forest, *Luke & Robertson 2397* (EA). Meru: Ngaia Forest, *Luke et al. 7168* (EA); Lower Imenti Forest, *Faden & Faden 74/895* (EA, K). Teita Taveta: Mbololo Hill, *Wakenene & Mwangangi 670* (EA); Ngangao Forest, *Faden et al. 523* (K).

3. **Tricalysia niamniamensis** Schweinf. ex Hiern, Fl. Trop. Afr. 3: 123. 1877; F.T.E.A. Rubiac. 2: 552. 1988; K.T.S.L.: 549. 1994. —Type: Sudan, S.W. Gunango, 7 Feb. 1870, *G.A. Schweinfurth 2883* [lectotype: P (P00072410), designated by E. Robbrecht in Bull. Jard. Bot. Natl. Belg. 57: 171. 1987; isolectotypes: BM (BM000903111), K (K000346894)]

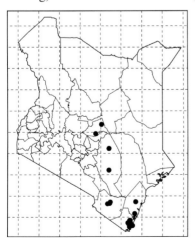
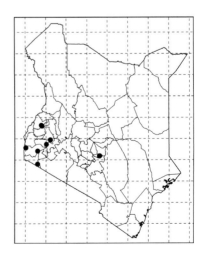

Plate 79 A–C. *Tricalysia microphylla*; D, E. *T. pallens*. Photo by GWH (A, B) and VMN (C–E).

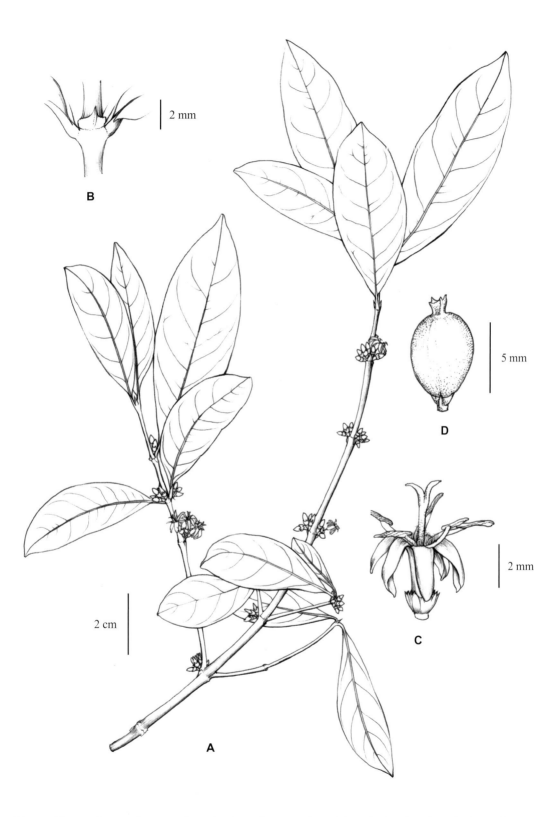

Figure 58 *Tricalysia pallens*. A. flowering branch; B. portion of branch showing the stipule; C. flower; D. fruit. Drawn by NJ.

Shrub or small tree, up to 6 m tall. Leaf-blades narrowly elliptic, ovate or obovate, 1–9 cm × 1–2.5 cm, rounded, acute or shortly acuminate at the apex, cuneate, rounded or cordate at the base, brownish above and green beneath, glabrescent to pubescent on both sides, with 4–6 pairs of main lateral nerves; domatia hairy; petioles 1–6 mm long; stipules sheathing, 1.5–2.5 mm long. Flowers 6-merous, subsessile, few to many in axillary inflorescences; bracts and bracteoles cupular, toothed. Corolla white; tube 6–7.5 mm long; lobes 4–5 mm long. Fruits orange, subglobose, glossy, 5–7 mm in diameter.

Distribution: Western and central Kenya. [East Africa].

Habitat: bushlands, woodlands, or rocky zones; 850–1650 m.

Bungoma: ca. 4 km from Webuye on road to Kitale, *Gilbert & Mesfin 6592* (EA). Embu: Seven Forks, *Robertson 2010* (EA, K). Homa Bay: Mfangano Island, *Paul 211* (EA). Kisumu: Kericho-Kisumu Road, *Gillett 19343* (EA, K). Migori: Bukeria, *Napier 5290* (EA).

4. **Tricalysia bridsoniana** Robbr., Bull. Jard. Bot. Natl. Belg. 56: 146. 1986; F.T.E.A. Rubiac. 2: 559. 1988; K.T.S.L.: 548. 1994. —Type: Kenya, Kilifi, Arabuko-Sokoke Forest, 24 Nov. 1978, *J.P.M. Brenan et al. 14685* [holotype: K (K000311614); isotypes: BR (BR0000008382603), EA (EA000001646), K (K000311615), WAG (WAG0003086)]

Shrub, up to 5 m tall. Leaf-blades obovate, elliptic or ovate, 5–10 cm × 1.5–3.5 cm, acute at the apex, cuneate at the base, coriaceous, with 3–4 pairs of main lateral nerves; domatia hairy; petioles 1–3 mm long; stipules sheathing, ca. 1 mm long, with awns up to 4 mm long. Flowers solitary, sessile; bracts and bracteoles 3–4 pairs, cupular. Calyx-limb ellipsoid, 4–5 mm long, with 4–5 minute lobes at the tip. Corolla white; tube 13–27 mm long; lobes 6–10 mm long, acute. Fruit red, ovoid, ca. 3.5 mm long.

Distribution: Central and coastal Kenya. [Tanzania].

Habitat: Submontane forests, lowland forests, or sandstone hills; 50–1350 m.

Kilifi: ca. 2 miles east of Jilore Forest Station, *Spjut & Ensor 2789* (EA); Arabuko-Sokoke Forest, *Brenan et al. 14685* (K). Kitui: Mutha Hill, *Luke & Stone 8213* (K). Kwale: Shimba Hills, *Luke & Luke 9463* (EA).

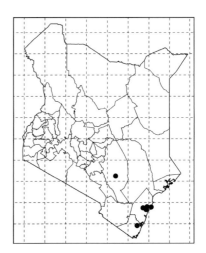

18. Trib. **Octotropideae** Bedd.

Shrubs, small trees or sometimes lianas, less often with supra-axillary spines. Leaves opposite, petiolate; domatia present or absent; stipules triangular, ovate or lanceolate, acuminate, sometimes keeled. Flowers (4–)5(–6)-merous, few to many in axillary clusters or cymes, or axillary and/or terminal panicles, rarely solitary. Corolla-tube short to cylindrical; lobes contorted. Stamens inserted at throat; anthers dorsifixed, exserted. Ovary 1–2-locular, with 1–2(–10) ovules in each locule. Fruits globose, subglobose or elliptic, (1–)2(–16)-seeded.

Seven genera occur in Kenya.

1a. Plants armed with spines .... 65. *Didymosalpinx*
1b. Plants unarmed 2
2a. Ovule single in each locule 3
2b. Ovules few in each locule 5
3a. Flowers in terminal and axillary panicles; corolla-lobes 6–9; style narrowly club-shaped, densely pubescent .... 66. *Lamprothamnus*
3b. Flowers usually sessile in bracteate axillary clusters; corolla-lobes 4–5; style slender or filiform, pubescent or just hairy at extreme base 4
4a. Fruits ovoid, crowned with persistent and sometimes beaked calyx .... 63. *Cremaspora*
4b. Fruits globose, not crowned with persistent calyx .... 69. *Polysphaeria*
5a. Flowers sessile and appearing terminal on short leafless spurs or sometimes pedicellate in pairs from the axils of young leaves .... 68. *Feretia*
5b. Flowers few to many in axillary pedunculate panicles or cymes 6
6a. Flowers pedicellate in pedunculate axillary cymes; stigmatic club deeply bifid, each lobe with 5 membranous ciliate wings .... 64. *Galiniera*
6b. Flowers many in axillary, lax, pedunculate panicles; stigmatic club long, fusiform, with 10 membranous ciliate wings... 67. *Kraussia*

63. **Cremaspora** Benth.

Shrubs, small trees or sometimes lianas Branches appearing supra-axillary and often subtended by small, rounded, often cordate leaves. Leaves opposite, shortly petiolate; stipules keeled, acuminate, deciduous. Inflorescences of axillary clusters, sessile, dense; bracts and bracteoles triangular-acuminate. Calyx-tube ovoid; limb with campanulate tube and 5 teeth. Corolla white, tube cylindrical or narrowly funnel-shaped; lobes 5, narrowly oblong, contorted. Stamens 5; anthers dorsifixed, linear, exserted. Ovary 2-locular; ovule single in each locule; style filiform, exserted. Fruits ovoid, indehiscent, 2-locular, (1–)2-seeded. Seeds half-ovoid, flattened.

A genus with two species in tropical Africa and the Comoros Islands; one species with two subspecies in Kenya.

1. **Cremaspora triflora** (Thonn.) K. Schum., Nat. Pflanzenfam. 4(4): 88. 1891; F.T.E.A. Rubiac. 2: 733. 1988; K.T.S.L.: 511. 1994. ≡ *Psychotria triflora* Thonn., Beskr. Guin. Pl.: 108. 1827. —Type: Ghana, near Asiama, *P. Thonning 299* [lectotype: C (C10004441), **designated here**; isolectotypes: C (C10004440), P-JU, S (S-G-5131)]

Shrub, small tree or sometimes a liana, up to 9 m tall. Branches opposite above the leaf axils often subtended by rounded, cordate leaves, up to 4 cm in diameter. Leaf-blades lanceolate, oblong-elliptic to obovate, 2–18 cm × 0.5–9 cm, acuminate at the apex, cuneate to broadly rounded at the base, glabrous above, glabrous to pubescent or hairy beneath; stipules narrowly triangular, 5–7 mm long. Flowers in opposite axillary clusters. Calyx-tube up to 2 mm long. Corolla-tube 3–5.5(–10) mm long; lobes up to 7 mm long. Fruits dark purple, ovoid or ellipsoid, 7–15 mm long, crowned with the persistent calyx.

1a. Calyx-lobes triangular, less than 1.5 mm long; fruits not drawn out into a beak ......a. subsp. *triflora*
1b. Calyx-lobes narrowly triangular, up to 2 mm long; fruits distinctly drawn out into a beak .... b. subsp. *confluens*

a. subsp. **triflora**

Calyx-lobes triangular, less than 1.5 mm long. Fruits not drawn out into a beak.
Distribution: Southern Kenya. [Tropical Africa].
Habitat: Evergreen forests and thickets; 800–2000 m.
Teita Taveta: Mwatate River Valley, *Faden et al. 70/444* (EA, K); Taita Hills, *Mungai et al. in EW 1761* (EA).

b. subsp. **confluens** (K. Schum.) Verdc., Kew Bull. 35: 132. 1980; F.T.E.A. Rubiac. 2: 735. 1988; K.T.S.L.: 511. 1994. ≡ *Cremaspora confluens* K. Schum., Pflanzenw. Ost-Afrikas, C: 383. 1895. —Types: Tanzania, Zanzibar, Mkokotoni, *F.L. Stuhlmann 607* [syntype: B (destroyed)]; Tanzania,

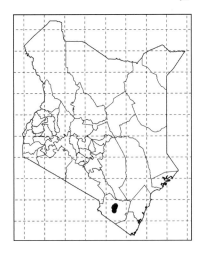
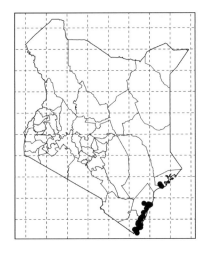

Plate 80 A, B. *Cremaspora triflora* subsp. *confluens*. Photo by GWH.

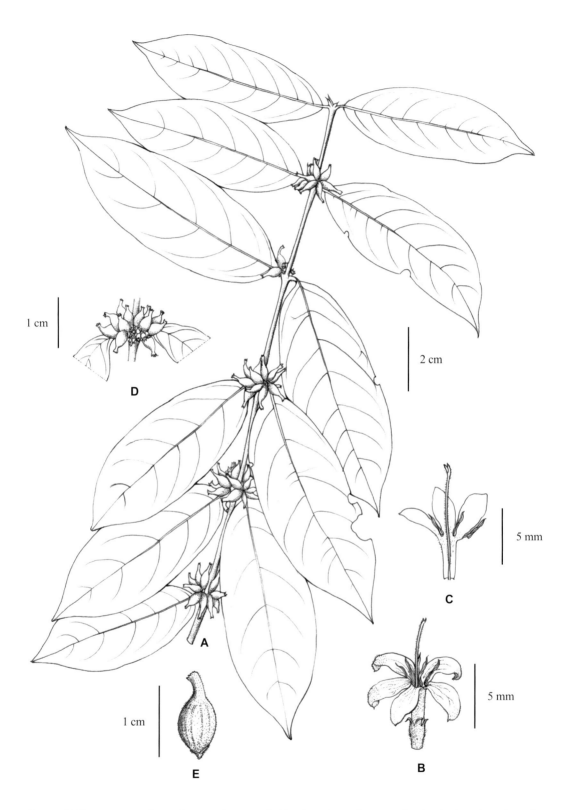

Figure 59 *Cremaspora triflora* subsp. *confluens*. A. fruiting branch; B. flower; C. longitudinal section of corolla, showing the stamens and style; D. infructescence; E. fruit. Drawn by NJ.

Uzaramo, Tambani, *F.L. Stuhlmann 6144* [syntype: B (destroyed)]; Tanzania, Zanzibar, Massazine, *H.G. Faulkner 2345* [neotype: K (K000286656), **designated here**] (Figure 59; Plate 80)

Calyx-lobes narrowly triangular, up to 2 mm long. Fruits distinctly drawn out into a beak.

Distribution: Coastal Kenya. [Tanzania].

Habitat: Coastal forests, bushlands, and thickets; up to 500 m.

Kilifi: Arabuko-Sokoke Forest, *Robertson 3710* (EA, K); Kaya Jibana, *Robertson & Luke 4506* (EA); Jilore-Mida, *Sangai in EA 15709* (EA). Kwale: Ukunda, Kaya Kinondo, *SAJIT 006087* (HIB); Shimba Hills, *Magogo & Glover 180* (EA, K); Mrima Hill, *Faden & Faden 77/694* (EA). Lamu: Witu Forest, *Robertson & Luke 5492* (EA). Tana River: Nairobi Ranch, *Festo & Luke 2490* (EA).

64. Galiniera Delile

Shrubs or small trees, with smooth and grey bark. Leaves opposite, petiolate; domatia present or absent; stipules triangular. Flowers sweet-scented, 5-merous, few to many in pedunculate axillary cymes. Calyx-tube short; limb-lobes triangular, acute. Corolla-tube very short; lobes ovate, obtuse. Stamens attached at the top of the corolla-tube, exserted and spreading; filaments short; anthers dorsifixed, narrowly oblong. Ovary 2-locular, with (1–)2 ovules in each locule; style short, subulate. Fruit a berry, red, globose, (2)4-seeded.

A small genus of two species, one species restricted to tropical Africa including Kenya.

1. **Galiniera saxifraga** (Hochst.) Bridson, F.T.E.A. Rubiac. 2: 696. 1988; K.T.S.L.: 513. 1994. ≡ *Pouchetia saxifraga* Hochst. ex A. Rich., Tent. Fl. Abyss. 1: 355. 1848. —Type: Ethiopia, Aber, near Addeselam, 9 Jan. 1840, *G.H.W. Schimper 863* [holotype: B (destroyed), lectotype: K (K000419929), **designated here**; isolectotypes: BR (BR0000008358721, BR 0000009812536), G (G00424954), HAL (HAL 0113667), K (K000419930), MPU (MPU 011688), NYBG (NY00132546, NY00133031), REG (REG000631, REG000632), TUB (TUB 004542, TUB004543, TUB004555)] (Figure 60; Plate 81)

Shrub or small tree, up to 14 m tall. Leaf-blades elliptic to oblong-elliptic, 5.5–20 cm × 2–8 cm, acuminate at the apex, cuneate at the base, glabrous above, glabrous to sparsely pubescent beneath, often with reddish veins; domatia conspicuous or obscure; stipules triangular, up to 1 cm long. Flowers few to many in pedunculate cymes, 5-merous. Calyx-tube very short; limb-lobes triangular, acute, up to 0.5 mm long. Corolla white and pink-tipped; tube very short; lobes ovate, up to

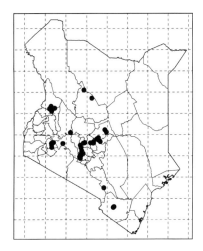

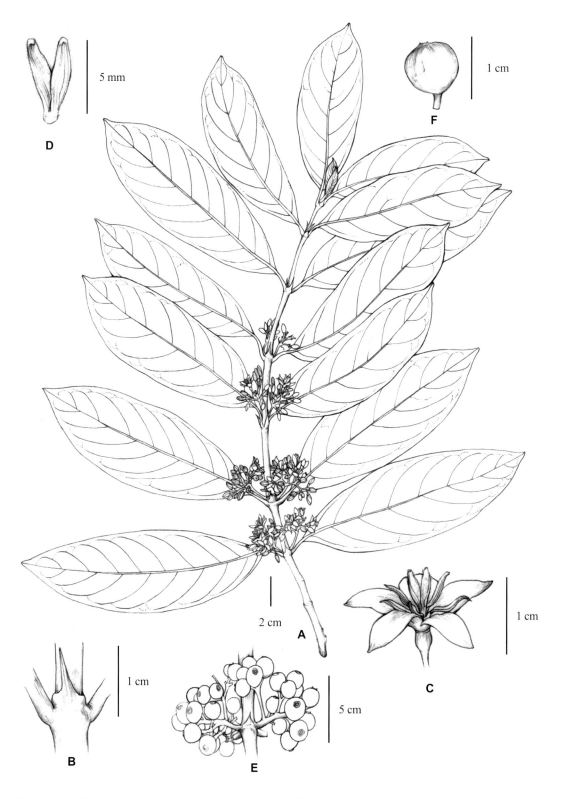

Figure 60 *Galiniera saxifraga*. A. flowering branch; B. portion of branch showing the stipule; C. flower; D. stigmas; E. infructescence; F. fruit. Drawn by NJ.

Plate 81 A–F. *Galiniera saxifraga*. Photo by YDZ (A), GWH (B–E) and CL (F).

1 cm long, obtuse. Berries red or purple-brown, up to 9 mm in diameter. Seeds 4–5 mm long.

Distribution: Western, central and southern Kenya. [East Africa, from Ethiopia to Angola and Zambia].

Habitat: Montane forests; 1700–3000 m.

Bomet: southwest Mau Forest Reserve, *Geesteranus 5747* (K). Elgeyo-Marakwet: Kapcherop Forest, *SAJIT 006924* (HIB). Kericho: southwest Mau Forest, Sambret, *Kerfoot 2910* (EA, K). Kiambu: Tigoni, Brackenridge Farm, *Luke & Luke 8289* (EA). Kirinyaga: Mount Kenya Forest, *Perdue & Kibuwa 8292* (EA, K). Makueni: Chyulu Hills, *Someren 8351* (K). Meru: Mount Kenya, *SAJIT 002922* (HIB). Murang'a: Kimakia Forest Reserve, *Agnew et al. 5617* (EA). Nakuru: Solai Forest, *Brasnett 1381* (EA, K). Nyeri: Aberdare Forest, *Gardner 2303* (EA, K). Samburu: Ndoto Mountains, *Bytebier & Kirika 30* (EA, K). Teita Taveta: Taita Hills, Yale Rock, *Beentje et al. 919* (K). Tharaka-Nithi: Chuka Forest, *SAJIT VK0051* (HIB). West Pokot: Cherangani Hills, *SAJIT Z0018* (HIB).

65. **Didymosalpinx** Keay

Shrubs or rarely small trees, often with pairs of supra-axillary spines. Leaves opposite, petiolate; stipules triangular, acute or acuminate-subulate. Flowers large, 5-merous, axillary or supra-axillary, solitary on each side of the node. Calyx-tube cylindric; limb-lobes linear, lanceolate to narrowly triangular, sometimes unequal. Corolla-tube funnel-shaped; lobes ovate, broadly ovate-elliptic or triangular, with contorted aestivation. Ovary unilocular, with several ovules embedded in 2 placentas. Fruits globose or ellipsoid, crowned with the persistent calyx-limb. Seeds 8–16, lenticular.

A small genus of four species restricted to tropical Africa; only one species in Kenya.

1. **Didymosalpinx norae** (Swynn.) Keay, Bull. Jard. Bot. État Bruxelles 28: 64. 1958; F.T.E.A. Rubiac. 2: 524. 1988; K.T.S.L.: 512. 1994. ≡ *Gardenia norae* Swynn., J. Linn. Soc., Bot. 40: 80. 1911. —Type: Zimbabwe, Chirinda Forest, 23 June 1906, *C.F.M. Swynnerton 11* [holotype: BM (BM000903070); isotypes: K (K000414936, K000414937)] (Figure 61; Plate 82)

Shrub or rarely small tree, up to 4(–10) m tall, with paired supra-axillary spines. Leaf-blades elliptic, 4.5–17 cm × 1.5–7 cm, acuminate at the apex, cuneate at the base; domatia small, hairy; petioles up to 1 cm long; stipules broad triangular, up to 4 mm long. Flowers white, solitary on each side of the node; peduncles up to 3.5 cm long. Calyx-tube 3–4 mm long; limb-lobes linear to narrowly linear-triangular, 3–6 mm long. Corolla-tube

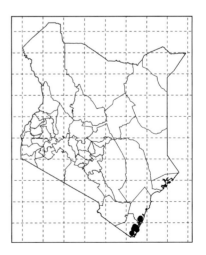

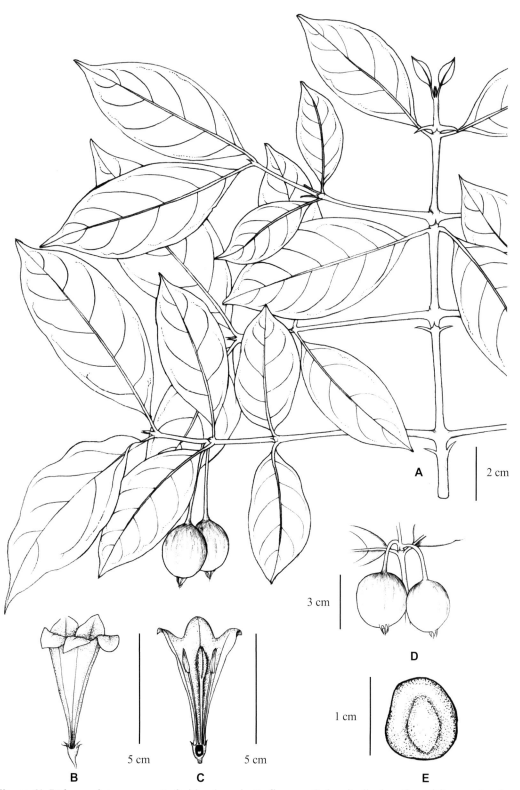

Figure 61 *Didymosalpinx norae*. A. fruiting branch; B. flowers; C. longitudinal section of flower, showing stamens, ovary and style; D. fruits; E. seed. Drawn by NJ.

Plate 82 *Didymosalpinx norae*. Photo by GWH.

funnel-shaped, 4.5–7.5 cm long; lobes broadly ovate-elliptic, (1–)2 cm long. Fruits green or white, with deep green longitudinal lines, globose or ellipsoid, 2–4 cm long, 10-ribbed. Seeds irregularly orbicular, compressed.

Distribution: Coastal Kenya. [East Tropical Africa].

Habitat: Forests; 100–900 m.

Kilifi: Kaya Jibana, *Luke & Robertson 2657* (EA, K). Kwale: Shimba Hills, *SAJIT 006031* (HIB); Mrima Hill, *Verdcourt 1854* (EA, K).

66. **Lamprothamnus** Hiern

Shrubs or small trees. Leaves coriaceous, ovate or elliptic, oblong, elliptic or ovate, subsessile, rounded to subacute at the apex, cordate at the base; stipules broadly triangular, apiculate. Flowers sweet-scented, in terminal and axillary panicles. Calyx-tube campanulate; limb-lobes 4–6, rounded-truncate. Corolla white; tube narrowly funnel-shaped; lobes 6–9, oblong-elliptic. Ovary 1–2-locular; ovule solitary in each locule; style narrowly club-shaped, densely pubescent. Fruits red, subglobose, crowned by persistent calyx-limb. Seeds ovoid.

A monotypic genus restricted to East Africa including Kenya.

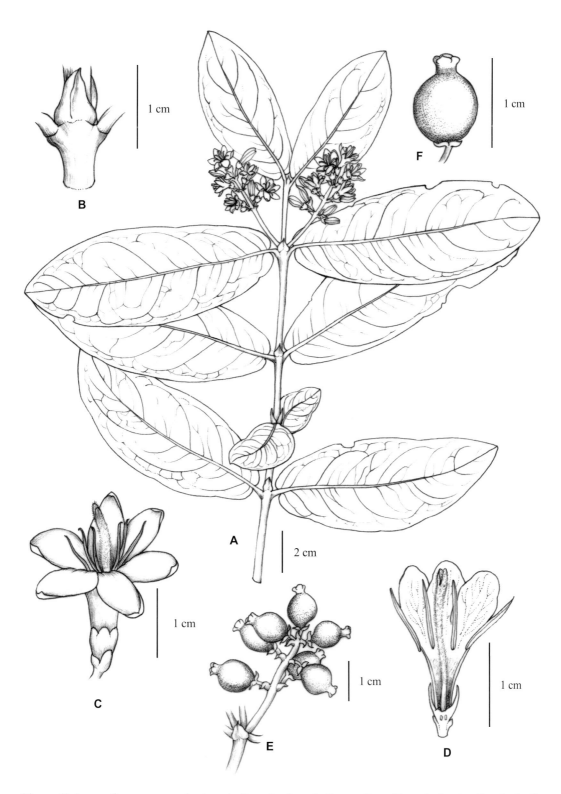

Figure 62 *Lamprothamnus zanguebaricus*. A. flowering branch; B. a portion of branch showing the stipule; C. flower; D. longitudinal section of flower, showing stamens, ovary and style; E. infructescence; F. fruit. Drawn by NJ.

Plate 83 *Lamprothamnus zanguebaricus*. Photo by GWH.

1. **Lamprothamnus zanguebaricus** Hiern, Hooker's Icon. Pl. 13: t. 1220. 1877; F.T.E.A. Rubiac. 2: 579. 1988; K.T.S.L.: 519. 1994. —Type: Tanzania, Dar es Salaam, Oct. 1868, *J. Kirk s.n.* [holotype: K (K000316235)] (Figure 62; Plate 83)

Shrub or small tree, up to 9 m tall, with grey, fissured bark. Leaf-blades coriaceous, glabrous, ovate or elliptic, subsessile, ovate or elliptic, oblong, elliptic or ovate, 2–16 cm × 1–6.5 cm, rounded to subacute at the apex, cordate at the base; stipules broadly triangular, apiculate, up to 8 mm long. Flowers sweet-scented, in terminal and axillary panicles. Peduncles up to 2.5 cm long. Calyx-tube campanulate; limb-lobes up to 2 mm long. Corolla-tube up to 1.2 cm long; lobes up to 1.1 cm long. Style densely pubescent, up to 2 cm long. Fruits red, subglobose, 6–11 mm in diameter.

Distribution: Coastal Kenya. [South Somalia to Tanzania].

Habitat: Moist forests, wooded grasslands, bushlands, thickets, and woodlands; up to 300 m.

Garissa: ca. 11 km south of Kulank W. P., *Gillett 20381* (EA, K). Kilifi: Mitunguni, *Robertson 4081* (EA, K); just below Jilore Forest Station, *Perdue & Kibuwa 10098* (EA). Kwale: Godoni Forest, *Kokwaro 3984* (EA); Chuna Forest, *Luke & Robertson 568* (EA); between Mwereni and Ndavya, *Robertson 4319* (EA, K). Lamu: Iwezo, *Power s.n.* (K). Makueni: Masalani, *Marsh 26* (EA). Mombasa: *Bally 5704* (K). Tana River: Tana River National Primate Reserve, *Luke et al. TPR160* (EA).

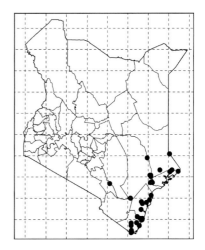

67. **Kraussia** Harv.

Shrubs or small trees. Leaves opposite, shortly petiolate; blades lanceolate or ovate-lanceolate, obtuse to acuminate, often with domatia; stipules triangular to lanceolate, sometimes apiculate. Flowers 5-merous, many in axillary, lax, pedunculate panicles. Calyx-tube ovate to turbinate; limb-lobes triangular, rounded to acute. Corolla-tube funnel-shaped, densely bearded at the throat; lobes oblong, with imbricate aestivation. Stamens attached at the mouth of the tube, exserted and spreading; anthers linear, dorsifixed near the base. Ovary 2-locular, with 1–3 impressed ovules. Style very short, swollen; stigmatic club long, fusiform, with 10 membranous ciliate wings. Fruit a berry, globose, crowned with the persistent calyx-limb, 2–6-seeded.

A small genus of four species, restricted to tropical and South Africa; two species in Kenya.

1a. Leaf-blades subcoriaceous; calyx-limb less than 1.5 mm long; fruits 2-seeded .... 1. *K. kirkii*
1b. Leaf-blades chartaceous; calyx-limb 3–4 mm long; fruits 4–6-seeded ....2. *K. speciosa*

1. **Kraussia kirkii** (Hook. f.) Bullock, Bull. Misc. Inform. Kew 1934: 231. 1934; F.T.E.A. Rubiac. 2: 701. 1988; K.T.S.L.: 519. 1994. ≡ *Rhabdostigma kirkii* Hook. f., Gen. Pl. 2: 109. 1873. —Type: Tanzania, Kilwa (Quiloa), 10 Jan. 1867, *J. Kirk 105* [holotype: K (K000319501)] (Plate 84)

Shrub up to 4.5(–8.5) m tall. Leaf-blades subcoriaceous, glabrous, elliptic to oblong-elliptic, 5–17 cm × 2–6.5 cm, obtuse to slightly acuminate at the apex, cuneate at the base; domatia present; petioles up to 1 cm long; stipules triangular, 2.5–6 mm long. Flowers several in a lax axillary panicle; peduncles up to 7 cm long; pedicels up to 1.2 cm long. Calyx-tube 1.5–2 mm long; limb-lobes triangular, up to 1.5 mm long, acute to rounded. Corolla-tube 2.5–4 mm long, bearded at throat; lobes oblong-lanceolate, 4–7 mm long, acute. Ovary 2-locular, with single ovule in each locule. Fruits subglobose or 2-lobed, 5–5.5 mm in diameter. Seeds 2, ca. 4 mm long.

Distribution: Coastal Kenya. [Tanzania].

Habitat: Coastal forests, bushlands, or wooded grasslands; up to 500 m.

Kilifi: Arabuko-Sokoke Forest, *SAJIT s.n.* (HIB); Mangea Hill, *Luke & Robertson 268*

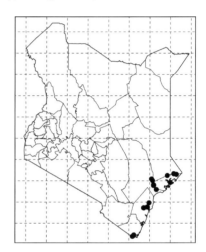

Plate 84 *Kraussia kirkii*. Photo by GWH.

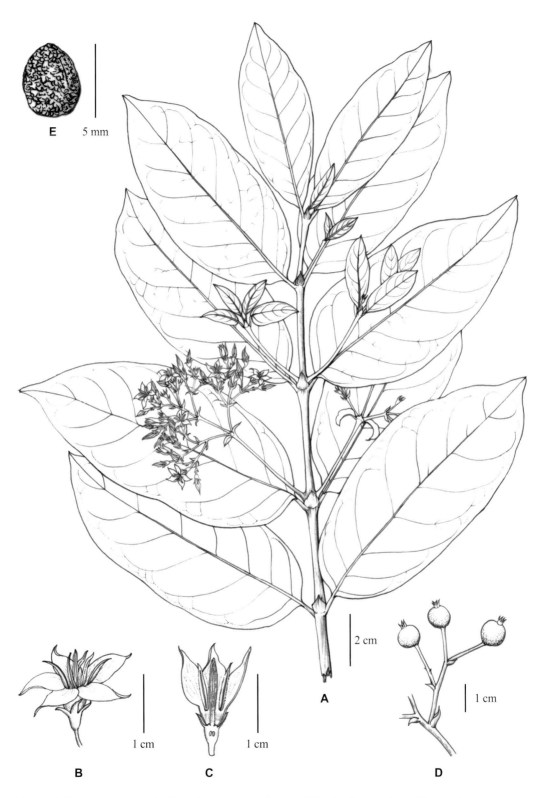

Figure 63 *Kraussia speciosa*. A. flowering branch; B. flower; C. longitudinal section of flower, showing stamens, ovary and style; D. infructescence; E. seed. Drawn by NJ.

(EA). Kwale: near Marenji Forest Reserve, *Luke & Luke 9034* (EA); Lungalunga-Vanga Road, *Faden & Faden 77/371* (EA, K). Lamu: Bodhei to Basuba, *Festo & Luke 2680* (EA). Tana River: Tana River National Primate Reserve, *Medley 401* (EA); Wema Forest, *Robertson & Gafo 6594* (EA, K).

2. **Kraussia speciosa** Bullock, Bull. Misc. Inform. Kew 1931: 256. 1931; F.T.E.A. Rubiac. 2: 701. 1988; K.T.S.L.: 519. 1994. —Type: Tanzania, Morogoro, Kimboza, Mikese-Kisaki road, 4 Sept. 1930, *P.J. Greenway 2516* [holotype: K (K000353158); isotypes: EA (EA000001688, EA000001689), K (K000353159)] (Figure 63)

Shrub or small tree, up to 9 m tall. Leaf-blades chartaceous, glabrous, oblong-elliptic, 8–18(–21) cm × 3–8 cm, acute to acuminate at the apex, rounded at the base; domatia occasionally present; petioles up 1.2 cm long; stipules triangular to ovate-triangular, up to 1 cm long, acute. Flowers several in a lax, axillary panicle; peduncles up to 5 cm long; pedicels up to 1 cm long. Calyx-tube 1.5–1.8 mm long; limb-lobes triangular, 3–4 mm long, acute. Corolla-tube 4.5–7 mm long; lobes lanceolate to narrowly ovate, 7–13 mm long, acute to acuminate. Ovary 2-locular, with 2–3 ovules in each locule. Fruits rounded or 2-lobed, 8–9 mm in diameter. Seeds 4–6, ca. 5 mm long.

Distribution: Coastal Kenya. [Tanzania].
Habitat: Forests; up to 900 m.

Kilifi: Kombeni River, Valley edge of Kaya Fimboni, *Robertson & Luke 5846B* (EA). Kwale: Shimba Hills, *Magogo & Glover 435* (EA); Base Titanium Dam, *Nyange & Luke 493* (EA). Lamu: Utwani Ndogo Forest, *Rawlins 318* (EA, K). Tana River: Tana River National Primate Reserve, *Luke et al. TPR538* (EA, K).

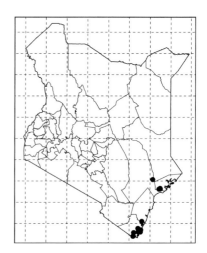

68. **Feretia** Delile

Shrubs or small trees. Leaves opposite; stipules usually ovate. Flowers fragrant, sessile and appearing terminal on short leafless spurs or sometimes pedicellate in pairs from the axils of young leaves, 5-merous; bracts and bracteoles papery when dry. Calyx-tube narrowly ovoid or campanulate; limb-lobes oblong or lanceolate, apiculate. Corolla-tube funnel-shaped; lobes ovate or oblong, obtuse or subapiculate. Stamens inserted at throat; anthers sessile, exserted. Ovary 2-locular, with 2–10 ovules. Fruits globose, fleshy, few-seeded. Seeds somewhat compressed.

A genus of two species restricted to tropical Africa; one species with two subspecies in Kenya.

1. **Feretia apodanthera** Delile, Ann. Sci. Nat., Bot., sér. 2, 20: 92, t. 1, 4. 1843; F.T.E.A. Rubiac. 2: 698. 1988; K.T.S.L.: 512. 1994. —Type: Ethiopia, Djeladjeranne on bank of R. Takazze, *J.G. Galinier s.n.* [holotype: MPU (MPU 07019)]

Shrub or small tree, up to 9 m tall. Leaf-blades elliptic to ovate, 1–8 cm × 0.5–5 cm, obtuse or rounded at the apex, cuneate at the base, glabrous to sparsely pubescent on both sides; domatia present; petioles up to 1.1 cm long; stipules ovate to triangular, 2–5 mm long, acuminate or apiculate. Flowers usually precocious, sessile to shortly pedicellate, on short leafless spurs or in leaf-axils; pedicels up to 3.6 cm long; bracts brown and papery when dry. Calyx-tube up to 2 mm long; lobes linear to lanceolate, up to 5 mm long, acute. Corolla-tube funnel-shaped, up to 1.4 cm long; lobes ovate or oblong, 0.5–1.5 cm long, obtuse to rounded or acute. Fruits red or white with purple streaks, globose, 0.3–1.7 cm in diameter. Seeds flattened, 3–7 mm long.

1a. Leaf-blades acute or less often obtuse at the apex; petioles less than 4 mm long; fruits less than 7 mm in diameter .... a. subsp. *apodanthera*
1b. Leaf-blades obtuse to rounded or sometimes acute at the apex; petioles up to 11 mm long; fruits up to 17 mm in diameter .... b. subsp. *keniensis*

a. subsp. **apodanthera**

Leaf-blades acute or less often obtuse at the apex. Petioles 2–4 mm long. Fruit 3–7 mm in diameter.

Distribution: Western Kenya. [North of Tropical Africa, from Mauritania to Eritrea].
Habitat: Riverine forests; 1000–1100 m.
Nakuru: ca. 14 km southwest of Mukutan, 1070 m, 29 July 1990, *Luke BFFP885* (EA).

b. subsp. **keniensis** Bridson, Kew Bull. 34: 368. 1979; W.F.E.A.: 151. 1987; F.T.E.A. Rubiac. 2: 699. 1988; K.T.S.L.: 519. 1994. —Type: Kenya, Kwale, north of Jadini, 3 Dec. 1959, *P.J. Greenway 9624* [holotype: K (K000319169); isotypes: EA (EA000001690), PRE (PRE0594504-0)] (Figure 64)

Leaf-blades obtuse to rounded or sometimes acute at the apex. Petioles 2–11 mm long; Fruits 5–17 mm in diameter.

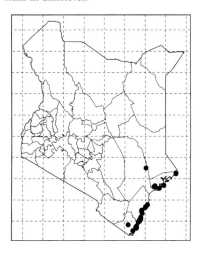

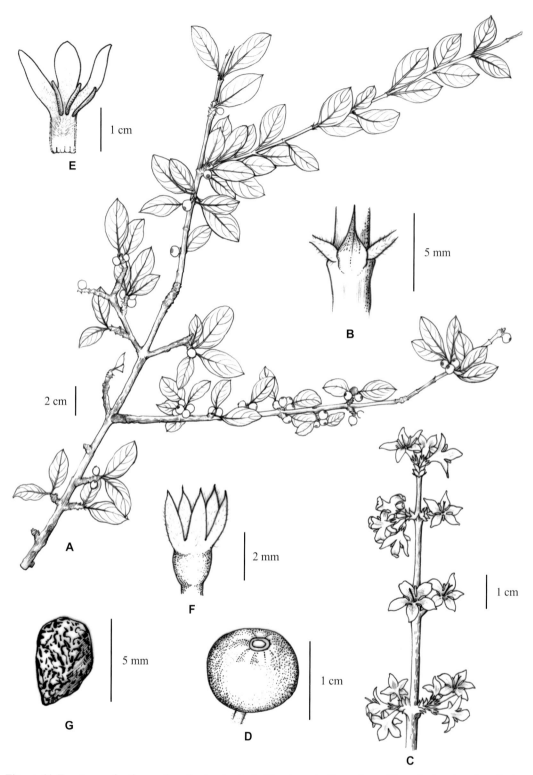

Figure 64 *Feretia apodanthera* subsp. *keniensis*. A. fruiting branch; B. portion of branch showing the stipule; C. a flowering branch; D. ovary and calyx; E. longitudinal section of the corolla, showing the stamens; F. fruit; G. seed. Drawn by NJ.

Distribution: Eastern and coastal Kenya. [Somalia].

Habitat: Coastal forests or bushlands; up to 50 m.

Kilifi: Watamu, *Kimeu et al. KEFRI644* (EA); Gede Forest, *Gerhardt & Steiner 32* (EA); Malindi, *Robertson 4212* (EA, K). Kwale: Diani Forest, *Robertson 4273* (EA, K); Jadini Hotel Forest, *Faden 70/209* (EA, K). Lamu: Witu Forest, *Robertson & Luke 5527* (EA); east edge of Dodori Reserve, *Kuchar 13694* (EA); Kiunga, *Michiri 540* (EA). Mombasa: Mtwapa, *Higgins EA13148* (EA); Bamburi, *Campbell 16138* (EA). Tana River: Kitwa Pembe Hill, *Faden & Faden 74/1106* (EA); Galole, *Makin in EA 14541* (EA).

69. **Polysphaeria** Hook. f.

Shrubs or small trees. Leaf-blades lanceolate to oblong-elliptic or less often rounded; stipules triangular, acuminate, with raised mid-line. Flowers usually sessile in bracteate axillary clusters, 4–5-merous, bracteoles often several per flower and usually joined to form a cup at the base of the calyx. Calyx-tube obconical; limb cup- or bowl-shaped, truncate or 4–5-toothed. Corolla-tube narrowly funnel-shaped or cylindrical, densely hairy at the throat; lobes 4–5, strictly contorted. Stamens inserted in the corolla-tube; anthers sessile or subsessile, included or partly exserted. Ovary 2-locular; ovule single in each locule, pendulous; style slender, pubescent, exserted. Fruits globose, rather tough, 1–2-seeded.

A genus of about 22 species constricted in tropical Africa, Madagascar and Comoros Islands; three species in Kenya.

1a. Calyx in young bud almost or quite enclosing the corolla .... 3. *P. cleistocalyx*
1b. Calyx never enclosing corolla even in young buds 2
2a. Leaf-blades usually narrowly elliptic; calyx usually truncate, glabrous; corolla usually glabrous .... 1. *P. multiflora*
2b. Leaf-blades usually elliptic or ovate; calyx short, usually distinctly toothed; corolla often densely pubescent .... 2. *P. parvifolia*

1. **Polysphaeria multiflora** Hiern, Fl. Trop. Afr. 3: 127. 1877; F.T.E.A. Rubiac. 2: 571. 1988; K.T.S.L.: 532. 1994. —Types: Mozambique, ca. 42 km up the R. Rovuma, 13 Mar. 1861, *J. Kirk s.n.* [lectotype: K (K000419934), **designated here**]; Tanzania, Mafia Is., *B. Frere s.n.* [syntype: K (K000419945)]; Mozambique, ca. 42 km up the R. Rovuma, *J. Kirk s.n.* [syntype: K]

= *Polysphaeria multiflora* Hiern subsp. *pubescens* Verdc., Kew Bull. 35: 127. 1980; F.T.E.A. Rubiac. 2: 573. 1988; K.T.S.L.: 532. 1994. —Type: Kenya, Tana River, Bura, 28 Sept. 1957, *P.J. Greenway 9239* [holotype: K (K000311609); isotypes: EA (EA000001631, EA000001632, EA000001633), PRE (PRE0594413-0)]

= *Polysphaeria lanceolata* Hiern, Fl. Trop. Afr. 3: 128. 1877; F.T.E.A. Rubiac. 2: 574. 1988; K.T.S.L.: 532. 1994. —Types: Mozambique, Shupanga, 2 Apr. 1860, *J. Kirk s.n.* [lectotype: K (K000419941), **designated here**]; ibid., 10 Jan. 1863, *J. Kirk s.n.* [syntype: K (K000419940)]

Shrub, up to 3.5(–4.5) m tall. Leaf-blades glabrous, lanceolate or narrowly elliptic to oblong-elliptic, 2–12 cm × 1–4.5 cm, acuminate to acute at the apex, cuneate to rounded at the base; petioles up 4–10 mm long; stipules triangular, 2.5–5 mm long, keeled. Inflorescences sessile or shortly to distinctly pedunculate, few- to many-flowered, globose; bracteoles cupular, ca. 1 mm long. Calyx shallowly cup-shaped, 1.5–2 mm long, glabrous or pubescent, truncate. Corolla white; tube ca. 4 mm long; lobes ovate, 1.5–2 mm long. Fruits globose, 6–9 mm in diameter, glabrous.

Distribution: Cental, southern, and coastal Kenya. [Aldabra Islands, Comoros, Malawi, Mozambique, Somalia, and Tanzania].

Habitat: Forests, woodlands, or thickets; up to 1400 m.

Kilifi: Pangani, *Robertson & Luke 4526* (EA); Lali Hills, *Adamson 26* (EA); Watamu, *Robertson 232* (EA). Kwale: Mwena River, *Faden et al. 77/382* (EA); Shimoni, *Ossent 707* (EA). Lamu: Baomo South Forest, *Kimberly 349* (EA, K). Machakos: Kindaruma Dam, *Gillet & Faden 18240* (EA). Teita Taveta: Msau River Valley, *Kabuye et al. 660* (EA); Mount Kasigau, *Medley 699* (EA). Tana River: Nairobi Ranch, *Festo & Luke 2370* (EA); Tana Delta, *Leauthaud 115* (EA); Tana River National Primate Reserve, *Luke et al. TPR26* (EA, K). Tharaka-Nithi: Kijegge Hill, *Beentje & Powys 4073* (EA).

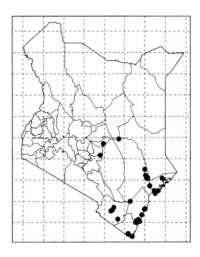

2. **Polysphaeria parvifolia** Hiern, Fl. Trop. Afr. 3: 128. 1877; F.T.E.A. Rubiac. 2: 573. 1988; K.T.S.L.: 532. 1994. —Types: Tanzania, Zanzibar, *J. Kirk 38* [lectotype: K (K000419947)], **designated here**; Tanzania, Zanzibar, Sept. 1873, *J.M. Hildebrandt 1181* [syntype: K (K000419948); isosyntypes: CORD (CORD00004481), L (L0057543), LE (LE00017377)] (Figure 65. Plate 85)

Shrub or small tree, up to 6(–9) m tall. Leaf-blades elliptic to ovate, 0.5–9 cm × 0.5–4.5 cm, subacute to acuminate at the apex, cuneate, truncate, rounded or subcordate at the base, glabrous to pubescent on both sides; petioles 3–5 mm long; stipules ca. 1 mm long, apiculate. Inflorescences sessile or rarely pedunculate, globose; bracts ovate-triangular, ca. 3 mm long. Calyx-tube ca. 0.5 mm long; limb with triangular teeth. Corolla white; tube funnel-shaped, 3–4.5 mm long; lobes ovate, 1.5–2 mm long. Fruits globose, red to purplish red, 7–10 mm in diameter, glabrous or slightly pubescent.

Distribution: Coastal Kenya. [Ethiopia, Somalia, South Sudan, Sudan, and Tanzania].

Habitat: Forests, woodlands, or thickets; up to 1400 m.

Kilifi: Arabuko-Sokoke Forest, *Moggridge 276* (EA); Marafa, *Polhill & Paulo 782* (EA); Kaya Kivara, *Robertson & Luke 4771* (EA). Kwale: Shimba Hills, *SAJIT 005468* (HIB); Kaya Kinondo Forest, *SAJIT 005446* (HIB); Shimoni, *Luke & Mbinda 5855* (EA). Lamu: Witu Forest, *Katende 1748* (EA); North of Hindi, *Kuchar 12834* (EA). Mombasa: Likoni, *Napier 3315* (EA). Tana River: Tana River National Primate Reserve,

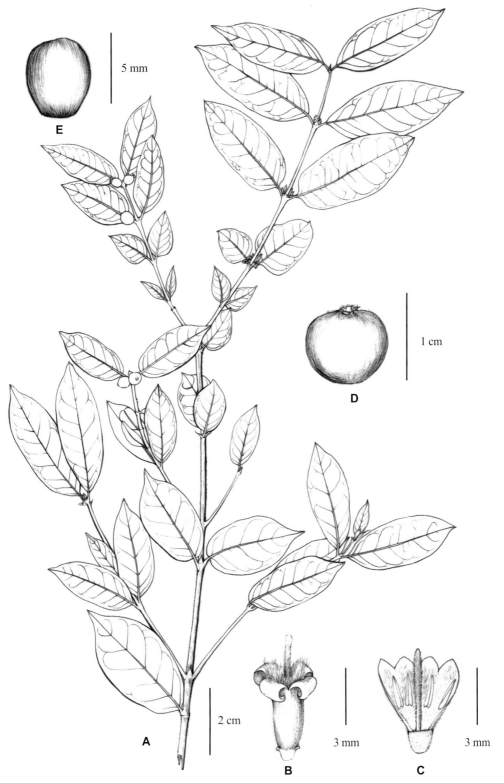

Figure 65 *Polysphaeria parvifolia*. A. a fruiting branch; B. flower; C. longitudinal section of corolla, showing the stamens and style; D. fruit; E. seed. Drawn by NJ.

Plate 85 A–C. *Polysphaeria parvifolia*. Photo by GWH.

Luke et al. TPR552 (EA); North of Wema, *Gillett & Kibuwa 19917* (EA, K).

Calyx in young bud almost or quite enclosing the corolla; tube ca. 1 mm long; limb ca. 4 mm

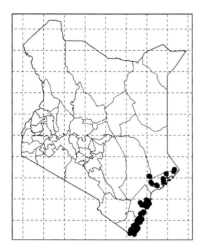

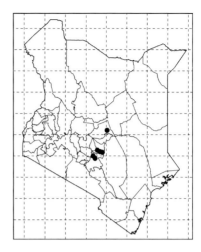

3. **Polysphaeria cleistocalyx** Verdc., Kew Bull. 35: 112. 1980; F.T.E.A. Rubiac. 2: 577. 1988; K.T.S.L.: 532. 1994. —Type: Tanzania, Kilosa, Mikumi Station, Dec. 1898, *W. Goetze 389* [holotype: K (K000316680)]

Shrub, up to 6 m tall, with glabrous stems. Leaf-blades glabrous, lanceolate or narrowly oblong-lanceolate, 5–14 cm × 2–4 cm, acute at the apex, cuneate at the base; petioles 4–8 mm long; stipules triangular, ca. 3 mm long, keeled. Inflorescences sessile or shortly pedunculate, globose; bracts and bracteoles 1.5–2.5 mm long.

long. Corolla cream white; tube funnel-shaped, ca. 4 mm long; lobes oblong-ovate, 2–2.5 mm long. Fruits white to purple-black, subglobose or depressed globose, 7–10 mm in diameter.

Distribution: Central Kenya. [Tanzania].
Habitat: Forests; 700–1200 m.

Embu: Thiba River near Mashamba, *Robertson 1893* (EA, K). Machakos: Mabaloni Hill, *Faden & Evans 70/126* (EA, K). Meru: Meru National Park, *Luke et al. 7297* (EA). Murang'a: Ithanga Hills, *Faden & Evans 69/936* (EA).

19. Trib. **Sherbournieae** Mouly & B.Bremer

Subshrubs, shrubs, or lianas. Stipules interpetiolar, entire. Raphides absent. Inflorescences always pseudo-terminal. Secondary pollen presentation present. Stigma club-shaped or capitate, stigmatic lobes fused over most of their length. Ovary 2-locular, placentas parietal with many ovules. Fruit a fleshy berry with seeds imbedded in the fleshy placenta. Seeds without adaxial excavation, exotestal, with secondary thickenings and folded testa.

Two genera occur in Kenya.

1a. Flowers few to several in subcapitulate cymes; bracts leaf-like, up to 2 cm long .... 70. *Mitriostigma*

1b. Flowers subsessile to pedicellate in axillary, compact to lax racemes or panicles; bracts not leaf-like, less than 1 cm long .... 71. *Oxyanthus*

70. **Mitriostigma** Hochst.

Small shrubs. Leaves opposite, petiolate; domatia absent or present as small hairy tufts; stipules triangular, scarcely connate at the base. Flowers 5-merous, subsessile to shortly pedicellate in small cymes. Calyx-tube turbinate to ellipsoid; limb-tube distinct, lobes narrowly triangular. Corolla white; tube cylindrical at the base, campanulate or funnel-shaped above; lobes erect or spreading. Stamens inserted in broad part of corolla-tube; anthers subsessile. Ovary 2-locular, with 6–12 ovules; style slender; stigmatic club winged. Fruits globose to ellipsoid, crowned by persistent calyx-limb. Seeds fleshy, compressed or not, always with folded testa.

A small genus of five species restricted to tropical and southern Africa; one species in Kenya.

1. **Mitriostigma greenwayi** Bridson, Kew Bull. 34: 127. 1979; F.T.E.A. Rubiac. 2: 539. 1988. —Type: Kenya, Kwale, Jadini, 7 Dec. 1959, *P.J. Greenway 9639* [holotype: K (K000311616); isotypes: EA (EA000001650, EA000001651), PRE] (Figure 66; Plate 86)

Small shrub, up to 1 m tall. Leaf-blades coriaceous, glabrous, obovate, 7.5–22 cm × 3.5–10.5 cm, obtuse at the apex, obtuse or cuneate at the base; domatia absent; petioles up to 1 cm long; stipules triangular, up to 1.5 cm long, acute or acuminate. Flower 5-merous, few to 15 in subcapitulate cymes; bracts leaf-like, up to 2 cm long; peduncles up to 1 cm long. Calyx-tube ovate, 2–3 mm long; limb-tube ca. 1 mm long, lobes narrowly triangular, acute, up to 2 mm long. Corolla white; tube slender, up to 2 cm long; lobes narrowly obovate, up to 1.1 cm long. Fruits black, globose or slightly fusiform, 2–2.5 cm long, crowned by persistent calyx-limb. Seeds red-black, ca. 4 mm in diameter.

Distribution: Coastal Kenya. [Endemic].

Habitat: Lowland and coastal forests; up to 250 m.

Kilifi: Kaya Kambe, *Luke & Robertson 4788* (EA); Pangani, *Faden et al. 77/531* (EA). Kwale: Kaya Waa, *SAJIT 005581* (HIB); Kaya Diani, *Luke & Luke 9019* (EA).

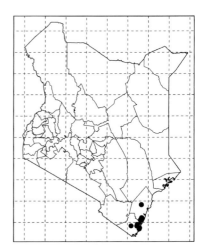

Plate 86 *Mitriostigma greenwayi*. Photo by GWH.

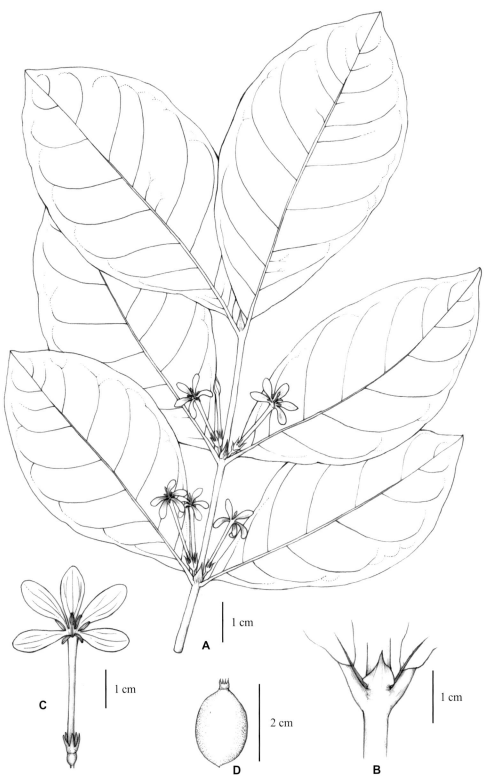

Figure 66 *Mitriostigma greenwayi*. A. a flowering branch; B. a portion of branch showing the stipule; C. flower; D. fruit. Drawn by NJ.

71. **Oxanthus** DC.

Shrubs, trees, or rarely lianas. Leaves opposite, shortly petiolate; domatia often present as hairy tufts; stipules ovate, persistent. Flowers 5-merous, subsessile to pedicellate in axillary, compact to lax racemes or panicles. Calyx-tube campanulate or turbinate-oblong; limb with distinct tube and 5 narrow lobes. Corolla white; tube slender, cylindrical; lobes 5, narrow, contorted, lanceolate to elliptic, spreading or reflexed. Stamens 5, inserted at the throat of corolla-tube; anthers sessile or subsessile, linear, exserted. Ovary (1–)2-locular with numerous ovules; style very slender, usually exserted. Fruit a berry, globose, ellipsoid or fusiform, with numerous seeds.

A genus of about 34 species restricted to tropical Africa; four species in Kenya.

1a. Inflorescences of a lax or compact panicle or cyme, 10–many-flowered 2
1b. Inflorescences of a compact or narrowly compact panicle, with less than 11 flowers 3
2a. Bracteoles conspicuous, narrowly lanceolate, 1.2–9 mm long; calyx-tube 1.5–2 mm long .... 1. *O. speciosus*
2b. Bracteoles inconspicuous, subulate, less than 2 mm long; calyx-tube 2–3 mm long .... 4. *O. pyriformis*
3a. Inland species; calyx-limb with lobes 2–3.5(–5) mm long; fruits ellipsoid to fusiform .... 2. *O. goetzei*
3b. Coastal species; calyx-limb with lobes (4–)5–8 mm long; fruits pyriform ... 3. *O. zanguebaricus*

1. **Oxyanthus speciosus** DC., Ann. Mus. Natl. Hist. Nat. 9: 218. 1807; F.T.E.A. Rubiac. 2: 527. 1988; K.T.S.L.: 524. 1994. —Type: Sierra Leone, *H. Smeathman s.n.* [lectotype: G-DC (G00666298), **designated here**; isolectotypes: BM (BM0009 03169, BM000903170), G (G00665913, G00665 929, G00665930, G00665931, G00665932, G00666 297, G00666298), TUB (TUB004541)]

Shrub or small tree, up to 6(–12) m tall. Leaf-blades elliptic to oblong, 7–25.5 cm × 3–12.5 cm, acuminate at the apex, cuneate to rounded at the base; petioles up to 1.5 cm long; stipules triangular or lanceolate, up to 2 cm long, acute. Inflorescences many-flowered, of a lax panicle; peduncles up to 1 cm long; bracts and bracteoles narrowly lanceolate, 1.2–9 mm long. Calyx-tube 1.5–2 mm long; limb-tube 0.7–1.8 mm long, lobes triangular 1–2.5 mm long, acute. Corolla-tube 1.8–7 mm long; lobes oblong-lanceolate, 0.7–1.8 cm long, acute. Fruits globose, ellipsoid to fusiform, 1.7–6.5 cm long, with persistent calyx-limb.

subsp. **stenocarpus** (K. Schum.) Bridson, Kew Bull. 34: 116. 1979; F.T.E.A. Rubiac. 2: 529. 1988; K.T.S.L.: 524. 1994. ≡ *Oxyanthus stenocarpus* K. Schum., Bot. Jahrb. Syst. 33: 345. 1903. —Type: Tanzania, E. Usambara Mts., Gonja, Sept. 1893, *C. Holst 4265* [lectotype: K (K000419887), designated by B. Sonké and E. Robbrecht in Bull. Jard. Bot. Nat. Belg. 65: 128. 1996; isolectotypes: HBG (HBG521359), KFTA (KFTA0000812), LECB (LECB0001543)] (Figure 67; Plate 87 A, B)

Fruits always fusiform, (2.1–)3–6.5 cm long.

Distribution: Northern, western, cental, southern, and coastal Kenya. [Eastern Africa from Ethiopia to northern South Africa].

Habitat: Forests; 1400–2300 m.

Note: There are four subspecies recorded in the species *Oxyanthus speciosus* DC.: subsp. *speciosus* occurs in tropical Africa, from West Africa to Tanzania, also in Ethiopia; subsp. *gerrardii* (Sond.) Bridson occurs in South Africa; subsp. *mollis* (Hutch.) Bridson occurs in central Africa from Tanzania to D.R. Congo and Angola; subsp. *stenocarpus* (K. Schum.) Bridson occurs in eastern Africa and also in Kenya..

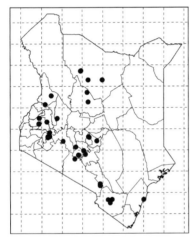

Baringo: Kabarnet, *Dale 2442* (EA, K). Embu: Kathita-Kiangombe, *Riley 776264* (EA). Kajiado: Ngong Forest, *Khayota 59* (EA). Kakamega: vicinity of Kakamega Forest, *Perdue & Kibuwa 9442* (EA, K). Kericho: Tinderet Forest Reserve, *Geesteranus 5364* (K). Kiambu: Kieni Forest Station, *Mulwa 117* (EA). Kilifi: Gede Forest, *Nash 7* (EA). Kirinyaga: Castle Forest Station, *Spjut & Ensor 3024* (EA, K). Makueni: Chyulu Hills, *Luke 851* (EA). Marsabit: Mount Kulal, *Synnott 1776* (EA). Nairobi: Karura Forest Station, *Muchai 135* (EA, K). Nakuru: Mau Forest, *Mutangah & Kamau 51* (EA). Nandi: *Wye 1842* (K). Samburu: Mathew's Range, *Ichikwa 858* (EA). Teita Taveta: Ngangao Forest, *SAJIT 004521* (HIB). Trans-Nzoia: Kapolet, *Tweedie 2710* (K). West Pokot: Cherangani Hills, *SAJIT 006898* (HIB).

2. **Oxyanthus goetzei** K. Schum., Bot. Jahrb. Syst. 28: 491. 1900; F.T.E.A. Rubiac. 2: 531. 1988; K.T.S.L.: 523. 1994. —Type: Tanzania, Iringa, Lofia River, 1898, *W. Goetze 445* [holotype: B (destoroyed); lectotype: K (K000419885), **designated here**]

Shrub, up to 4.5 m tall. Leaf-blades coriaceous, glabrous, elliptic to broadly elliptic, 5–19 cm × 2.5–8 cm, acute to acuminate at the apex, acute to obtuse at the base; petioles up to 1.2 cm long; stipules narrowly triangular, up to 3.5 cm long, acute. Inflorescences 1–8-flowered, compact; bracts and bracteoles lanceolate, 4–9 mm long. Calyx-tube 2–5 mm long; limb-tube 1.7–2.5 mm long, lobes 2–3.5(–5) mm long. Corolla white to yellowish; tube 5.8–11.2 cm long; lobes oblong, up to 3 cm long, acuminate. Fruits brown, ellipsoid to fusiform, 3–4.5 cm long, with persistent calyx-limb.

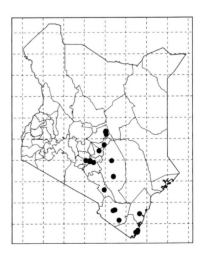

subsp. **keniensis** Bridson, Kew Bull. 34: 120.

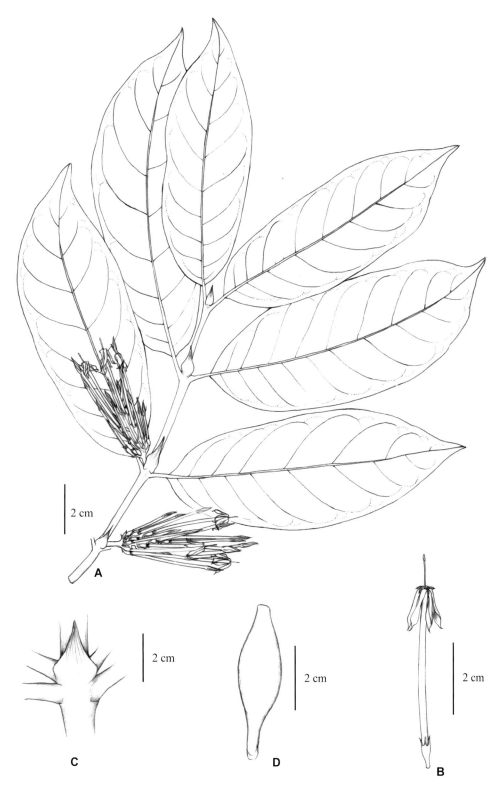

Figure 67 *Oxyanthus speciosus* subsp. *stenocarpus*. A. a flowering branch; B. a portion of branch showing the stipule; C. flower; D. fruit. Drawn by NJ.

1979; F.T.E.A. Rubiac. 2: 531. 1988; K.T.S.L.: 523. 1994. —Type: Kenya, Teita Taveta, Mount Kasigau, 600 m, Oct. 1938, *B. Joanna in C.M. 8883* [holotype: K (K000419885); isotype: EA (EA000001659)] (Plate 87C)

Stipules 0.5–1.6(–2.4) cm long. Bracteoles 4–6 mm long. Calyx-tube 2–3 mm long.

Distribution: Central and coastal Kenya. [Tanzania].

Habitat: Forests; up to 1700 m.

Note: *Oxyanthus goetzei* subsp. *goetzei* has longer bracteoles and calyx-tube and occurs in eastern Africa from Tanzania to Zimbabwe and Mozambique.

Embu: Kiangombe Mount, *Smith et al. 282* (K). Kiambu: Thika Falls, *Napier et al. 2366* (EA, K). Kilifi: west of Ganze, *Reitsma et al. 200* (K). Kitui: Nuu Hill, *Mwachala et al. 545* (EA). Kwale: Gongoni Forest, *Chidzinga & Luke 009* (EA); Kaya Kinodo, *Malombe et al. K11646* (EA). Machakos: Mabaloni Hill, *Faden & Evans 70/129* (EA). Makueni: Kibwezi, *Scheffler 388* (K). Meru: Ngaya Forest, *SAJIT Z0219* (HIB). Teita Taveta: Mount Kasigau, *Joana in C.M. 8883* (EA); Mbale, above Youth Polytechnic, *Mwachala 111* (EA). Tharaka-Nithi: Kijegge Hill, *Beentje & Powys 4087* (EA, K).

3. Oxyanthus zanguebaricus (Hiern) Bridson, Kew Bull. 34: 119. 1979; F.T.E.A. Rubiac. 2: 530. 1988; K.T.S.L.: 524. 1994. ≡ *Gardenia zanguebarica* Hiern, Fl. Trop. Afr. 3: 105. 1877. —Type: Tanzania, Bagamoyo, Ruvu (Kingani), May 1874, *J.M. Hildebrandt 1268* [holotype: K (K000419886); isotypes: BM, CORD (CORD00004464)] (Plate 87D–F)

Shrub or small tree, up to 8.5 m tall. Leaf-blades coriaceous, glabrous, elliptic to narrowly ovate, 6–19 cm × 2–7 cm, acute or shortly acuminate at the apex, obtuse or cuneate at the base; petiole up to 1 cm long; stipules narrowly to broadly triangular, up to 1.1 cm long. Inflorescences (3–)5–11-flowered, compact; bracts and bracteoles lanceolate, up to 1(–1.2) cm long. Calyx-tube 2.5–3 mm long; limb-tube ca. 2 mm long, lobes linear-subulate, 4–8 mm long. Corolla-tube 7–11.5 cm long; lobes oblong-lanceolate, 1–2.2 cm long, acuminate. Fruit ovoid, 2.5–5.5 cm long.

Distribution: Coastal Kenya. [Coastal areas of East Africa, from Somalia to Mozambique].

Habitat: Coastal forests, woodlands, or bushlands; up to 500 m.

Garissa: Boni Reserve, *Kuchar 13575* (EA). Kilifi: Arabuko-Sokoke Forest, *SAJIT 006457* (HIB); Kaya Rabai, *Luke & Robertson 2282* (EA); Mida, *Donald 453* (EA). Kwale: Shimba Hills, *SAJIT V0214* (HIB); Marenji Forest Reserve, *Luke & Robertson 1743* (EA); Shimoni, *Luke & Robertson 2354* (EA). Lamu: Witu Forest, *Robertson & Luke 5482* (EA, K). Tana River: Nairobi Ranch, *Festo & Luke 2506* (EA).

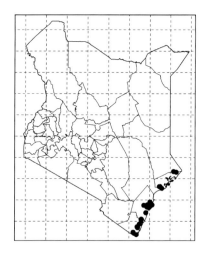

4. Oxyanthus pyriformis (Hochst.) Skeels, U.S.D.A. Bur. Pl. Industr. Bull. 248: 56. 1912; F.T.E.A. Rubiac. 2: 534. 1988; K.T.S.L.: 524. 1994. ≡ *Megacarpha pyriformis* Hochst., Flora 27:

551. 1844. —Type: South Africa, Port Natal, Jule 1839, *C.F.F. Krauss 110* [holotype: B (destroyed); lectotype: K (K000419890), **designated here**; isolectotypes: K (K000419889), TUB (TUB004539, TUB004540)]

Shrub or small tree, up to 10 m tall. Leaf-blade coriaceous, glabrous, ovate to broadly elliptic, 8–30 cm × 3.5–17 cm, obtuse or shortly acuminate at the apex, cuneate to truncate or rounded and usually asymmetric at the base; petioles up to 1.5 cm long; stipules ovate to triangular, up to 2.8 cm long, acute to acuminate. Inflorescence a lax 10–30-flowered cyme, with a short peduncle; bracteoles inconspicuous. Calyx-tube ovoid, 2–3 mm long; limb-tube up to 2.5 mm long, lobes filiform to subulate, up to 4 mm long. Corolla-tube 3.4–13.6 cm long; lobes oblong to narrowly obovate, 1–2.6 cm long, acute to acuminate. Fruits green, obovoid to fusiform, 1.8–4 cm long, without calyx-limb.

1a. Inland subspecies; corolla-tube less than 4 cm long; calyx-lobes 1–2 mm long........................
....a. subsp. *brevitubus*
1b. Coastal subspecies; corolla-tube more than 11 cm long; calyx-lobes (1–)2–4 mm long ....
....b. subsp. *longitubus*

a. subsp. **brevitubus** Bridson, Kew Bull. 34: 126. 1979; F.T.E.A. Rubiac. 2: 536. 1988; K.T.S.L.: 524. 1994. —Type: Kenya, Kajiado, Emali Forest, Aug. 1940, *V.G.L. van Someren 92* [holotype: K (K000311620); isotypes: EA (EA000001653, EA000001654, EA000001655), K (K000311621, K000311622)]

Corolla-tube 3.4–4 cm long. Calyx-lobes 1–2 mm long.

Distribution: Central and southern Kenya. [Tanzania].

Habitat: Montane forests; 1200–1800 m.
Kajiado: Emali Hill, *Luke & Robertson 119* (EA). Kitui: Mutito Hill, *Gardner K3616* (EA, K). Teita Taveta: Mount Kasigau, *Medley 666* (EA).

b. subsp. **longitubus** Bridson, Kew Bull. 34: 126. 1979; F.T.E.A. Rubiac. 2: 536. 1988; K.T.S.L.: 524. 1994. —Type: Kenya, Kwale, Shimba Hills, Mwele Mdogo Forest, 23 Aug. 1953, *R.B. Drummond & J.H. Hemsley 3961* [holotype: K; isotypes: BR (BR0000008849670), EA

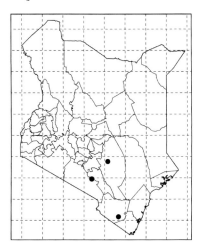
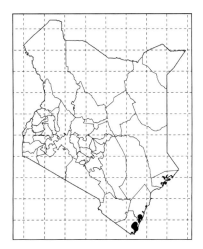

Plate 87 A, B. *Oxyanthus speciosus* subsp. *stenocarpus*; C. *O. goetzei* subsp. *keniensis*; D–F. *O. zanguebaricus*; G–I. *O. pyriformis* subsp. *longitubus*. Photo by SWW (A), GWH (B, D, F, I), YDZ (C), VMN (E) and BL (G, H).

(EA000001652)] (Plate 87G–I)

Corolla-tube 11–13.6 cm long. Calyx-lobes (1–)2–4 mm long.

Distribution: Coastal Kenya. [Endemic].
Habitat: Coastal forests; up to 500 m.
Kilifi: Pangani Rocks, *Luke 1839* (EA, K).
Kwale: Shimba Hills, *SAJIT 006151* (HIB); Gongoni Forest, *Luke & Robertson 2375* (EA, K).

20. Trib. **Pavetteae** A. Rich. ex Dumort.

Shrubs, small trees, or lianas. Stipules interpetiolar, entire or rarely fimbriate. Raphides absent. Inflorescences terminal or sometimes pseudo-axillary. Aestivation contorted to the left. Secondary pollen presentation usually present. Stigmatic lobes fused over most of their length. Ovary with two locular, axile placentas with one to many ovules. Fruits fleshy. Seeds with adaxial excavation. Embryo-radicle inferior or lateral. Endosperm entire or slightly to deeply ruminate. Seeds exotestal, with thickenings (as a continuous plate) along outer tangential wall, or absent. Pollen grains in monads, 3–4(–5)-colporate.

Seven genera occur in Kenya.

1a. Scandent shrubs or lianas 2
1b. Shrubs or small trees 3
2a. Stems 4-angled, with recurved spines .... 74. *Cladoceras*
2b. Stems not 4-angled, unarmed .... 72. *Rutidea*
3a. Ovule single in each locule .... 75. *Pavetta*
3b. Ovules usually more than one, sometimes many in each locule 4
4a. Corolla-tube over 1 cm long .... 73. *Leptactina*
4b. Corolla-tube usually less than 6 mm long 5
5a. Flowers 5–6-merous, in sessile few-flowered clusters at the ends of short shoots; ovules 3(–4) in each locule .... 76. *Tennantia*
5b. Flowers (4–)5-merous, in terminal or pseudoaxillary, few- to many-flowered corymbose, sessile or pedunculate; ovules 1 to many in each locule 6
6a. Fruits never crowned by the persistent calyx-limb; seed 1(–2), reticulate ..... 78. *Coptosperma*
6b. Fruits mostly crowned by the persistent calyx-limb; seeds (1–)several to numerous, not reticulate .... 77. *Tarenna*

72. **Rutidea** DC.

Scandent shrubs, with hairy stems and opposite lateral branches. Leaves opposite, shortly petiolate; domatia absent or present as hairy tufts; stipules with broad base and several fimbriae or a single lobe. Flowers 4–5(–6)-merous, sessile or pedicellate in sessile or pedunculate panicles;

bracts and bracteoles present. Calyx-tube ovoid to campanulate; limb-tube short, lobes ovate subulate or filiform. Corolla white or cream, salver-shaped; tube cylindrical to funnel-shaped; lobes contorted dextrorsely in the bud, spreading or reflexed. Stamens inserted at the mouth of the corolla, exserted; filaments short; anthers oblong, apiculate. Ovary 2(–3)-locular; ovule solitary in each locule; style slender; stigma fusiform clavate or globose, entire or rarely 2–3-lobed. Fruit a drupe, spherical, 1-locular, usually with persistent calyx-limb. Seed 1, globose.

A genus of about 21 species restricted to tropical Africa; three species in Kenya.

1a. Stipules with (3–)5–9 fimbriae .... 1. *R. orientalis*
1b. Stipules with a single linear to subulate lobe 2
2a. Flowers 5-merous; stigma fusiform to clavate .... 2. *R. smithii*
2b. Flowers 4-merous; stigma ellipsoid to broadly ellipsoid .... 3. *R. fuscescens*

1. **Rutidea orientalis** Bridson, Kew Bull. 33: 253. 1978; F.T.E.A. Rubiac. 2: 606. 1988; K.T.S.L.: 542. 1994. —Type: Tanzania, Mpanda, Kungwe-Mahali Peninsula, near head of Ntali R., below Mt. Kungwe, 9 Sept. 1959, *R.M. Harley 9582* [holotype: K (K000412153); isotypes: BR (BR0000008856081), EA (EA000001684), K (K000412154)] (Figure 68; Plate 88A–D)

Scandent shrub or liana, up to 6 m tall; young branches usually densely covered with golden to rust-coloured hairs. Leaf-blades elliptic to oblanceolate, 4–15.5 cm × 2–7(–8.5) cm, acuminate at the apex, obtuse to rounded and usually asymmetrically at the base, sparsely strigose above, sparsely to densely hairy beneath; petioles up to 1.8 cm long, pubescent; stipules up to 1.5 cm long, with (3–)5–9(–12) fimbriae, hairy outside. Flowers 5-merous, several in terminal dense panicles. Calyx-tube up to 1 mm long; limb-tube 0.2–1 mm long, lobes narrowly triangular, up to 4 mm long. Corolla-tube up to 9 mm long, lobes ovate to rounded, up to 3.5 mm long. Stigma well-exserted, fusiform. Fruits yellow to red, globose, 5–9 mm in diameter. Seed 1, globose, 4–5 mm in diameter.

Distribution: Western Kenya. [Eastern and southern Africa from Kenya to Zimbabwe].

Habitat: Forest edges or thickets; 800–2300 m.

Bomet: Sotik, *Bally B4645* (EA). Kakamega: Kakamega Forest, *SAJIT 006744* (HIB). Kericho: southwest Mau Forest, *Kerfoot 2888* (EA, K). Nakuru: Endabarra, *Bally B4988* (EA, K). Nandi: North Nandi Forest, *SAJIT 006634* (HIB). Trans Nzoia: Cherangani Hills, *SAJIT 006891* (HIB).

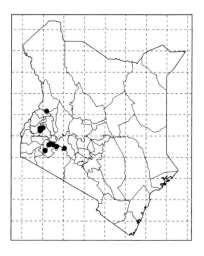

2. **Rutidea smithii** Hiern, Fl. Trop. Afr. 3: 189. 1877; F.T.E.A. Rubiac. 2: 608. 1988; K.T.S.L.: 542. 1994. —Type: D.R. Congo, *C. Smith s.n.* [holotype: K (K000412161)] (Plate 88E–G)

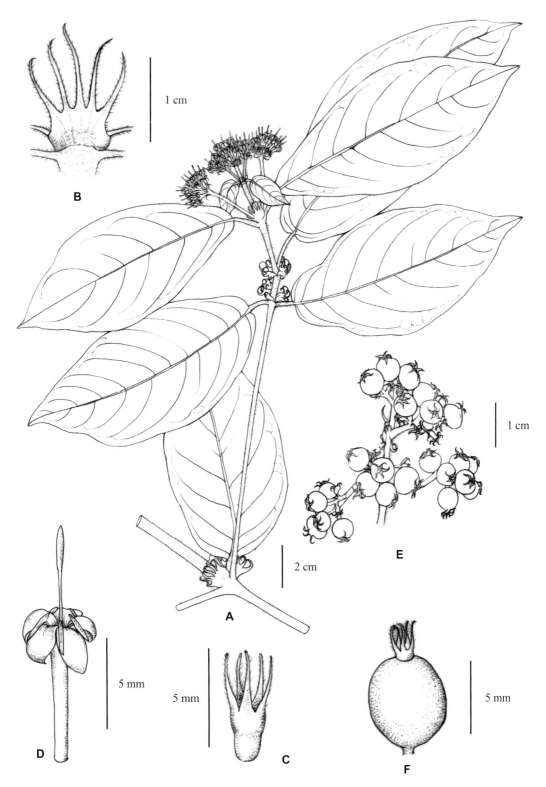

Figure 68 *Rutidea orientalis*. A. flowering branch; B. stipule; C. calyx; D. corolla; E. infructescence; F. fruit. Drawn by NJ.

Plate 88 A–D. *Rutidea orientalis*; E–G. *R. smithii*. Photo by YDZ (A, B) and GWH (C–G).

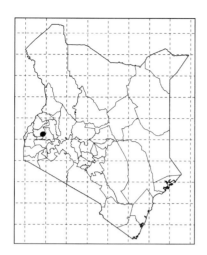

Shrub or liana, up to 7 m tall, with pubescent young branches. Leaf-blades elliptic to obovate, 3–15 cm × 1.5–9 cm, acuminate at the apex, cuneate, obtuse, rounded or subcordate at the base, glabrous to sparsely pubescent above and beneath; petioles up to 3 cm long; stipules oblong to triangular, up to 2(–4) mm long, with a subulate lobe up to 1.2 cm long. Flowers 5(–6)-merous, in dense or lax panicles. Calyx-tube up to 1 mm long; limb-lobes ovate. Corolla-tube up to 5 mm long; lobes ovate, up to 3 mm long, rounded. Stigma fusiform, well exserted. Fruits yellow to orange, globose, 5–7 mm in diameter.

Habitat: Forest edges; 1000–2000 m.

Distribution: Western Kenya. [Tropical Africa].

Kakamega: Kakamega Forest, *Gilbert 6873* (EA).

3. **Rutidea fuscescens** Hiern, Fl. Trop. Afr. 3: 191. 1877; F.T.E.A. Rubiac. 2: 609. 1988; K.T.S.L.: 542. 1994. —Type: Mozambique, Moramballa, 30 Dec. 1858, *J. Kirk s.n.* [holotype: K (K000412145)]

Shrub or liana, up to 9 m tall, with puberulous to finely pubescent young branches. Leaf-blades elliptic to ovate, 2–11(–16) × 1.5–7 cm, acute to acuminate at the apex, obtuse to rounded at the base, glabrous or sparsely pubescent on both sides; petioles up to 1.5(–3.5) cm long; stipules oblong, 1–2 mm long, with a linear to subulate lobe up to 1.2 cm long. Flowers 4-merous, in compact panicles up to 5 cm in diameter. Calyx-tube up to 1 mm long; limb-tube very short, truncate or slightly lobed. Corolla-tube 4–9 mm long; lobes broadly ovate, up to 3 mm long, rounded. Stigma ellipsoid to broadly ellipsoid, exserted. Fruits yellow to red, globose, 4.5–6 mm in diameter.

1a. Bracteoles always absent from the pedicels; pedicels 1–3 mm long .... a. subsp. *fuscescens*
1b. Bracteoles present on the pedicels; pedicels less than 1.5 mm long .... b. subsp. *bracteata*

a. subsp. **fuscescens**

Bracteoles always absent from the pedicels. Pedicels 1–3 mm long.

Distribution: Coastal Kenya. [Tropical East Africa].

Habitat: Coastal forest edges; 350–400 m.
Kwale: Shimba Hills, *Luke 1626* (EA, K).

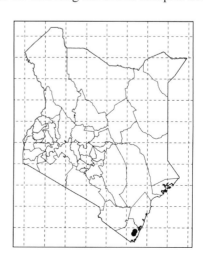

b. subsp. **bracteata** Bridson, Kew Bull. 33: 276. 1987; F.T.E.A. Rubiac. 2: 610. 1988; K.T.S.L.: 542. 1994. —Type: Uganda, Mengo, near Entebbe, May 1935, *P. Chandler 1224* [holotype: K (K000412149); isotypes: BR (BR0000008856050), EA (EA000001685), K (K000412150), MHU (MHU000021)]

Siaya: Yala Town, *Kokwaro 1780* (EA).

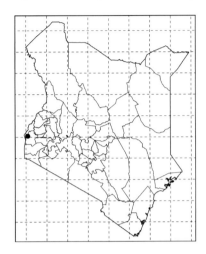

Bracteoles present on the pedicels. Pedicels 1–1.5 mm long.

Distribution: Western Kenya. [East Africa, from Kenya to Zimbabwe].

Habitat: Inland forest edges; 1200–1300 m.

73. **Leptactina** Hook. f.

Shrubs or small trees. Leaves opposite, petiolate, sometimes with domatia; stipules usually conspicuous. Flowers bisexual, 4–6-merous in cymes, usually terminal on lateral shoots. Calyx-tube ellipsoid or obconic; lobes large, leaf-like, (sub-)equal, persistent. Corolla-tube very long, narrowly cylindrical; lobes linear, lanceolate to elliptic, long, pointed, spreading, contorted in bud. Stamens sessile. Ovary 2-locular, with numerous ovules; stigma 2-lobed, lobes linear, slightly exserted. Fruits subglobose or oblong to oblong-conic, often longitudinally ribbed, numerous-seeded.

A genus of 28 species restricted to tropical Africa; only one species in Kenya.

1. **Leptactina platyphylla** (Hiern) Wernham, Bot. Jahrb. Syst. 51: 278. 1913; F.T.E.A. Rubiac. 2: 689. 1988; K.T.S.L.: 520. 1994. ≡ *Mussaenda platyphylla* Hiern, Fl. Trop. Afr. 3: 70. 1877. —Type: D.R. Congo, Monbuttu-land, Bongwa's village, 15 Apr. 1870, *G. Schweinfurth 3626* [holotype: K (K000411784); isotype: BM (BM000528539)] (Figure 69; Plate 89)

Shrub or small tree, up to 7.5 m tall. Leaf-blades elliptic, 7–30(–38) cm × 4–15 cm, acuminate at the apex, rounded to cuneate at the base, glabrous to sparsely or densely pubescent on both sides; petioles up to 2 cm long; stipules broad, leaf-like, up to 2.3(–3) cm long, reflexed. Flowers white, 5–6-merous, in

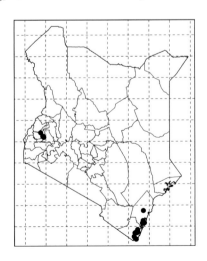

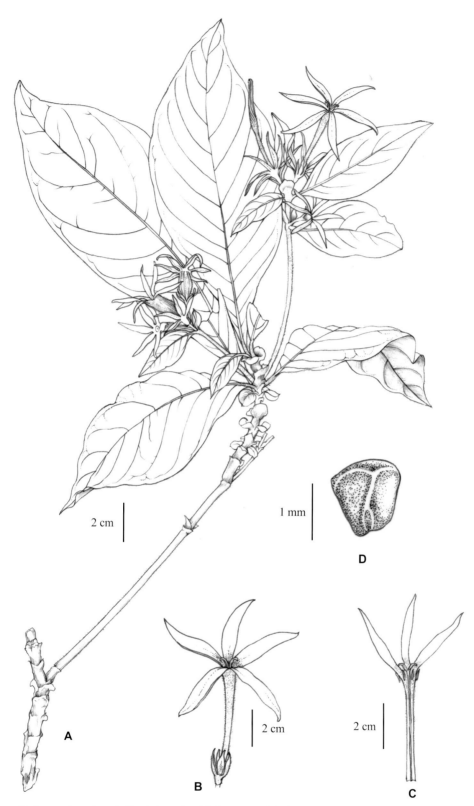

Figure 69 *Leptactina platyphylla*. A. branch with flowers and fruits; B. flower; C. longitudinal section of corolla, showing the stamens and style; D. seed. Drawn by NJ.

Plate 89 A–C. *Leptactina platyphylla*. Photo by BL (A, C) and ZXZ (B).

dense cymes, terminal on lateral shoots, with peduncles up to 2 cm long. Calyx-tube 2–5 mm long; lobes lanceolate, equal, leaf-like, up to 2 cm long, acute. Corolla-tube narrowly long, up to 11 cm long; lobes lanceolate, up to 5.2 cm long, acute. Fruits ellipsoid, 1–2 cm long, longitudinally ribbed.

Distribution: Western and coastal Kenya. [Central, eastern and southern Africa, from Cameroon to Kenya and Mozambique].

Habitat: Forests, woodlands, or secondary bushlands; up to 1700 m.

Kakamega: Kakamega Forest, *Kokwaro 4310* (EA). Kilifi: Kambe Kaya, *Hawthorne 409* (EA); Chasimba Kaloleni, *Faden et al. 70/948* (EA); Mangea Hill, *Luke & Robertson 274* (EA). Kwale: Shimba Hills, *SAJIT V0146* (HIB); Mwagandi Forest, *Magogo & Glover 1145* (EA).

74. **Cladoceras** Bremek.

Scandent shrubs with quadrangular stems; some axillary branchlets reduced to recurved spines. Leaves opposite, shortly petiolate; stipules connate into a short sheath at the base. Flowers white, 5-merous, scented, in dense corymbs or subcapitate, terminal on short axillary branchlets. Calyx-tube ovoid; lobes ovate-triangular. Corolla-tube narrowly cylindrical; lobes lanceolate to elliptic, obtuse. Stamens inserted at the throat; anthers subsessile, exserted. Style filiform; stigma 2-lobed, lobes lanceolate, flattened, included. Ovary 2-locular; ovules 4–5 in each locule. Fruits globose, crowned by the persistent calyx-limb.

A monotypic genus restricted to coastal regions of East Africa including Kenya.

1. **Cladoceras subcapitatum** (K. Schum. & K. Krause) Bremek., Hooker's Icon. P. 35: t. 3411. 1940; F.T.E.A. Rubiac. 2: 605. 1988; K.T.S.L.: 509. 1994. ≡ *Chomelia subcapitata* K. Schum. & K. Krause, Bot. Jahrb. Syst. 39: 525. 1907. —Types: Tanzania, Rufiji, Mafia I., 26 Nov. 1900, *W. Busse 426* [syntype: B (destroyed); lectotype: K (K000353086), **designated here**; isolectotype: EA (EA000001687)]; Tanzania, Uzaramo, near Dar es Salaam, Mogo Forest, *A. Engler 3241 & F.L. Stuhlmann 155* [syntype: B (destroyed)] (Figure 70)

Scandent shrub, up to 2 m tall, with quadrangular stems. Leaf-blades glabrous, elliptic, 5–12 cm × 2–3.5 cm, acute to shortly acuminate at the apex, cuneate at the base; petioles 2–4.5 mm long; stipules 2.5–3.5 mm long. Flowers white, 9–15 in dense corymbs, terminal on axillary branchlets up to 13 cm long. Calyx-tube up to 1.5 mm long; lobes ca. 2.5 mm long. Corolla-tube 2.5–3.5 cm long; lobes ovate, ca. 7 mm long. Style up to 7.5 mm long. Fruits globose ca. 8 mm in diameter.

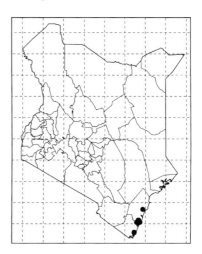

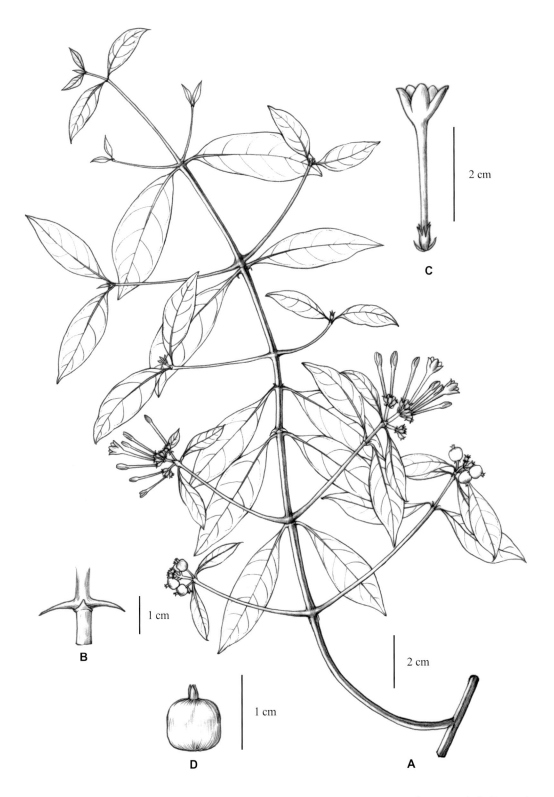

Figure 70 *Cladoceras subcapitatum*. A. branch with flowers and fruits; B. spines; C. flower; D. fruit. Drawn by NJ.

Distribution: Coastal Kenya. [Tanzania].
Habitat: Coastal forests or bushlands; up to 900 m.

Kilifi: Arabuko-Sokoke Forest, *Robertson et al. 6882* (EA). Kwale: Gongoni Forest, *Luke 2424* (EA).

75. **Pavetta** L.

Unarmed shrubs or small trees. Raphides absent. Leaves opposite or rarely whorled, sometimes with bacterial nodules scattered on the blade or arranged along the mid-rib and domatia in the nerve axils; stipules generally persistent, shortly connate, triangular, often aristate or with a short mucro or a keeled lobe. Inflorescences cymose to corymbiform, few- to many-flowered, sessile to pedunculate, terminal on main and lateral stems or axillary. Flowers pedicellate or sessile, bisexual, hermaphrodite, always 4-merous. Calyx-tube turbinate to campanulate; limb truncate or lobed. Corolla-tube slender, cylindrical to funnel-shaped, inside glabrous or pubescent in the throat; lobes 4, contorted in bud. Stamens 4(–5); filaments short; anthers dorsifixed near the base. Ovary 2-locular; ovules 1(–2) in each locule; stigma very shortly 2-lobed. Fruit a drupe, globose to ovoid, with calyx limb persistent or deciduous; pyrenes 2.

A big genus with about 360 species, palaeotropical, widespread in Africa, tropical Asia, Australia, and Pacific islands; 19 species in Kenya.

1a. Corolla-tube shorter than or as long as the lobes, densely bearded at the throat 2
1b. Corolla-tube usually slender, much longer than or rarely as long as the lobes, never bearded at the throat 3
2a. Western Kenya species; leaves paired or 3-whorled; nodules absent .... 2. *P. ternifolia*
2b. Central Kenya species; leaves paired; nodules linear, restricted to the nerves beneath .... 1. *P. hymenophylla*
3a. Leaves on short spurs; flowers solitary .... 19. *P. uniflora*
3b. Leaves spaced on branches; flowers many in dense or lax corymbs 4
4a. Leaves linear .... 4. *P. linearifolia*
4b. Leaves narrowly elliptic to ovate or obovate 5
5a. Inflorescences terminal on short leafless branchlets 6
5b. Inflorescences terminal on branches bearing one or more pairs of leaves 10
6a. Leaves linear to lanceolate or oblanceolate, 8–30 cm long .... 14. *P. crassipes*
6b. Leaves never linear, less than 12 cm long 7
7a. Leaves pubescent to tomentose beneath; calyx pubescent or tomentose ...... 17. *P. dolichantha*
7b. Leaves glabrous to pubescent beneath; calyx glabrous to slightly pubescent 8
8a. Inflorescences dense, sub-umbellate .... 18. *P. subcana*
8b. Inflorescences lax 9
9a. Flowers cream to yellow; corolla-tube 1–2 mm wide at the top; calyx-lobes less than 1 mm long .... 15. *P. gardeniifolia*

9b. Flowers white or cream; corolla-tube ca.1 mm wide at the top; calyx-lobes 1–2 mm long .... 16. *P. sepium*
10a. Calyx-lobes transversely oblong; corolla-tube 7–9 mm long .... 3. *P. teitana*
10b. Calyx-lobes absent, small toothed, subulate, narrowly triangular, linear or lanceolate; corolla-tube less than 3.4 cm long 11
11a. Leaves pubescent above, pubescent to tomentose beneath; calyx-tube densely pubescent to tomentose 12
11b. Leaves glabrous above, glabrous to sparsely pubescent beneath; calyx-tube glabrous to sparsely pubescent 14
12a. Inflorescences dense, less than 3 cm across .... 6. *P. elliottii*
12b. Inflorescences lax to very lax, up to 10 cm across 13
13a. Coastal species; stipules with stiff hairs inside .... 5. *P. sansibarica*
13b. Inland species; stipules without silky hairs inside .... 7. *P. oliveriana*
14a. Inflorescences compact; calyx-lobes filiform .... 10. *P. stenosepala*
14b. Inflorescences compact to lax corymbs or subumbellate; calyx-lobes various 15
15a. Inland species, found above 1000 m 16
15b. Coastal species, found below 1000 m 17
16a. Leaf-blades pubescent beneath; calyx-lobes linear, 3.5–5 mm long .... 11. *P. aethiopica*
16b. Leaf-blades glabrous beneath; calyx-lobes subulate, 0.5–4 mm long .... 12. *P. abyssinica*
17a. Corolla-tube less than 1 cm long... 8. *P. sphaerobotrys*
17b. Corolla-tube 1–2.2 cm long 18
18a. Inflorescences compact, 2–4 cm across; nodules dot-like, scattered .... 9. *P. crebrifolia*
18b. Inflorescences lax to very lax, 6–13 cm across; nodules linear and confined to the mid-rib ... . 13. *P. tarennoides*

1. **Pavetta hymenophylla** Bremek., Repert. Spec. Nov. Regni Veg. 37: 68. 1934; F.T.E.A. Rubiac. 2: 632. 1988; K.T.S.L.: 528. 1994. —Type: Tanzania, Lushoto, Amani, 2 Aug. 1911, *M. Grote* in *Herb. Amani 3541* [holotype: B (destroyed); lectotype: K (K000286957), **designated here**; isolectotype: EA (EA000 001680)] (Plate 90A, B)

Shrub or small tree, up to 12 m tall. Leaf-blades elliptic to obovate, 6.5–23.5 cm × 2–11 cm, acute to acuminate at the apex, cuneate at the base; petioles up to 3 cm long; stipule-limbs truncate, 1–3 mm long, keeled. Corymbs terminal on main or lateral branches, lax; pedicels 5–9 mm long. Calyx-tube 1–2 mm

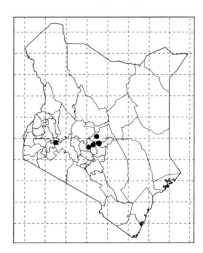

long; limb-tube 1–2 mm long, lobes ovate-oblong, 1–3 mm long. Corolla-tube 3–8 mm long, bearded at the throat; lobes oblong, 6–10 mm long. Fruits black, globose, 6–10 mm in diameter. Seeds brown-black.

Distribution: Western and central Kenya. [East Africa, from Kenya to Mozambique].

Habitat: Montane forests or forest edges; 1750–2400 m.

Kericho: Mau Forest, *Mwangangi & Kamau M7* (EA). Kirinyaga: Kamweti, *Agnew 5862* (EA). Meru: Marania, *Bally B3536* (K). Nyeri: Keria, *Brunt 1483* (K). Tharaka-Nithi: Chogoria Forest, *SAJIT 003976* (HIB).

2. **Pavetta ternifolia** Hiern, Fl. Trop. Afr. 3: 177. 1877; F.T.E.A. Rubiac. 2: 633. 1988; K.T.S.L.: 530. 1994; C.P.K.: 382. 2016. —Type: Tanzania, Bukoka, Karagwe, Feb. 1862, *J.H. Speke & J.A. Grant 422* [holotype: K (K000728922)] (Plate 90C–E)

= *Pavetta yalaensis* Bremek., Kew Bull. 9: 501. 1954. —Type: Kenya, N. of Yala R., 31 May 1951, *G.S. Rogers 741* [holotype: K; isotypes: BM (BM000903482), BR (BR0000008851581), EA, S (S-G-4609), PRE (PRE0594756-0), U (U0122096)]

Shrub or small tree, up to 4.5(–7) m tall. Leaves ternate or opposite, elliptic-obovate, 4–15.5 cm × 1–5 cm, acuminate at the apex, cuneate at the base; petioles up to 1 cm long; stipule-limbs truncate, up to 2.5 mm long. Corymbs terminal on main or lateral branches; pedicels up to 6 mm long. Calyx with tube ca. 1 mm long; limb-tube 1.5–3 mm long; lobes ovate-oblong, 1.5–4 mm long. Corolla white or pale yellow, with tube 4–6 mm long; lobes ovate-oblong, 4.5–7 mm long. Fruits black, 2-lobed, 6–8 mm in diameter. Seeds dark red-brown, smooth.

Distribution: Western Kenya. [D.R. Congo and tropical East Africa].

Habitat: Forests or hills; 1350–1900 m.

Bomet: Sotik, *Tumpnell 60/459* (EA). Kakamega: Kakamega Forest, *Faden 77/838* (EA). Kericho: Cheptuiyet, *Kerfoot 2183* (EA). Kisii: near Magombo Market, *Vuyk 517* (EA). Kisumu: Morongiot, *SAJIT 006979* (HIB). Nandi: *McDonald 1349* (EA, K). Trans Nzoia: near Tyack's Bridge, *Tweedie 4314* (EA).

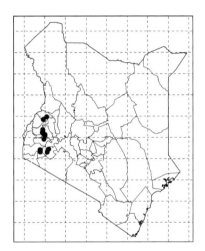

3. **Pavetta teitana** K. Schum., Pflanzenw. Ost-Afrikas C: 389. 1895; F.T.E.A. Rubiac. 2: 637. 1988; K.T.S.L.: 530. 1994. —Types: Kenya, Taita, *J.M. Hildebrandt 2555* [holotype: B (destroyed)]; Taita Taveta, Tsavo National Park, Ngulia Safari Camp, 6 Dec. 1980, *M.G. Gilbert 6006* [neotype: EA, **designated here**; isoneotype: K] (Plate 91A–E)

= *Pavetta kaessneri* S. Moore, J. Bot. 43: 250. 1905. —Type: Kenya, Machakos, Mukaa, 4 June 1902, *T. Kässner 920* [holotype: BM; isotype: K (K000311698)]

Shrub or small tree, up to 7 m tall. Leaf-blades lanceolate, lanceolate-oblong, oblong or rarely ovate, 9–18 cm × 2.5–5.5 cm, acute

Plate 90 A, B. *Pavetta hymenophylla*; C–E. *P. ternifolia*. Photo by YDZ (A, B), GWH (C, E) and QFW (D).

to slightly acuminate at the apex, acute to cuneate at the base, glabrous above, tomentose beneath; nodules small, scattered; petioles up to 1.5 cm long; stipule-limbs truncate, 3–5 mm long. Corymbs terminal on the main branches, compact to lax; pedicels 1–5 mm long. Calyx-tube 1–1.3 mm long; limb-tube 1–1.5 mm long, lobes transversely oblong, rounded to obtuse, ca. 0.5 mm long. Corolla-tube 7–9 mm long; lobes oblong-lanceolate, 5–7 mm long. Fruits black, globose, 7–8 mm in diameter.

Distribution: Central and southern Kenya. [Endemic].

Habitat: Evergreen riverine forests, dry forests, bushlands, hillsides, or wooded grasslands; 700–1850 m.

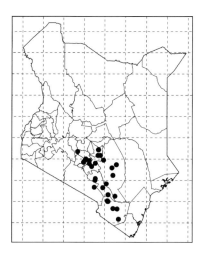

Embu: Kiangombe, *Dyson 425* (K). Kiambu: Fourteen Falls, *SAJIT-Z001-19* (HIB); Kuraiha Farm, *Malombe & Kirika 1* (EA). Kitui: Mutha Hill, *Luke & Stone 8230* (EA). Machakos: Matuu, *Fukuoka K-202* (EA). Makueni: Mbuinzau, *Luke 2155A* (EA). Teita Taveta: Kasigau Mount, *Faden et al. 69/430* (EA); Ngulia Hills, *Luke et al. 11196* (EA).

4. **Pavetta linearifolia** Bremek., Repert. Spec. Nov. Regni Veg. 37: 128. 1934; F.T.E.A. Rubiac. 2: 637. 1988; K.T.S.L.: 528. 1994. —Type: Kenya, Tana River, Kinakomba, 1889, *Leroy 1039* [holotype: P (P03916403)]

Shrub or small tree up to 4.5 m tall. Leaf-blades linear, 3.5–12.5 cm × 0.8–2 cm, acute at the apex, obtuse to rounded at the base, margins always revolute; petioles 2–3 mm long; stipule-limbs truncate, 1.5–4 mm long. Corymbs terminal on main or lateral branches, moderately compact; pedicels 1–2.5 mm long. Calyx-tube ca. 1 mm long; limb-tube ca. 1 mm long; lobes ovate-triangular, 0.5–1 mm long. Corolla-tube 7–8.5 mm long; lobes oblong-

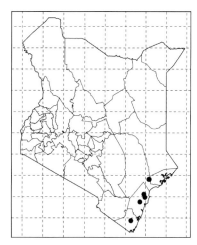

lanceolate, 4–5 mm long. Fruits black, globose, 5–6 mm in diameter.

Distribution: Coastal Kenya. [Tanzania].

Habitat: Coastal forest, thickets, dry bush on hillslopes; up to 450 m.

Kilifi: Mangea Hill, *Luke & Robertson 628* (EA). Kwale: Kaya Puma, *Luke et al. 6339* (EA). Tana River: Wema, *Robertson & Luke 1254* (EA).

5. **Pavetta sansibarica** K. Schum., Bot.

Jahrb. Syst. 28(1): 79. 1899; F.T.E.A. Rubiac. 2: 641. 1988; K.T.S.L.: 529. 1994. —Types: Tanzania, Zanzibar, *Marseeler 49* [syntype: B (destroyed)] & *31* [syntype: B (destroyed); lectotype: BR (BR0000008851123), **designated here**]

Shrub 1.5–5 m tall. Leaf-blades lanceolate, oblanceolate or elliptic, 7.5–20 cm × 1–8 cm, acuminate at the apex, cuneate at the base, sparsely strigose or pubescent; nodules present or absent; petioles up to 4 cm long, densely pubescent; stipule-limbs subtruncate to triangular, 1–3 mm long. Corymbs terminal on lateral branches, very lax, few–many-flowered; pedicels 2.5–11 mm long, densely pubescent. Calyx-tube 1.5 mm, pubescent; limb-tube very short, lobes narrowly triangular to linear-subulate, 1–3 mm long. Corolla-tube 1.2–2 cm long, pubescent outside, sparsely pubescent inside; lobes oblong, 8–11 mm long. Fruit black, globose, 5–8 mm in diameter, sparsely to densely pubescent.

subsp. **trichosphaera** (Bremek.) Bridson, Kew Bull. 32: 614. 1978. —Type: Kenya, Kwale, Shimba Hills, 1927, *H.M. Gardner in F.D. 1445* [holotype: K (K000352536)] (Plate 91F)

= *Pavetta shimbensis* Bremek., Kew Bull. 11: 172. 1956. —Type: Kenya, Kwale, Shimba Hills, Mwele Mdogo Forest, 5 Feb. 1953, *R.B. Drummond & J.H. Hemsley 1131A* [holotype: K (K000352535)]

Leaves narrowly to broadly elliptic, 9–20 cm × 4.5–8 cm; calyx-lobes 1.5–3 mm long; calyx-tube covered with long hairs.

Distribution: Coastal Kenya. [Tanzania].
Habitat: Lowland forest; 250–400 m.

Note: *Pavetta sansibarica* subsp. *sansibarica* has smaller leaves and shorter pedicels and just occurs in Tanzania.

Kilifi: Kaya Jibana, *Luke et al. 4305* (K). Kwale: Shimba Hills, *SAJIT 006055* (HIB).

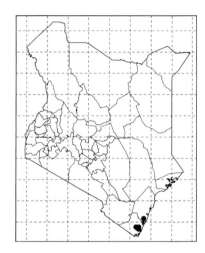

6. **Pavetta elliottii** K. Schum. & K. Krause, Bot. Jahrb. Syst. 39: 551. 1907; F.T.E.A. Rubiac. 2: 643. 1988; K.T.S.L.: 528. 1994. —Type: Kenya, Nairobi, *G.F. Elliott 76* [holotype: B (destroyed); lectotype: K (K000311697), **designated here**] (Plate 92A)

= *Pavetta trichocalyx* Bremek., Repert. Spec. Nov. Regni Veg. 47: 90. 1939. ≡ *Pavetta elliottii* K. Schum. & K. Krause var. *trichocalyx* (Bremek.) Bridson, Kew Bull. 32: 616. 1978. —Type: Kenya, Embu, Chuka, 26 Feb. 1922, *R.E. & T.C.E Fries 1996* [holotype: S (S-G-4607)]

Shrub, sometimes scrambling, up to 2 m tall. Leaf-blades ovate, elliptic or rarely obovate, 2.5–10 cm × 1.5–7.5 cm, obtuse to acuminate at the apex, rounded or cordate at the base, tomentose beneath; nodules small, scattered; petioles short, tomentose; stipule-limbs truncate, up to 5 mm long. Corymbs terminal on main and short lateral branches,

Plate 91 A–E. *Pavetta teitana*; F. *P. sansibarica* subsp. *trichosphaera*. YDZ (A–E) and GWH (F).

compact. Calyx-tube 1.5–2 mm long; limb-tube 0.7–1 mm long, lobes narrowly triangular to lanceolate, 2–5 mm long. Corolla-tube 1.5–3 cm long; lobes lanceolate, 5–8 mm long. Fruits black, globose, 8–9 mm in diameter.

Distribution: Western and central Kenya. [Endemic].

Habitat: Forests or bushlands; 1650–2150 m.

Kiambu: Komo Kuraiha Farm, *Malombe & Kirika 21* (EA). Machakos: Ol Donyo Sabuk, *Bally 7460* (EA). Makueni: Kithembe Hill, *Mwangangi 2136* (EA). Murang'a: Kangure Forest, *Mwachala et al. 218* (EA). Nairobi: near Rowallan Scout Camp, *Gilbert 5830* (EA); Langata Forest, *Mathenge 804* (EA). Narok: Intona Area, *Msafiri 950* (EA). Nyeri: Kampi ya Farasi, *Mbnui & Kiglanya 473* (EA). Tharaka-Nithi: Chuka, *Fries et al. 1996* (K).

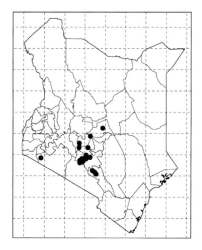

7. **Pavetta oliveriana** Hiern, Fl. Trop. Afr. 3: 174. 1877; F.T.E.A. Rubiac. 2: 644. 1988; K.T.S.L.: 528. 1994. —Type: Tanzania, Bukoba, Karagwe, Dec. 1861, *J.H. Speke & J.A. Grant 160* [holotype: K (K000728923)] (Plate 92B–E)

Shrub or small tree, sometimes scrambling, up to 4.5(–7) m tall. Leaf-blades elliptic to broadly elliptic or ovate, 3–21.5 cm × 1.5–9.5 cm, acute to acuminate at the apex, cuneate, rounded or cordate at the base, finely pubescent above, tomentose beneath; nodules scattered; petioles up to 6 cm long; stipules triangular, 5–8 mm long. Corymbs terminal on main and lateral branches, lax; pedicels 1–8 mm long, tomentose. Calyx-tube 1–2 mm long; limb-lobes triangular, 2–5 mm long. Corolla-tube 1.5–3.5 cm long; lobes oblong, 4–6 mm long. Fruits black, globose, 8–10 mm in diameter.

Distribution: Northwestern, western, and southern Kenya. [Tropical Africa, from Eritrea, Sudan to D.R. Congo and Tanzania].

Habitat: Forests, forest edges, wooded bushlands, thicket mosaics, or grazed grasslands; 1000–2150 m.

Baringo: Tugen Hills, *Kuchar 8559* (EA). Bomet: Sotik, *Bally 7451* (K). Bungoma: southwest of Mount Elgon, *Symes 306* (EA). Elgeyo-Marakwet: Marakwet, *Lindsay 152* (EA, K). Homa Bay: Soklo, *Kirika 223* (EA). Kajiado: Olenyamu, *Glover & Samuel 2902* (EA). Kakamega: Kakamega Forest, *SAJIT 006685* (HIB). Kisii: south Mucrango Hedge, *Vuyk 124* (K). Kisumu: Ishiulu, *Kamau 244*

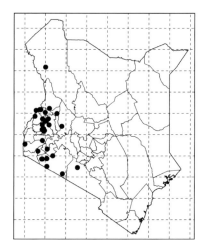

Plate 92 A. *Pavetta elliottii*; B–E. *P. oliveriana*. Photo by YDZ (A) and GWH (C–E).

(EA). Nandi: Kaimosi, *Williams & Piers 606* (EA). Narok: Endama, *Glover et al. 1973* (EA); Entasekera, *Glover et al. 2060* (EA). Trans Nzoia: Kitale, *Bogdan 4281* (EA). Turkana: Moruassigar, *Newbould 7088* (EA). Uasin Gishu: Kapsaret, *Birchford 17906* (EA). West Pokot: Cherangani Hills, *SAJIT 005069* (HIB).

8. **Pavetta sphaerobotrys** K. Schum., Bot. Jahrb. Syst. 28: 494. 1900; F.T.E.A. Rubiac. 2: 648. 1988; K.T.S.L.: 529. 1994. —Type: Tanzania, Kilosa, Luhembe (Ruhembe), *W. Goetze 403* [holotype: B (destroyed); lectotype: K (K000287138), **designated here**]

Shrub or small tree, up to 9 m tall. Leaf-blades elliptic to oblanceolate, 3.5–15 cm × 1.5–5.5 cm, acuminate or acute at the apex, cuneate at the base; nodules along the midrib and lateral nerves; petioles 1–1.6 cm long; stipule-limbs truncate, 2–7 mm long. Corymbs terminal on main or lateral branches, compact; pedicels up to 6 mm long. Calyx-tube 1–1.2 mm long; limb-lobes lanceolate, 2–7 mm long. Corolla-tube 5–10 mm long; lobes oblong, 3–6.5 mm long. Fruits greenish black, globose, 6–7 mm in diameter.

subsp. **tanaica** (Bremek.) Bridson, Kew Bull. 32(3): 623. 1978; F.T.E.A. Rubiac. 2: 649. 1988; K.T.S.L.: 529. 1994. ≡ *Pavetta tanaica* Bremek., Repert. Spec. Nov. Regni Veg. 47: 27. 1939. —Type: Kenya, Tana River, near rapids, 17 Apr. 1934, *Sampson 80* [holotype: K (K000287144)]

= *Pavetta manamoca* Bremek., Kew Bull. 8: 502. 1954. —Type: Kenya, Tana River, Garissa, *J. Adamson in Bally 5851* [holotype: K (K000311731); isotype: EA]

Stipules thinly membranous. Corolla-tube 6.5–10 mm long; lobes 4–6.5 mm long.

Distribution: Eastern Kenya. [Somalia].
Habitat: Lowland forests; up to 50 m.
Note: There are three subspecies recorded in this species *Pavetta sphaerobotrys* K. Schum., with only subsp. *tanaica* (Bremek.) Bridson distributed in Kenya. The other two subspecies, subsp. *sphaerobotrys* and subsp. *lanceisepala* (Bremek.) Bridson, have longer corolla-tube, and only occur in Tanzania.

Tana River: Tana River Primate Reserve, *Robertson & Luke 4573* (EA).

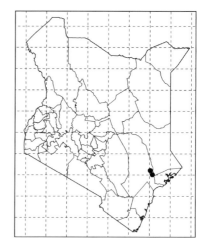

9. **Pavetta crebrifolia** Hiern, Fl. Trop. Afr. 3: 172. 1877; F.T.E.A. Rubiac. 2: 649. 1988; K.T.S.L.: 527. 1994. —Type: Kenya, Kilifi, Malindi, Sabaki R., Oct. 1873, *J. Kirk s.n.* [holotype: K (K000311730)]

Shrub, up to 4 m tall. Leaf-blades elliptic to obovate, 3.5–13.5 cm × 1.5–6 cm, acute to rounded and sometimes shortly apiculate at the apex, cuneate at the base, subcoriaceous; nodules scattered; petioles up to 2 cm long; stipule-limbs 2–7 mm long, subtruncate to broadly ovate. Corymbs terminal on leafy lateral branches, subumbellate, compact;

pedicels 2–9 mm long. Calyx-tube 1–1.2 mm long; limb-tube 0.7–1 mm long; lobes linear to lanceolate or narrowly triangular, 3.5–12 mm long. Corolla-tube 10–22 mm long, glabrous outside; lobes 5–8 mm long, acuminate. Fruits greenish black, globose, 8–11 mm in diameter.

1a. Plants entirely glabrous..a. var. *crebrifolia*
1b. Plants with young branches, leaves beneath, pedicels and calyx and fruits pubescent ....
....b. var. *pubescens*

a. var. **crebrifolia** (Plate 93A, B)

Plants entirely glabrous.
Distribution: Southern and coastal Kenya. [East Africa, from Somalia to Tanzania].
Habitat: Forests, bushlands, woodlands, coastal thickets, or wooded grasslands; up to 1550 m.
Kilifi: Kaya Jibana, *Luke & Luke 4328* (EA); Arabuko-Sokoke Forest, *Musyoki & Hansen 1011* (EA); *Polhill & Paulo 801*

b. var. **pubescens** Bridson, Kew Bull. 32: 624. 1978; F.T.E.A. Rubiac. 2: 650. 1988. —Type: Kenya, Lamu, Mararani, Boni Forest, 9 Sept. 1961, *Gillespie 327* [holotype: K (K000287215)] (Plate 93C)

Distribution: Central and coastal Kenya. [Endemic].
Habitat: Forest edges; up to 600 m.
Kitui: near Endau, *SAJIT MU0131* (HIB). Lamu: Boni Forest, *Gillespie 327* (K).

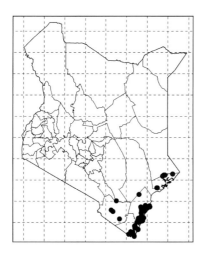

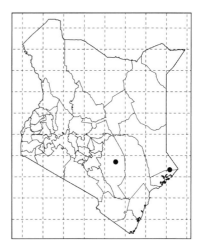

(EA). Kwale: Dzombo Hill, *Robertson et al. MDE336* (EA); Mrima Hill, *Verdcourt 5279* (K); Shimoni, *Drummond & Hemsley 3920* (EA). Lamu: Bodhei, *Festo & Luke 2641* (EA); Witu Forest, *Robertson & Luke 5483* (EA). Teita Taveta: Marimbongonyi Valley, *Mungai et al. EW3088/A* (K).

10. **Pavetta stenosepala** K. Schum., Abh. Preuss. Akad. Wiss. 1894: 26. 1894; W.F.E.A.: 155. 1987; F.T.E.A. Rubiac. 2: 653. 1988; K.T.S.L.: 529. 1994. —Type: Tanzania, Lushoto, Deremai, Feb. 1893, *C. Holst 2268* [holotype: B (destroyed); lectotype: K (K000728924), **designated here**; isolectotype: HBG (HBG521318)] (Plate 94)

Plate 93 A, B. *Pavetta crebrifolia* var. *crebrifolia*; C. *P. crebrifolia* var. *pubescens*. Photo by GWH (A, B) and NW (C).

Plate 94 A–C. *Pavetta stenosepala*. Photo by GWH (A, B) and VMN (C).

Shrub, up to 4.5 m tall. Leaves mostly crowded on the top of the branches; blades narrowly oblanceolate to obovate, 3–13.5 cm × 1–6 cm, acute to shortly acuminate at the apex, cuneate at the base; nodules mostly occurring on the mid-rib; petioles up to 3 cm long; stipules triangular to ovate and gradually caudate, up to 1.4 cm long. Corymbs terminal on lateral branches, very compact; pedicels reduced. Calyx-tube very short; limb-lobes filiform to linear, up to 1.2 cm long. Corolla-tube 1–2 cm long; lobes oblong-elliptic, 3–5.5 mm long. Fruits black, globose, 5.5–7 mm in diameter.

Distribution: Coastal Kenya. [Tanzania].
Habitat: Lowland forests; up to 500 m.

Kilifi: Gede Forest, *Gerhardt & Steiner 16* (EA); Arabuko-Sokoke Forest, *Trump 85* (EA); Kaya Kivara, *Robertson & Luke 4752* (EA). Kwale: Shimba Hills, *SAJIT 005502* (HIB);

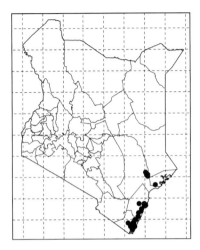

Buda Mafisini Forest, *Luke & Rorbertson 1711* (EA); Pengo Forest, *Drummond & Hemsley 1207* (EA). Lamu: Witu, *Rawlins 1256* (EA). Tana River: Tana River Primate Reserve, *Luke & Robertson 1155* (EA).

11. **Pavetta aethiopica** Bremek., Kew Bull. 11(1): 174. 1956; W.F.E.A.: 154. 1987; F.T.E.A. Rubiac. 2: 659. 1988; K.T.S.L.: 527. 1994. —Type: Ethiopia, Sidamo, Mega Mt., 7 Sept. 1953, *P.R.O. Bally 9397* [holotype: K (K00 0287329); isotypes: BR (BR0000008849960), EA (EA000001670)]

Shrub, up to 2 m tall. Leaf-blades elliptic to narrowly obovate, 5–9 cm × 3–5 cm, shortly acuminate at the apex, acute to cuneate at the base; nodules concentrated near the mid-rib; petioles up to 1.4 cm long; stipules lanceolate, 8–10 mm long. Corymbs terminal on lateral shoots, compact; pedicels 2–5 mm long. Calyx-tube ca. 1.5 mm long; lobes linear, 3.5–5 mm long. Corolla-tube ca. 1.5 cm long; lobes lanceolate-oblong, ca. 8 mm long. Fruits black, globose, ca. 7 mm in diameter.

Distribution: Northern Kenya. [Ethiopia].

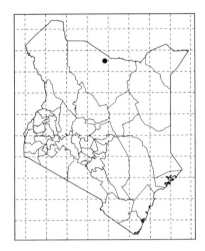

Habitat: Thickets; ca. 1000 m.
Marsabit: Huri hills, *Synnott 1879* (EA).

12. **Pavetta abyssinica** Fresen., Mus. Senckenberg. 2: 166. 1837; F.T.E.A. Rubiac. 2: 660. 1988; K.T.S.L.: 527. 1994. —Type: Ethiopia, between Halei and Temben, *E. Rüppel s.n.* [holotype: FR (FR0030070); isotype: FR (FR0030071)]

Shrub or small tree, up to 9 m tall. Leaf-blades narrowly elliptic to elliptic or obovate, 3.5–15.5 cm × 1–6.5 cm, acuminate at the apex, cuneate at the base; nodules along the nerves or scattered; petioles up to 1.5 cm long; stipules truncate, up to 1 cm long. Corymbs terminal on lateral branches, compact to lax; pedicels up to 1 cm long. Calyx-tube 1–2 mm long; lobes subulate, 0.5–5 mm long. Corolla-tube 0.8–2 cm long; lobes oblong, 0.4–1 cm long. Fruits black, globose, 5–10 mm in diameter.

1a. Young branches and petioles glabrous or sparsely pubescent; calyx and corolla-tube glabrous outside..a. var. *abyssinica*
1b. Young branches and petioles densely pubescent; calyx and corolla-tube pubescent outside .....
....b. var. *lamurensis*

a. var. **abyssinica** (Plate 95)

= *Pavetta kenyensis* Bremek., Repert. Spec. Nov. Regni Veg. 37: 149. 1934. —Type: Kenya, Nyeri, Liki R., 18 Aug. 1934, *E. Battiscombe 1312* [holotype: K; isotype: EA (EA000001668)]

= *Pavetta trichotropis* Bremek., Kew Bull. 3: 354. 1949. —Type: Kenya, Elgeyo, Kapsowar, Marakwet Hills, June 1932, *Gardner 611* [holotype: EA]

Young branches and petioles glabrous or sparsely pubescent; calyx and corolla-tube glabrous outside.

Distribution: Northern, northwestern, western, central, and southern Kenya. [Tropical East Africa, from Eritrea to Tanzania].

Habitat: Forests or hillsides; 1300–2350 m.

Baringo: Katimok Forest, *Dale K2438* (EA). Bomet: Tinderet Forest Reserve, *Geesteranus 5365* (K). Bungoma: Mount Elgon, *Bridson 77* (EA). Elgeyo-Marakwet: Marakwet, *Lindsay 160* (EA). Kajiado: Ngong Hills, *Khayota 148* (EA). Kericho: Namanga Hill Forest, *Simon et al. 529* (K). Kiambu: Kikuyu, *Battiscombe 44* (K). Kirinyaga: south Mount Kenya, *SAJIT 003038* (HIB). Makueni: Kithembe Hill, *Mwangangi 2138* (EA). Marsabit: Marsabit Forest, *Faden 68/626* (EA); Mount Kulal, *Luke & Luke 10819* (EA). Meru: Nyambene Hills, *SAJIT Z0168* (HIB); Ngaia Forest, *Luke et al. 10277* (EA). Nakuru: Elburgon, *Baker K179* (EA). Nandi: near Kapsabet, *Dale 3123* (EA). Narok: Loita Hills, *Beentje 2559* (K). Nyeri: near Ontulili Forest Station, *SAJIT 003139* (HIB). Samburu: Matthew's Range, *Luke 14236* (EA). Turkana: Loima Forest, *Mbonge 29* (EA, K). West Pokot: Lelan Forest, *Bridson 100* (K).

b. var. **lamurensis** (Bremek.) Bridson, Kew Bull. 32: 636. 1978; F.T.E.A. Rubiac. 2: 662. 1988; K.T.S.L.: 527. 1994. ≡ *Pavetta lamurensis* Bremek., Repert. Spec. Nov. Regni Veg. 37: 148. 1934. —Type: Kenya, Kiambu, Limuru, 6 July 1909, *G. Scheffler 314* [holotype: B (destroyed); lectotype: K (K000311728), **designated here**; isolectotypes: BM, E (E00193689), G (G00359390), S (S-G-4593)]

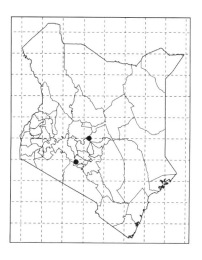

Plate 95 A–E. *Pavetta abyssinica* var. *abyssinica*. Photo by GWH (A–C) and SWW (D, E).

= *Pavetta abyssinica* Fresen. var. *prescottii* Bridson, Kew Bull. 32(2): 636. 1978. —Type: Kenya, Nyeri, Nanyuki, 1926, *Prescott Decie s.n.* [holotype: BM (BM000903459)]

Distribution: Central Kenya. [Endemic].
Habitat: Upland forests; ca. 2200 m.
Kiambu: Limuru, *Scheffler 314* (K). Nyeri: Nanyuki, *Prescott Decie s.n.* (BM).

13. **Pavetta tarennoides** S. Moore, J. Bot. 43: 353. 1905; F.T.E.A. Rubiac. 2: 663. 1988; K.T.S.L.: 530. 1994. —Type: Kenya, Kwale, Mwele Mdogo, 2 Mar. 1902, *T. Kässner 230* [holotype: BM (BM000903465)] (Plate 96)

Shrub up to 5 m tall. Leaf-blades elliptic to ovate, 4–17 cm × 2–7.5 cm, acute to slightly acuminate at the apex, cuneate at the base; nodules linear, confined to the mid-rib; petioles 1–3.5 cm long; stipule-limbs truncate to slightly triangular, 2–7 mm long. Corymbs terminal on main and lateral leafy branches, very lax; pedicels 5–18 mm long. Calyx-tube 1–1.3 mm long; limb-tube 1–2 mm long, lobes subulate to linear, 2–4 mm long. Corolla-tube 1.2–2.2 cm long; lobes oblong-lanceolate, 0.7–1.2 cm long. Fruits black, globose, 5–7 mm in diameter.

Distribution: Coastal Kenya. [Endemic].
Habitat: Lowland forests; 300–450 m.
Kwale: Shimba Hills, *Luke & Robertson 2719* (EA).

14. **Pavetta crassipes** K. Schum., Pflanzenw. Ost-Afrikas, C: 389. 1895; F.T.E.A. Rubiac. 2: 670. 1988; K.T.S.L.: 527. 1994; U.T.S.K.: 343. 2005. —Types: Kenya, Kavirondo, Karachonyo, *A. Fischer 313* [holotype: B (destroyed)]; Bungoma, Broderick Falls, Mar. 1959, *E.M. Tweedie 1804* [neotype: K, **designated here**] (Figure 71; Plate 97A, B)

Shrub or small tree, up to 8 m tall. Leaves crowded at the top of branches, paired or 3–4-whorled; blades narrowly oblong or oblanceolate, 8–30 cm × 2–7.5 cm, rounded to obtuse at the apex, obtuse to attenuate at the base; petioles up to 1.5 cm long; stipule-limbs truncate, 2–6 mm long. Corymbs terminal on short leafless lateral branches, lax; pedicels up to 7 mm long. Calyx-tube 1–1.5 mm long; limb-tube 1–2 mm long, truncate or undulate to slightly dentate. Corolla-tube 0.8–1.5 cm long; lobes oblong, 4–6 mm long. Fruit black,

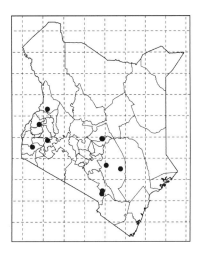

Plate 96 *Pavetta tarennoides*. Photo by GWH.

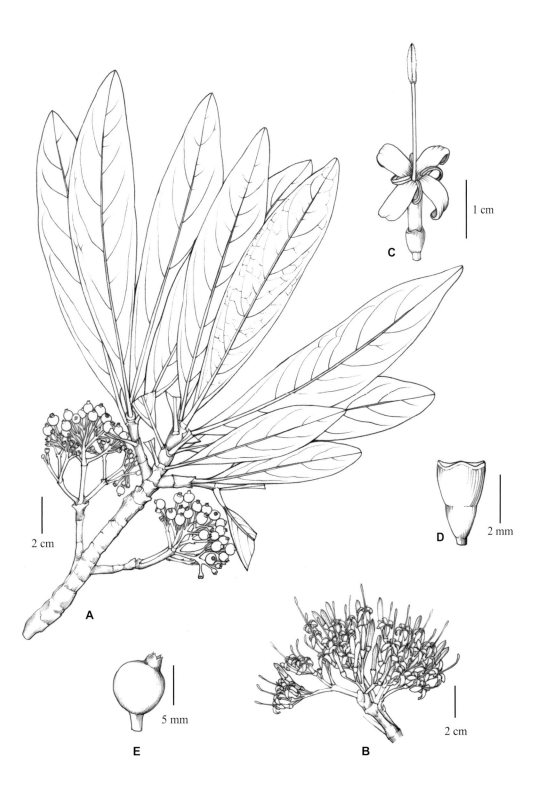

Figure 71. *Pavetta crassipes*. A. fruiting branch; B. inflorescences; C. flower; D. calyx; E. fruit. Drawn by NJ.

globose, 6–8 mm in diameter.

Distribution: Western and central Kenya. [Tropical Africa].

Habitat: Hills or open grasslands; 850–1850 m.

Bungoma: Broderick Falls, *Tweedie 1804* (K). Homa Bay: *Katz 28* (EA). Kajiado: Chyulu Hills, *Luke & Luke 9270* (EA). Kisumu: Muhoroni, *Brich 60/363* (EA). Kitui: Kabonge, *Edwards 172* (EA). Meru: Makundune, *Gillett 20116* (EA). West Pokot: Chapereria, *Nyamwaya 28* (EA).

15. **Pavetta gardeniifolia** Hochst. ex A. Rich., Tent. Fl. Abyss. 1: 351. 1848; F.T.E.A. Rubiac. 2: 676. 1988; K.T.S.L.: 528. 1994. —Type: Ethiopia, Maundet, 16 July 1842, *G.W. Schimper 1141* [holotype: P; isotypes: BR (BR 0000008850300), HAL (HAL0113829), JE (JE00000337), K (K000728915), L (L005 7930, L0057931), LG (LG0000090029547), MO (MO-391812), MPU (MPU021996), REG (REG 000629), TUB (TUB004584)] (Plate 97C–E)

Shrub or small tree, up to 7 m tall. Leaf-blades elliptic or oblanceolate to obovate, 1.6–12.5 cm × 1–6 cm, acute, obtuse, rounded or occasionally slightly emarginate at the apex, cuneate at the base; nodules present, scattered; petioles up to 20 mm long; stipule-limbs subtruncate to triangular, 2–5 mm long, shortly acuminate, hairy inside. Corymbs moderately compact to very lax, terminal on leafless spurs or short lateral branches; pedicels up to 22 mm long. Flowers sweet-scented, cream to yellow. Calyx-limb 0.5–1.2 mm long, truncate, shallowly dentate or shortly lobed. Corolla cream to bright yellow; tube 0.4–1.5 cm long; lobes narrowly oblong, 3–7 mm long. Fruits black, globose, 6–8 mm in diameter.

Distribution: Northern, northwestern, western, central, eastern, and southern Kenya. [Tropical and southern Africa].

Habitat: Upland forests, thickets, woodlands, open bushlands on rocky places, or wooded grasslands; 700–2250 m.

Baringo: Along the road from Nakuru to Lake Hannington, *Faden & Napper 69/070* (EA, K). Kajiado: Oloitokitok, *Muasya et al. GBK 02/54/07* (K). Kiambu: Muguga, *Fukuoka K-96* (EA). Kisii: ca. 3 km west of Ikonge, *Vuyk 1975* (EA). Kitui: Kibwezi, *Luke 15021* (EA). Laikipia: Maisor Ranch, *Cameron 206* (EA). Machakos: Kyamutheke Hill, *Luke et al. 3854* (EA). Makueni: Kilima Kiu, *Kokwaro 3020* (EA). Mandera: Dandu, *Gilbert 13423* (EA). Marsabit: Mount Nyiru, *Luke et al. 13987* (EA). Meru: Lewa, *Linsen & Giesen 143* (EA). Narok: Olarro Camp, *Luke et al. 7318* (EA); Loita Hills, *Mwangangi 4017* (EA). Samburu: Mathew's Range, *Ichikawa 410* (EA); Ol Doinyo Sabachi, *Gilbert 5372* (EA). Teita Taveta: Kasigau Mount, *Luke & Luke 4104* (EA). Turkana: Moruassigar, *Newbould 7043* (EA). Wajir: close to Wajir, *Ali 16338* (EA). West Pokot: Kogh Mount, *Caufield 221* (EA).

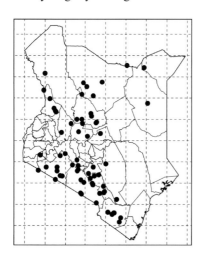

Plate 97 A, B. *Pavetta crassipes*; C–E. *P. gardeniifolia*. Photo by GWH (A, B) and YDZ (C–E).

16. Pavetta sepium K. Schum., Pflanzenw. Ost-Afrikas, C: 389. 1895; F.T.E.A. Rubiac. 2: 678. 1988; K.T.S.L.: 529. 1994. —Type: Tanzania, Kilimanjaro, Apr. 1894, *G. Volkens 2193* [holotype: B (destroyed); lectotype: BM (BM000903426), **designated here**]

Shrub or small tree, up to 5.5 m tall. Leaf-blades lanceolate, elliptic to ovate, 3.5–12.5 cm × 1–5.5 cm, acute to acuminate at the apex, cuneate at the base; nodules present, scattered; petioles up to 1.5 cm long; stipule-limbs translucent, truncate, 2–5 mm long. Corymbs terminal on short spurs or short leafless lateral branches, lax; pedicels 3–11 mm long. Flowers white or cream. Calyx-tube 0.5–1 mm long; limb-tube 0.5–1 mm long, lobes narrowly triangular to linear, 1–2 mm long. Corolla-tube slender, up to 12 mm long; lobes oblong to oblanceolate, 3–6 mm long. Fruits black, globose, ca. 5 mm in diameter.

1a. Leaves, pedicels, calyx and corolla-tube pubescent .... a. var. *sepium*
1b. Leaves, pedicels, calyx and corolla-tube sparsely pubescent to glabrous 2
2a. Leaves sparely pubescent; corolla-tube less than 6 mm long .... b. var. *glabra*
2b. Leaves glabrous; corolla-tube 6–12 mm long .... c. var. *merkeri*

a. var. **sepium**

Leaves, pedicels, calyx and corolla-tube pubescent.
Distribution: Southern Kenya. [Tanzania].
Habitat: Hillsides; 1200–1250 m.
Kajiado: Loitoktok, *Rauh 235* (EA). Teita Taveta: Taita Hills, *Napier 1139* (EA, K).

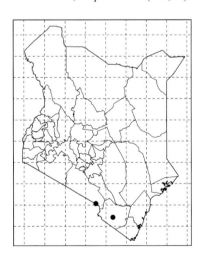

b. var. **glabra** Bremek., Repert. Spec. Nov. Regni Veg. 37: 178. 1934; F.T.E.A. Rubiac. 2: 679. 1988; K.T.S.L.: 529. 1994. —Type: Kenya, Machakos, Ukambani, 28 Jan. 1906, *G. Scheffler 103* [holotype: B (destroyed); lectotype: K (K000311724), **designated here**; isolectotypes: AMD (AMD.116079), BM, E (E00193683), P (P00553374), PRE (PRE0594406-0), S (S-G-4602)]

Leaves sparely pubescent; corolla-tube 4.5–6 mm long.
Distribution: Southern Kenya. [Endemic].

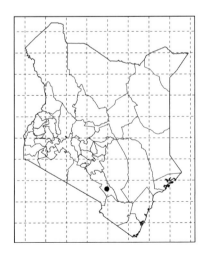

Habitat: Bushland; ca. 1000 m.

Makueni: Kibwezi, *Scheffler 103* (K).

c. var. **merkeri** (K.Krause) Bridson, Kew Bull. 32(3): 645. 1978; F.T.E.A. Rubiac. 2: 680. 1988; K.T.S.L.: 529. 1994. ≡ *Pavetta merkeri* K. Krause, Bot. Jahrb. Syst. 43: 146. 1909. —Type: Tanzania, "Massaisteppe", near the East African Rift Valley, 1904, *M. Merker 822* [holotype: B (destroyed)]

Leaves glabrous; corolla-tube 6–12 mm long.

Distribution: Central and southern Kenya. [Tanzania].

Habitat: Forests, bushlands, or woodlands on rocky places; 550–1250 m.

Kajiado: Rombo Chala, *Luke 3944* (EA). Kitui: Ikutha, *Bally 1605* (EA). Makueni: Kibwezi Forest, *Luke 14789* (EA). Meru: Meru National Park, *Luke et al. 7306* (EA). Narok: Losuate Hills, *Kuchar 13748* (EA). Samburu: Ndoto Mountains, *Gilbert et al. 5624* (EA, K). Teita Taveta: Mzima Springs, *Gillett 21016* (EA).

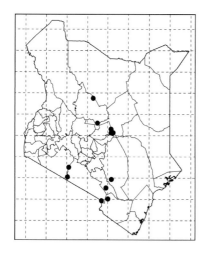

17. **Pavetta dolichantha** Bremek., Repert. Spec. Nov. Regni Veg. 37: 175. 1934; F.T.E.A. Rubiac. 2: 680. 1988; K.T.S.L.: 528. 1994.

—Type: Tanzania, Mpwapwa, 5 Apr. 1929, *H.E. Hornby 96* [holotype: K (K000352610); isotype: EA (EA000001697)]

Shrub, up to 2 m tall. Leaves always crowded on the top of branches; blades oblanceolate to obovate, 1.5–6 cm × 0.6–3 cm, obtuse, rounded or emarginate at the apex, cuneate at the base, pubescent to tomentose; petioles 2–4 mm long. Corymbs terminal on short leafless branches, few-flowered, compact; pedicels 2–5 mm long, tomentose. Calyx-tube 1–1.5 mm long; limb-tube 0.5–1 mm long; lobes triangular, 0.7–2 mm long. Corolla-tube 1.2–3.1 cm long; lobes ovate, 3–4.5 mm long. Fruits black, globose, 5–7 mm in diameter.

Distribution: Central, eastern, and southern Kenya. [Tanzania].

Habitat: Hills or bushlands; 300–1650 m.

Garissa: Mado Gashi-Garissa Road, *Gilbert & Thulia 1705* (EA, K). Kajiado: Olenyamu, *Glover & Samule 2902* (K); Emart Olkimpai Hills, *Pearce & Vollesen 947* (K). Narok: Suswa Hill, *Glover et al. 4006* (EA). Samburu: Matakweni Hill, *Faden et al.*

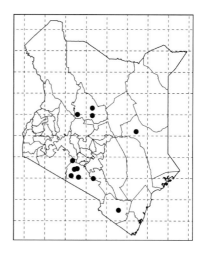

74/941 (K). Teita Taveta: Sagala Hills, *Polhill & Paulo 966* (EA).

18. Pavetta subcana Hiern, Fl. Trop. Afr. 3: 172. 1877; F.T.E.A. Rubiac. 2: 681. 1988; K.T.S.L.: 529. 1994. —Type: Sudan, Bahr el Ghasal, Jur (Ghattas Zeriba), 4 June 1871, *G. Schweinfurth 3250* [holotype: K (K000728911); isotypes: BM (BM000903429), P]

Shrub, up to 3.5 m tall. Leaf-blades narrowly elliptic to elliptic, 0.5–8 cm × 0.4–3 cm, obtuse or sometimes acute at the apex, cuneate at the base, glabrous or pubescent on both sides; petioles up to 1 cm long; stipule-limbs 2–5 mm long, truncate. Corymbs dense, sub-umbellate, few–several-flowered, terminal on leafless spurs or branches; pedicels 1–4 mm long. Calyx-tube ca. 1 mm long; limb-tube 1 mm long, lobes triangular, 1–2 mm long. Corolla-tube 1–2.2 cm long; lobes oblong-elliptic, 4–7.5 mm long. Fruit black, globose, 5–8 mm in diameter.

var. **longiflora** (Vatke) Bridson, Kew Bull. 32(3): 646. 1978; F.T.E.A. Rubiac. 2: 681. 1988; K.T.S.L.: 529. 1994. ≡ *Pavetta gardeniifolia* Hochst. ex A. Rich. var. *longiflora* Vatke, Oesterr. Bot. Z. 25: 231. 1875. —Types: Ethiopia, Habab, *J.M. Hildebrandt 436* [syntype: B (destroyed)]; Eritrea, Keren, Jule 1870, *O. Beccari 148* [syntype: B (destroyed); lectotype: FT (FT003451), **designated here**]

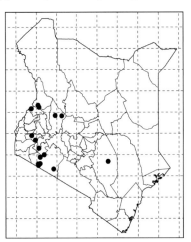

Young branches and leaves always glabrous; corolla-tube always glabrous outside.

Distribution: Western and central Kenya. [Central Africa to Ethiopia and D.R. Congo].

Habitat: Forests, thickets, riparian bushlands, or grasslands of riversides; 300–1750 m.

Note: We did not find the specimens *Welby s.n.* in Kew, which collected from Lake Rudolf of Turkana District in Kenya in 1899, so we are not sure whether *Pavetta subcana* var. *subcana* is found in Northern Kenya.

Baringo: Vtya, *Luke 210* (EA). Bomet: Sotik, *Dale 1047* (EA). Kisumu: Kisumu Airport, *Kokwaro 1824* (EA). Kitui: Endau, *SAJIT MU0131* (HIB). Narok: Mara River-Ngerendei, *Glover et al. 146* (EA); Masai Mara Reserve, *Kuchar 9740* (EA); Loita Plains, *Kuchar 14155* (EA). Trans Nzoia: Suam River, *Tweedie 3029* (K). West Pokot: Suk Escarpment, *Napier 2049* (EA).

19. Pavetta uniflora Bremek., Hooker's Icon. Pl. 32: *t*. 3194. 1933; F.T.E.A. Rubiac. 2: 686. 1988; K.T.S.L.: 530. 1994. —Type: Kenya, Kilifi, Arabuko-Sokoke Forest, 1929, *R.M. Graham* in *F.D. 1856* [holotype: K; isotypes: EA (EA000001693, EA000001694, EA000001695), G (G00014627)]

Shrub or small tree, up to 7 m tall. Leaves confined to very short spurs; blades narrowly ovate to obovate, 1–4 cm × 0.5–1.8 cm, rounded to obtuse at the apex, narrowly cuneate at the base; petioles up to 3 mm long; stipule-limbs with subulate tip up to 2 mm long. Flowers sessile on the leafy spurs, solitary or occasionally paired. Calyx-tube 1–2 mm long; limb-tube 1.8–2 mm long, lobes narrowly triangular, linear, 4–9 mm long. Corolla-tube slender, 2.6–4.3 cm long; lobes narrowly oblong, 4.5–6 mm long, acute to acuminate.

Fruits black, globose, 5–6 mm in diameter, with persistent calyx.

Distribution: Coastal Kenya. [East Africa, from Somalia to Mozambique].

Habitat: Coastal forests, dry bushlands, or hills; up to 350 m.

Kilifi: Arabuko-Sokoke Forest, *Polhill & Paulo 859* (EA); Mangea Hill, *Roberson & Luke 1779* (EA). Kwale: Dzombo Forest, *Robertson et al. MDE150* (EA); Shimba Hills, *Robertson & Luke 2810* (EA). Tana River: Galana Ranch, *Bally 16739* (EA).

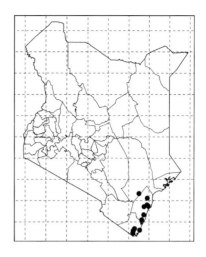

76. **Tennantia** Verdc.

Shrubs with leaves and flowers on lateral short shoots. Stipules narrowly triangular, keeled. Flowers 5–6-merous, in sessile few-flowered clusters. Calyx-tube cup-shaped; limb with distinct tube, lobes ovate. Corolla-tube very short; lobes oblong, contorted. Ovary 2-locular; ovules 3(–4) in each locule; style clavate, winged. Fruits small, globose, crowned with the persistent calyx-limb. Seeds few, segment-shaped, angular.

A monotypic genus restricted in Kenya and Somalia.

1. **Tennantia sennii** (Chiov.) Verdc. & Bridson, Kew Bull. 36: 511. 1981; F.T.E.A. Rubiac. 2: 582. 1988; K.T.S.L.: 547. 1994. ≡ *Tricalysia sennii* Chiov., Fl. Somalia 2: 240, f. 141. 1932. —Type: Somalia, Saar-Tumai, 1 Jule 1929, *L. Senni 437* [holotype: FT (FT003416)] (Figure 72; Plate 98)

= *Xeromphis keniensis* Tennant, Kew Bull. 22: 435. 1968. —Type: Kenya, Machakos, Mtito Andei, 16 Jan. 1961, *P.J. Greenway 9740* [holotype: K (K000286858); isotype: EA (EA000002963)]

Shrub, up to 3 m tall. Leaf-blades elliptic to narrowly obovate, 1–6.5 cm × 0.5–2.5 cm, rounded at the apex, cuneate at the base, glabrous or puberulous; stipules narrowly triangular, up to 2 mm long. Flowers white, pink or yellow, 3–5 in sessile clusters on lateral short shoots. Calyx-tube ca. 1.2 mm long; limb-tube ca. 1.2 mm long; lobes very short. Corolla-tube up to 2 mm long; lobes oblong, 4–5 mm long. Fruits yellow-green to

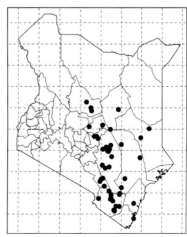

Figure 72 *Tennantia sennii*. A. a fruiting branch; B. flower; C. dissected corolla, showing the stamens; D. seed. Drawn by NJ.

Plate 98 *Tennantia sennii*. Photo by NW.

black, rounded, 5–6 mm in diameter, crowned by persistent calyx-limb. Seeds segment-shaped, ca. 4 mm long.

Distribution: Central, eastern and coastal Kenya. [Somalia].

Habitat: Dry bushlands or woodlands; up to 1100 m.

Embu: *Gillett & Mathew 19086* (EA, K). Garissa: south of Hagadera, *Kuchar & Msafiri 6436* (EA). Isiolo: Ngare Ndare, *Schultka 217* (EA). Kilifi: Ndara Ranch to Kajire, *Robertson 6529* (EA, K). Kitui: Mutomo Hill, *SAJIT MU0207* (HIB); Sukyangwa, *Kuchar 15059* (EA). Kwale: Twiga Beach, *Ekkens 649* (EA). Makueni: Kibwezi Forest Reserve, *Luke et al. 15324* (EA, K). Meru: Meru National Park, *Hamilton 737* (EA). Tharaka-Nithi: *Gilbert et al. 5718* (EA, K). Samburu: Mount Lolokwe, *Gillett 18970* (EA, K). Teita Taveta: Mount Kasigau, *Medley 960* (EA); Tsavo National Park, *Hucks 1965* (EA). Tana River: north of Hola, *Robertson 1769* (EA, K).

77. **Tarenna** Gaertn.

Shrubs, small trees or rarely lianas. Leaves opposite, subsessile to petiolate, sometimes with domatia; stipules interpetiolar, triangular, often aristate. Inflorescences of terminal corymbs, few- to many-flowered, sessile or pedunculate; bracteate or bracts reduced. Flowers often fragrant, (4–)5-merous, pedicellate or less often sessile. Calyx-tube turbinate to ovoid; limb 5-lobed. Corolla white to yellowish; tube cylindrical to funnel-shaped; lobes oblong, contorted. Stamens 5, inserted in corolla throat, exserted; filaments short or reduced; anthers dorsifixed. Ovary 2-locular with 1 to many in each locule; style slender; stigma fusiform or linear, sulcate or striate, shortly 2-lobed. Fruits berry-like, globose to ellipsoid, crowned with persistent calyx-limb, (1–)2- to many-seeded. Seeds hemispherical to variously shaped with hilar cavity.

A large genus of ca. 200 species widespread in the tropics and subtropics of the Old World; three species in Kenya.

1a. Corolla densely covered with adpressed hairs outside, with spreading lobes; seeds 2–4 per fruit .... 1. *T. trichantha*
1b. Corolla glabrous outside, with reflexed lobes; seeds usually more than 5 per fruit 2
2a. Stipules gradually narrowed or with subulate acumen up to 6 mm long; corolla-lobes oblong, less than 6 mm long .... 2. *T. pavettoides*
2b. Stipules obtuse to acute, with or without an apiculum up to 1 mm long; corolla-lobes narrowly oblong, 6–7 mm long .... 3. *T. drummondii*

1. **Tarenna trichantha** (Baker) Bremek., Repert. Spec. Nov. Regni Veg. 37: 207. 1934; F.T.E.A. Rubiac. 2: 602. 1988; K.T.S.L.: 547. 1994. ≡ *Pavetta trichantha* Baker, Bull. Misc. Inform. Kew 1894: 148. 1894. —Type: Seychelles, Aldabra I., *Abbott 23/26* [holotype: K]

Shrub or small tree, up to 4.5 m tall. Leaf-blades elliptic, ovate or rounded, 3–10 cm × 1.5–6 cm, acute to shortly acuminate at the apex, cuneate to rounded at the base, glabrous to pubescent on both sides; domatia present as tufts of whitish hairs; petioles up to 1.6 cm long; stipules triangular, 1.5–3 mm long, shortly apiculate, caducous. Flowers (4–)5-merous, many in dense corymbs, terminal on short lateral branches. Calyx-tube 1–1.3 mm long; limb-lobes triangular to lanceolate, up to 1.3 mm long. Corolla white, salver-shaped, densely covered with adpressed hairs outside; tube 4–5 mm long; lobes oblong, ca. 2 mm long, rounded. Fruits green, 4–5 mm in diameter, 2-locular, usually 2–4-seeded.

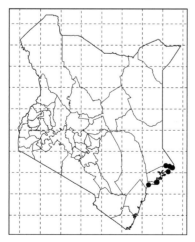

Distribution: Coastal Kenya. [Aldabra Islands, Comoros, Mozambique, and Tanzania].

Habitat: Coastal bushlands or thickets; up to 100 m.

Lamu: Sankuri Hill, *Festo et al. 2787* (EA); Ras Tenewi, *Luke & Robertson 1436* (EA). Tana River: Shekiko, *Luke & Robertson 1376* (EA).

2. **Tarenna pavettoides** (Harv.) Sim, Forest Fl. Cape: 239. 1907; F.T.E.A. Rubiac. 2: 588. 1988; K.T.S.L.: 547. 1994. ≡ *Kraussia pavettoides* Harv., Fl. Cap. 3: 22. 1865. —Type: South Africa, Natal, Durban [Port Natal], Field Hill, *J. Sanderson 656*

[holotype: TCD; isotypes: GAR (GRA 0002795-0), K (K000411741), MELB]

Shrub or small tree, up to 10 m tall. Leaf-blades elliptic, obovate to oblanceolate, 5–20 cm × 2–9 cm, acuminate to acute at the apex, cuneate at the base; glabrous or glabrescent to pubescent at the beneath nerves; domatia present; petiole up to 3 cm long; stipule-limbs deltoid, 2–6 mm long, gradually narrowed or with subulate acumen up to 6 mm long. Flowers (4–)5-merous, many in lax or congested corymbs; pedicels up to 8 mm long. Calyx-tube 1–1.5 mm long; limb 1–2.2 mm long, with ovate, deltoid or oblong lobes. Corolla white; tube 3–6.5 mm long; lobes oblong 3.5–6 mm long, rounded or sometimes apiculate or emarginate. Fruits green to black, 4–8 mm in diameter, 2–15-seeded.

1a. Young stems glabrous to sparsely pubescent; fruits ca. 15-seeded .... a. subsp. *friesiorum*
1b. Young stems always densely covered with adpressed hairs; fruits 4–13-seeded ....
.... b. subsp. *gillmanii*

a. subsp. **friesiorum** (K. Krause) Bridson, Kew Bull. 34: 386, fig. 2A & 3A,B. 1979; F.T.E.A. Rubiac. 2: 588. 1988; K.T.S.L.: 547. 1994. ≡ *Pavetta friesiorum* K. Krause, Notizbl. Bot. Gart. Berlin-Dahlem 10: 606. 1929. —Type: Kenya, Meru, between Meru and Nithi River, 25 Feb. 1922, *R.E. & T.C.E. Fries 1967* [holotype: B (destroyed); lectotype: K (K000353241), **designated here**; isolectotype: BR (BR0000008010933)]

Young stems glabrous to sparsely pubescent. Fruit ca.15-seeded.
Distribution: Central Kenya. [Tanzania].

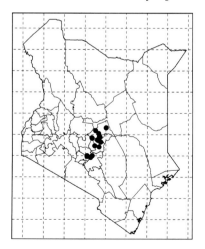

Habitat: Mountane forests; 1200–1600 m.
Embu: *Bally B2205* (K). Kiambu: Fourteen Falls, *Faden 68/740* (EA). Meru: Ngaia Forest, *Luke et al. 7160* (EA, K). Murang'a: south side of Thika River, *Faden 68/802* (EA, K). Tharaka-Nithi: Chuka, *Singh 3620* (EA).

b. subsp. **gillmanii** Bremek. ex Bridson, Kew Bull. 34(2): 386. 1979; F.T.E.A. Rubiac. 2: 590. 1988. —Type: Tanzania, Bukoba, Nshamba, Sept. 1935, *H. Gillman 565* [holotype: K (K000411749); isotype: EA (EA000001630)] (Plate 99A–C)

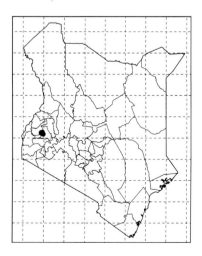

Plate 99 A–C. *Tarenna pavettoides* subsp. *gillmanii*; D–F. *T. drummondii*. Photo by GWH.

Figure 73 *Tarenna drummondii*. A. a flowering branch; B. longitudinal section of corolla, showing the stamens; C. infructescence; D. fruit. Drawn by NJ.

Young stems always densely covered with adpressed hairs; fruits 4–13-seeded.

Distribution: Western Kenya. [Tropical East Africa from Sudan to east D.R. Congo and Zambia].

Habitat: Upland forest edges or thickets; 1500–1800 m.

Kakamega: Kakamega Forest, *SAJIT 006715* (HIB). Nandi: west of Kaimosi Police Post, *Gilbert & Mesfin 6699* (EA, K).

3. **Tarenna drummondii** Bridson, Kew Bull. 34: 379. 1979; F.T.E.A. Rubiac. 2: 590. 1988; K.T.S.L.: 546. 1994. —Type: Kenya, Kwale, Mwasangombe Forest, 27 Aug. 1953, *R.B. Drummond & J.H. Hemsley 4024* [holotype: K (K000311612); isotypes: B (B100295999), BR (BR0000008007131), EA (EA000001629), FT (FT003383), K (K000311613)] (Figure 73; Plate 99D–F)

Shrub or small tree, up to 9 m tall. Leaf-blades elliptic to narrowly obovate, 8–19 cm × 2.5–9 cm, acute to acuminate at the apex, cuneate to obtuse or rounded and sometimes unequal at the base; domatia occasionally present; petioles up to 2 cm long; stipules triangular or oblong, 3–4 mm long, obtuse to acute, with or without an apiculum up to 1 mm long. Flowers many in terminal, subsessile, lax corymbs; pedicels up to 1.2 cm long. Calyx-tube 1–1.2 mm long; limb-lobes triangular to oblong, apiculate. Corolla white or cream; tube cylindrical, 4–6 mm long; lobes narrowly oblong, 6–7 mm long, rounded or irregularly emarginate. Fruits black, rounded, 5–7 mm in diameter.

Distribution: Coastal Kenya. [Tanzania].

Habitat: Coastal forests; up to 500 m.

Kilifi: Cha Simba, *Luke & Robertson 1879* (EA); Kambe Kaya, *Hawthorne 439* (EA). Kwale: Shimba Hills, *SAJIT 006072* (HIB); Dzombo Hill, *Robertson et al. MDE339* (EA, K); Mrima Hill, *Faden et al. 77/688* (K).

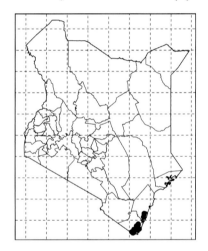

78. **Coptosperma** Hook. f.

Small trees or shrubs. Leaves coriaceous, less often chartaceous; stipules ovate to triangular, without darkened central area, acute to aristate. Inflorescences corymbose, generally terminal on main and leafy lateral branches, sometimes on short spurs or axillary. Flowers (4–)5-merous. Calyx-limb tubular, truncate, toothed or lobed. Corolla-tube glabrous outside; lobes contorted in bud. Ovary 2-locular, with 1–8 ovules impressed in large placenta or pendulous from small placenta; style exserted. Fruits fleshy, brown or red, drying blackish, 1(–2)-seeded. Seeds blackish-brown, reticulate.

A genus with about 20 species in eastern tropical Africa, Comoros Islands, Madagascar, Seychelles and the Mascarenes; six speices in Kenya.

1a. Stipules hairy within; inflorescences markedly lax; calyx-limb truncate to shortly toothed....... ....6. *C. nigrescens*
1b. Stipules not hairy within; inflorescences congested to moderately lax; calyx-limb distinctly lobed 2
2a. Inflorescences axillary; bracteoles somewhat cupular, surrounding the base of the calyx-tube.. .... 5. *C. supra-axillare*
2b. Inflorescences terminal; bracteoles not cupular 3
3a. Leaves subsessile, often very small, less than 3.5 cm × 1.2 cm 4
3b. Leaves distinctly petiolate, often larger, up to 13.5 cm × 6.3 cm 5
4a. Leaf-blades elliptic to oblong-elliptic, glabrous; corolla-tube shorter than lobes .... 1. *C. kibuwae*
4b. Leaf-blades very narrowly elliptic, puberulous on both sides; corolla-tube equaling the lobes.. .... 2. *C. wajirense*
5a. Domatia absent; stipules ovate, 3–7 mm long; calyx-tube 1–1.5 mm long....... 3. *C. graveolens*
5b. Domatia conspicuous; stipules ovate to triangular, 5–10 cm long .... 4. *C. littorale*

1. **Coptosperma kibuwae** (Bridson) Degreef, Syst. & Geogr. Pl. 71: 375. 2002. ≡ *Tarenna kibuwae* Bridson, Kew Bull. 34: 395. 1979; F.T.E.A. Rubiac. 2: 597. 1988; K.T.S.L.: 546. 1994. —Type: Kenya, Garissa, Garissa-Dadaab Road, 19 km from Garissa, 230 m, 11 May 1974, *J.B. Gillett & F.N. Gacathi 20598* [holotype: K (K000311611); isotypes: EA (EA000001624, EA000001625)]

Shrub, up to 3 m tall. Leaves mostly on short spurs; blades subsessile, subcoriaceous, glabrous, elliptic to oblong-elliptic, 0.8–2.4 cm × 0.2–1.2 cm, obtuse at the apex, cuneate at the base; domatia absent; stipules triangular, 1–1.5 mm long. Inflorescences of terminal cymes on short spurs, few-flowered. Calyx-tube ca. 1 mm long; limb ca. 0.8 mm long, lobes broadly triangular. Corolla greenish cream; tube funnel-shaped, ca. 3.5 mm long; lobes oblong, ca. 6 mm long. Fruits green, rounded, ca. 8 mm in diameter, 1-seeded.

Distribution: Eastern Kenya. [Endemic].
Habitat: Dry bushlands; 200–400 m.

Garissa: Garissa-Dadaab Road, *Gillett & Gachathi 20598* (EA, K). Tana River: Korokora, *Paulo 473* (EA, K).

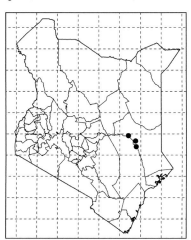

2. **Coptosperma wajirense** (Bridson) Degreef, Syst. & Geogr. Pl. 71: 381. 2002. ≡ *Tarenna wajirensis* Bridson, Kew Bull. 34: 398. 1979; F.T.E.A. Rubiac. 2: 597. 1988; K.T.S.L.: 547. 1994. —Type: Kenya, Wajir, Dadaab-Wajir Road, 9 km south of Sabale, 180 m, 12 May 1974, *J.B. Gillett & F.N. Gachathi 20649* [holotype: K (K000311610); isotype: EA

(EA000001623)]

Shrub, up to 2 m tall. Leaves crowded on short spurs or spaced on short lateral branches; blades lanceolate or narrowly elliptic, 1.5–3.5 cm × 0.5–1 cm, obtuse at the apex, cuneate or obtuse at the base, sessile or shortly petiolate, domatia absent; stipules triangular-ovate, 1.2–2 mm long. Inflorescences of dense cymes, terminal on short lateral branches, 5–25-flowered. Calyx-tube ca. 1 mm long, limb ca. 1 mm long, with rounded lobes. Corolla white or cream; tube ca. 4 mm long; lobes oblong, ca. 4 mm long, rounded. Fruit black, rounded, ca. 4.5 mm in diameter, 1-seeded.

Distribution: Eastern Kenya. [Endemic].
Habitat: Dry bushlands; 150–400 m.

Wajir: ca. 40 km north of Wajir on road to Tarbaj, *Gillett 19723* (EA, K); ca. 9 km south of Sabale, *Gillett & Gachathi 20649* (EA, K).

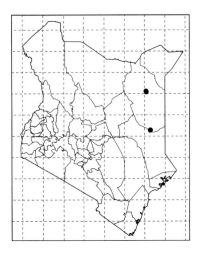

3. **Coptosperma graveolens** (S. Moore) Degreef, Syst. & Geogr. Pl. 71: 374. 2002. ≡ *Pavetta graveolens* S. Moore, J. Bot. 45: 267. 1907. ≡ *Tarenna graveolens* (S. Moore) Bremek., Repert. Spec. Nov. Regni Veg. 37: 193. 1934; F.T.E.A. Rubiac. 2: 598. 1988; K.T.S.L.: 546. 1994. —Type: Uganda, Toro, near mouth of Mpanga R., 21 Aug. 1906, *A.G. Bagshawe 1173* [holotype: BM (BM000528122)] (Figure 74; Plate 100A–C)

= *Tarenna graveolens* (S. Moore) Bremek. var. *impolita* Bridson, Kew Bull. 34: 395. 1979; F.T.E.A. Rubiac. 2: 599. 1988. ≡ *Coptosperma graveolens* (S. Moore) Degreef var. *impolitum* (Bridson) Degreef, Syst. Geogr. Pl. 71: 375. 2002. —Type: Kenya, South Horr, 15 Aug. 1944, *J. Adamson 25* in *Bally 3573* [holotype: K (K000353282)]

= *Tarenna boranensis* Cufod., Nuovo Giorn. Bot. Ital., n.s., 55: 86. 1948. —Types: Ethiopia, Pozzi di El Banno, 2 May 1939, *R. Corradi 2765* [lectotype: FT (FT003388), **designated here**] & *2766* [isolectotype: FT (FT003389)] & *2767* [isolectotype: FT (FT003390)]; Ethiopia, Tra El Banno ed El Dire, 14 May 1939, *R. Corradi 2776* [isolectotype: FT (FT003391)]; Ethiopia, Strada Atana-Murlè, 13 July 1939, *R. Corradi 2778* [syntype: FT (FT003392)]; Ethiopia, Fiume Sagan, 26 May 1939, *R. Corradi 2779* [syntype: FT (FT003393)]; Ethiopia, Rive del Ghizo, 28 July 1939, *R. Corradi 2780* [syntype: FT (FT003394)] & *2781* [syntype: FT (FT003395)] & *2782* [syntype: FT (FT003396)]; Ethiopia, Rive del Caschei, 4 July 1939, *R. Corradi 2783* [syntype: FT (FT003397)]; Ethiopia, Ruscello di El Dire, 15 May 1939, *R. Corradi 2784* [syntype: FT (FT003398)]; Ethiopia, El Meti bassa, 24 May 1939, *R. Corradi 2785* [syntype: FT (FT003399)]; Ethiopia, Strada di Amar Cocche, 22 June 1939, *R. Corradi 2786* [syntype: FT (FT003400)] & *2787* [syntype: FT (FT003401)] & *2788* [syntype: FT (FT003402)]; Kenya, Kitui, May 1877, *J.M. Hildebrandt 2755* [syntype: WU; isosyntypes: BM (BM000528130), K (K000411738), M (M0106267), P (P00546226, P00546227), W (W0010252)]; Tanzania, Lake Victoria, *Conrads 287* [syntype: WU]; Tanzania,

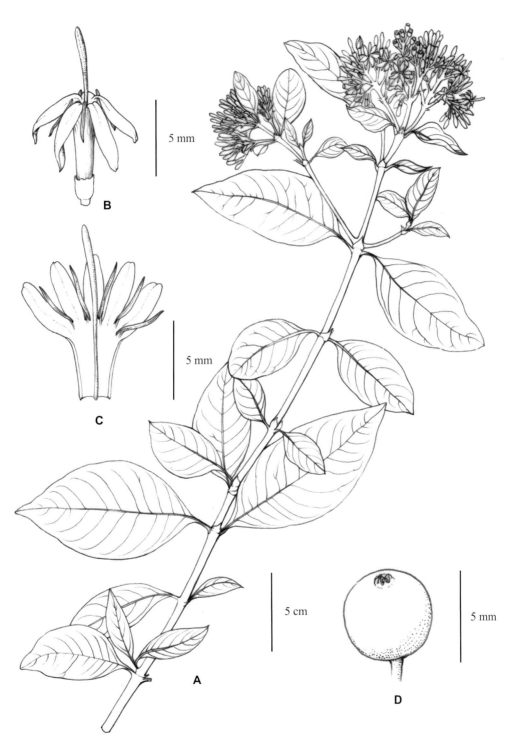

Figure 74 *Coptosperma graveolens*. A. branch with inflorescence; B. flower; C. longitudinal section of the corolla, showing the stamens and style; D. fruit. Drawn by NJ.

Usambara Mts., Mashewa, July 1893, *C. Holst 3562* [syntype: W; isosyntypes: K (K000411739), KFTA (KFTA0000631), M (M0106312)]

Shrub or tree, up to 9(–14) m tall. Leaf-blades coriaceous, lanceolate to narrowly elliptic, 3–13.5 cm × 1–6.5 cm, obtuse to acuminate at the apex, cuneate at the base; domatia absent; petioles up to 3 cm long; stipules ovate, 3–7 mm long, caducous. Inflorescences of rather compact corymbs, terminal, sessile; pedicels up to 3(–5) mm long. Calyx-tube 1–1.5 mm long; limb-tube up to 1.5 mm long; lobes up to 1.5 mm long. Corolla whitish to yellowish; tube cylindrical, (2–)3–6 mm long; lobes oblong, 3–6 mm long. Fruits black, rounded, 3.5–7 mm in diameter, with persistent calyx-limb, 1(–2)-seeded.

Distribution: Widespread in Kenya. [Tropical East Africa, also in Ethiopia and Somalia].

Habitat: Bushlands, grasslands, forest margins, or other rocky places; up to 2200 m.

Baringo: Mount Tiati, *Beentje 3130* (EA); near Chepkesin, *Bonnefille & Riollet 75/55* (EA); near Lake Hannington, *Faden & Napper 69/067* (EA, K). Busia: Port Victoria, *Glasgow 45/41* (EA). Homa Bay: Mfangano Island, *Paul 212* (EA). Kajiado: southwest of Ngong Hills, *Gillett 18686* (EA, K). Kericho: Ternan Fossil site, *Collinson 11* (EA). Kiambu: behind Blue Posts Hotel, *Faden 66/216* (EA, K). Kilifi: Watamu Forest, *Robertson et al. 5223* (EA); Mnarani, *Faden & Faden 71/815* (EA); Marereni, *Graham 1669* (K). Kirinyaga: Thiba River, *Ossent 679* (K). Kitui: Muumoni Hill Forest, *Mwachala et al. 612* (EA); Makongo Forest Reserve, *Mwangangi 4312* (EA); Mutomo Hill, *Gillett 19142* (EA, K). Kwale: Gandini, *Robertson & Luke 4856* (EA). Laikipia: Mugie Ranch, *Luke et al. 6503* (EA); Ol Ari Nyiro Ranch, *Robertson 4106* (EA, K). Lamu: Kiunga, *Kuchar 13598* (EA); Kiunga Marine National Park, *Taiti 25* (EA). Machakos: Lukenya Hill, *Bally 8990* (EA). Makueni: Kibwezi Hill, *Gatheri et al. 79/133* (EA, K); Kilima Kiu, *Gillett 18377* (K). Mandera: Dandu, *Gillett 12763* (EA, K). Marsabit: Marsabit National Reserve, *Luke & Luke 10876* (EA). Meru: Ngaya Forest, *SAJIT Z0194* (HIB); near Lake Nkuga, *SAJIT 003821* (HIB); Leopard Rock, *Hamilton 127* (EA). Mombasa: Port Tudor, *Gardner 2619* (EA). Nairobi: Nairobi Arboretum, *Dale 2885* (EA); Langata, *Ngweno 65* (EA); Karura Forest, *Bally 9216* (EA). Nakuru: Crater Lake Sanctuary, *Luke 3264* (EA); Lake Nakuru National Park, *Kutilek 105* (EA, K). Narok: Mount Suswa, *Glover & Samuel 3321* (EA); Chepkorobotik Forest, *Glover et al. 24* (EA); Soit Ololol Escarpment, *Sikes 251* (EA). Nyandarua: Karati Forest, *Gardner 1146* (EA). Nyeri: Zawadi Estate, *Faden et al. 74/571* (EA, K). Samburu: Mathew's Range, *Fratkin 14* (EA, K); Barsaloi, *Shikano 53* (EA); Maralal Leroki Hills, *Leakey 939* (EA). Teita Taveta: Sagala Forest, *SAJIT 005362* (HIB); Ngulia Hill, *Gilbert 2730* (EA); Mount Kasigau, *Gilbert & Gilbert 6115* (EA). Tana River: near Dalu, *Luke & Robertson 1296* (EA); Kora National

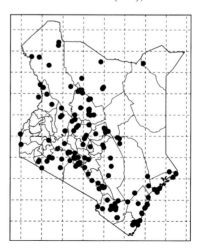

Reserve, *Hemming 83/113* (EA). Turkana: Kakuma, *Itani 78/19* (EA); Kuwalathe, *Paulo 1065* (EA); Lokitaung Gorege, *Carter & Stannard 128* (EA). West Pokot: Lomasei, *Mathew 6822* (K).

4. **Coptosperma littorale** (Hiern) Degreef, Syst. & Geogr. Pl. 71: 376. 2002. ≡ *Enterospermum littorale* Hiern, Fl. Trop. Afr. 3: 92. 1877. ≡ *Tarenna littoralis* (Hiern) Bridson, Kew Bull. 34: 397. 1979; F.T.E.A. Rubiac. 2: 600. 1988; K.T.S.L.: 546. 1994. —Type: Mozambique, Luame R., 3 June 1858, *J. Kirk s.n.* [lectotype: K (K000411723), designated by D.M. Bridson in Kew Bull. 34: 397. 1979]

Shrub or small tree, up to 10 m tall. Leaf-blades coriaceous, ovate to obovate or oblanceolate, 3–10 cm × 1.3–6 cm, rounded to obtuse at the apex, cuneate at the base; domatia conspicuous; petioles up to 1.5 cm long; stipules ovate to triangular, 5–10 mm long. Inflorescences of dense corymbs, terminal, usually sessile. Calyx-tube up to 1 mm long; limb with rounded teeth. Corolla whitish; tube narrowly funnel-shaped, 3–4 mm long; lobes oblong, 1.7–3 mm long. Fruits black, rounded, 5–7 mm in diameter, 1-seeded.

Distribution: Coastal Kenya just near Tanzania. [East Africa, from Kenya to northern South Africa].

Habitat: Coastal bushlands or forest margins; up to 10 m.

Kwale: Funzi Island, *Luke 5737* (EA); Mukurumudzi River, *Nyange & Luke 562* (EA, K).

5. **Coptosperma nigrescens** Hook. f., Gen. Pl. 2: 87. 1873; F.T.E.A. Rubiac. 2: 602. 1988; K.T.S.L.: 547. 1994. ≡ *Tarenna nigrescens* (Hook. f.) Hiern, Fl. Trop. Afr. 3: 92. 1877. —Type: Mozambique, Moramballa, 31 Dec. 1858, *J. Kirk s.n.* [lectotype: K (K000411724), designated by W.P. Hiern in Fl. Trop. Afr. 3: 92. 1877] (Plate 100D)

Shrub or tree, up to 6 m tall. Leaf-blades coriaceous, lanceolate, narrowly elliptic to narrowly ovate, 3–11 cm × 1–4 cm, acuminate or acute at the apex, cuneate or acute at the base; domatia absent; petioles 5–10 mm long; stipules 0.5–2(–3) mm long, obtuse or acute, hairy within, caducous. Inflorescences of lax corymbs, terminal, sessile or with peduncle up to 2 cm long, often supported by rudimentary leaves; pedicels up to 14 mm long. Calyx-

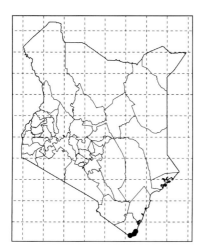
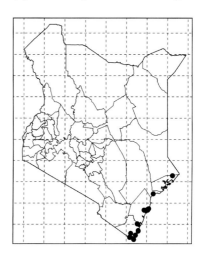

tube 1.5–2 mm long; limb-tube up to 2 mm long, truncate or with very short teeth. Corolla white or yellowish, glabrous outside, pubescent at throat; tube (1.5–)3–5 mm long; lobes 5–6 mm long. Fruits black when ripe, rounded, 6–8 mm in diameter, with persistent calyx-limb, 1-seeded.

Distribution: Coastal Kenya. [East Africa from Somalia to Mozambique, also in Madagascar and Comoros Islands].

Habitat: Lowland forests or bushlands; up to 400 m.

Kilifi: Arabuko-Sokoke Forest, *Simpson 386* (EA); near Malindi, *Robertson & Luke 5279* (EA, K); Gede Forest, *Omondi & Obunyali KEFRI308* (EA, K). Kwale: Kisite-Mpunguti Marine Park, *Mwadime et al. 227* (EA); Dzombo Hill, *Robertson et al. MDE264* (EA, K); Diani Forest, *Gillett & Kibuwa 19857* (EA, K). Lamu: Boni Forest Reserve, *Robertson & Luke 5592* (EA, K). Tana River: Shekiko, *Luke & Robertson 1369* (EA).

6. **Coptosperma supra-axillare** (Hemsl.) Degreef, Syst. & Geogr. Pl. 71: 379. 2002. ≡ *Pavetta supra-axillaris* Hemsl., J. Bot. 54(Suppl. 2): 19. 1916. ≡ *Tarenna supra-axillaris* (Hemsl.) Bremek., Repert. Spec. Nov. Regni Veg. 37: 206. 1934; F.T.E.A. Rubiac. 2: 599. 1988; K.T.S.L.: 547. 1994. —Type: Seychelles, Aldabra I., Ile Esprit, 1908, *J.C.F. Fryer s.n.* [holotype: K (K000172503)] (Plate 100E)

Shrub or small tree, up to 7 m tall. Leaf-blades coriaceous, glabrous, narrowly elliptic to elliptic, 3–12 cm × 1–4 cm, acute to subacuminate at the apex, cuneate at the base; domatia absent; petioles up to 10 mm long; stipules ovate or sometimes triangular, 2–5 mm long, caducous. Inflorescences of axillary or terminal cymes, with peduncles up to 2 cm long; bracts scale-like or with triangular lobes; bracteoles cupular. Calyx-tube 0.7–1.2 mm long; limb 0.7–1 mm long, lobed. Corolla white to cream-white; tube 2.5–5 mm long; lobes oblong, 2–3.5 mm long. Fruits black, rounded, 4–5 mm in diameter, 1-seeded.

Distribution: Coastal Kenya. [East Africa from Tanzania to Mozambique and Zimbabwe].

Habitat: Lowland forests or woodlands; up to 300 m.

Kilifi: Arabuko-Sokoke Forest, *Faden et al. 71/637* (EA); Kaya Kivara, *Robertson & Luke 4694* (EA); Mangea Hill, *Luke & Robertson 1783* (EA, K). Kwale: Shimba Hills, *Magogo & Glove 590B* (EA); Maluganji Forest Reserve, *Robertson & Luke 6008* (EA). Lamu: Mundane Range, *Robertson & Luke 5588* (EA, K); Boni Forest Reserve, *Luke & Robertson 1520* (EA, K). Mombasa: *Boivin s.n.* (BR).

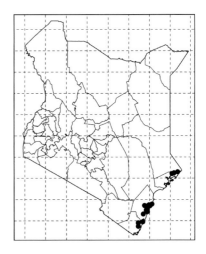

Plate 100 A–C. *Coptosperma graveolens*; D. *C. nigrescens*; E. *C. supra-axillare*. Photo by YDZ (A, B), SWW (C), BL (D) and VMN (E).

21. Trib. **Gardenieae** A. Rich. ex DC.

Shrubs, trees, or lianas. Raphides absent. Stipules interpetiolar, entire. Inflorescences terminal or pseudoaxillary. Aestivation contorted to the left, rarely to the right. Secondary pollen presentation predominantly present. Stigma club-shaped, capitate or linear, stigmatic lobes fused over most of their length, rarely divided. Ovary of 2–9(–16) carpels, uni- to plurilocular, placentas axile or parietal with one to many ovules. Fruit a berry with a dense fleshy mesocarp and a pulpy placenta embedding the seeds. Seeds with or without adaxial excavation.

Five genera occur in Kenya.

1a. Plants always armed with spines .... 83. *Catunaregam*
1b. Plants unarmed 2
2a. Ovary 1-locular, with numerous ovules on 2–9 parietal placentas; fruits usually with a thick fibrous, woody or leathery wall 3
2b. Ovary 2(–6)-locular, with single to many ovules in each locule; fruits not as above 4
3a. Flowers 5–12-merous; solitary or in few-flowered clusters; ovary with numerous ovules on 2–9 parietal placentas .... 79. *Gardenia*
3b. Flowers 5-merous, often solitary and pendent; ovary with numerous ovules on 2 parietal placentas .... 81. *Rothmannia*
4a. Flowers usually very large, with corolla-tube over 1 cm long .... 80. *Heinsenia*
4b. Flowers small, with corolla-tube usually less than 6 mm long .... 82. *Aidia*

79. **Gardenia** J. Ellis

Shrubs or small trees, unarmed or spinescent. Leaves opposite or rarely verticillate, petiolate; stipules connate, persistent to caducous. Flowers white to cream, fading to yellow, often fragrant, 5–12-merous, isostylous, solitary or in few-flowered clusters, terminal or pseudo-axillary. Calyx-tube globose, ovoid or oblong-ellipsoid; limb well-developed. Corolla salver-shaped, funnel-shaped or campanulate; lobes 5–12, contorted, obtuse to rounded at the apex. Stamens as many as the corolla-lobes, sessile or subsessile; anthers dorsifixed. Ovary 1-locular, with numerous ovules on 2–9 parietal placentas; style elongate; stigma exserted, clavate or 2-lobed. Fruits globose or ellipsoid, usually with a thick fibrous or woody wall. Seeds numerous, ellipsoid, compressed, embedded in pulp.

A genus of about 130 species widespread in tropical and subtropical regions of Africa, Asia, Madagascar, and Pacific islands; five species in Kenya.

1a. Desert shrub; leaf-blades 1–4 cm × 0.5–1.5 cm .... 2. *G. fiorii*
1b. Not a desert shrub; leaf-blades more than 4 cm × 2 cm 2
2a. Leaf-blades with almost transverse lateral nerves; corolla-tube 1.2–1.5 cm long; fruits 1–1.2 cm in diameter.. 1. *G. transvenulosa*
2b. Leaf-blades with normal lateral nerves; corolla-tube usually more than 2.5 cm long; fruit more than 1.6 cm in diameter...3
3a. Ovary with 3 placentas .... 3. *G. posoquerioides*
3b. Ovary with 6–9 placentas 4
4a. Fruits ellipsoid, usually smooth and without coarse ribs .... 5. *G. ternifolia*
4b. Fruits globose to ellipsoid, usually coarsely 8–11-ribbed .... 4. *G. volkensii*

1. **Gardenia transvenulosa** Verdc., Kew Bull. 34: 347. 1979; F.T.E.A. Rubiac. 2: 502. 1988; K.T.S.L.: 514. 1994. —Type: Kenya, Kilifi, Arabuko-Sokoke Forest, 8 June 1973, *B.M. Musyoki & O.J. Hansen 997* [holotype: K (K000311624); isotypes: C (C10001253), EA (EA000001609)]

Shrub or small tree, up to 4(–9) m tall. Leaf-blades coriaceous, elliptic to broadly elliptic, 4–12.5(–14) cm × 2.5–8 cm, acuminate at the apex, cuneate to rounded at the base, with almost transverse lateral nerves, glabrous or nearly so except for domatia; petioles up to 1.8 cm long; stipules sheathing at the base, soon splitting and becoming cup-like. Flowers white, solitary or paired, 5-merous, pseudo-axillary. Calyx-tube subglobose or ovoid, 2–3 mm long; limb-tube 2–3 mm long, with narrowly lanceolate lobes up to 10 mm long. Corolla-tube narrowly campanulate, 1–1.5 cm long; lobes rounded, 5–6 mm long. Fruits globose, 1–1.2 cm in diameter, crowned by the persistent calyx-limb.

Distribution: Coastal Kenya. [Mozambique and Tanzania].

Habitat: Lowland forest; up to 150 m.

Kilifi: Arabuko-Sokoke Forest, *Musyoki & Hansen 997* (EA, K); Nyari between Dida and Tezo, *Luke 1938* (EA, K).

2. **Gardenia fiorii** Chiov., Result. Sci. Miss. Stefan.-Paoli Somal. Ital. 1: 90. 1916; F.T.E.A. Rubiac. 2: 504. 1988; K.T.S.L.: 513. 1994. —Types: Somalia, Iscia Baidoa, 17 Oct. 1913, *G. Paoli 1220* [lectotype: FT (FT003409), **designated here**]; Golonle, 29 July 1913, *G. Paoli 783* [syntype: FT (FT003407)]; between Audinle and Berdale, 16 Oct. 1913, *G. Paoli 976* [syntype: FT (FT003408)]

Shrub, up to 4 m tall. Leaves always 3-whorled, often crowded on lateral short shoots, sessile to subsessile; blades elliptic, obovate to oblanceolate, 1–4 cm × 0.5–1.5 cm, acute at the apex, cuneate to narrowly subcordate at the base, densely pubescent on both sides; stipules

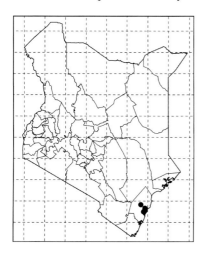

2–4.5 mm long. Flowers white, solitary, sessile to subsessile at tips of short branchlets. Calyx-tube ca. 1.5 mm long; limb-tube ca. 1 mm long, lobes linear to linear-spathulate, ca. 4 mm long, enlarged in fruit. Corolla-tube slender, 1.3–2.5 cm long; lobes 5–6, elliptic-oblong, 0.7–1.6 cm long. Ovary with 2 placentas. Fruits subglobose, 9–10 mm in diameter, hairy, crowned by the persistent calyx-limb.

Distribution: Central and eastern Kenya. [Somalia and Ethiopia].

Habitat: Open bushland; up to 750 m.

Garissa: Lebugombisso, *Adamson 6008* (EA). Kitui: Kasiokoni, *Edwards 182* (EA). Mandera: south of El Wak on Wajir Road, *Gilbert & Thulin 1650* (EA, K). Tana River: Garissa-Garsen Road, *Faden & Faden 74/1003* (EA, K).

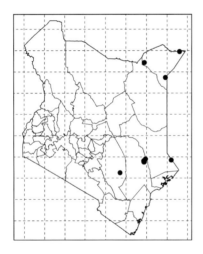

3. **Gardenia posquerioides** S. Moore, J. Linn. Soc., Bot. 40: 81. 1911; F.T.E.A. Rubiac. 2: 504. 1988; K.T.S.L.: 513. 1994. —Type: Zimbabwe, Chirinda Forest, 1128–1219 m, 31 Jan. 1906, *C.F.M. Swynnerton 71* [lectotype: BM (BM000903077), **designated here**; isolectotypes: K (K000419841), US (02496632)]; Chirinda Forest, *C.F.M. Swynnerton 6504* [syntype: BM] (Plate 101A)

Shrub or small tree, up to 6 m tall. Leaves glabrous, opposite or 3-wholed; blades ovate, elliptic, oblong or obovate, 10–22 cm × 4.5–9.5 cm, acuminate at the apex, cuneate at the base; petioles up to 1.5 cm long; stipules ovate to triangular, 1–1.5 cm long, obtuse. Flowers white, solitary, 5–6-merous, terminal, shortly pedicellate. Calyx-tube cylindric-fusiform, ca. 1.5 cm long; limb-tube 1.2–2 cm long, lobes linear-oblong, 2.5–6.5 cm long. Corolla-tube 9–12.5 cm long, lobes broadly oblong, 4–5.5 cm long, obtuse. Ovary with 3 placentas. Fruits fusiform, 4–6 cm long, markedly 6-ribbed, crowned by the persistent calyx-limb.

Distribution: Coastal Kenya. [South Africa, Tanzania, and Zimbabwe].

Habitat: Lowland forests; up to 400 m.

Kilifi: Ribe Kaya, *Hawthorne 304* (EA); Kwale: Shimba Hills, *SAJIT 006124* (HIB); Mrima Hill, *Robertson et al. 16* (EA, K).

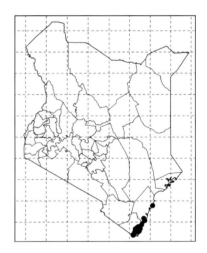

4. **Gardenia volkensii** K. Schum., Bot. Jahrb. Syst. 34: 332. 1904; W.F.E.A.: 152. 1987; F.T.E.A. Rubiac. 2: 507. 1988; K.T.S.L.: 514. 1994; U.T.S.K.: 249. 2005. —Type: Tanzania, Zanzibar coast area, 8 Nov. 1902, *A. Engler 2199* [holotype: B (destroyed)] (Plate 101B, C)

Shrub or small tree, up to 4.5(–10) tall. Leaves sessile to shortly petiolate, 3-whorled at ends of short shoots; blades obovate, 2–

9.5 cm × 1–5.5 cm, rounded at the apex, cuneate at the base, glabrous to finely puberulous or slightly scabrid; stipules semicircular to ovate-triangular, 2–5 mm long. Flowers white, fading to yellow or orange, solitary, sweet-scented, subsessile. Calyx-lobes oblong, up to 15 mm long; limb-tube 5–14 mm long, lobes variable. Corolla-tube narrowly cylindrical, 2.5–12.5 cm long; lobes elliptic to narrowly obovate, 2–5 cm long. Placentas 6–9. Fruit ellipsoid to subglobose, 4–11 cm long, always markedly coarsely 8–11-ribbed.

Distribution: Widespread in Kenya. [Tropical Africa, from Ethiopia and Somalia to northern South Africa].

Habitat: Grasslands, bushlands, or woodlands; up to 1800 m.

Baringo: north Loruk, *Leippert 5222* (EA). Elgeyo-Marakwet: Marakwet, *Lindsay 5* (EA). Homa Bay: Mfangano Island, *Paul 488* (EA). Isiolo: Samburu National Reserve, *Mwangangi et al. 55* (EA). Kilifi: Arabuko-Sokoke Forest, *Robertson 5213* (EA, K); Kaloleni, *Luke & Luke 9043* (EA); Dida, *Kimeu et al. KEFRI584* (K). Lamu: Mokowe to Bodhei, *Festo & Luke 2634* (EA). Makueni: Yatta Plateau, *Robertson 5016* (K). Marsabit: Lag Manzili Valley, *Gillett 13509* (EA, K); Marsabit Forest, *Muasya et al. GBK10* (EA). Meru: Meru National Park, *Hamilton 143* (EA). Migori: Suna, *Rayner s.n.* (EA). Narok: Masai Mara Game Reserve, *Kokwaro & Mathenge 2775* (EA); Ol Choro Orogwe Ranch, *Glover et al. 2011* (EA, K); Loita Hills, *Kuchar 7072* (EA). Samburu: Mount Ndoto, *Hepper & Jaeger 7250* (K); Barceloi, *Shikano 10* (EA). Teita Taveta: Mbololo River, *Greenway & Kanuri 13026* (EA, K). Tana River: southeast Golbanti, *Robertson & Luke 5433* (EA, K); Kurawa, *Polhill & Paulo 642* (EA, K); Kipendi West Forest, *Medley 355* (EA). Trans-Nzoia: Suam River, *Tweedie 3851* (K). Turkana: Lokori, *Newbould 7314* (EA); Mailongol Mountains, *Mathew 6808* (K). Uasin Gishu: Turbo, *Bally 2463* (K). West Pokot: Cherangani Hills, *SAJIT 006878* (HIB).

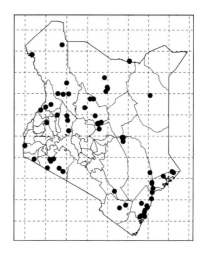

5. Gardenia ternifolia Schumach. & Thonn., Beskr. Guin. Pl.: 147. 1827; F.T.E.A. Rubiac. 2: 508. 1988; K.T.S.L.: 514. 1994. —Type: Ghana, Accra (Gah) and Adampi, *P. Thonning 140* [holotype: C (C10003878); isotypes: C (C10003877), FT, G-DC]

Shrub or small tree, up to 6(–10) m tall. Leaves subsessile, 3-whorled at the ends of short ternate shoots, 4–18 cm × 2–11 cm, blades oblanceolate to obovate, rounded or obtuse at the apex, attenuate to obtuse at the base; stipules broadly ovate, 2–4 mm long. Flowers white, fading to yellow, solitary, terminal. Calyx-tube oblong, up to 1 cm long; limb-tube 0.6–1.5 cm long, lobes very variable, subulate, linear to spathulate, up to 1.5 cm long. Corolla-tube cylindric, 4.5–11 cm long; lobes elliptic, 2–5.5 cm long. Fruits ellipsoid, 3.5–8 cm long, usually smooth and without coarse ribs.

subsp. **jovis-tonantis** (Welw.) Verdc., Kew Bull.34: 354. 1979; W.F.E.A.: 152. 1987; F.T.E.A. Rubiac. 2: 509. 1988; K.T.S.L.: 514. 1994. ≡ *Decameria jovis-tonantis* Welw., Apont.: 579, nota: 12. 1859. ≡ *Gardenia ternifolia* Schumach. & Thonn. var. *jovis-tonantis* (Welw.) Aubrév., Fl. For. Soud.-Guin.: 460. 1950. —Type: Angola, Golungo-Alto, Serra de Alto Queta, Dec. 1854, *F.M.J. Welwitsch 2573* [lectotype: LISU (LISU208562), designated by Verdcourt, Kew Bull. 34: 354. 1979; isolectotypes: BM (BM000924151), K, LISU (LISU208563)] (Figure 75; Plate 101D–F)

= *Gardenia lutea* Fresen., Mus. Senckenberg. 2: 167. 1837; K.T.S.: 441. 1921. —Type: Ethiopia, 2 days north of Gondar, *E. Rüppell s.n.* [holotype: FR (FR0030049)]

= *Gardenia goetzei* Stapf & Hutch., J. Linn. Soc., Bot. 38: 427. 1909. ≡ *Gardenia ternifolia* Schumach. & Thonn. var. *goetzei* (Stapf & Hutch.) Verdc., Kew Bull. 34: 355. 1979, **syn. nov.**; F.T.E.A. Rubiac. 2: 509. 1988. —Type: Tanzania, Kissaki Steppe, 1898, *G. Goetze 44* [holotype: K (K000419846)]

Young branches sparsely to densely strigose-pubescent. Flowers 6–9-merous; calyx-tube slightly pubescent.

Distribution: Widespread in Kenya. [Tropical Africa, from Ethiopia, Sudan, to Zambia and Angola, also in northern South Africa].

Habitat: Grasslands, bushlands, or woodlands; up to 2100 m.

Note: *Gardenia ternifolia* subsp. *ternifolia* with usually 6-merous flowers occurs in tropical Africa from Senegal to Ethiopia. Besides, two varieties were recorded in the subspecies *Gardenia ternifolia* subsp. *jovis-tonantis* (Welw.) Verdc. However, we believe the trait of indumentum of leaves is a continuous variation.

Baringo: Salawa Forest, *SAJIT 006593* (HIB). **Embu**: Mwea National Reserve, *Mwangangi 4702* (EA). **Garissa**: Boni Forest, *Adamson 5992* (EA). **Kajiado**: Chyulu Reserve, *Gardner 2573* (K). **Kitui**: *Bally 1553* (K). **Kwale**: Shimba Hills, *Magogo & Glover 4* (EA, K). **Laikipia**: Ol Ari Nyiro Ranch, *Muasya 2054* (EA). **Lamu**: Utwani Forest, *Rawlins 269* (EA, K). **Machakos**: Donyo Sabuk, *Lind et al. 5514A* (EA). **Makueni**: Chyulu Hills, *Luke & Luke 5230* (EA). **Masabit**: Masabit Forest, *SAJIT Z0285* (HIB). **Migori**: Kuja, *Glasgow 46/60* (K). **Nairobi**: Langata, *Archer 407* (EA). **Narok**: Masai Mara National Reserve, *Reinhard & Radhe 36* (EA). **Teita Taveta**: Bura, *Mwachala 311* (EA). **Trans-Nzoia**: Saiwa Swamp National Park, *Kirika 179* (EA). **Uasin Gishu**: Turbo, *Bally 2463* (K).

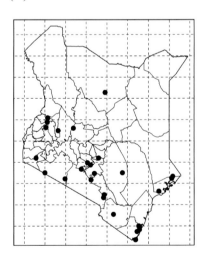

Plate 101 A. *Gardenia posoquerioides*; B, C. *G. volkensii*; D–F. *G. ternifolia* subsp. *jovis-tonantis*. Photo by GWH (A–C, F) and YDZ (D, E).

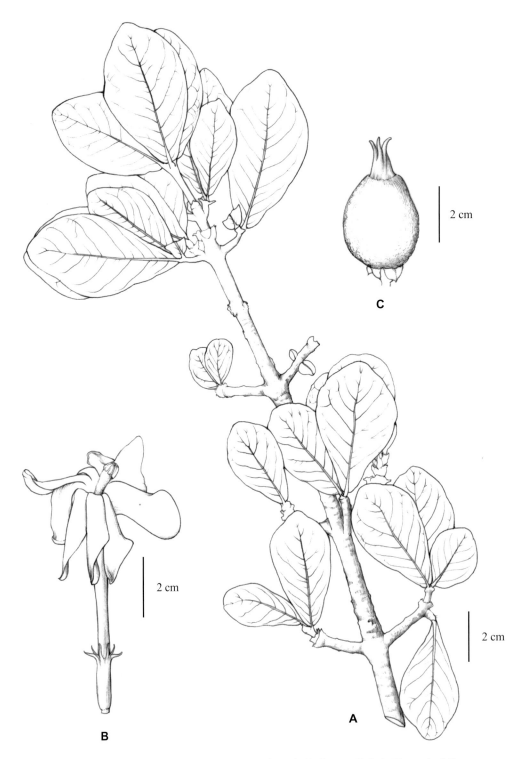

Figure 75 *Gardenia ternifolia* subsp. *jovis-tonantis*. A. a branch; B. flower; C. fruit. Drawn by NJ.

80. **Heinsenia** K. Schum.

Shrubs or trees. Leaves opposite, petiolate, with or without domatia; stipules triangular, acute to acuminate. Inflorescences several-flowered on very short terminal branchlets subtended by a single leaf. Calyx-tube ovoid to tubular; limb shortly cylindrical, with 5 triangular to lanceolate teeth. Corolla-tube narrowly campanulate; lobes 5, ovate to lanceolate. Ovary 2-locular, with 2–10 ovules in each locule; style filiform or elongate-clavate, with upper part 10-ribbed. Fruits subglobose, crowned by the persistent calyx-limb, 1–several-seeded.

A monotypic genus restricted to tropical Africa including Kenya.

1. **Heinsenia diervilleoides** K. Schum., Bot. Jahrb. Syst. 23: 454. 1897; F.T.E.A. Rubiac. 2: 730. 1988; K.T.S.L.: 515. 1994. —Type: Tanzania, E. Usambara Mts., Derema (Nderema), *E. Heinsen 23* [holotype: B (destroyed); lectotype: K (K000172772), **designated here**; isolectotypes: BR (BR0000008846686, BR0000008846693)] (Figure 76; Plate 102)

Shrub or small tree, up to 12(–15) m tall. Leaf-blades often red when young or with the nerves red beneath, glabrous to puberulous, elliptic, oblong to oblanceolate, 5–16.5 cm × 1–8 cm, acuminate at the apex, cuneate at the base; petioles up to 1 cm long; stipules triangular, 4–7 mm long. Flowers white, spotted pink inside, 2–6 in cymes on very

Plate 102 *Heinsenia diervilleoides*. Photo by SWW.

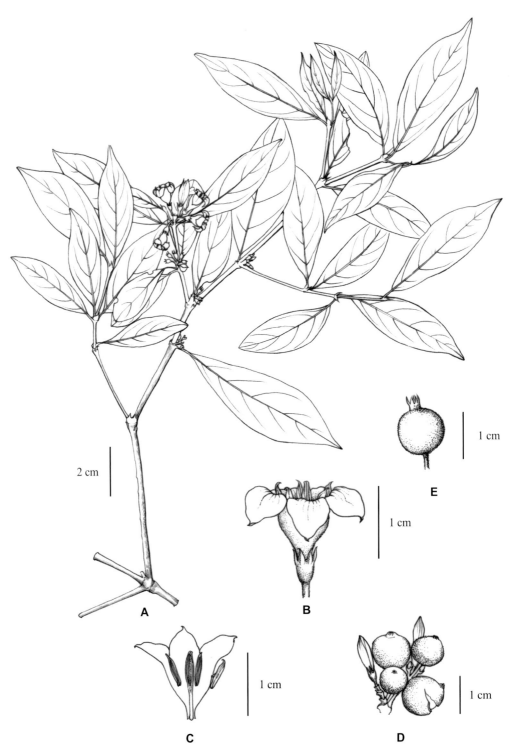

Figure 76 *Heinsenia diervilleoides*. A. flowering branch; B. flower; C. longitudinal section of corolla, showing the stamens and style; D. infructescence; E. fruit. Drawn by NJ.

short terminal branchlets. Calyx-tube 1–1.5 mm long; limb 2–4 mm long. Corolla-tube up to 1.2 cm long; lobes ovate, 3.5–8.5 mm long, obtuse. Fruits greenish purple, subglobose, 1.1–1.3 cm in diameter.

Distribution: Northern, western, central, southern, and coastal Kenya. [Tropical Africa].

Habitat: Forests; 350–2400 m.

Baringo: Katimok Forest, *Dale 2421* (EA). Bungoma: near Forester's House, *Drummond & Hemsley 4764* (EA, K). Homa Bay: Nyanza Basin, *Moon 588* (EA). Kakamega: Kakamega Forest, *SAJIT 006689* (HIB). Kericho: Kimugu Tea Estate, *Perdue & Kibuwa 9264* (EA, K). Kiambu: Kanyan Tea Estate, *Luke 442* (EA). Kisumu: Morongiot, *SAJIT 006977* (HIB). Kwale: Shimba Hills, *Luke & Robertson 2734* (EA). Machakos: Ol Doinyo Sapuk National Park, *Faden et al. 74/1305* (K). Makueni: Chyulu Hills, *Luke & Luke 10304* (EA). Marsabit: Marsabit Forest, *Verdcourt 2223* (K). Meru: Meru Forest, *SAJIT 002371* (HIB). Migori: Bukuria, *Graham 8* (K). Nandi: South Nandi Forest, *SAJIT 006674* (HIB). Nyeri: Lower Sagana Forest, *Gardner 18566* (K). Samburu: Mathew's Range, *Luke 14140B* (EA, K). Teita Taveta: Ngangao Forest, *Faden et al. 209* (K).

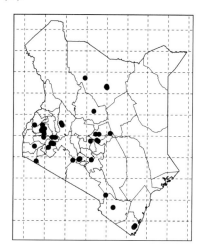

81. **Rothmannia** Thunb.

Shrubs or small trees. Leaves opposite or ternate, petiolate, sometimes with domatia; stipules triangular, acuminate. Flowers terminal on very reduced branches, large, usually 5-merous, often solitary and pendent. Calyx-tube obconical to cylindrical; limb-tube velvety inside; lobes present or absent, erect. Corolla-tube funnel-shaped or campanulate; lobes contorted. Stamens inserted in the corolla-tube; anthers included or partly exserted. Ovary 1-locular, with 2 parietal placentas; ovules numerous; style slender. Fruits globose or ellipsoid, with a thick or leathery wall. Seeds numerous.

A genus of about 43 species widespread in tropical Africa, Madagascar and tropical Asia; six species in Kenya.

1a. Corolla-tube over 11 cm long 2
1b. Corolla-tube less than 8 cm long 3
2a. Inland species; calyx glabrous outside, tube usually less than 1 cm long .... 4. *R. longiflora*
2b. Coastal species; calyx gloden or reddish-brown pubescent, tube 9–14 mm long ....
 .... 5. *R. macrosiphon*
3a. Flowers solitary on short branches, or 2–3 in fascicles; peduncles 2–8 mm long; corolla pubescent outside .... 6. *R. manganjae*
3b. Flowers solitary, subsessile; corolla glabrous or pubescent outside 4

4a. Leaves chartaceous, glabrous to pubescent; calyx and corolla pubescent outside .... 3. *R. urcelliformis*

4b. Leaves coriaceous, glabrous; calyx and corolla glabrous outside 5

5a. Stipules triangular, up to 6 mm long; calyx-limb often splitting; corolla-lobes 1.3–3.5 cm long; fruits globose .... 1. *R. fischeri*

5b. Stipules triangular, less than 2.5 mm long; calyx-limb never splitting; corolla-lobes 2.3–4.5 cm long; fruits ellipsoid to broadly ellipsoid .... 2. *R. ravae*

1. **Rothmannia fischeri** (K. Schum.) Bullock, Ann. Transvaal Mus. 17: 224. 1937; F.T.E.A. Rubiac. 2: 512. 1988; K.T.S.L.: 541. 1994. ≡ *Randia fischeri* K. Schum., Pflanzenw. Ost-Afrikas C: 380. 1895. —Type: Tanzania, Mwanza, Kayenzi, *A. Fischer 318* [syntype: B (destroyed); lectotype: K (K000414841), **designated here**]; ibid., *A. Fischer 296* [syntype: B (destroyed)]

Shrub or small tree, up to 8(–10) m tall. Leaves coriaceous, glabrous, elliptic, narrowly obovate to obovate, 2–11(–14) cm × 1–6 cm, obtuse, acute or subacuminate at the apex, acute, cuneate or obtuse at the base, with domatia; petioles up to 1 cm long; stipules triangular, up to 6 mm long, acuminate. Flowers solitary; peduncle very abbreviated. Calyx glabrous; tube 3–7 mm long; limb-tube 4–11 mm long, lobes absent or filiform to subulate, up to 18 mm long. Corolla cream with red spots, tube funnel-shaped or narrowly cylindrical below then funnel-shaped, 3–8 cm long; lobes ovate or lanceolate, 1.3–3.5 cm long. Stamens slightly exserted. Fruits green with pale spots, rounded, 3–6.2 cm in diameter.

subsp. **verdcourtii** Bridson, Kew Bull. 39: 68. 1984; F.T.E.A. Rubiac. 2: 513. 1988; K.T.S.L.: 541. 1994. —Type: Kenya, Teita Taveta, Bura near Voi, 914 m, Jule 1937, *I.R. Dale* in *F.D. 3787* [holotype: K; isotypes: BR (BR0000008855237), EA (EA000001663, EA000001664)] (Plate 103A–D)

Leaf-blades elliptic to broadly elliptic, acute to subacuminate at the apex, obtuse to acute at the base. Calyx-limb truncate or repand, with minute lobes. Corolla-tube gradually funnel-shaped, 5.5–7.2 cm long; lobes 1.3–1.8 cm long.

Distribution: Central and southern Kenya. [Tanzania].

Habitat: Bushlands; 900–2100 m.

Note: There are three subspecies recorded in the species *Rothmannia fischeri* (K. Schum.) Bullock.: subsp. *fischeri* occurs in eastern Africa from Tanzania to northern part of South Africa; subsp. *moramballae* (Hiern) Bridson just occurs in eastern Africa from Zimbabwe to northeast South Africa.

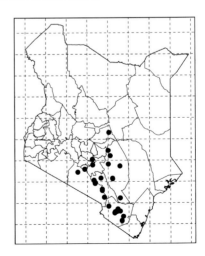

Kajiado: Kampi ya Ndege, *Bally B9458* (K). Kitui: Mutomo Hill, *SAJIT MU0075* (HIB); Endau forest, *Owino & Mathenge 214* (EA). Machakos: Matuu, *Fukuoka K-200* (EA). Makueni: Kampi ya Mawe, *Katumani 1039* (EA); Chyulu Plains, *Luke et al. 8363* (EA); Keo Hill, *Bally 2590* (EA). Meru: Ngaia Forest, *Luke et al. 7141* (EA). Murang'a: Nairobi-Mwingi Road to Embu Road, *KSCP/PGRWG 003/91/2000* (K). Teita Taveta: Kasigau Forest, *SAJIT 005396* (HIB); west slope of Bura Valley, *Mwachala et al. EW1181* (EA); Manda Hill, *Gillett 2367* (EA).

2. **Rothmannia ravae** (Chiov.) Bridson, Kew Bull. 39: 71. 1984. ≡ *Randia ravae* Chiov., Fl. Somala 2: 237, f. 139. 1932. —Type: Somalia, Middle Juba, Badadda (Baddada), *L. Senni* 272 [holotype: FT]

Shrub or small tree, up to 6(–10) m tall. Leaf-blades narrowly to broadly elliptic, 5–12 cm × 1.7–5 cm, acuminate at the apex, acute to cuneate at the base, with domatia; petioles 4–10 mm long; stipules triangular, 1–2.5 mm long. Flowers solitary, peduncles very abbreviated. Calyx glabrous; tube 5–8 mm long; limb-tube 6–10 mm long, distinctly ribbed, lobes filiform, 4–15 mm long. Corolla white to greenish-yellow, with red to purple spots, glabrous outside; tube narrowly cylindrical at the base, campanulate above, 3–7 cm long; lobes ovate, 2–3 cm long. Stamens slightly exserted. Fruits ellipsoid to globose, 4–8 cm long, green with pale spots.

Distribution: Southern and coastal Kenya. [Somalia and Tanzania].

Habitat: Forests or thickets; up to 1000 m.

Kilifi: Mangea Hill, *Luke & Robertson 1805* (EA, K); Malindi, *Robertson 7720* (EA); Bamba, *Bally B8550* (K). Kwale: Chuna Forest, *Luke & Robertson 570* (EA); Shimba Hills, *Luke 2932* (EA). Lamu: Mangai, *Festo & Luke 2774* (EA, K). Teita Taveta: Maungu Hills, *Faden et al. 70/177* (EA, K); Ngangao Forest, *Block et al. 318* (K).

3. **Rothmannia urcelliformis** (Hiern) Bullock ex Robyns, Fl. Spermatophyt. Parc Nat. Albert 2: 340. 1947; F.T.E.A. Rubiac. 2: 514. 1988; K.T.S.L.: 541. 1994. ≡ *Gardenia urcelliformis* Schweinf. ex Hiern, Fl. Trop. Afr. 3: 104. 1877. —Types: Sudan, Nabambissoo, 19 Feb. 1870, *G. Schweinfurth 3034* [lectotype: K (K000414785), **designated here**]; D.R. Congo, Niamniam, 5 Mar. 1870, *G. Schweinfurth 3292* [syntype: BM (BM000903105)] (Plate 103E, F)

Shrub or small tree, up to 9(–15) m tall. Leaves opposite or occasionally ternate; blades elliptic to obovate-elliptic, 6–18 cm × 2–10 cm, distinctly acuminate at the apex, cuneate at the base, glabrous, glabrescent or rarely pubescent on both sides; petioles up to 1.1 cm long; stipules triangular, up to 1.2(–1.5) cm long, long-acuminate. Flowers solitary; peduncles very abbreviated; bracteoles filiform. Calyx

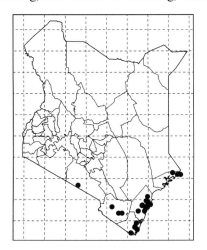

Plate 103 A–D. *Rothmannia fischeri* subsp. *verdcourtii*; E, F. *R. urcelliformis*. Photo by GWH (A, E, F), NW (B, D) and VMN (C).

pubescent; tube 3–6(–11) mm long, 5-angled; limb-tube up to 2.3 cm long, lobes linear, up to 2(–2.5) cm long. Corolla white to cream with reddish-purple markings; tube narrowly cylindrical below, funnel-shaped above, 3–7.7 cm long; lobes lanceolate, 1.2–4.5 cm long, acuminate. Stamens slightly exserted. Fruits globose to ellipsoid, 2.5–7.6 cm long.

Distribution: Western, central, and southern Kenya. [Tropical Africa].

Habitat: Forests; 1200–2000 m.

Kajiado: Ol Doinyo Orok Hill, *Nyakundi 231* (EA). Kakamega: Kakamega Forest, *SAJIT 006725* (HIB). Kericho: Cheptuiyet South Belgut, *Kerfoot 4866* (EA). Kiambu: Chania Falls, *Faden 68/784* (EA). Kirinyaga: Njukini Forest, *Brunt 1481* (K). Kisumu: northwest Kakamega Forest, *Mabberley & Tweedie 1095* (EA, K). Kitui: Endau Forest, *Daughty H8/38/3* (EA). Makueni: Chyulu Hills, *Luke et al. 10319B* (EA). Meru: Ngaia Forest, *Luke et al. 9579* (EA, K). Nairobi: Karura Forest, *Mwangangi & Abdalla 268* (EA, K); Nairobi City Park, *Kahurananga & Kiilu 2970* (EA). Nyamira: Ngoina Tea Estate, *Spjut & Ensor 3135* (EA, K). Samburu: Mathew's Range, *Ichikawa 885* (EA). Teita Taveta: Kasigau Hills, *Luke et al. 5396* (K). Tharaka-Nithi: Mount Kenya, *Luke 693* (EA).

4. **Rothmannia longiflora** Salisb., Parad. Lond.: t. 65. 1807; F.T.E.A. Rubiac. 2: 515. 1988; K.T.S.L.: 541. 1994. —Type: Ghana, Cape Coast, *L.J. Brass s.n.* [holotype: BM (BM000903045)] (Plate 104A)

Shrub, small tree or rarely liana, up to 9 m tall. Leaf-blades glabrous, elliptic or broadly elliptic, 6–15(–18) cm × 2.5–6(–8) cm, acuminate at the apex, cuneate at the base, with domatia; petioles up to 0.8(–1) cm long; stipules triangular, ca. 3 mm long, acuminate. Flowers solitary, terminal on short branches above a single leaf; peduncles up to 1 cm long; bracteoles 5–9, scale-like. Calyx glabrous outside; tube 0.5–1 cm long; limb-tube 0.7–1 cm long, lobes shortly triangular or linear, 1–2(–4) mm long. Corolla purple-green, finely pubescent outside; tube narrowly cylindrical below, funnel-shaped above, 14–18(–24) cm long; lobes ovate, 1.3–2.5 cm long, obtuse. Stamens included or slightly exserted. Fruits globose or ellipsoid, 3.5–5.2(–7) cm long, 10-ribbed.

Distribution: Western Kenya. [Tropical Africa].

Habitat: Forests; 1600–1700 m.

Kakamega: Kakamega Forest, *Mwangangi & Gliniars 34* (EA).

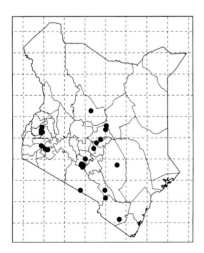
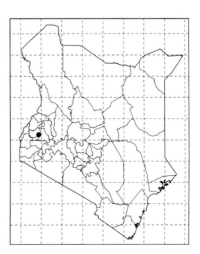

5. **Rothmannia macrosiphon** (K. Schum. ex Engl.) Bridson, Kew Bull. 31: 180. 1976; F.T.E.A. Rubiac. 2: 518. 1988; K.T.S.L.: 541. 1994; C.P.K.: 387. 2016. ≡ *Randia macrosiphon* K. Schum. ex Engl., Abh. Königl. Akad. Wiss. Berlin: 28. 1894. —Type: Tanzania, Tanga, Duga, June 1893, *C. Holst 3179* [holotype: B (destroyed); lectotype: K (K000352652), **designated here**; isolectotypes: G (G00359435), M (M0106394), P (P03934776)] (Figure 77; Plate 104B)

Shrub or small tree, up to 8 m tall, with young branches finely velutinous. Leaf-blades coriaceous, glabrous or nearly so, oblanceolate to obovate, 4.5–14.5 cm × 2–7 cm, acuminate at the apex, cuneate at the base, without domatia; petioles up to 1.5 cm long; stipules triangular, 3–5 mm long, acuminate. Flowers solitary, terminal above a single leaf; peduncles up to 1.6 cm long; bracteoles triangular, 2–6 mm long. Calyx golden or reddish-brown pubescent; tube 9–14 mm long; limb-tube up to 11 mm long, with linear lobes. Corolla white with reddish purple spots inside, pubescent; tube narrowly cylindrical below, campanulate above, 11–24 cm long; lobes ovate, 1.3–2.6 cm long. Stamens included. Fruits ovoid or rounded, 2.5–3.2 cm long, slightly ribbed, pubescent.

Distribution: Coastal Kenya. [Tanzania].
Habitat: Coastal forests; up to 500 m.

Kilifi: Mangea Hill, *Luke & Robertson 312* (EA). Kwale: Shimba Hills, *SAJIT 005928* (HIB); Buda Mafisini Forest Reserve, *Luke & Robertson 1676* (EA, K); Kaya Muhaka, *Makokha et al. 1568* (EA).

6. **Rothmannia manganjae** (Hiern) Keay, Bull. Jard. Bot. État 28: 56. 1958; F.T.E.A. Rubiac. 2: 517. 1988; K.T.S.L.: 541. 1994. ≡ *Gardenia manganjae* Hiern, Fl. Trop. Afr. 3: 103. 1877. —Type: Zimbabwe, Manganja Hills, Sept. 1861, *C.J. Meller s.n.* [holotype: K (K000414833)] (Plate 104C–E)

= *Randia fratrum* K. Krause, Notizbl. Bot. Gart. Berlin-Dahlem 10: 604. 1929. —Type: Kenya, Meru, 19 Feb. 1922, *R.E. & T.C.E. Fries 1631* [holotype: B (destroyed); lectotype: K (K000414840), **designated here**; isolectotype: S (S15-61821)]

Shrub or small tree, up to 12(–15) m tall. Leaves opposite, narrowly elliptic to oblanceolate, (4.3–)8–15.2 cm × 2.1–5.9 cm, base cuneate, apex acuminate or acute; petioles

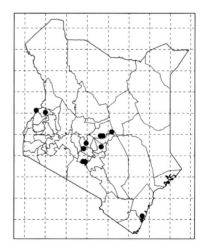

Plate 104 A. *Rothmannia longiflora*; B. *R. macrosiphon*; C–E. *R. manganjae*. Photo by GWH.

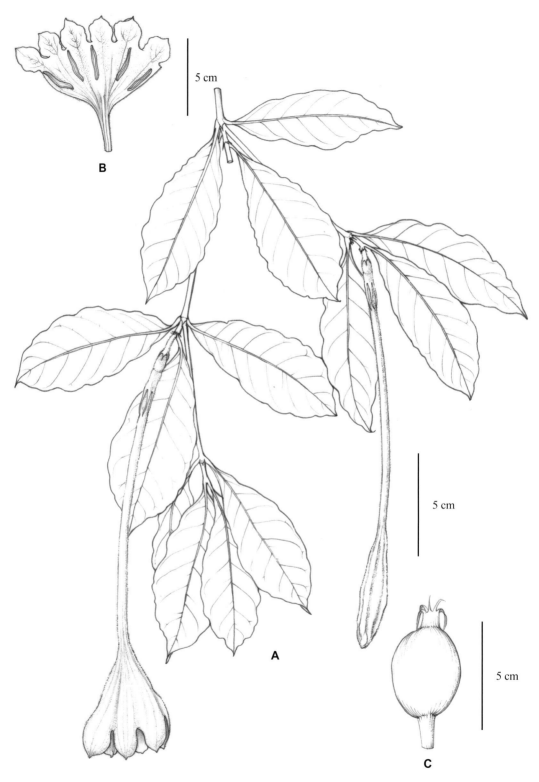

Figure 77 *Rothmannia macrosiphon*. A. flowering branch; B. longitudinal section of the upper part of corolla, showing the stamens; C. fruit. Drawn by NJ.

0.3–1.6 cm long; stipules triangular, 1–3 mm long, caducous. Flowers solitary, terminal on short branches above a single leaf; peduncles 2–8 mm long; pedicels absent. Calyx glabrous, with tube 0.4–1 cm long; limb-tube 0.4–1.2 cm long, with shortly triangular lobes up to 3 mm long. Corolla white with reddish marks inside, campanulate, 2.5–7.5 cm long; lobes ovate, 1–3.1(–4.3) cm long. Style 1.5–3 cm long. Fruits brown to black, globose, 1.8–4 cm in diameter.

Distribution: Western, central and coastal Kenya. [Malawi, Mozambique, Tanzania, and Zimbabwe].

Habitat: Forests; 200–1900 m.

Embu: Manyatta, *Graham 1720* (EA, K). Kilifi: Kaya Jibana, *Luke & Robertson* 2653 (EA, K). Meru: Meru Forest, *SAJIT 003892* (HIB); Lower Imenti Forest, *Faden & Faden 74/884* (EA, K). Murang'a: Tuso, *Napier 2314* (K). Nairobi: Nairobi Arboretum, *Quene & Wege 89* (EA). Nyeri: Nyeri Railway Station Forest, *Gardner 2477* (EA, K). Trans-Nzoia: Kwa Muthoni Shopping Center, *Maundu et al. 7* (EA).

82. **Aidia** Lour.

Small trees or shrubs, sometimes scandent. Leaves opposite, petiolate, usually with domatia; stipules persistent or more often caducous, triangular. Inflorescences pseudoaxillary, lateral at nodes with subtending scalelike leaves, cymose, few- to many-flowered, sessile to pedunculate. Flowers bisexual, 5-merous, sessile or pedicellate. Calyx-tube ovoid or turbinate; limb cupular or campanulate, slightly lobed or denticulate. Corolla-tube cylindrical, hairy at the throat; lobes contorted, acute, usually strongly reflexed. Stamens exserted; anthers sagittate. Ovary 2(–3)-locular; ovules several to many in each locule; style exserted; stigma clavate, fusiform or 2-lobed. Fruits globose; calyx-limb persistent or deciduous. Seeds several to numerous, angled to compressed.

A genus of about 55 species widespread in tropical Africa, South and Southeast Asia; only one species in Kenya.

1. **Aidia abeidii** S.E. Dawson & Gereau, Novon 20: 257. 2010. —Type: Tanzania, Bagamoyo, Zaraninge Forest, 260 m, 1 July 1998, *Y.S. Abeid 264* [holotype: K (K 000771307 & K000771310); isotypes: BR, M, MO (MO-2292748), NHT, UPS] (Figure 78)

Tree or somewhat scandent shrub, up to 12 m tall. Leaf-blades coriaceous, glabrous, elliptic to elliptic-lanceolate, (3–)7–13 cm × 1–4.2 cm, acuminate at the apex, narrowly cuneate and sometimes slightly asymmetric at the base; domatia present. Inflorescence

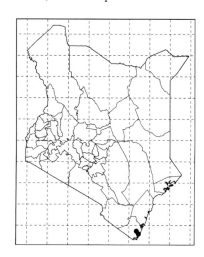

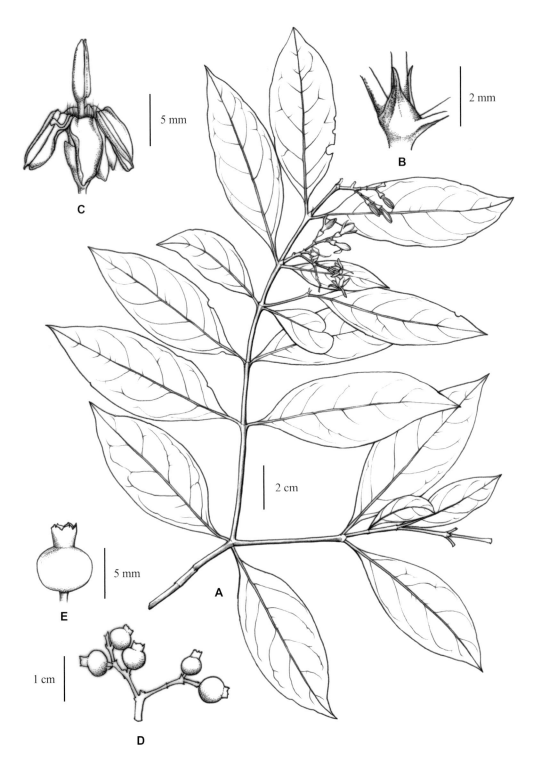

Figure 78 *Aidia abeidii*. A. a flowering branch; B. a portion of branch showing the stipule; C. flower; D. inflorescences; E. fruit. Drawn by NJ.

lateral at nodes with subulate subtending scale-like leaf, 7–35-flowered; bracts and bracteoles rounded. Calyx-tube turbinate, 3.7–4 mm long; limb funnel-shaped, 2–2.5 mm long, lobes rounded, ca. 0.5 mm long, short-mucronulate. Corolla-tube funnel-shaped, ca. 5 mm long, hairy at the throat; lobes ovate, 6.3–6.8 mm long, margins revolute. Ovary 2-locular; ovules 12–22; style ca. 12 mm long. Fruits subglobose, 5–6 mm in diameter, with persistent calyx-limb.

Distribution: Coastal Kenya. [Tanzania].
Habitat: Lowland forests; up to 400 m.
Kwale: Shimba Hills, *Luke 8317* (EA, K); Gongoni Forest, *Luke 2948* (EA, K).

83. **Catunaregam** Wolf

Shrubs or small trees, often armed with spines or spinescent short shoots. Leaves opposite or often clustered on short shoots; stipules interpetiolar, ovate-acuminate to triangular, apiculate, deciduous. Inflorescences terminal on short shorts, fasciculate or cymose, sessile to pedunculate; bracteate or bracts reduced. Flowers 1–6, 5-merous, subsessile to pedicellate. Calyx-tube ovoid or campanulate; limb-tube shortly cylindrical, lobes short, spathulate to obovoid. Corolla white to cream or pale green; tube campanulate to subrotate; lobes oblong, obovate to rounded. Fruits globose, ellipsoid, or ovoid-globose, with persistent calyx-limb. Seeds numerous, ellipsoid, angled or reniform.

A genus of 12 species widespread in Africa and Asia; two species in Kenya.

1a. Leaf-blades obovate, obtuse at the apex; inflorescences fasciculate; fruits 1.3–2.5 cm long .... 1. *C. nilotica*
1b. Leaf-blades elliptic, often acuminate at the apex; inflorescences often cymose; fruits 1.8–3.3 cm long .... 2. *C. spinosa*

1. **Catunaregam nilotica** (Stapf) Tirveng., Taxon 27: 515. 1978; F.T.E.A. Rubiac. 2: 497. 1988; K.T.S.L.: 506. 1994. ≡ *Randia nilotica* Stapf, J. Linn. Soc., Bot. 37: 519. 1906. —Type: Ethiopia, Sennar, *K.G.T. Kotschy 400* [lectotype: K (K000414808), designated by D.D. Tirvengadum in Taxon 27: 515. 1978] (Plate 105A, B)

Subshrub, shrub or small tree, up to 6 m tall, often armed with solitary alternate spines up to 3.5 cm long. Leaves clustered on short lateral shoots; blades glabrous or densely pubescent, obovate to oblanceolate, 1–7 cm × 0.5–3 cm, rounded or emarginate at the apex, attenuate at the base; petioles up to 1 cm long; stipules ovate to triangular, 1–1.5 mm long. Flowers 1–2 per spur shoot; pedicels up to

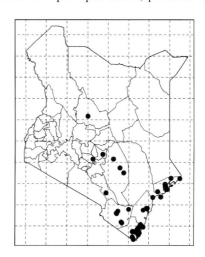

1.6 cm long. Calyx-tube cup-shaped, 1.5–2 mm long; limb-tube somewhat longer than the calyx; lobes oblong to spathulate, up to 3.5 mm long. Corolla white or cream, fading to yellow; tube short; lobes obovate, spathulate or rounded, 5–10 mm long. Fruits yellow-brown, ellipsoid to subglobose, 1.3–2.5 cm in diameter.

Distribution: Central and coastal Kenya. [Tropical Africa, from Somalia to Nigeria].

Habitat: Wooded grasslands, woodlands, bushlands, or thickets; up to 1200 m.

Embu: Mbeere, *Kirika et al. NMK571* (K). Kilifi: near Jilore Forest Station, *Perdue & Kibuwa 10146* (EA, K); Mangea Hill, *Moonrow 993* (K). Kitui: Endau, *Archer 579* (EA). Kwale: Shimba Hills, *Block et al. 415* (K); Muhaka Forest, *Faden & Faden 77/712* (EA, K); Mwaluganje Elephant Sanctuary, *Mwadime et al. 164* (EA). Lamu: Baragoni Hindi, *Gilbert & Kuchar 5844* (EA, K); Lamu Island, *Sangai 964* (EA). Machakos: Mananja, *Kirika et al. KEFRI203* (EA, K). Makueni: Kibwezi Forest Reserve, *Luke & Luke 14371* (EA). Mombasa: Nyali, *Birch 62/93* (EA). Samburu: *Moonrow 1053* (K). Teita Taveta: Mbololo Forest, *CPG s.n.* (PE); Rong'e Juu, *Mwachala et al. in EW3309* (EA). Tana River: Kurawa, *Polhill & Paulo 651* (EA, K).

2. **Catunaregam spinosa** (Thunb.) Tirveng., Bull. Mus. Hist. Nat., Sér. 3, Bot. 35: 13. 1978; F.T.E.A. Rubiac. 2: 499. 1988; K.T.S.L.: 507. 1994. ≡ *Gardenia spinosa* Thunb., Gardenia: 16 (1780). —Type: China, Macao, *P.J. Bladh s.n.* [holotype: UPS-Thunb (V-006111); isotype: BM (BM000945258)] (Figure 79; Plate 105C)

Shrub or small tree, up to 7.5 m tall, often armed with opposite or alternate spines up to 3 cm long. Leaves opposite on main branches or mostly clustered on lateral shoots, elliptic, obovate to spathulate, 1.2–13 cm × 0.7–6 cm, glabrous to densely velvety-woolly on both sides, acute to obtuse at the apex, attenuate at the base; petioles up to 1.2 cm long; stipules caducous, triangular, ca. 5 mm long. Inflorescences terminal on lateral short shoots, 1–3-flowered. Calyx-tube ovoid to ellipsoid, 3.5–7 mm long; lobes broadly elliptic to oblanceolate, 4–8 mm long, acute to rounded. Corolla white, fading to yellow; tube campanulate, 4–6 mm long; lobes oblong, obovate or round, 0.6–1(–1.5) cm long. Fruits yellow-green to brown, ellipsoid to globose, 1.8–3.5 cm long; seeds 4–5 mm long.

Distribution: Eastern and coastal Kenya. [Asia and Africa].

Habitat: Lowland forests; up to 300 m.

Kilifi: Mangea Hill, *Luke 1613* (EA). Kwale: Shimba Hills, *SAJIT 006018* (HIB); Mrima Hill, *Verdcourt 1859* (K). Lamu: Boni Forest, *Katz 54* (EA); near Mangai, *Festo & Luke 2756* (EA). Tana River: Tana River Primate Reserve, *Luke & Robertson 1166* (EA); northeast Garsen, *Gillett & Kibuwa 19930* (EA, K).

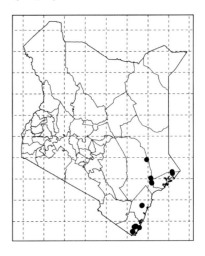

Plate 105 A, B. *Catunaregam nilotica*; C. *C. spinosa*. Photo by BL (A, B) and VMN (C).

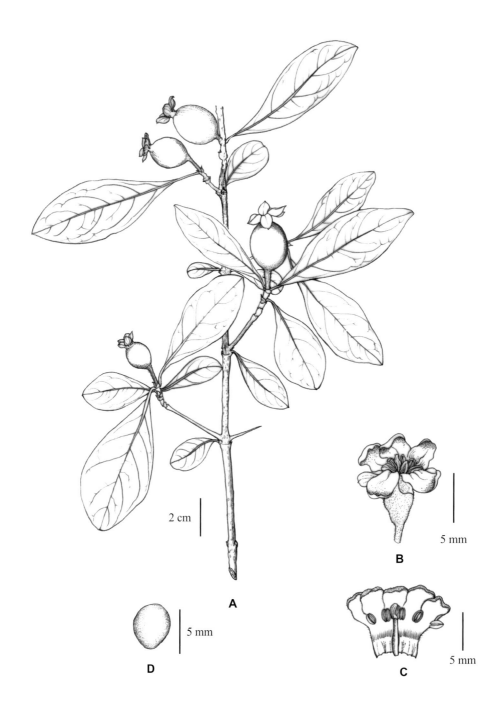

Figure 79 *Catunaregam spinosa*. A. a fruiting branch; B. flower; C. longitudinal section of corolla, showing the stamens and style; D. seed. Drawn by NJ.

Glossary

acumen: a slender, tapering, and pointed tip or apex of a leaf or other plant structure.

aestivation: arrangement or folding pattern of floral parts, such as petals or sepals, in a flower bud before it opens; imbricate/contorted aestivation.

albumen: a nutrient-rich tissue in plant seeds that provides nourishment to the developing embryo.

bracteole: a small, modified leaf or bract that is found on the stalk of a flower or inflorescence.

caducous: shed or fall off at an early growing stage.

calyculus (pl. calyculi): a small cup-shaped structure.

calyx-limb: the expanded, non-joining part as distinct from the tube in a joined calyx, usually leaf-like.

calyx-tube: the tube in a joined calyx.

capitate: like a head or a knob.

chartaceous: thick, stiff, and papery, resembling the characteristics of parchment or cardboard.

club-shaped: having a thickened, rounded base that gradually narrows towards the tip, resembling the shape of a club.

clavate: club-like or gradually thickened towards the apex.

coccus (pl. cocci, adj. coccous): a part of a fruit that is a dry, one-seeded structure resulting from the splitting of a compound ovary.

coarse ribs: prominent, thick, and noticeable raised lines or ridges on the surfaces.

corneous endosperm: a hard, horn-like tissue found in the seed of certain plants, serving as a nutrient storage for the developing embryo.

dextrorsely: in a right-handed or clockwise direction.

domatia: specialized structures that provide shelter or housing for other organisms, such as ants or mites.

dorsifixed: describing an anther attachment where the base is attached to the filament along its back or dorsal side.

embryo-radicle: embryonic root of the plant, which develops into the future root of the plant.

endosperm: nutrient-rich tissue found in seeds that nourishes the developing embryo.

excavation: hollowing or concave shape.

exotestal: describing a seed coat that is formed by a single layer of cells derived from the outer integument.

fissured: describing the presence of deep, narrow cracks or grooves on the surface of plant tissues, such as bark or stems.

funnel-shaped: referring to a flower with a tubular corolla that gradually widens into a broad, open, and often flared limb resembling a funnel.

hilar cavity: a small indentation or depression located near the point where the seed is attached to the seed stalk or funiculus; it is often visible as a small concave area on the seed surface.

invagination: the process of being turned inside out or folded back on itself to form a cavity or pouch.

lenticel (adj. lenticellate): a small, raised pore or opening on the surface of stems or fruits that allow for gas exchange.

limb-tube: fused or united portion of a corolla or a calyx that forms a tube-like structure, excluding any expanded or lobed parts.

locule (pl. loculi, adj. locular): a chamber or a compartment within an ovary where seeds are formed and contained.

loculicidal: describing the way a capsule opens by splitting along the walls of its individual compartments or locules.

medifixed: referring the base of an anther is attached at the middle of the stalk.

nodule: a small, rounded or irregularly shaped swelling or bump on a plant structure.

papyraceous: referring to a thin and papery texture, resembling paper.

parietal placenta: arrangement of ovules to the inner wall, or parietal region, of the ovary.

pit-like: referring to a feature or structure that resembles a small, shallow depression or cavity on the surface of plant tissues.

pyriform: pear-like or pear-shaped.

placenta: a specialized tissue found on the inner walls of the ovary, serving as the attachment point for one or multiple ovules or seeds.

precocious: referring to plants or plant parts that exhibit early development or flowering ahead of the typical schedule for their species.

pseudoaxillary: appearing to arise from the axil of a leaf, but actually located slightly above or below it.

puberulous: with fine, short, and soft hairs or pubescence on the surface.

pulp: a soft, fleshy, and often juicy tissue found within fruits, typically surrounding the seeds.

putamen: hard, woody shell or outer layer of certain plant seeds.

pyrene: (of a fruit) the seed plus a hard layer of endocarp (often sculptured) surrounding the seed.

raphide: a needle-like crystal of calcium oxalate in the cells.

ribbon-like: long and thin, resembling a ribbon.

ruminate endosperm: an intricately folded and convoluted tissue found in certain seeds, resembling the appearance of a brain.

salver-shaped: referring to a flower with a tubular corolla that abruptly widens into a flat, open, and usually symmetrical limb.

scrambling: referring to a growth habit where stems or branches sprawl and climb over other plants or structures for support.

secondary pollen presentation: In Rubiaceae, it refers to a unique mechanism where the stamens are initially held in a lowered position within the flower. After a pollinator, typically a bee, lands on the flower and triggers specific movements or vibrations, the stamens rapidly move upward and release the pollen onto the visiting pollinator. This secondary presentation of pollen increases the chances of successful pollination.

septicidal: describing the manner in which a capsule opens along the lines of junction of the carpels, i.e. along the septa.

seta (pl. setae): a bristle or stiff hair on a stipule.

spinescent: terminating in a spine.

spur: a short, stubby branch or shoot.

stigmatic knob: a rounded or knob-like structure located at the tip of the stigma, which is a part of the female reproductive organ of a flower.

stipule-limb: expanded or flattened part of a stipule.

streak: a linear or elongated marking or pattern often found on petals, leaves, fruits, or stems of plants.

striate: of fine, longitudinal lines or stripes on the surface of a plant structure.

sulcate: of longitudinal grooves or furrows on plant structures.

tangential wall: a specific layer within a seed that runs parallel to the surface, providing structural support and protection to the seed.

ternate: arranged in a whorl or cluster of three.

testa: the outer protective layer or coat of a seed, providing it with physical protection.

theca (adj. thecous): a protective covering or case of an anther which encloses pollens.

velutinous: soft and fuzzy, typically caused by the presence of dense, short, and fine hairs or trichomes on the surface.

velvety: having a smooth, soft appearance.

Abbreviations

F.T.E.A. (FTEA): *Flora of Tropical East Africa.*
K.T.S.L.: *Kenya Trees, Shrubs, and Lianas.*
U.K.W.F.F.: *Upland Kenya Wild Flowers and Ferns.*
U.T.S.K.: *Usefull Tress and Shrubs for Kenya.*
W.F.E.A.: *Collins Photo Guide to Wild Flowers of East Africa.*
C.P.K.: *Common Plants of Kenya.*

New Typifications

Agathisanthemum bojeri Klotzsch, **lectotype designation**
Agathisanthemum globosum (Hochst. ex A. Rich.) Klotzsch, **lectotype designation**
Anthospermum usambarense K. Schum., **second-step neotype designation**
Breonadia salicina (Vahl) Hepper & J.R.I. Wood, **lectotype designation**
Calycosiphonia spathicalyx (K. Schum.) Robbr., **lectotype designation**
Canthium sarogliae Chiov., **lectotype designation**
Chassalia cristata (Hiern) Bremek., **lectotype designation**
Chassalia discolor K. Schum., **lectotype designation**
Chassalia parvifolia K. Schum., **neotype designation**
Chassalia umbraticola Vatke, **lectotype designation**
Cladoceras subcapitatum (K. Schum. & K. Krause) Bremek, **lectotype designation**
Cremaspora triflora (Thonn.) K. Schum., **lectotype designation**
Cremaspora triflora (Thonn.) K. Schum. subsp. *confluens* (K.Schum.) Verdc., **neotype designation**
Diodia aulacosperma K. Schum., **lectotype designation**
Dirichletia asperula K. Schum., **lectotype designation**
Galiniera saxifraga (Hochst.) Bridson, **lectotype designation**
Galium chloroionanthum K. Schum., **lectotype designation**
Galium ossirwaense K. Krause, **neotype designation**
Gardenia fiorii Chiov., **lectotype designation**
Gardenia posoquerioides S. Moore, **lectotype designation**
Geophila obvallata (Schumach.) Didr. subsp. *ioides* (K. Schum.) Verdc., **lectotype designation**
Heinsenia diervilleoides K. Schum., **lectotype designation**

Heinsia crinita (Widmark) G. Taylor subsp. *parviflora* (K. Schum. & K. Krause) Verdc., **lectotype designation**
Keetia zanzibarica (Klotzsch) Bridson, **neotype designation**
Kohautia caespitosa Schnizl. var. *amaniensis* (K. Krause) Bremek., **lectotype designation**
Mussaenda keniensis K. Krause, **neotype designation**
Mussaenda monticola K. Krause, **neotype designation**
Oldenlandia herbacea (L.) Roxb. var. *holstii* (K. Schum. ex Engl.) Bremek., **lectotype designation**
Oldenlandia scopulorum Bullock var. *lanceolata* Bremek., **lectotype designation**
Oxyanthus goetzei K. Schum., **lectotype designation**
Oxyanthus pyriformis (Hochst.) Skeels, **lectotype designation**
Oxyanthus speciosus DC., **lectotype designation**
Paederia pospischilii K. Schum., **neotype designation**
Pavetta abyssinica Fresen. var. *lamurensis* (Bremek.) Bridson, **lectotype designation**
Pavetta crassipes K. Schum., **neotype designation**
Pavetta elliottii K. Schum. & K. Krause, **lectotype designation**
Pavetta hymenophylla Bremek, **lectotype designation**
Pavetta sansibarica K. Schum., **lectotype designation**
Pavetta sepium K. Schum., **lectotype designation**
Pavetta sepium K. Schum. var. *glabra* Bremek., **lectotype designation**
Pavetta sphaerobotrys K. Schum., **lectotype designation**
Pavetta stenosepala K. Schum., **lectotype designation**
Pavetta subcana Hiern var. *longiflora* (Vatke) Bridson, **lectotype designation**
Pavetta teitana K. Schum., **neotype designation**
Phyllopentas elata (K. Schum.) Kårehed & B. Bremer, **lectotype designation**
Plectronia telidosma K. Schum., **lectotype designation**
Polysphaeria lanceolata Hiern, **lectotype designation**
Polysphaeria multiflora Hiern, **lectotype designation**
Polysphaeria parvifolia Hiern, **lectotype designation**
Psychotria amboniana K. Schum., **lectotype designation**
Psychotria tanganyicensis Verdc., **lectotype designation**
Psydrax schimperianus (A. Rich.) Bridson, **lectotype designation**
Randia fratrum K. Krause, **lectotype designation**
Rhodopentas bussei (K. Krause) Kårehed & B. Bremer, **lectotype designation**
Rothmannia fischeri (K. Schum.) Bullock, **lectotype designation**
Rothmannia macrosiphon (K. Schum. ex Engl.) Bridson, **lectotype designation**
Rothmannia urcelliformis (Hiern) Bullock ex Robyns, **lectotype designation**
Rytigynia acuminatissima (K. Schum.) Robyns, **neotype designation**
Rytigynia bugoyensis (K. Krause) Verdc., **lectotype designation**
Rytigynia celastroides (Baill.) Verdc., **lectotype designation**
Rytigynia decussata (K. Schum.) Robyns, **lectotype designation**
Rytigynia eickii (K. Schum. & K. Krause) Bullock, **lectotype designation**

Rytigynia uhligii (K. Schum. & K. Krause) Verdc., **lectotype designation**
Spermacoce filituba (K. Schum.) Verdc., **lectotype designation**
Spermacoce minutiflora (K. Schum.) Verdc., **lectotype designation**
Spermacoce subvulgata (K. Schum.) J.G. García, **lectotype designation**
Tarenna boranensis Cufod., **lectotype designation**
Tarenna pavettoides (Harv.) Sim subsp. *friesiorum* (K. Krause) Bridson, **lectotype designation**
Triainolepis africana Hook. f. subsp. *hildebrandtii* (Vatke) Verdc., **lectotype designation**
Urophyllum holstii K. Schum., **lectotype designation**
Vangueria loranthifolia K. Schum., **lectotype designation**
Vangueria microphylla K. Schum., **lectotype designation**
Vangueria schumanniana (Robyns) Lantz, **lectotype designation**
Vangueria volkensii K. Schum., **lectotype designation**

New Synonyms and Nomenclatural Novelties

Gardenia goetzei Stapf & Hutch., **syn. nov.** = **Gardenia ternifolia** Schumach. & Thonn. subsp. **jovis-tonantis** (Welw.) Verdc.

Gardenia ternifolia Schumach. & Thonn. var. *goetzei* (Stapf & Hutch.) Verdc., **syn. nov.** = **Gardenia ternifolia** Schumach. & Thonn. subsp. **jovis-tonantis** (Welw.) Verdc.

Lasianthus kilimandscharicus K. Schum. subsp. *glabrescens* Jannerup, **syn. nov.** = **Lasianthus kilimandscharicus** K. Schum.

Pentodon pentandrus Vatke var. *minor* Bremek, **syn. nov.** = **Pentodon pentandrus** Vatke

Edrastima goreensis (DC.) Neupane & N.Wikstr var. **trichocarpa** (Bremek.) Y.D. Zhou, **comb. nov.**

Scleromitrion lancifolium (Schumach.) Y.D. Zhou, **comb. nov.**

Scleromitrion lancifolium var. **scabridulum** (Bremek.) Y.D. Zhou, **comb. nov.**

Index

A

Adina microcephala 211
Adina rubrostipulata 209
Adina rubrostipulata var. *discolor* 209
Afrocanthium 6, 244, **250**
Afrocanthium keniense 251, **254**
Afrocanthium kilifiense 251, **254**, 255
Afrocanthium lactescens 250, **251**, 255
Afrocanthium peteri 251, **253**
Afrocanthium pseudoverticillatum ....
.... 250, **251**, 252, 255
Agathisanthemum 3, 106, **120**
Agathisanthemum bojeri **120**, 121
Agathisanthemum globosum 120, **122**
Aidia 8, 414, **431**
Aidia abeidii **431**, 432
Anthospermeae 9, **177**
Anthospermum 2, 4, **177**
Anthospermum aberdaricum 180
Anthospermum herbaceum **177**, 178, 181
Anthospermum herbaceum var. *villosicarpum*
.... 179
Anthospermum holtzii 54
Anthospermum usambarense .. 177, **180**, 181
Anthospermum villosicarpum 177, **179**
Anthospermum welwitschii 177, **182**

B

Borreria filituba 165
Borreria minutiflora 169
Borreria princeae 170
Borreria radiata 165

Borreria subvulgata 165
Breonadia 4, 208, **211**
Breonadia microcephala 211
Breonadia salicina **211**, 212, 213
Bullockia 6, 243, **265**
Bullockia dyscritos 265, **267**, 269
Bullockia fadenii 265, **268**
Bullockia mombazensis 265, 266, 269
Bullockia pseudosetiflora 265, **270**
Bullockia setiflora 265, **268**, 269

C

Calycosiphonia 7, 314, **320**
Calycosiphonia spathicalyx **321**, 322
Canthium 5, **275**
Canthium celastroides 288
Canthium crassum 293
Canthium dyscriton 267
Canthium fadenii 268
Canthium glaucum 275, **278**, 279
Canthium gueinzii 244
Canthium indutum 300
Canthium inopinatum 265
Canthium kaessneri 263
Canthium keniense 254
Canthium kilifiense 254
Canthium lactescens 251
Canthium lividum 258
Canthium mombazense 265
Canthium neglectum 283
Canthium oligocarpum **275**
Canthium oligocarpum subsp. **friesiorum** ....
.... 275, **276**, 277, 279

Canthium oligocarpum subsp. **intermedium**
.... 275, **276**
Canthium oligocarpum subsp. **oligocarpum**.
.... 275, **276**
Canthium peteri 253
Canthium phyllanthoideum 271
Canthium pseudosetiflorum 270
Canthium pseudoverticillatum 251
Canthium recurvifolium 264
Canthium robynsianum 251
Canthium rubrocostatum 257
Canthium sarogliae 283
Canthium schimperianum 260
Canthium setiflorum 268
Canthium setiflorum subsp. *telidosma* 268
Canthium tetraphylla 275
Canthium tetraphyllum **278**
Canthium tetraphyllumucum 279
Canthium urophyllum 289
Canthium zanzibaricum 248
Carphalea glaucescens 69
Catunaregam 5, 414, **433**
Catunaregam nilotica **433**, 435
Catunaregam spinosa 433, **434**, 435, 436
Cephaelis peduncularis var. *suaveolens* 61
Cephaelis suaveolens 61
Chamaepentas 3, 63, **64**
Chamaepentas hindsioides **64**, 65, 66
Chassalia 2, 4, 20, **23**
Chassalia cristata 2, **24**, 25, 27
Chassalia discolor **26**
Chassalia discolor subsp. *discolor* 28
Chassalia discolor subsp. *grandifolia* 28
Chassalia discolor subsp. **taitensis**. 28, 29, 30
Chassalia kenyensis 24, **31**
Chassalia parvifolia 24, **28**, 30
Chassalia subochreata 24, **31**
Chassalia umbraticola 24, **26**, 27
Chazaliella abrupta 22
Chazaliella abrupta var. *parvifolia* 23

Chomelia subcapitata 373
Cinchonoideae **203**
Cladoceras 2, 5, 365, **373**
Cladoceras subcapitatum **373**, 374
Coffea 6, 314, **315**
Coffea arabica 315, **316**, 317
Coffea arabica var. *intermedia* 316
Coffea eugenioides 315, **316**, 318
Coffea fadenii **315**
Coffea pseudozanguebariae 315, **319**
Coffea rhamnifolia 315, **320**
Coffea sessiliflora 315, **319**
Coffea spathicalyx 321
Coffeeae 223, **314**
Conostomium 3, 106, **114**
Conostomium camptopodum 114
Conostomium longitubum **114**, 115
Conostomium microcarpum 116
Conostomium quadrangulare 114, **116**
Coptosperma 8, 365, **406**
Coptosperma graveolens ..407, **408**, 409, 413
Coptosperma graveolens var. *impolitum* ... 408
Coptosperma kibuwae **407**
Coptosperma littorale 407, **411**
Coptosperma nigrescens 407, **411**, 413
Coptosperma supra-axillare 407, **412**, 413
Coptosperma wajirense **407**
Cordylostigma 3, 106, **125**
Cordylostigma longifolium 125, **127**, 128
Cordylostigma obtusilobum 125, **126**
Cordylostigma prolixipes **125**
Cordylostigma virgatum 125, **126**
Craterispermeae 9, **17**
Craterispermum 4, **17**
Craterispermum schweinfurthii **17**, 18, 19
Cremaspora 7, **333**
Cremaspora confluens 334
Cremaspora triflora **334**
Cremaspora triflora subsp. **confluens** ....
.... **334**, 335, 336

Cremaspora triflora subsp. **triflora** **334**
Crossopterygeae 223, **236**
Crossopteryx 5, **236**
Crossopteryx febrifuga **236**, 237, 238

D

Debia 134
Decameria jovis-tonantis 418
Dibrachionostylus 3, 106, **123**
Dibrachionostylus kaessneri **123**, 124
Didymosalpinx 5, 333, **340**
Didymosalpinx norae **340**, 341, 342
Dimetia 134
Diodella 3, 106, **160**
Diodella sarmentosa **160**, 161, 162
Diodia 3, 106, **155**
Diodia aulacosperma **155**
Diodia aulacosperma var. **angustata** ....
.... 155, **157**
Diodia aulacosperma var. **aulacosperma** ......
.... **155**, 156, 157
Diodia sarmentosa 160
Diodia senensis 170
Dirichletia 4, 63, **69**
Dirichletia asperula 69
Dirichletia ellenbeckii 69
Dirichletia glaucescens **69**, 70, 71
Dolichopentas 3, 63, **74**
Dolichopentas decora 75, **77**
Dolichopentas decora var. **decora** **77**
Dolichopentas decora var. **triangularis** **77**
Dolichopentas longiflora **75**, 76

E

Edrastima 3, 106, **128**, 134
Edrastima goreensis **129**
Edrastima goreensis var. **goreensis** .. 129, **130**
Edrastima goreensis var. **trichocarpa** ....
.... 129, **130**, 131
Empogona 6, 314, **323**
Empogona kirkii var. *glabrata* 324
Empogona ovalifolia **323**
Empogona ovalifolia var. **glabrata** ....
.... **324**, 326, 327
Empogona ovalifolia var. **ovalifolia** . **324**, 326
Empogona ovalifolia var. **taylorii** ....
.... 324, **325**, 326
Empogona ruandensis 323, **325**
Empogona taylorii 325
Enterospermum littorale 411
Eumachia 4, **20**
Eumachia abrupta **22**
Eumachia abrupta var. **abrupta** 21, 22, **23**
Eumachia abrupta var. **parvifolia** **23**
Exallage 134

F

Fadogia 2, 243, **291**
Fadogia cienkowskii **291**, 292
Feretia 7, 333, **348**
Feretia apodanthera **349**
Feretia apodanthera subsp. **apodanthera** .. **349**
Feretia apodanthera subsp. **keniensis** ....
.... **349**, 350
Fleroya rubrostipulata 209

G

Galiniera 7, 333, **337**
Galiniera saxifraga **337**, 338, 339
Galium 1, 185, **188**
Galium acrophyum 189, **197**
Galium afroalpinum 194
Galium aparinoides 189, **191**, 192
Galium brenanii 189, **194**, 196
Galium chloroionanthum 189, **190**, 192
Galium glaciale 189, **195**

Galium glaciale var. **glaciale** **195**, 200
Galium glaciale var. **satimmae** 195, **197**
Galium hochstetteri 197
Galium kenyanum 189, **198**
Galium mollicomum 199
Galium mollicomum var. *friesiorum* 199
Galium ossirwaense 188, **199**, 200
Galium rotundifolium var. *hirsutum* .. 189, 190
Galium ruwenzoriense 189, **194**, 196
Galium scioanum 189, **198**
Galium scioanum var. **glabrum** 198, **199**
Galium scioanum var. **scioanum** **198**, 200
Galium simense 189, **191**
Galium simense var. *hysophilum* 197
Galium simense var. *keniense* 197
Galium spurium 189, **193**
Galium spurium var. **africanum** **193**
Galium spurium var. *spurium* 193
Galium thunbergianum 188, **189**
Galium thunbergianum var. **hirsutum** ....
.... **189**, 190, 192
Galium thunbergianum var. *thunbergianum* ....
.... 190
Gardenia 5, **414**
Gardenia crinita 224
Gardenia fiorii 414, **415**
Gardenia goetzei 418
Gardenia lutea 418
Gardenia manganjae 428
Gardenia norae 340
Gardenia posoquerioides 415, **416**, 419
Gardenia spinosa 434
Gardenia ternifolia 415, **417**
Gardenia ternifolia subsp. **jovis-tonantis** ......
.... **418**, 419, 420
Gardenia ternifolia subsp. *ternifolia* 418
Gardenia ternifolia var. *goetzei* 418
Gardenia ternifolia var. *jovis-tonantis* 418
Gardenia transvenulosa **415**
Gardenia urcelliformis 425

Gardenia volkensii 415, **416**, 419
Gardenia zanguebarica 362
Gardenieae 223, **414**
Geophila 2, 20, **32**
Geophila herbacea **32**, 33, 34
Geophila ioides 35
Geophila obvallata 32, **33**
Geophila obvallata subsp. *involucrata* 35
Geophila obvallata subsp. **ioides** 34, **35**
Geophila obvallata subsp. *obvallata* 35
Geophila obvallata subsp. *pilosa* 35
Geophila repens 33
Grumilea bequaertii var. *pubescens* 46
Grumilea lauracea 48
Grumilea riparia 37
Guettarda 4, **219**
Guettarda speciosa **220**, 221
Guettardeae 203, **219**

H

Hallea rubrostipulata 209
Hedyotis 134
Hedyotis affinis 142
Hedyotis aspera 107
Hedyotis corymbosa 152
Hedyotis fugax 143
Hedyotis globosa 122
Hedyotis goreensis 129
Hedyotis herbacea 146
Hedyotis johnstonii 139
Hedyotis lancifolia 131
Hedyotis monanthos 136
Hedyotis rupicola 140
Hedyotis virgata 126
Heinsenia 7, 414, **421**
Heinsenia diervilleoides **421**, 422
Heinsia 5, **224**
Heinsia crinita **224**
Heinsia crinita subsp. *crinita* 225

Heinsia crinita subsp. **parviflora** **225**, 226, 227
Heinsia parviflora 225
Heinsia zanzibarica.................. 224, **225**, 227
Hymenodictyeae............................... 203, **204**
Hymenodictyon 4, **204**
Hymenodictyon floribundum.. **204**, 205, 207
Hymenodictyon kurria...............................204
Hymenodictyon parvifolium ... 204, **206**, 207
Hypodematium sphaerostigma.................. 171

I

Involucrella 134
Ixora 5, **239**
Ixora narcissodora 239, **240**, 241, 242
Ixora scheffleri..239
Ixora scheffleri subsp. **keniensis**...............239
Ixora scheffleri subsp. *scheffleri* 240
Ixoreae 223, **238**
Ixoroideae..**223**

K

Keetia.. 2, 243, **244**
Keetia gueinzii......................... **244**, 245, 246
Keetia lukei 244, **248**
Keetia venosa 244, **247**, 249
Keetia zanzibarica 244, **248**, 249
Knoxia longituba....................................... 101
Knoxieae..**9**, **63**
Kohautia.. 3, **106**
Kohautia aspera..**107**
Kohautia caespitosa 107, **109**
Kohautia caespitosa var. **amaniensis**... **110**, 111
Kohautia caespitosa var. **caespitosa** **110**
Kohautia caespitosa var. **kitaliensis**......... **110**
Kohautia coccinea............................ **107**, 108
Kohautia longifolia 127
Kohautia obtusiloba 126

Kohautia prolixipes 125
Kohautia virgata.. 126
Kraussia....................................... 7, 333, **345**
Kraussia kirkii 345, **346**
Kraussia pavettoides 402
Kraussia speciosa..................... 345, 347, **348**
Kurria floribunda 204

L

Lagynias pallidiflora.................................303
Lamprothamnus 7, 333, **342**
Lamprothamnus zanguebaricus 343, 344, **345**
Lasiantheae 9, **13**
Lasianthus 4, **13**
Lasianthus kilimandscharicus **14**, 15, 16
Lasianthus kilimandscharicus subsp. *glabrescens* 14
Leptactina.............................. 7, 365, **370**
Leptactina platyphylla.............**370**, 371, 372
Logania capensis...37

M

Megacarpha pyriformis 362
Meyna comorensis 278
Meyna tetraphylla 278
Meyna tetraphylla subsp. *comorensis*........278
Mitracarpus 2, 105, **157**
Mitracarpus hirtus **158**, 159
Mitracarpus villosus.................................158
Mitragyna...4, **208**
Mitragyna rubrostipulata **209**, 210
Mitratheca richardsonioides 143
Mitriostigma...7, **356**
Mitriostigma greenwayi **356**, 357, 358
Multidentia 6, 243, **293**
Multidentia crassa **293**
Multidentia sclerocarpa........... **293**, 294, 295

Mussaenda 2, 5, 224, **228**
Mussaenda arcuata 230, 232
Mussaenda erythrophylla 230, **231**
Mussaenda keniensis 233
Mussaenda microdonta 230, **233**
Mussaenda microdonta subsp. *microdonta* 233
Mussaenda microdonta subsp. **odorata** **233**, 234, 235
Mussaenda monticola 230, **231**, 234
Mussaenda odorata 233
Mussaenda platyphylla 370
Mussaenda schimperi 78
Mussaenda zanzibarica 225
Mussaendeae 223, **224**

N

Nauclea 4, 208, **217**
Nauclea latifolia **217**, 218
Nauclea microcephala 211
Naucleeae 203, **208**
Nerium salicinum 211

O

Octotropideae **333**
Oldenlandia 3, 106, 128, **134**
Oldenlandia acicularis 135, **148**
Oldenlandia affinis 135, **142**
Oldenlandia affinis subsp. *affinis* 143
Oldenlandia affinis subsp. **fugax** **143**
Oldenlandia amaniensis 110
Oldenlandia borrerioides 140
Oldenlandia bullockii 130
Oldenlandia caespitosa 152
Oldenlandia capensis 152
Oldenlandia corymbosa 135, **151**
Oldenlandia corymbosa var. **caespitosa** 151, **152**
Oldenlandia corymbosa var. **corymbosa** **151**, 153
Oldenlandia corymbosa var. **linearis** 151, **153**
Oldenlandia corymbosa var. **nana** ... 151, **154**
Oldenlandia cryptocarpa 134, **140**
Oldenlandia fastigiata 134, 135, **148**
Oldenlandia fastigiata var. **fastigiata** **148**, 150
Oldenlandia fastigiata var. **pseudopenton** 148, **150**
Oldenlandia fastigiata var. **somala** .. 148, **150**
Oldenlandia filipes 144
Oldenlandia friesiorum 135, **138**, 141
Oldenlandia goreensis 129, 130, 134
Oldenlandia goreensis var. *trichocarpa* 130
Oldenlandia herbacea 135, **146**
Oldenlandia herbacea var. *caespitosa* 152
Oldenlandia herbacea var. **herbacea** 146, **147**, 149
Oldenlandia herbacea var. **holstii** 146, **147**, 149
Oldenlandia holstii 147
Oldenlandia ichthyoderma 135, **146**
Oldenlandia johnstonii 135, **139**
Oldenlandia kaessneri 123, 145
Oldenlandia lancifolia 131, 131
Oldenlandia lancifolia var. *scabridula* 132
Oldenlandia linearis 154
Oldenlandia linearis var. *nana* 154
Oldenlandia longifolia 127
Oldenlandia longituba 114
Oldenlandia monanthos 134, **136**, 137
Oldenlandia obtusiloba 126
Oldenlandia pedunculata 125
Oldenlandia prolixipes 125
Oldenlandia richardsonioides 135, **143**
Oldenlandia richardsonioides var. *gracilis* ... 143
Oldenlandia rosulata 135, **142**
Oldenlandia rosulata var. **littoralis** **142**
Oldenlandia rupicola 135, **140**, 141

Oldenlandia s.s. 134
Oldenlandia scopulorum 135, **144**
Oldenlandia scopulorum var. *lanceolata*... 144
Oldenlandia somala 150
Oldenlandia subg. *Anotidopsis* 128
Oldenlandia subtilis 152
Oldenlandia verticillata var. *trichocarpa* .. 130
Oldenlandia violacea 134, **136**, 137
Oldenlandia wiedemannii 135, **145**
Oldenlandia wiedemannii var. *glabricaulis*.....
 145
Ophiorrhiza lanceolata 87
Otomeria 3, 63, **101**
Otomeria elatior **102**
Otomeria oculata **102**, 103, 104
Oxyanthus 7, 356, **359**
Oxyanthus goetzei 359, **360**
Oxyanthus goetzei subsp. *goetzei* 362
Oxyanthus goetzei subsp. **keniensis**. **360**, 364
Oxyanthus pyriformis 359, **362**
Oxyanthus pyriformis subsp. **brevitubus**......
 .... **363**
Oxyanthus pyriformis subsp. **longitubus** ......
 .... **363**, 364
Oxyanthus speciosus **359**
Oxyanthus speciosus subsp. *gerrardii* 360
Oxyanthus speciosus subsp. *mollis* 360
Oxyanthus speciosus subsp. *speciosus* 360
Oxyanthus speciosus subsp. **stenocarpus** ......
 .... **360**, 361, 364
Oxyanthus stenocarpus 360
Oxyanthus zanguebaricus 359, **362**, 364

P

Pachystigma decussatum 285
Pachystigma gillettii 300
Pachystigma loranthifolium 299
Pachystigma loranthifolium subsp. *salaense*...
 299

Pachystigma schumannianum 302
Pachystigma schumannianum subsp. *mucronulatum*
 303
Paederia 2, **183**
Paederia pospischilii **183**, 184, 185
Paederieae 9, **182**
Paraknoxia parviflora 96
Parapentas 3, 63, **72**
Parapentas battiscombei **72**, 73, 74
Pauridiantha 4, **10**
Pauridiantha holstii 10
Pauridiantha paucinervis **10**, 11, 12
Pauridiantha paucinervis subsp. *holstii* 10
Pavetta 4, 7, 365, **375**
Pavetta abyssinica 376, **388**
Pavetta abyssinica var. **abyssinica** ... **389**, 390
Pavetta abyssinica var. **lamurensis** **389**
Pavetta abyssinica var. *prescottii* 391
Pavetta aethiopica 376, **388**
Pavetta crassipes 375, **391**, 393, 395
Pavetta crebrifolia 376, **384**
Pavetta crebrifolia var. **crebrifolia** .. **385**, 386
Pavetta crebrifolia var. **pubescens** ... **385**, 386
Pavetta dolichantha 375, **397**
Pavetta elliottii 376, **380**, 383
Pavetta elliottii var. *trichocalyx* 380
Pavetta friesiorum 403
Pavetta gardeniifolia 375, **394**, 395
Pavetta gardeniifolia var. *longiflora* 398
Pavetta graveolens 408
Pavetta hymenophylla 375, **376**, 378
Pavetta kaessneri 377
Pavetta kenyensis 389
Pavetta lamurensis 389
Pavetta linearifolia 375, **379**
Pavetta manamoca 384
Pavetta merkeri 397
Pavetta oliveriana 376, **382**, 383
Pavetta parviflora 257
Pavetta sansibarica 376, **380**

Pavetta sansibarica subsp. **trichosphaera** **380**, 381
Pavetta sepium 376, **394**
Pavetta sepium var. **glabra** **396**
Pavetta sepium var. **merkeri** 396, **397**
Pavetta sepium var. **sepium** **396**
Pavetta shimbensis 380
Pavetta sphaerobotrys 376, **384**
Pavetta sphaerobotrys subsp. *lanceisepala* 384
Pavetta sphaerobotrys subsp. *sphaerobotrys* 384
Pavetta sphaerobotrys subsp. **tanaica** **384**
Pavetta stenosepala 376, **385**, 387
Pavetta subcana 375, **398**
Pavetta subcana var. **longiflora** **398**
Pavetta subcana var. *subcana* 398
Pavetta supra-axillaris 412
Pavetta tanaica 384
Pavetta tarennoides 376, **391**, 392
Pavetta teitana 376, **377**, 381
Pavetta ternifolia 375, **377**, 378
Pavetta trichantha 402
Pavetta trichocalyx 380
Pavetta trichotropis 389
Pavetta uniflora 375, **398**
Pavetta yalaensis 377
Pavetteae 223, **365**
Pentanisia 2, 63, **96**
Pentanisia foetida 96, **97**, 100
Pentanisia longituba 96, 101
Pentanisia ouranogyne 96, **98**, 99, 100
Pentanisia parviflora **96**
Pentanisia schweinfurthii 96, **97**
Pentanisia zanzibarica 84
Pentanopsis 4, 106, **112**, 134
Pentanopsis fragrans **112**, 113
Pentas 3, 63, **83**
Pentas ainsworthii 87
Pentas arvensis 83, **86**
Pentas bussei 92

Pentas decora 77
Pentas decora var. *triangularis* 77
Pentas elata 81
Pentas hindsioides 64
Pentas lanceolata 83, **87**
Pentas lanceolata var. **lanceolata** ... **87**, 90, 91
Pentas lanceolata var. **leucaster** 87, **88**, 91
Pentas lanceolata var. **nemorosa** 87, **89**, 91
Pentas leucaster 88
Pentas longiflora 75
Pentas longiflora f. *glabrescens* 75
Pentas micrantha **83**
Pentas micrantha subsp. *micrantha* 84
Pentas micrantha subsp. **wyliei** **83**
Pentas parvifolia 93
Pentas parvifolia f. *spicata* 93
Pentas parvifolia var. *nemorosa* 89
Pentas pubiflora 83, **85**, 86
Pentas quadrangularis 116
Pentas schimperi 78
Pentas schimperiana 78
Pentas suswaensis 83, **89**
Pentas thomsonii 78
Pentas triangularis 77
Pentas wyliei 83
Pentas zanzibarica 83, **84**
Pentas zanzibarica var. **tenuifolia** 84, **85**
Pentas zanzibarica var. **zanzibarica** **84**, 86
Pentodon 3, 106, **117**
Pentodon pentandrus **117**, 118, 119
Pentodon pentandrus var. *minor* 117
Phyllopentas 3, 63, **78**
Phyllopentas elata 78, **81**, 82
Phyllopentas schimperi **78**, 79, 80
Phyllopentas schimperiana 78
Plectronia bibracteata 272
Plectronia bugoyensis 289
Plectronia eickii 285
Plectronia microterantha 251
Plectronia rhamnifolia 320
Plectronia sclerocarpa 293

Plectronia telidosma 268	**Psychotria mahonii** var. **puberula** **46**, 47
Plectronia venosa 247	**Psychotria mahonii** var. **pubescens** **46**, 47
Polysphaeria 7, 333, **351**	*Psychotria megistosticta* var. *puberula* 46
Polysphaeria cleistocalyx 351, **355**	**Psychotria mildbraedii** 36, **61**
Polysphaeria lanceolata 351	*Psychotria nairobiensis* 57
Polysphaeria multiflora **351**	*Psychotria obvallata* 33
Polysphaeria multiflora subsp. *pubescens* 351	**Psychotria orophila** 36, **41**, 42
Polysphaeria parvifolia 351, **352**, 353, 354	*Psychotria peduncularis* var. *ciliato-stipulata*. 61
Pouchetia saxifraga 337	
Pseudomussaenda 5, 224, **228**	*Psychotria peduncularis* var. *suaveolens* 61
Pseudomussaenda flava **228**, 229	**Psychotria petitii** 36, **45**
Psychotria 4, 20, **35**	**Psychotria pseudoplatyphylla** 36, **39**, 42
Psychotria abrupta 22	**Psychotria punctata** 36, **54**
Psychotria albidocalyx 52	**Psychotria punctata** var. **minor** **56**
Psychotria albidocalyx var. *velutina* 48	**Psychotria punctata** var. **punctata** **56**, 59
Psychotria alsophila 36, **60**	**Psychotria punctata** var. **tenuis** **56**
Psychotria amboniana 36, **48**, 50, 51	*Psychotria riparia* 37
Psychotria amboniana var. *velutina* 48	*Psychotria riparia* var. *puberula* 38
Psychotria bagshawei 36, **39**, 40	**Psychotria schliebenii** 37, **53**
Psychotria capensis 36, **37**	**Psychotria schliebenii** var. **parvipaniculata** . 53
Psychotria capensis subsp. *riparia* 37	**Psychotria schliebenii** var. **sessilipaniculata** . 53
Psychotria capensis var. **capensis** **37**, 40	*Psychotria subochreata* 31
Psychotria capensis var. **puberula** . 37, **38**, 40	**Psychotria taitensis** 36, **41**
Psychotria ceratoloba 36, **61**, 62	**Psychotria tanganyicensis** 36, **60**, 62
Psychotria crassipetala 36, **38**	*Psychotria tarambassica* 58
Psychotria cristata 24	*Psychotria triflora* 334
Psychotria faucicola 36, **57**, 59	*Psychotria volkensii* 58
Psychotria fractinervata **43**, 44, 47	**Psychotrieae** 9, **20**
Psychotria herbacea 32	**Psydrax** 2, 6, 243, **256**
Psychotria hirtella 58	**Psydrax faulknerae** 256, **260**, 262
Psychotria holtzii 36, **54**, 55	**Psydrax kaessneri** 257, **263**
Psychotria holtzii var. *pubescens* 54	**Psydrax lividus** 256, **258**, 259
Psychotria kirkii 36, **57**, 59	**Psydrax parviflora** subsp. **rubrocostata** .. 262
Psychotria kirkii var. *hirtella* 58	**Psydrax parviflorum** **257**
Psychotria kirkii var. *nairobiensis* 57	**Psydrax parviflorum** subsp. **parviflorus** . 257
Psychotria kirkii var. *tarambassica* 58	**Psydrax parviflorum** subsp. **rubrocostatus** **257**
Psychotria kirkii var. *volkensii* 57	
Psychotria lauracea 36, **48**, 49, 50	**Psydrax parviflorus** 256
Psychotria leucopoda 37, **52**, 55	**Psydrax polhillii** 257, **263**
Psychotria macrophylla 60	**Psydrax recurvifolius** 256, **264**
Psychotria mahonii 36, **45**	**Psydrax robertsoniae** 256, 257, **261**, 262

Psydrax schimperiana 262
Psydrax schimperianus 257, **260**
Pyrostria 6, 243, **271**
Pyrostria bibracteata 271, **272**, 273, 274
Pyrostria phyllanthoidea **271**, 273

R

Randia fischeri 424
Randia fratrum 428
Randia macrosiphon 428
Randia nilotica 433
Randia ravae 425
Rhabdostigma kirkii 346
Rhodopentas 3, 63, **92**
Rhodopentas bussei **92**, 94, 95
Rhodopentas parvifolia 92, **93**, 95
Richardia 2, 105, **173**
Richardia brasiliensis 173, **174**, 176
Richardia scabra **173**, 175, 176
Rondeletia febrifuga 236
Rothmannia 5, 414, **423**
Rothmannia fischeri **424**
Rothmannia fischeri subsp. **fischeri** 424
Rothmannia fischeri subsp. **moramballae** .....
.... 424
Rothmannia fischeri subsp. **verdcourtii** ....
.... **424**, 426
Rothmannia longiflora 423, **427**, 429
Rothmannia macrosiphon **428**, 429, 430
Rothmannia manganjae 423, **428**
Rothmannia ravae 424, **425**
Rothmannia urcelliformis 424, **425**, 426
Rubia 1, 185, **186**
Rubia conotricha 186
Rubia cordifolia **186**
Rubia cordifolia subsp. **conotricha** .. **186**, 187
Rubia cordifolia subsp. *cordifolia* 186
Rubia ruwenzoriensis 194
Rubiaceae **1**

Rubieae 9, **185**
Rubioideae **9**
Rutidea 2, 7, **365**
Rutidea fuscescens 366, **369**
Rutidea fuscescens subsp. **bracteata** ....
.... 369, **370**
Rutidea fuscescens subsp. **fuscescens** **369**
Rutidea orientalis **366**, 367, 368
Rutidea smithii **366**, 368
Rytigynia 5, 6, 244, **280**
Rytigynia acuminatissima **281**, 282
Rytigynia bugoyensis 281, **289**, 290
Rytigynia celastroides 280, 281, **288**
Rytigynia decussata 281, **285**, 287
Rytigynia eickii 281, **285**, 287
Rytigynia friesiorum 276
Rytigynia gillettii 300
Rytigynia induta 300
Rytigynia microphylla 288
Rytigynia mrimaensis 281, **286**
Rytigynia neglecta 281, **283**, 284
Rytigynia parvifolia 280, 281, **288**, 290
Rytigynia uhligii 281, **283**, 284

S

Sarcocephalus latifolius 217
Scleromitrion 3, 106, **131**, 134
Scleromitrion lancifolium **131**
Scleromitrion lancifolium var. **scabridulum** .
.... **132**, 133
Sherbournieae 223, **355**
Spermacoce 2, 105, **163**
Spermacoce chaetocephala 163, **165**, 166
Spermacoce filituba 164, **165**, 166
Spermacoce hirta 158
Spermacoce laevis 163, **167**
Spermacoce minutiflora 163, **169**
Spermacoce princeae 164, **170**, 172
Spermacoce princeae var. *pubescens* 170

Spermacoce pusilla 163, **167**, 168	*Tricalysia ovalifolia* 323
Spermacoce radiata 163, **165**	*Tricalysia ovalifolia* var. *glabrata* 324
Spermacoce senensis 164, **170**, 172	*Tricalysia ovalifolia* var. *taylorii* 325
Spermacoce sphaerostigma 163, **171**	**Tricalysia pallens** 328, **329**, 330, 331
Spermacoce subvulgata 163, **165**	*Tricalysia ruandensis* 325
Spermacoce villosa 158	*Tricalysia sennii* 399
Spermacoceae 9, **105**, 134	

T

Tapiphyllum mucronulatum 303
Tapiphyllum schumannianum 302
Tardavel kaessneri 165
Tarenna 8, 365, **401**
Tarenna boranensis 408
Tarenna drummondii 402, 404, 405, **406**
Tarenna graveolens 408
Tarenna graveolens var. *impolita* 408
Tarenna kibuwae 407
Tarenna littoralis 411
Tarenna nigrescens 411
Tarenna pavettoides **402**
Tarenna pavettoides subsp. **friesiorum** ... **403**
Tarenna pavettoides subsp. **gillmanii** ....
.... **403**, 404
Tarenna supra-axillaris 412
Tarenna trichantha **402**
Tarenna wajirensis 407
Tennantia 7, 365, **399**
Tennantia sennii **399**, 400, 401
Triainolepis 4, 63, **66**
Triainolepis africana **67**
Triainolepis africana subsp. *africana* 69
Triainolepis africana subsp. **hildebrandtii** ....
.... **67**, 68
Triainolepis hildebrandtii 67
Tricalysia 7, 314, **328**
Tricalysia bridsoniana 328, **332**
Tricalysia microphylla **328**, 330
Tricalysia niamniamensis 328, **329**

U

Uncaria 2, 208, **214**
Uncaria africana **214**
Uncaria africana subsp. **africana** **215**
Uncaria africana subsp. **lacus-victoriae** ......
.... 215, 216, **217**
Uragoga ceratoloba 61
Uragoga macrophylla 60
Uragoga mildbraedii 61
Urophylleae 9, **10**
Urophyllum holstii 10
Urophyllum paucinerve 10

V

Vangueria 6, 243, 244, **298**
Vangueria acuminatissima 281
Vangueria apiculata 298, **310**, 311
Vangueria campanulata 307
Vangueria gillettii 298, **300**
Vangueria induta 298, **300**, 301
Vangueria infausta var. *campanulata* 307
Vangueria infausta 298, **307**
Vangueria infausta subsp. *campanulata* 307
Vangueria infausta subsp. *infausta* 307
Vangueria infausta subsp. **rotundata** ....
.... **307**, 309, 308
Vangueria infausta subsp. *rotundata* var.
  campanulata 307
Vangueria infausta subsp. *rotundata* var. *rotundata* ..
.... 307
Vangueria infausta var. *rotundata* 309

Vangueria kyimbilensis 312
Vangueria loranthifolia **299**
Vangueria loranthifolia subsp. **loranthifolia**.
.... **299**, 301
Vangueria loranthifolia subsp. **salaensis** . **299**
Vangueria loranthifolium 298
Vangueria madagascariensis ... 298, **305**, 306
Vangueria microphylla 288
Vangueria pallidiflora ....
.... 5, 243, 298, **303**, 304, 306
Vangueria randii 298, **309**
Vangueria randii subsp. **acuminata**. **309**, 310
Vangueria randii subsp. *chartacea* 310
Vangueria randii subsp. *randii* 309
Vangueria randii subsp. *vollesenii* 310
Vangueria rotundata 307
Vangueria schumanniana 298, **302**
Vangueria schumanniana subsp. **mucronulatum**
.... 302

X

Xeromphis keniensis 399